NEW VIEWS OF THE MAGELLANIC CLOUDS
IAU SYMPOSIUM NO. 190

COVER ILLUSTRATION: A neutral hydrogen view of the Magellanic system taken with the Parkes multi-beam receiver. The Large Magellanic Cloud is on the left and the Small Magellanic Cloud is on the right. Other features include the Bridge of gas between the Clouds as well as the start of the Magellanic Stream (seen above the SMC).

Data courtesy of: Haynes, Staveley-Smith, Mebold, Kalberla, Bruens, Filipović, Putman and Muller (see papers by: Putman, Gibson & Staveley-Smith and Haynes et al.). The Australia Telescope is funded by the Commonwealth of Australia for operation as a National Facility managed by CSIRO.

Published on behalf of the
INTERNATIONAL ASTRONOMICAL UNION

by
Astronomical Society of the Pacific
First published 1999

Managing Editor, D. H. McNamara
Production Manager, Enid Livingston

EDITORIAL/PUBLISHING OFFICE:
Managing Editor
PO Box 24463
211 KMB Brigham Young University
Provo, UT 84602-4463
USA

(801) 378-2298 Phone
(801) 378-2265 Fax
pasp@astro.byu.edu E-mail

CATALOG/BOOK ORDERS:
IAU Publications
390 Ashton Avenue
San Francisco, CA 94112-1722
USA

(415) 337-1100 Phone
(415) 337-5205 Fax
catalog@aspsky.org E-mail
www.aspsky.org Web Site

Printed by Sheridan Books, Inc.
Chelsea, Michigan

Library of Congress Catalog Card Number: 99-68552
ISBN: 1-58381-021-8

INTERNATIONAL ASTRONOMICAL UNION
UNION ASTRONOMIQUE INTERNATIONALE

NEW VIEWS OF THE MAGELLANIC CLOUDS

Proceedings of the 190th Symposium
of the International Astronomical Union
held in Victoria, Canada
12-17 July 1998

Edited by:

YOU-HUA CHU
Astronomy Department, University of Illinois, Urbana, Illinois, USA

NICHOLAS B. SUNTZEFF
Cerro Tololo Inter-American Observatory, NOAO, La Serena, Chile

JAMES E. HESSER
National Research Council of Canada, Herzberg Institute of Astrophysics, Victoria, British Columbia, Canada

and

DAVID A. BOHLENDER
National Research Council of Canada, Herzberg Institute of Astrophysics, Victoria, British Columbia, Canada

Publisher

ASTRONOMICAL SOCIETY OF THE PACIFIC

Information on other IAU Symposium proceedings is given at the back of this volume

Table of Contents

Section B. Poster Papers

Part 3. Massive Stars and 30 Doradus

Section A. Invited Reviews

Part 4. Detailed Chemical Abundances

Section A. Invited Reviews

Section B. Poster Papers

Part 5. Stellar Populations and Surveys

Section A. Invited Reviews

Section B. Poster Papers

Part 6. Stellar Clusters

Section A. Invited Reviews

Section B. Poster Papers

Part 7. Tidal Interactions

Section A. Invited Reviews

Section B. Poster Papers

Part 8. Cepheids and Distances

Section A. Invited Reviews

Section B. Poster Papers

Part 9. Summary

Preface

International Astronomical Union Symposia on the Magellanic Clouds have achieved a pattern of being held somewhat more freqently than once per decade in alternating hemispheres. IAU Symposium 190, New Views of the Magellanic Clouds, held in Victoria, British Columbia, Canada on 12-17 July, 1998 brought together nearly 200 people from 21 countries for intense scrutiny of the many new developments bearing on our understanding of the Milky Way's major satellite galaxies. The meeting was co-sponsored by the IAU, the Canadian Institute of Theoretical Astrophysics, the Department of Physics and Astronomy of the University of Victoria, and the National Research Council (NRC) of Canada's Herzberg Institute of Astrophysics (HIA), whose headquarters are at the Dominion Astrophysical Observatory (DAO) in Victoria.

The organizers' hopes for a meeting brimming with new science and stimulating discussions were fully satisfied, thereby fulfilling two other goals. The first was to honour Dr. Sidney van den Bergh in the year of his official retirement from the National Research Council, where for nine years he was Director of the DAO. The second was to celebrate the 80th anniversary of the commissioning of the Plaskett Telescope, whose first spectrogram was secured on 1918 May 6. Although the Plaskett Telescope has never observed the Magellanic Clouds, its users have made major contributions to knowledge of the Milky Way to which the Clouds are bound.

The Scientific Organizing Committee was co-chaired by You-Hua Chu (USA) and Nicholas Suntzeff (USA), who were assisted by Enzo Brocato (Italy), Anne Cowley(USA), Kenneth Freeman (Australia), Paul Hodge (USA), Monica Rubio (Chile), Monique Spite (France), Lister Staveley-Smith (Australia), Nolan Walborn (USA), Douglas Welch (Canada) and Hans Zinnecker (Germany). The Local Organizing Committee was co-chaired by David Bohlender (NRC) and Jim Hesser (NRC), with assistance from David Crampton (NRC), Jay Gallagher (U. Wisconsin), David Hartwick (U. Victoria), Simon Morris (NRC), Brenda Parrish (NRC) and Don VandenBerg (U. Victoria).

The meeting format incorporated long poster sessions and various opportunities for informal discussions and socializing. These began with an opening reception hosted by NRC on a beautiful summer evening at the University of Victoria Faculty Club. The Sophysticates String Quartet composed of Sophia Darwin, Stephanie Côté, Anka Venema and Frank Younger welcomed participants with lovely classical music. To celebrate the 80th anniversary of the DAO, a mid-week luncheon was held at the Observatory followed by an open house. The HIA staff demonstrated their activities in submm, optical and far-ultraviolet instrumentation, data archiving and distribution, and operations of the Plaskett Telescope. In contrast to the weather throughout the week, this event was marked by unseasonably cool temperatures and a low cloud cover that blocked the spectacular views staff claimed are regularly enjoyed from Little Saanich Mountain! Within hours of completing the Observatory tour, excellent weather returned for the remainder of the meeting, including the Faculty Club banquet. Lively eastern European music provided by the Dumka Quartet comprised of Tony Booker, Tom Marcaccini, Alex Olson and Kate Rhodes preceded friends, former students and colleagues who paid tribute to Sidney van den Bergh for his many achievements in international and Canadian astronomy.

We are saddened that two active participants in IAU Symposium 190, Jurij Frantsman and Rebecca Elson, are no longer with us; their strong presence helped make the meeting a scientific success.

We thank the staff of the Department of Physics and Astronomy for provision of Internet connectivity, and the HIA staff for their hospitality during the luncheon and open house at DAO. The Editors apologize for the delay in producing these proceedings. While some important papers indeed arrived late, other responsibilities prevented the prompt attention the proceedings deserved.

In addition to the sponsoring organizations, the organizers thank our volunteers, Rob Cardinal, Brad Cavanagh, Mike Letawsky, John Ouellette, Russell Robb, Carolyn Swerdlyk and James Taylor, for their services as audio-visual technicians, distributors of discussion sheets, and general sources of information for our visitors. We also thank Wolfgang Brandner, who contributed to meeting planning, and to Eva Grebel for her numerous contributions pre- and post-meeting. Brenda Parrish (NRC) assisted with many aspects of meeting planning and registration, and converted (often illegible) discussion sheets into text. Pat McGuire, Teresa McGuire, Victoria Emery and Ian Wong from the University's Conference Management Services provided efficient, friendly and professional support for the conference. Stéphane Courteau took the photographs included in these proceedings. It was our pleasure to have worked with such enthusiastic, capable, dedicated individuals.

The editors: You-Hua Chu, Nicholas Suntzeff, Jim Hesser, and David Bohlender
October 1999

Participant List

Abe Rihei, Japan, ⟨abe@a.phys.nagoya-u.ac.jp⟩
Alves David, USA, ⟨alves@igpp.llnl.gov⟩
Ambrocio-Cruz Patricia, Mexico, ⟨patricia@astroscu.unam.mx⟩
Andersen Johannes, Denmark, ⟨ja@astro.ku.dk⟩
Anguita Claudio, Chile, ⟨canguita@das.uchile.cl⟩
Ardeberg Arne, Sweden, ⟨arne@astro.lu.se⟩
Arias Lorena, Mexico, ⟨lorena@astroscu.unam.mx⟩
Azzopardi Marc, France, ⟨azzopardi@obmara.cnrs-mrs.fr⟩
Barbá Rodolfo, USA, ⟨rbarba@stsci.edu⟩
Battinelli Paolo, Italy, ⟨battinel@oarhp1.rm.astro.it⟩
Beaulieu Jean-Philippe, France, ⟨beaulieu@iap.fr⟩
Beaulieu Sylvie, United Kingdom, ⟨beaulieu@ast.cam.ac.uk⟩
Bergbusch Peter, Canada, ⟨bergbush@deneb.phys.uregina.ca⟩
Bohannan Bruce, USA, ⟨bruce@noao.edu⟩
Bohlender David, Canada, ⟨david.bohlender@hia.nrc.ca⟩
Boily Christian, Germany, ⟨cmb@ari.uni-heidelberg.de⟩
Bolatto Alberto, USA, ⟨bolatto@bu.edu⟩
Bomans Dominik, Germany, ⟨bomans@astro.ruhr-uni-bochum.de⟩
Bono Giuseppe, Italy, ⟨bono@oat.ts.astro.it⟩
Brandner Wolfgang, USA, ⟨brandner@ifa.hawaii.edu⟩
Bregman Joel, USA, ⟨jbregman@umich.edu⟩
Brocato Enzo, Italy, ⟨brocato@astrte.te.astro.it⟩
Caulet Adeline, USA, ⟨caulet@astro.uiuc.edu⟩
Cavanagh Brad, Canada, ⟨cavanagh@uvic.ca⟩
Chin Yi-nan, Taiwan (ROC), ⟨einmann@biaa.sinica.edu.tw⟩
Chu You-Hua, USA, ⟨chu@astro.uiuc.edu⟩
Cioni Maria Rosa, Netherlands, ⟨mrcioni@strw.leidenuniv.nl⟩
Clayton Geoffrey, USA, ⟨gclayton@fenway.phys.lsu.edu⟩
Cole Andrew, USA, ⟨cole@astro.wisc.edu⟩
Cook Kem, USA, ⟨kcook@mordred.llnl.gov⟩
Côté Stéphanie, Canada, ⟨stephanie.cote@hia.nrc.ca⟩
Courteau Stéphane, Canada, ⟨courteau@astro.ubc.ca⟩
Cowley Anne, USA, ⟨anne.cowley@asu.edu⟩
Crampton David, Canada, ⟨david.crampton@hia.nrc.ca⟩
Crotts Arlin, USA, ⟨arlin@astro.columbia.edu⟩
Cunha Katia, USA, ⟨katia@baade.physics.utep.edu⟩

Da Costa Gary, Australia, ⟨gdc@mso.anu.edu.au⟩
de Boer Klaas, Germany, ⟨deboer@astro.uni-bonn.de⟩
de Oliveira Marcio, Brazil, ⟨marcio@if.ufrgs.br⟩
Demers Serge, Canada, ⟨demers@astro.umontreal.ca⟩
Dennefeld Michel, France, ⟨dennefel@iap.fr⟩
DePoy Darren, USA, ⟨depoy@astronomy.ohio-state.edu⟩
Di Benedetto G, Italy, ⟨pdibene@ifctr.mi.cnr.it⟩
Dickel John, USA, ⟨johnd@astro.uiuc.edu⟩
Dickel Hélène, USA, ⟨lanie@astro.uiuc.edu⟩
Dickey John, USA, ⟨john@ast1.spa.umn.edu⟩
Dieball Andrea, Germany, ⟨adieball@astro.uni-bonn.de⟩
Dopita Michael, Italy, ⟨mad@madras.anu.edu.au⟩
Durrell Patrick, Canada, ⟨durrell@impatiens.physics.mcmaster.ca⟩
Dusterberg Christian, Germany, ⟨cduester@astro.uni-bonn.de⟩
Ecker George, Israel,
Efremov Yuri, Russia, ⟨efremov@sai.msu.su⟩
Elson Rebecca, England, ⟨deceased⟩
Feast Michael, South Africa, ⟨mwf@uctvax.uct.ac.za⟩
Filipović Miroslav, Germany, ⟨fica@mpe.mpg.de⟩
Frantsman Jurij, Latvia, ⟨deceased⟩
Fukui Yasuo, Japan, ⟨fukui@a.phys.nagoya-u.ac.jp⟩
Gallagher Jay, USA, ⟨jsg@astro.wisc.edu⟩
Gallart Carme , Chile, ⟨gallart@das.uchile.cl⟩
Gardiner Lance, Korea (South), ⟨ltg@omega.sunmoon.ac.kr⟩
Garnett Don, USA, ⟨dgarnett@as.arizona.edu⟩
Gaustad John, USA, ⟨jgaustal@swarthmore.edu⟩
Geisler Doug, Chile, ⟨doug@phys.cfm.udec.cl⟩
Gieren Wolfgang, Chile, ⟨wgieren@coma.cfm.udec.cl⟩
Gilmore Gerry , United Kingdom, ⟨gil@ast.cam.ac.uk⟩
Girardi Leo, Germany, ⟨leo@mpa-garching.mpg.de⟩
Gower Ann, Canada, ⟨gower@uvastro.phys.uvic.ca⟩
Graff David, USA, ⟨david.graff@osu.edu⟩
Grebel Eva, USA, ⟨grebel@astro.washington.edu⟩
Gusev Alexander, Russia, ⟨gusev@sai.msu.su⟩
Haberl Frank, Germany, ⟨fwh@mpe.mpg.de⟩
Hanes Dave, Canada, ⟨hanes@astro.queensu.ca⟩
Harris Gretchen, Canada, ⟨glharris@astro.uwaterloo.ca⟩

Harris William, Canada, ⟨ harris@physics.mcmaster.ca ⟩
Harris Jason, USA, ⟨ jharris@ucolick.org ⟩
Hartwick David, Canada, ⟨ hartwick@uvastro.phys.uvic.ca ⟩
Hasegawa Tetsuo, Japan, ⟨ tetsuo@ghz.mtk.ioa.s.u-tokyo.ac.jp ⟩
Hatzidimitriou Despina, Greece, ⟨ dh@physics.uch.gr ⟩
Haynes Raymond, Australia, ⟨ rhaynes@atnf.csiro.a ⟩
Heikkilä Arto, Sweden, ⟨ arto@oso.chalmers.se ⟩
Hesser James, Canada, ⟨ jim.hesser@hia.nrc.ca ⟩
Hill Vanessa, Germany, ⟨ vhill@eso.org ⟩
Hodge Paul, USA, ⟨ hodge@astro.washington.edu ⟩
Holtzman Jon, USA, ⟨ holtz@nmsu.edu ⟩
Hunter Deidre, USA, ⟨ dah@lowell.edu ⟩
Hutchings John, Canada, ⟨ john.hutchings@hia.nrc.ca ⟩
Hwang Una, USA, ⟨ hwang@rosserv.gsfc.nasa.gov ⟩
Innanen Kimmo, Canada, ⟨ kimimn@yorku.ca ⟩
Irwin Alan, Canada, ⟨ irwin@beluga.phys.uvic.ca ⟩
Johansson Lars, Sweden, ⟨ leb@oso.chalmers.se ⟩
Johnson Jennifer, USA, ⟨ jennifer@ucolick.org ⟩
Jones Paul, Australia, ⟨ p.jones@nepean.uws.edu.au ⟩
Kawamura Akiko, Japan, ⟨ kawamura@a.phys.nagoya-u.ac.jp ⟩
Keller Stefan, Australia, ⟨ stefan@mso.anu.edu.au ⟩
Kim Sungeun, USA, ⟨ sek@astro.uiuc.edu ⟩
Kontizas Evangelos, Greece, ⟨ ekonti@astro.noa.gr ⟩
Kontizas Mary, Greece, ⟨ mkonti@astro.noa.gr ⟩
Korn Andreas, Germany, ⟨ a.korn@lsw.uni-heidelberg.de ⟩
Kravtsov Valery, Russia, ⟨ scorpi@sai.msu.su ⟩
Kunkel William, Chile, ⟨ kunkel@ociw.edu ⟩
Langer Norbert, Germany, ⟨ ntl@astro.physik.uni-potsdam.de ⟩
Letawsky Mike, Canada, ⟨ michael.letawsky@hia.nrc.ca ⟩
Lehner Nicolas, Northern Ireland, ⟨ n.lehner@qub.ac.uk ⟩
Li Pak Shing, USA, ⟨ pakshing@ncsa.uiuc.edu ⟩
Loup Cecile, France, ⟨ loup@iap.fr ⟩
Majewski Steven, USA, ⟨ srm4n@virginia.edu ⟩
Marconi Marcella, Italy, ⟨ marcella@cerere.na.astro.it ⟩
Marquette Jean-Baptiste, France, ⟨ marquett@iap.fr ⟩
Marx-Zimmer Monika, Germany, ⟨ mmarx@astro.uni-bonn.de ⟩
Massey Philip, USA, ⟨ massey@noao.edu ⟩

Maurice Eric, France, ⟨ maurice@observatoire.cnrs-mrs.fr ⟩
Mighell Kenneth, USA, ⟨ mighell@noao.edu ⟩
Milsztajn Alain, France, ⟨ amilsztajn@cea.fr ⟩
Mizuno Akira, Japan, ⟨ mizuno@a.phys.nagoya-u.ac.jp ⟩
Mizuno Norikazu, Japan, ⟨ norikazu@a.phys.nagoya-u.ac.jp ⟩
Moneti Andrea, Spain, ⟨ moneti@iso.vilspa.esa.es ⟩
Morgan David, United Kingdom, ⟨ dhm@roe.ac.uk ⟩
Morris Simon, Canada, ⟨ simon.morris@hia.nrc.ca ⟩
Niemela Virpi, Argentina, ⟨ virpi@fcaglp.fcaglp.unlp.edu.ar ⟩
Nordstrom Birgitta, Denmark, ⟨ birgitta@astro.ku.dk ⟩
Oke Bev, Canada, ⟨ oke@dao.nrc.ca ⟩
Oey Sally, USA, ⟨ oey@stsci.edu ⟩
Olsen Knut, USA, ⟨ kolsen@noao.edu ⟩
Olszewski Edward, USA, ⟨ edo@as.arizona.edu ⟩
Onishi Toshikazu, Japan, ⟨ ohnishi@a.phys.nagoya-u.ac.jp ⟩
Ostheimer James, USA, ⟨ jco9w@umbriel.astro.virginia.edu ⟩
Ouellette John, Canada, ⟨ ouellet@uvastro.phys.uvic.ca ⟩
Pak Soojong , Korea (South), ⟨ soojong@mpe-garching.mpg.de ⟩
Palouš Jan, Czech Republic, ⟨ palous@ig.cas.cz ⟩
Panagia Nino, USA, ⟨ panagia@stsci.edu ⟩
Parker Joel, USA, ⟨ joel@boulder.swri.edu ⟩
Parrish Brenda, Canada, ⟨ brenda.parrish@hia.nrc.ca ⟩
Petre Robert, USA, ⟨ rob@lheavx.gsfc.nasa.gov ⟩
Points Sean, USA, ⟨ points@astro.uiuc.edu ⟩
Pritchard John, New Zealand, ⟨ j.pritchard@phys.canterbury.ac.nz ⟩
Probst Ronald, Chile, ⟨ rprobst@noao.edu ⟩
Putman Mary, Australia, ⟨ putman@mso.anu.edu.au ⟩
Rajagopal Jayadev, India, ⟨ jaydev@rri.ernet.in ⟩
Richter Philipp, Germany, ⟨ prichter@astro.uni-bonn.de ⟩
Ripepi Vincenzo, Italy, ⟨ ripepi@cerere.na.astro.it ⟩
Rosado Margarita, Mexico, ⟨ margarit@astroscu.unam.mx ⟩
Rubio Mónica, Chile, ⟨ mrubio@das.uchile.cl ⟩
Ruotsalainen Robert, USA, ⟨ ruotsalainen@ewu.edu ⟩
Saha Abhijit, USA, ⟨ saha@noao.edu ⟩
Saito Hiro, Japan, ⟨ saito@a.phys.nagoya-u.ac.jp ⟩
Sawa Takeyasu, Japan, ⟨ sawa@auephyas.aichi-edu.ac.jp ⟩
Schmidtke Paul, USA, ⟨ paul.schmidtke@asu.edu ⟩

Scowen Paul, USA, ⟨ paul.scowen@asu.edu ⟩
Seitzer Patrick, USA, ⟨ seitzer@astro.lsa.umich.edu ⟩
Shelton Robin, USA, ⟨ shelton@spots.gsfc.nasa.gov ⟩
Sirianni Marco, USA, ⟨ sirianni@pha.jhu.edu ⟩
Skillman Evan, USA, ⟨ skillman@ast1.spa.umn.edu ⟩
Smecker-Hane Tammy, USA, ⟨ tsmecker@uci.edu ⟩
Smith R. Chris, USA, ⟨ csmith@noao.edu ⟩
Smith Verne, USA, ⟨ verne@balmer.physics.utep.edu ⟩
Snowden Steve, USA, ⟨ snowden@lheavx.gsfc.nasa.gov ⟩
Sonneborn George, USA, ⟨ sonneborn@stars.gsfc.nasa.gov ⟩
Stanimirović Snezana, Australia, ⟨ sstanimi@atnf.csiro.au ⟩
Staveley-Smith Lister, Australia, ⟨ lstavele@atnf.csiro.au ⟩
Stecher Therdore, USA, ⟨ stecher@uit.gsfc.nasa.gov ⟩
Suntzeff Nicholas, Chile, ⟨ nsuntzeff@noao.edu ⟩
Sweigart Allen, USA, ⟨ sweigart@bach.gsfc.nasa.gov ⟩
Swerdlyk Caroline, Canada, ⟨ caroline@uvastro.phys.uvic.ca ⟩
Tavarez Maritza, USA, ⟨ tavarez@astro.lsa.umich.edu ⟩
Taylor James, Canada, ⟨ taylor@uvic.ca ⟩
Testor Gerard, France, ⟨ gerard.testor@obspm.fr ⟩
Thornley Michele, USA, ⟨ mthornle@nrao.edu ⟩
Turtle Tony, Australia, ⟨ turtle@physics.usyd.educ.au ⟩
Valle Giada, Italy, ⟨ giada@astr16pi.difi.upipi.it ⟩
Van Dyk Schuyler, USA, ⟨ vandyk@ipac.caltec.edu ⟩
van Loon Jacco, United Kingdom, ⟨ jacco@ast.cam.ac.uk ⟩
Vandenberg Don, Canada, ⟨ davb@uvvm.uvic.ca ⟩
Vanlandingham Karen, USA, ⟨ karenv@suelee.la.asu.edu ⟩
Venn Kim, USA, ⟨ venn@clare.physics.macalester.edu ⟩
Wakker Bart, USA, ⟨ wakker@astro.wisc.edu ⟩
Walborn Nolan, USA, ⟨ walborn@stsci.edu ⟩
Walker Alistair, Chile, ⟨ awalker@noao.edu ⟩
Wang Q. Daniel, USA, ⟨ wqd@dorado.astro.umass.edu ⟩
Weis Kerstin, Germany, ⟨ kweis@ita.uni-heidelberg.de ⟩
Welch Doug, Canada, ⟨ welch@physics.mcmaster.ca ⟩
Welty Daniel, USA, ⟨ welty@oddjob.uchicago.edu ⟩
Westerlund Bengt, Sweden, ⟨ bengt.westerlund@astro.uu.se ⟩
Williams Rosa, USA, ⟨ rosanina@milkyway.gsfc.nasa.gov ⟩
Winkler P. Frank, USA, ⟨ winkler@middlebury.edu ⟩

Yamaguchi Reiko, Japan, ⟨reiko@a.phys.nagoya-u.ac.jp⟩
Zaritsky Dennis, USA, ⟨dzaritsky@as.arizona.edu⟩
Zinnecker Hans, Germany, ⟨hzinnecker@aip.de⟩

Part 1. Overview and Cosmological Context of the Magellanic Clouds

Section A. Invited Reviews

Sidney van den Bergh with his wife Paulette and the guest speakers at the Symposium banquet. From left to right, Kim Innanen, Paul Hodge, Paulette van den Bergh, Sidney van den Bergh, David Crampton, Gretchen Harris, Yuri Efremov, and Jim Hesser. Paulette van den Bergh is holding an honourary membership in the Euro-Asian Astronomical Society presented by Prof. Efremov, while Sidney holds a framed Symposium poster signed by participants. A "rebellious action" foretold by Gretchen Harris during her banquet reminiscences as Sidney's graduate student is illustrated on page 568.

New Views of the Magellanic Clouds
IAU Symposium, Vol. 190, 1999
Y.-H. Chu, N.B. Suntzeff, J.E. Hesser, and D.A. Bohlender, eds.

Magellanic Cloud Studies, Past and Future

Paul Hodge

Department of Astronomy, Box 351580, University of Washington, Seattle, WA 98195, USA

Abstract. This introduction reviews early observations of the Magellanic Clouds, briefly describes the "Golden Age of Magellanic Discoveries", and looks ahead at promises for dramatic new results from the powerful telescopes of the 21st century.

1. Magellanic Cloud Studies, 1200-1950

Although the Magellanic Clouds were observed for centuries by the peoples of the southern hemisphere, it was not until about the thirteenth century that they came to the attention of northern hemisphere inhabitants. According to von Humbolt (1848), Arab explorers and traders, traveling up and down the east coast of Africa, were well aware of the Clouds. They named the Large Cloud the "White Ox".

Among the first Europeans to comment on the Clouds was the sixteenth century scientist Peter Martyr, who said (Eden 1555):

"But it is most certeyne, that it is not gyuen to anye one man to knowe all thynges...Neuerthelesse, the Portugales of owre tyme haue sayled to the fyue and fyftie degree of the south pole: Where, coompasinge abowte the poynt thereof, they myght see throughowte al the heauen about the same, certeyne shynynge whyte cloudes here and there among the starres, lyke vnto theym whiche are seene in the tracte of heauen cauled *Lactea via*, that is, the mylke whyte waye."

The Clouds first became associated with the Great Circumnavigator when Antonio Pigafetta, who accompanied Magellan on his famous voyage, wrote about them in his published account of the trip and proposed that they be named for Magellan as a memorial, Magellan having lost his life in the Philippines before the circumnavigation was complete. It is popular history that Pigafetta was Magellan's navigator. However, this is not the case; Pigafetta was a free-lance adventurer who joined Magellan in the hopes of having a unique experience (Dennefeld 1998).

By the nineteenth century serious astronomical studies of the Clouds was being done. One such investigation was that of James Dunlop, who observed them from Parramatta (near Sydney, Australia) and published particularly handsome and accurate sketches of each Cloud (Dunlop 1828). At about the same time Sir John Herschel visited Capetown in southern Africa to map the southern hemisphere skies. Setting up his 20-ft telescope (named for its focal length), Herschel identified large numbers of star clusters and nebulae, all to be included in his comprehensive "General Catalog" (Herschel 1864), which was later subsumed into the "New General Catalog" (Dreyer 1888), which, of course, is still in use and which lists dozens of Magellanic Cloud objects. Herschel recognized the Clouds as unusual objects and stated (Shapley 1957):

"The two Magellanic Clouds are very remarkable objects. The larger of the two is an accumulated mass of stars, and consists of clusters of stars of irregular form, either conical masses or nebulae of different magnitudes and degrees of condensation.....In no other portion of the heavens are so many nebulous and stellar masses thronged together in an equally small space."

Herschel's African observations were reported widely, even in the popular press (e.g., Locke 1835). Among the users of the General Catalog was Cleveland Abbe, Director of the Cincinnati Observatory (in Ohio, USA). Abbe apparently was impressed by the large density of star clusters and nebulae recorded in the Magellanic Clouds and correctly realized that this implied that the Clouds must be separate, distant systems similar to the Milky Way Galaxy. He went further, extrapolating this conclusion to include smaller and fainter nebulae as separate galaxies, thus anticipating Hubble's work by about 50 years. He wrote (Abbe 1867):

"The visible universe is composed of systems, of which the *Via Lactea*, the two Nubeculae (the Magellanic Clouds - ed.), and the Nebulae, are the individuals, and which are themselves composed of stars and clusters and of gaseous bodies..."

The advent of astronomical photography helped to transfer astronomy from a descriptive to a more quantitative science. The first application of photography to the Magellanic Clouds probably was for purposes of providing accurate stellar positions (Gould 1889). This work was carried out at the great observatories at Cordoba (Argentina), Capetown (South Africa) and Melbourne (Australia). Joining these in the late nineteenth century was Harvard Observatory's southern station at Arequipa, Peru, where the start was made on the magnificent and still valuable plate collection covering the Magellanic Clouds.

The beginnings of modern astrophysical studies of Magellanic Cloud stars can probably be traced to Wilhelmina Fleming's study of S Doradus, the first recognized "luminous blue variable" (or "Hubble-Sandage variable"). She derived its unusual light curve and noted its spectrum as having a P Cygni profile (Shapley 1957).

Another Harvard astronomer, Henrietta Leavitt, began studying the fainter variables in the Magellanic Clouds in 1904. Three years later she made an epochal discovery: while listing the Cepheid variables in order of their periods, she discovered that the list also was nearly in order of magnitudes, thus discovering the Cepheid period-luminosity relation (Wetzel 1955).

In the meantime Herschel's star clusters were to some extent neglected for nearly a hundred years. Shapley (1957) discussed them briefly, listing 10 objects that he considered to be true globular clusters in the Clouds. However, only 2 of Shapley's 10 clusters are known to be true old globular clusters (e.g., Olsen et al. 1999; Johnson 1999), the rest being populous clusters of younger age (see, for example, Hodge 1961). Shapley had suggested that Magellanic globulars might be peculiar, as he pointed out that long ago Annie Cannon had assigned A-type spectra to some of them.

Progress in Magellanic Cloud research in the first half of the twentieth century was slow, which should not be too surprising considering the small number of astronomers in the world at that time. Most work was concentrated in the South American observatories in Argentina and Chile and in South Africa and Australia, where positional astronomy dominated the staffs' efforts. Harvard's Boyden Station, which moved to Bloemfontein, South Africa, during this period, continued its photographic survey of the variable stars in the Clouds, but interest in other Cloud problems was modest.

2. The 1950's and 1960's: The Golden Age of Magellanic Discoveries

By mid-century developments in stellar and galactic astronomy helped to precipitate a strong new interest in the Magellanic Clouds. Stellar evolution and the meaning of stellar populations was finally becoming understandable and the Clouds offered a myriad of objects, all at the same distance, which could be compared to the new theoretical models. At the same time large telescopes, especially the Palomar 5-m, were probing nearby galaxies and the distant cosmos, making the Clouds important as the foundation of the extragalactic distance scale.

A few of the exciting new Magellanic studies of that era are listed here, though there were many others of equal importance:

- Gascoinge and Kron's use of photoelectric photometry to derive a more precise Cepheid period-luminosity relation (Gascoigne & Kron 1953).
- Arp's derivation of color-magnitude diagrams for clusters and field stars in the SMC, allowing comparison with stellar evolutionary models and age determinations (e.g., Arp 1958).
- Feast, Thackeray, and Wesselink's survey of the spectra, colors and luminosities of the Cloud's brightest stars, providing the first good extragalactic HR diagram (Feast et al. 1960).
- Thackeray's and Wesselink's discovery of RR Lyrae variables in the Clouds, giving a distance determination independent of that from Cepheids (Thackeray 1952).
- Payne-Gaposchkin and Gaposchkin's massive study of Cloud Cepheids, permitting analysis of light curve characteristics and exploring the question of age as a function of spatial distribution (e.g., Payne-Gaposchkin 1971).
- Westerlund's explorations of OB assications and other groups of young stars, illuminating the relationship between these objects and neutral and ionized gas clouds (e.g., Westerlund 1961).
- Kerr and his radio astronomy colleagues' pioneering observations of HI in the Clouds and their discovery of the gaseous bridge connecting them (Kerr et al. 1954).
- Mathewson et al.'s discovery of the Magellanic Stream (Mathewson et al. 1974).

As the second half of the century moved on, tremendous progress was achieved in exploring and exploiting the Magellanic Clouds. I will not describe the current status of this research, as it is amply covered by the papers given at this meeting.

3. Magellanic Cloud Research: 2000 –.

If the late twentieth century saw great strides in Magellanic astronomy, the twenty-first century promises even more marvelous developments. These will largely result from the introduction of several amazingly powerful new telescopes, both on the ground and in space.

Of the ground-based telescopes the most impressive is the VLT (the "Very Large Telescope"), being installed on a northern Chilean desert peak and consisting of four 8-m optical telescopes that will eventually be connected to be used interferometrically. Nearby is the site of a giant millimeter array, which will be the short radio wavelength equivalent of the Australia Telescope Compact Array. Two other giant new telescopes in Chile are the Gemini 8-m and the Magellan 6.5-m, both located in the dry mountains near La Serena. Far to the east will be another large optical instrument, the proposed South African Large Telescope.

The first 10 or so years of the new century also will see several powerful space observatories that will make possible many marvelous new possibilities for Magellanic Cloud research. Already HST has proved an extremely useful instrument, as several papers in this volume attest, and the NGST (the "Next Generation Space Telescope") is expected to do much more. Less well-known, perhaps, is the kind of remarkable new measurements that some of the other spacecraft will make possible. To take one example from the rich alphabet soup of telescope names, which includes AXAF, DIVA, GAIA, SIRTF and others, I will point out some things that will be possible to do with SIM (the "Space Interferometry Mission"). For astrometry SIM is planned to have a global positional accuracy of 4μ arcsec and a narrow-angle accuracy of 1μ arcsec. This kind of phenomenal precision will make it possible not only to determine the transverse motions of the Clouds with great ease, but also to measure proper motions of individual stars and clusters in the Clouds. With an expected parallax for the Clouds of 20μ arcsec, SIM will even be able to measure their distances geometrically.

When we combine this new astrometry with the powerful new optical and radio telescopes' results, it will be possible to do many things unthinkable not long ago. Here is a short list of Magellanic Cloud possibilities:

- construct a three-dimensional map of the cluster systems (and, of course, of stars).
- derive the dynamics of stars and clusters.
- determine the evolutionary and kinematic histories of stars.
- examine low mass star formation.
- explore the bottom of the stellar luminosity function.
- reconstruct the details of tidal encounters and possible mergers.
- discover planets.

Acknowledgments. The first part of this article owes a great debt to a review by Shapley (1957), which introduced me to much of the history of the Clouds and which pointed me to the remarkable book by Eden (1555), from which the Peter Martyr quote was taken.

References

Abbe, C. 1867, MNRAS, 27, 262

Arp, H. C. 1958, AJ, 63, 273

Dennefeld, M. 1998, personal communication

Dreyer, J. 1888, MemRAS, 49, Part 1

Dunlop, J. 1828, *A Catalog of Nebulae and Clusters*, Phil. Trans. Roy. Soc. London, 118, 113

Eden, R. 1555, *The First Three English Books on America*, (ed.) E. Arbor, (Birmingham, 1885), p.142

Feast, M., Thackeray, A. D., & Wesselink, A. J. 1960, MNRAS, 121, 337

Gascoigne, B., & Kron, G. 1953, PASP, 65, 32

Gould, B. A. 1889, AJ, 9, 36

Herschel, J. 1864, *A General Catalogue of Nebulae and Clusters of Stars*, Phil. Trans. Roy. Soc., Part 1

Hodge, P. W. 1961, ApJ, 133, 413

Johnson, J. 1999, this volume

Kerr, F., Hindman, J., & Robinson, B. 1954, Austr. J. Phys., 7, 297

Locke, R. A. 1835, *British Scientist Finds Life on Moon*, New York Sun, NY, Aug. 28 edition

Mathewson, D. S., Cleary, M. N., & Murray, J. D. 1974, ApJ, 190, 291

Olsen, K., et al. 1999, MNRAS, 300, 665

Payne-Gaposchkin, C. H. 1971, Smithsonian Contr. Astrophys., No. 13

Shapley, H. 1957, *The Inner Metagalaxy* (Yale U. Press, New Haven)

Thackeray, A. D. 1952, Trans. IAU, 8, 397

von Humbolt, A. 1948, *Cosmos* (Otte, New York), p. 279

Westerlund, B. 1961, Uppsala Ann. 5, No. 1

Wetzel, M. 1955, Annals Harvard College Obs., 109, No. 12

New Views of the Magellanic Clouds
IAU Symposium, Vol. 190, 1999
Y.-H. Chu, N.B. Suntzeff, J.E. Hesser, and D.A. Bohlender, eds.

Cosmic Star Formation from the Milky Way and its Satellites

F.D.A. Hartwick

Department of Physics and Astronomy, University of Victoria, Victoria, BC, Canada, V8W 3P6

Abstract. We use observations and evolutionary models of local objects to interpret a recent determination of the star-formation history of the universe. By fitting the global star-formation rate, the model predicts the ratio of spheroid to disk mass of ~1, an intergalactic medium (IGM) whose mass is ~2.3 times the mass in stars, and whose metallicity is ~0.1 $Z_\odot$.

1. Introduction

It is conventional to interpret observations of high-redshift galaxies using sophisticated galaxy evolution models. However, unless the observed spectra are of sufficiently high S/N, the resulting interpretations may not be unique (c.f. O'Connell 1996). On the other hand, deep color-magnitude diagrams of objects like globular clusters, dwarf spheroidals, and gas-rich systems such as the Galactic disk and the Magellanic Clouds in principle show us their complete star formation history (c.f. contributions in Leitherer et al. 1996 and this volume). To the extent that these systems can be considered representative of the universe as a whole (the mean density of the local group is close to that of the universe), we should be able to use them to deduce a global star-formation history.

2. The Cosmic Star-Formation History

The starting point for this work is the diagram of Madau et al. (1996) and an updated version of this diagram from Madau et al. (1998) which shows the cosmic star-formation rate (SFR) in units of $M_\odot$ $yr^{-1}vol^{-1}$ as a function of redshift z. Very similar results had been derived previously by Pei & Fall (1995) by modeling the observed column densities of H I and their metallicities in damped Lyα systems. Because we will use the results of stellar evolution calculations as input, it is convenient to transform the abscissa from z to lookback time by assuming a specific cosmological model. (The ordinate must also be modified slightly to take into account the model change.) The model assumed has H_0=70 km $sec^{-1}Mpc^{-1}$, Ω_0=0.2, Ω_Λ=0.8, and t_0=14.91 Gyr. This transformed cosmic SFR with a rescaled ordinate is shown in Fig. 1.

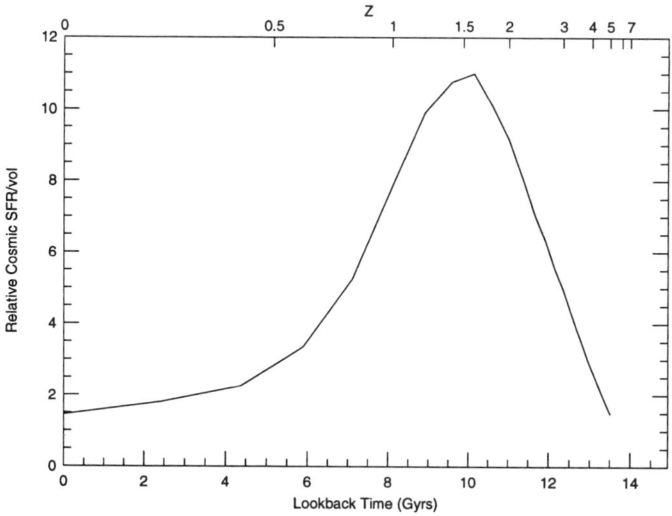

Figure 1. The relative star-formation rate density by Madau et al. (1998) plotted as a function of look-back time.

3. Modeling the Cosmic Star-Formation Rate

Results from the CFRS (Lilly et al. 1998) tell us that while large disks show increases in SFR by factors of ~3 by z=0.7, the red (spheroid) components show no evolution. We shall interpret the above observations as 'spheroids form early and disks form late' (i.e. we consider the hump at ~10 Gyrs in Fig. 1 as predominately due to evolution of the spheroid component while the plateau at shorter lookback times to be due predominately to disk evolution).

In order to model the spheroid component we shall use the one-zone mass-loss chemical evolution model of Hartwick (1976) which has been shown to reproduce the metal abundance distribution of Galactic halo components quite well (e.g. Fig. 1 of Hartwick 1983). The model predicts a metal abundance (Z) distribution given by

$$\mathrm{dS/dlogZ} = \ln 10 \times \mathrm{S}_{tot} \times (\mathrm{Z}/\mathrm{P}_{eff}) \times \mathrm{e}^{-Z/P_{eff}} \qquad (1)$$

where P_{eff} is the effective yield and varies inversely as the star formation (mass loss) rate, S is the cumulative number of stars formed and $S_{tot} = S_{Z=\infty}$.

For a given age-metallicity relation (AMR), we can calculate the star-formation rate by taking the product of eqn 1 with the derivative of the AMR, dlogZ/dt. An AMR was constructed by assuming an age of 13 Gyrs for M92 ([Fe/H]= −2.3), 11 Gyrs for M5 ([Fe/H]= −1.4), and passing through [Fe/H]=0 at $t = 0$. The resulting expression (shown dashed in Fig. 5) is

$$\log Z/Z_{\odot} = 0.117531 \times (1 - \mathrm{e}^{t_9/4.3}) \qquad (2)$$

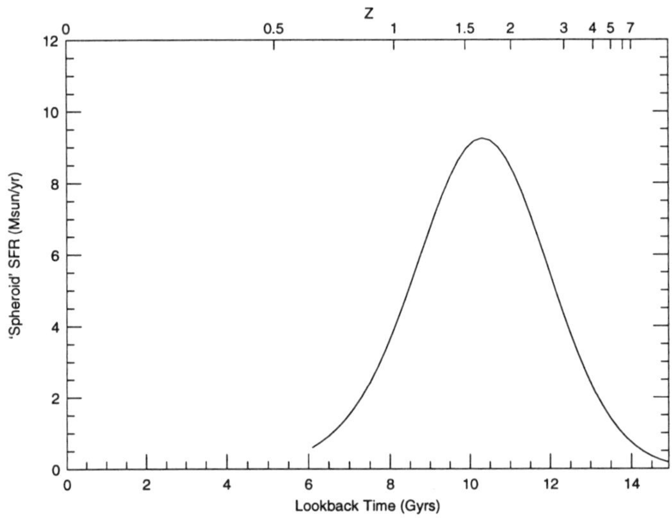

Figure 2. The star-formation rate for the spheroid component of the model.

where t_9 is the lookback time in Gyrs.

For an assumed value of $P_{eff} = -1.0$ and $S_{tot}=3.9\times10^{10}$ $M_\odot$, the resulting SFR is shown in Fig. 2. Note that the maximum SFR of $\sim$10 $M_\odot$ yr^{-1} (which depends only on the value of S_{tot} and the AMR slope) is in the range of values deduced by Steidel et al. (1996) for the $z > 3$ 'UV-dropout' galaxies.

For the disk component we are guided by the SFR derived recently for the Galactic disk by Rocha-Pinto & Maciel (1997) which is shown in Fig. 3. For ease of computation we have adopted the relation shown by the two straight line segments. (Note the increase in SFR by a factor of $\sim$3.4 between $t = 0$ and $t = 9$ Gyrs.)

The disk SFR of Fig. 3 can now be added to the spheroid SFR component of Fig. 2 and compared with the cosmic SFR of Fig. 1. This is shown in Fig. 4. Given the simplicity of the model, the agreement seems quite remarkable! We should emphasize that the composite galaxy modeled above is not expected to exist as a single object in nature. Rather, if the universe contained only one galaxy that is what it might look like!

4. Discussion and Predictions

As indicated above, the star-formation rate of each component was chosen to fit the relative SFR of Fig. 1. In doing so the model makes several predictions. The value of S_{tot} required to match the observed SFR implies a total baryon mass of 18.5×10^{10} $M_\odot$ and a total mass of long-lived stars and remnants of 3.1×10^{10} $M_\odot$. Integrating the assumed disk SFR of Fig. 3 yields a stellar mass of 2.5×10^{10} $M_\odot$. Allowing an additional 10% of the disk mass in gas we calculate the ratio of spheroid to disk mass of 1.1. Schechter & Dressler (1987) have determined this ratio observationally and find it to be $\sim$1. Further, the ratio of total

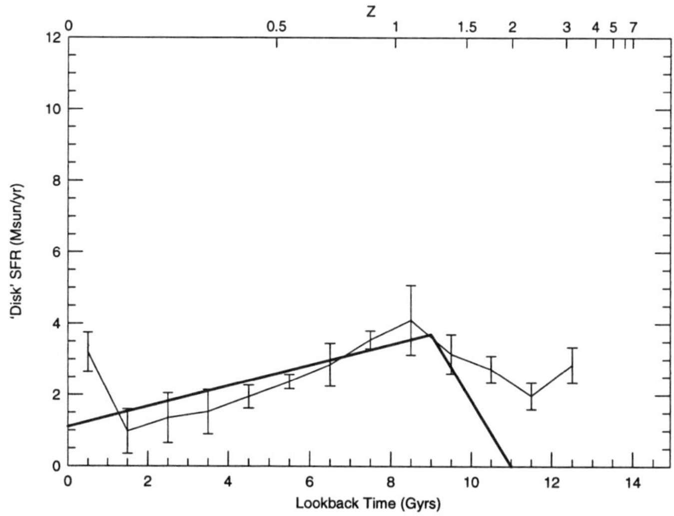

Figure 3. The adopted star-formation rate for the disk component of the model (heavy lines) plotted on top of the derived rate for the Galactic disk from Rocha-Pinto & Maciel (1997).

baryon mass to mass in stars is 3.3. Equating this ratio to $\Omega_{baryon}/\Omega_{stars}$ and assuming Ω_{stars}=0.003 (Persic & Salucci 1992) we find Ω_{baryon}=0.01 compared to Ω_{BBN}=0.03. Thus, the model predicts an alleviation of the 'dark baryon problem' by putting 2.3 times the mass in stars into the IGM as metal-enriched gas. Fig. 5 shows how the enrichment of the IGM is expected to vary as a function of lookback time. Note that the present metallicity of the IGM is predicted to be 0.1 $Z_{\odot}$ (assuming that it is well mixed).

How robust are these results? The cosmological model dependence is expected to be small, amounting basically to an expansion or contraction of the abscissae in Figs. 1-5. However drastic changes to the original Madau et al. (1998) cosmic star-formation rate will have an appreciable effect. For example, if recent suggestions (Hughes et al. 1998) of a factor of 5 increase in SFR at redshift z$\sim$3 are confirmed, then the spheroid contribution would have to rise substantially and increase the derived spheroid to disk ratio upsetting the already good agreement with available observations. However, if the dust-enshrouded star-formation component is more nearly evenly distributed in redshift, then our results should remain valid, requiring only a change in normalization.

5. How Do the Milky Way's Satellites Fit into This Picture?

The above model does not differentiate between star formation taking place in one large object or in many smaller ones. Indeed, star formation followed by mass loss may also be effective if it takes place during collisions between large numbers of smaller objects. It is easily shown that if the collisions which suddenly heat and/or sweep out the gas are Poisson distributed about mean

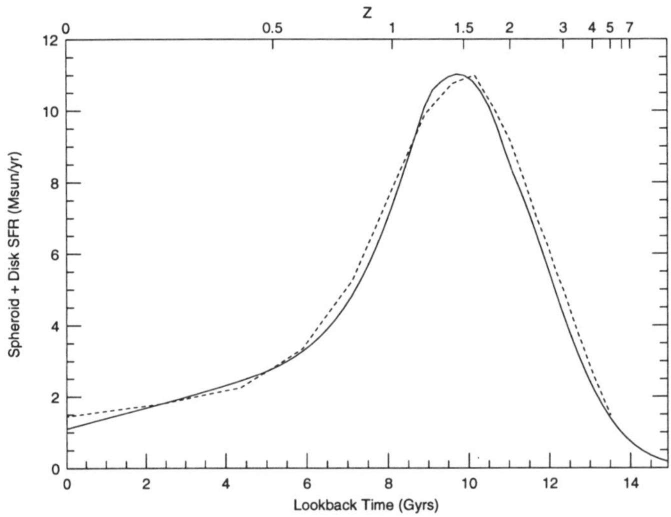

Figure 4. The model spheroid component of Fig. 2 and the model disk component of Fig. 3 combined (solid line) compared to the observed relation (dashed) from Madau et al. (1998) in Fig. 1.

$Z=P_{eff}$, then the expected metallicity distribution is virtually identical to eqn (1). Furthermore, collisions are likely to initiate bursts of star formation as well as eventually to quench them. This picture appears to favor a hierarchical galaxy formation process (e.g., Searle & Zinn 1978) whereby numerous small sub-units collapsed, began forming stars, then collided during assembly over an extended period of time. The gas removed by collisions between the building blocks would then either be lost to the IGM and/or settled down later to form the disk. The removal of gas might then be expected to leave an object resembling a dwarf spheroidal galaxy, if it could survive the sudden mass loss and potential tidal disruption. The large age spread observed from deep color-magnitude diagrams of individual dwarf spheroidal galaxies (e.g., Da Costa 1997) could be taken as evidence for the prolonged spheroid assembly seen in Fig. 2.

The Large Magellanic Cloud would appear to be a candidate for a massive leftover building block given that its most metal-poor clusters appear to be identical to those of the Galaxy (Olsen et al. 1999; Johnson et al. 1999). Such a picture would also allow a rationalization of the disk-like motions of the LMC clusters (Schommer et al. 1992) with the pressure-supported kinematics of the most metal-poor Galactic globular clusters.

Apparently the Magellanic Clouds have so far managed to avoid major mass loss by collision/star formation so that in the above scenario they would be considered relics, having remained relatively self-contained from the time of formation of the first stars. The other gas-rich dwarf irregular galaxies in the Local Group may be similar examples.

Acknowledgments. The author is most pleased to have participated in this celebration of Sidney van den Bergh's monumental contributions to astronomy. He also wishes to thank colleagues S. Gwyn, M. Hudson, J. Navarro,

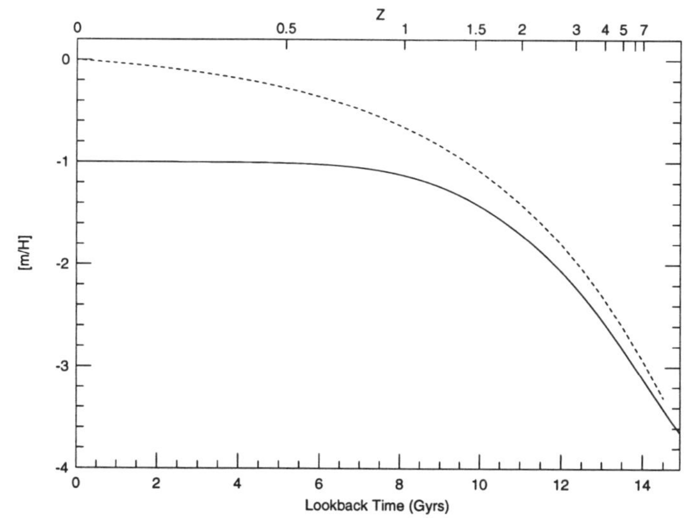

Figure 5. The predicted metallicity of the gas lost during spheroid star formation (solid line). A fixed fraction of this gas forms the disk while the rest becomes the IGM. Dashed line: the assumed age-metallicity relation of eqn 2.

and D. Vandenberg for many useful discussions. Finally the author gratefully acknowledges continued support from NSERC.

References

Da Costa, G. S. 1997, in "The Nature of Elliptical Galaxies", eds. M. Arnaboldi, G. S. Da Costa, & P. Saha, San Francisco: ASP, 116, 270

Hartwick, F. D. A. 1976, ApJ, 209, 418

Hartwick, F. D. A. 1983, Mem. del. Soc. Ast. It., 54, 51

Hughes, D., Serjeant, S., Dunlop, J., et al. 1998, Nature, 394, 241

Johnson, J. A., Bolte, M., Bond, H. E., et al. 1999, this volume

Leitherer, C., Fritze-v. Alvensleben, U., & Huchra, J., (eds) 1996, "From Stars to Galaxies", San Francisco: ASP, 98

Lilly, S., Schade, D., Ellis, R., et al. 1998, ApJ, 500, 75

Madau, P., Ferguson, H. C., Dickinson, M. E., et al. 1996, MNRAS, 283, 1388

Madau, P., Pozzetti, L., & Dickinson, M. 1998, ApJ, 498, 106

O'Connell, R. W. 1996, in "From Stars to Galaxies", eds. C. Leitherer, U. Fritze-v. Alvensleben, & J. Huchra, San Francisco: ASP, 98, 3

Olsen, K., Hodge, P. W., Mateo, M., et al. 1999, this volume

Pei, Y. C., & Fall, S. M. 1995, ApJ, 454, 69

Persic, M., & Salucci, P. 1992, MNRAS, 258, 14P

Rocha-Pinto, H. J., & Maciel, W. J. 1997, MNRAS, 289, 882

Schechter, P. L., & Dressler, A. 1987, AJ, 94, 563
Schommer, R. A., Suntzeff, N. B., et al. 1992, AJ,103, 447
Searle, L., & Zinn, R. 1978, ApJ, 225, 357
Steidel, C. C., Giavalisco, M., Pettinni, M., et al. 1996, ApJ, 462, L17

Discussion

Rebecca Elson: A couple of comments. We have a similar project underway in Cambridge to consider the Local Group in a cosmological context. We find that the dwarf galaxies contribute less than 10% of the total star formation at any redshift, and that none of the dwarf galaxies would be visible in the Hubble Deep Field at redshifts greater than approximately 0.5.

Hans Zinneckeer: In its starkest form, can we claim that all galactic globular clusters were made in Magellanic Cloud type systems which later collided and merged to build up the Milky Way Galaxy?

Hartwick: Possibly, but I also could imagine that when gas lost from spheroid formation eventually collapsed to build up the disk, another generation of more metal-rich clusters might have formed with more disk-like kinematics and spatial distributions.

William Harris: In response to Zinnecker's question, the Galaxy's globular clusters can't all have pre-formed in LMC-like things because a lot of them are very metal-rich by comparision with any clusters in the LMC or other dwarf galaxies. The bigger the galaxy, the more the globular cluster metallicity distribution is weighted to the high end, so the more you have that problem. I agree that you need to form clusters inside dwarf-sized systems, but the bottom line seems to me that you can't do it all in Magellanic-sized things beforehand and then just combine them later - you would get the wrong metallicity distribution.

Simon Morris: The CNOC2 survey is finding that the SFR at $z = 0.2$ to $z = 0.5$ has a substantial contribution from Magellanic (LMC) luminosity objects. This seems to differ from the situation in the Local Group.

David Graff: Since the Madau plot is based on observations of an ensemble of galaxies, it seems natural that its peak should be wider than that of any single galaxy. I don't think your model should be pushed to provide a detailed fit with the Madau plot.

Hartwick: The model I have described is a composite one. I would not expect any individual galaxy to resemble it.

Dominik Bomans: Recent results form JCMT indicate a large population of dusty star-forming galaxies at high redshift, flattening the downturn of the Madau diagram beyond $z = 2$. What effect would this have on predictions of your model?

Hartwick: As long as the dust-enshrouded component contributes a similar amount at all redshifts, then all that is required is an increase in model normalization and all predictions remain the same.

Part 1. Overview and Cosmological Context of the Magellanic Clouds

Section B. Poster Papers

(Top, from left) Stephanie Côté, Frank Younger, Sophia Darwin, and Anka Venema – the Sophysticates String Quartet – entertained guests at the opening reception, while (bottom, from left) Pat McGuire, Teresa McGuire, Brenda Parrish and Victoria Emery cheerfully registered participants and provided local information throughout IAU Symposium 190.

New Views of the Magellanic Clouds
IAU Symposium, Vol. 190, 1999
Y.-H. Chu, N.B. Suntzeff, J.E. Hesser, and D.A. Bohlender, eds.

SMC 1 or What's in a Name?

Hélène R. Dickel, TG Chair

Astronomy Department, University of Illinois, Urbana, IL 61801, USA

Task Group on Designations

Commission 5, International Astronomical Union

Abstract. What's in a name? Everything! SMC 1 is a planetary nebula in the **Large** Magellanic Cloud! URLs of Web documents are given to aid astronomers in designating their astronomical sources so as to avoid confusion in the literature.

1. Introduction to Designation Problems

What's in a name? everything! SMC 1 is a planetary nebula in the **Large** Magellanic Cloud! This new planetary nebula near the LMC was noted by Savage, Murdin, & Clark (1982); it is also known as SMP LMC 104A (Sanduleak, MacConnell, & Philip 1978).

A look at current literature reveals that nonstandard, confusing designations of astronomical sources remain a problem. In an effort to promote clear and unambiguous identification of all astronomical objects outside the solar system, the Task Group on Designations of the International Astronomical Union attempts to clarify existing astronomical designations and the TG reviews, updates, and advertises the IAU Recommendations for Nomenclature. These guidelines and other resources from the IAU are described in the following sections.

2. Designation Guidelines from the IAU

The following documents on the Web are provided as a service to astronomers to help them with designating astronomical sources of radiation outside the solar system:

- How to refer to a source or designate a new one
  `http://cdsweb.u-strasbg.fr/how.html`

- IAU Recommendations for Nomenclature
  `http://cdsweb.u-strasbg.fr/iau-spec.html`

- Second Reference Dictionary of Nomenclature of Celestial Objects
  `http://cdsweb.u-strasbg.fr/cgi-bin/Dic`

The "How to ..." document briefly outlines the steps for creating a proper designation while the "IAU Recommendations ..." gives the full details. The interactive "Second Reference Dictionary" is an electronic look-up which is updated on a regular basis. It provides full references and usages for more than 4000 different acronyms.

3. Additional Resources

The IAU recently developed a "registry" for new acronyms which is part of the Second Reference Dictionary. It may be found at:

- Registry of New Acronyms

  `http://cdsweb.u-strasbg.fr/cgi-bin/DicForm`

Registering an acronym, especially for large, on-going surveys where images and source lists may be produced in stages, helps insure the availability of a suitable, unique acronym for the survey and that the proposed designation conforms to the IAU recommendations.

The Task Group in collaboration with editors of several astronomical journals and managers of large data archives is now studying the feasibility of an automated system to detect nonconforming designations when an article and/or survey data are submitted for publication and/or to an electronic archive.

Members of the Task Group (who are listed at the end of the IAU Recommendations ...) may be contacted regarding general questions about astronomical nomenclature or for specific help in creating appropriate designations for their sources.

Acknowledgments. H. Dickel gratefully acknowledges the continued support and assistance of the IAU Task Group on Designations.

References

Savage, A., Murdin, P.G., & Clark, D.H. 1982, Observatory, 102, 229

Sanduleak, N., MacConnell, D.J., & Philip, A.G.D. 1978, PASP, 90, 621

Part 2. Interstellar Medium

Section A. Invited Reviews

Mary Putman and Bart Wakker are enjoying a poster session.

New Views of the Magellanic Clouds
IAU Symposium, Vol. 190, 1999
Y.-H. Chu, N.B. Suntzeff, J.E. Hesser, and D.A. Bohlender, eds.

New Magellanic Cloud Interstellar Matters

K.S. de Boer

Sternwarte, University of Bonn, D-53121 Bonn, Germany

Abstract.
Novel instrumentation has always led to new insights in the nature of the universe. Novel astronomical instrumentation is no exception. In numerous areas new instruments led to the detection of hitherto unknown phenomena. I will review some notable studies of the recent past with some emphasis on those dicoveries not extensively covered by others at this conference. Finally, very recent work based on ORFEUS echelle spectra is presented.

1. Introduction

New telescopes and new measuring techniques have always led to fast developments of knowledge and understanding in the particular research field. The research on the ISM of the Magellanic Clouds (MCs) is no different. The new satellites like ROSAT (X-ray maps), ISO (near and far IR photometry and spectroscopy), CGRO (Gamma rays), HST (imaging and spectroscopy in the visual and the UV), the space shuttle instruments HUT and ORFEUS (UV-photometry and spectroscopy), new ground based instruments like Hα interferometric mapping devices, IR-CCDs, the Australia Telescope (radiosynthesis), the NANTEN mm telescope (CO), and many others have shown what new windows the new techniques have opened. On many of these, presentations are made in this conference.

In this contribution I will review some notable studies from the bygone years. The choice is personal, thus biased, in part towards those pieces of research which otherwise might not be mentioned in this meeting. Before that I will review *what* we are actually looking at when we observe the interstellar medium in the Magellanic Clouds (and elsewhere). Finally, I will present the new results from ORFEUS echelle spectra of Magellanic Cloud stars.

2. WHAT is Actually Detected?

When we look at the Magellanic Clouds, the most conspicuous part of the ISM is formed by H II regions. We normally say we observe ionized gas, but what we see is, in fact, the *recombining* part of the ionised gas. This is gas in which T is low enough for recombination to really take place and in which the electron and proton density are large (relatively speaking). The optical (and also the radio) recombination lines therefore show us $\int n_e^2\, dl$ of gas at $T \simeq 10^4$ K. Clearly, we preferentially see the densest parts (n_e^2) of the H II gas, thus essentially the *edges* of the H II regions.

The intensities of the forbidden optical and UV emission lines, like [O III], [N II], C III], etc., are based on collisional excitation. In principle one sees $\int n_e^2\, dl$ at the temperature where the particular ion can exist, but matters are complicated because of excitation saturation, so that in part one sees $\int n_{\rm ion} f_{\rm ul}\, dl$ (see Osterbrock 1988, p. 132 ff). Models of the emission spectrum from diffuse gas are available from Domgörgen & Mathis (1994).

In the radio continuum one sees the radiation produced by electrons moving through gas and scattered by the ions. Again the dense portions of the gas radiate best, we see thus n_e^2.

When we make H I 21 cm measurements, we detect $\int n_{\rm H}\, dl$ so H I 21 cm data give us basically $n_{\rm H}$. However, we need to know the length of the sight line, l, over which emission takes place. This is impossible in almost all situations.

The studies of CO (see also the nice maps obtained with NANTEN; this conference) show the existence of that molecule, a molecule easily destroyed by photons, therefore present only in gas shielded from photons. Moreover, the CO emission comes from an excited state. Thus we see basically the very dense and dusty neutral gas, $n_{\rm H}^2$.

Mochizuki et al. (1994) recently surveyed the LMC in the cooling line of [C II] at 158 μm. Their map shows us the gas able to populate the fine structure state of C II, an excitation which in dusty cool clouds is based on $n_{\rm C} \times n_e$ and thus $n_{\rm H} \times n_e$. On the surface of such clouds the excitation is rather based on the part of the radiation field with $\lambda \leq 1335$ Å being able to excite C II as well. So which of the two locations dominates the emission at the positions measured?

The X-ray maps (Einstein, ROSAT, ASCA) show the emission by very hot ($T \geq 10^6$ K) gas. Also here the emission is dominated by the densest parts of the gas, thus on n_e^2, at those high temperatures where the cooling is somewhat more efficient than at neigbouring. However, these X-rays are easily absorbed by neutral gas. The absence of X-rays may therefore mean either of two things: that there is no million degree gas at all or that there is an absorbing cloud in front of the X-ray emitting gas.

Finally, absorption lines measure just the column density $\int n_{\rm X} dl$, but always related to the temperature at which the given absorber can exist. Now we know the line of sight l, but the gas is surely not evenly spread. Another problem is the lack of background sources of radiation, both in the visual and the UV as well as in the radio domain. This then oftentimes is compounded by lack of sensitivity of the detector (or by our wish to observe fainter than possible!).

All these marvelous measurements lure us into combining them to derive nice and clean parameters for the gas. Alas, all these measurements are made with telescopes having different beam sizes for each type of measurement. Thus all the problems of beam-filling factors come into play. Furthermore, the spatial resolution is different, so that a position bright in two kinds of data need not mean it is the same gas making the two kinds of detected radiation.

In all these studies the distance information is meagre, in the sense that depth information is almost impossible to obtain.

In spite of all these limitations, very nice and important work has been carried out in past years.

3. Notable Studies Since 1990

A very compact paper giving important results from EUV spectroscopy is that by Clayton et al. (1996). With the HUT they observed stars in the LMC to study their UV extinction. They could show that the UV-extinction continues to rise from 1500 toward 950 Å as expected from IUE measurements. This is also in line with models for dust extinction (e.g. Aanestad 1989). Using WUPPE the UV-polarization was measured and was found to lie *above* the Serkowsi law. Although the spectra were only of modest spectral resolution, Clayton et al. (1996) could fit the broader spectral depressions with absorption in the Lyman and Werner H_2 absorption bands. Toward 2 stars they find $N(H_2)$ of 1-2 $\cdot 10^{20}$ cm^{-2}.

Of particular relevance for the physics of cool clouds are the 21 cm H I absorption line studies carried out by Mebold et al. (1991) in the Magellanic Stream and by Dickey et al. (1994) and Marx-Zimmer et al. (1999) in the LMC itself. In the Magellanic Stream no absorption was detected, indicating the Stream gas is warm. In the LMC a large number of cold absorption components has been discovered. The average temperature is as low as $T_c \simeq 30$ K, compared to about 60 K for galactic gas seen in 21 cm absorption (Marx-Zimmer et al. 1999).

The SMC has been completely mapped in Hα with the Marseille 2-D Fabry-Perot interferometer (le Coarer et al. 1993) providing us with velocity and intensity information. Also in the LMC various fields have been studied (see e.g. Rosado et al. 1996). Much more of this kind of work needs to be done in the LMC, in particular to get velocities for Hα to be related with the kinematics of shells and with the kinematics in the LMC on a large scale.

A further notable study deals with the comparison of all the wonderful measurements in the different emission lines in the 30 Dor region (Poglitsch et al. 1995). These emissions include those of Br γ, CO, [O I] 63 μm, [C II] 158 μm, H_2 in the near IR, along with the broad band IRAS fluxes. The observational problems of spatial coverage, spatial resolution, as well as of uncertainties in the depth structure become acutely visible in this study.

Depth structure can, however, be found using interstellar absorption lines. For that one needs a large number of stars, closely packed in the sky, and spread appropriately in distance. Such a study was carried out by Vladilo et al. (1993) in the field of SN 1987A in the LMC. Stars at the front side of a spatial structure show few absorption components, stars in the back all. The task is just the one of sequencing, since in reality the distances of the stars are only poorly known, in particular their relative distances at 50 kpc from us. In this way Vladilo et al. (1993) could derive some of the depth structure of the material in the field around SN 1987A.

Another possibility to find depth structure utilises the supernova light echos. SN 1987A was used for that by Spyromilio et al. (1995) and by Xu et al. (1995). Here the dust is mapped in 3 dimensions based on the time sequence of the clouds lighting up in the sky projected around SN 1987A. More details about the 3-D structure were derived from emission line spectra of the surrounding gas (Xu & Crotts 1999).

Time dependent *changes* in the ISM are seldom seen. The gas in the vicinity of SN 1987A showed such variations, which were analysed by Panagia et al. (1991).

The reality of an outer layer of gas at coronal temperatures around the LMC (de Boer & Savage 1980) is now confirmed. Wakker et al. (1998) used better data than available at that earlier epoch (see de Boer 1984) and demonstrated that abundant C IV absorption absolutely unrelated to the background stars is present indeed.

A large number of studies revolved around solving the nature of the supergiant shells in the LMC. In particular hot gas was demonstrated to exist inside LMC 4 (Bomans et al. 1994, 1996) and this structure did not originate in the collision with a high-velocity cloud (Domgörgen et al. 1995). Given the ages of the stars inside LMC 4 a large scale trigger for the star formation is required.

At the SE edge of the LMC Blondiau et al. (1996), using ROSAT data, detected X-ray emitting gas as well as shadowing of X-ray gas by dust clouds detected in IRAS maps. This is suggestive of a bow shock at the leading edge of the LMC and dense star forming gas in the interior. Including the LMC motion and the need to have a large scale star formation trigger for the making of the LMC supershells, the bow-shock induced star formation model was developed (de Boer et al. 1998a).

4. ORFEUS

Let me now turn to the new results from the ORFEUS experiment. The ORFEUS telescope for far-UV spectroscopy flew on the ASTRO-SPAS platform during two space shuttle missions. The ORFEUS telescope feeds the UCB intermediate dispersion spectrograph, working between 300 and 1400 Å with $\lambda/\Delta\lambda \simeq 3 \cdot 10^3$, and the Tübingen-Heidelberg echelle spectrograph, working between 900 and 1400 Å with $\lambda/\Delta\lambda \simeq 10^4$. The attached IMAPS produces high resolution spectra ($\lambda/\Delta\lambda \simeq 1.5 \cdot 10^5$) of bright (and thus galactic) stars (see Jenkins & Peimbert 1997). During the second mission of Nov.-Dec. 1996 the echelle spectrograph functioned well and five Magellanic Cloud stars have been measured (4 in the LMC, 1 in the SMC).

4.1. ORFEUS Echelle Spectra and H_2

The spectra, with a resolution of about 30 $\mathrm{km\,s^{-1}}$, have been searched for the presence of absorption by H_2. The line of sight to one LMC star, LH 10:3120 (see de Boer et al. 1998b), and to the SMC star, HD 5980 (see Richter et al. 1998), both contain H_2 in quantities measurable in absorption in the Lyman and Werner bands. The molecule was found in all spectra at velocities pertaining to the Milky Way as well.

In the MC gas the population of the various J levels of H_2 does not follow a simple rule. As with galactic lines of sight, the lowest rotational levels are mainly populated through collisional excitation (thus showing $n_e \times n_{\mathrm{H}}$). The higher rotational levels are populated through the work of the radiation field. UV photons redistribute the level population to one related to the strength and shape of the radiation field convolved with the probabilities of the electronic transitions (Spitzer & Zweibel 1974).

LMC: Toward the star LH 10:3120, associated with N 11, the kinetic temperature derived from the lowest levels is $T \leq 50$ K (Richter et al. 1998a; de Boer et al. 1998b). Such a low temperature has not been seen in galactic H_2 gas (Spitzer et al. 1974). However, it seems to be in line with the very cold gas mentioned above as discovered in H I absorption studies (see Dickey et al. 1994, Marx-Zimmer et al. 1998). The total column density in front of LH 10:3120 is $N(H_2) \simeq 7 \cdot 10^{18}$ cm^{-2}.

SMC: Toward the star HD 5980, associated with NGC 346, two absorbing clouds are well resolved in the spectrum, *Cloud A* near +120 km s^{-1}, and *Cloud B* near 160 km s^{-1}. The kinematic temperature in *Cloud A*, which is seen in absorption only in levels $J \leq 2$, is $T \simeq 70$ K. In *Cloud B*, seen in levels $J \geq 5$, the UV pumping appears to be extremely strong leading to a level-population distribution equivalent to > 2350 K (Richter et al. 1998b), much higher than any found in the Milky Way (see Spitzer et al. 1974). Both clouds are seen in weak and blended absorption in levels $J = 3, 4$. The total column density in front of HD 5980 is $N(H_2) \simeq 5 \cdot 10^{16}$ cm^{-2}.

4.2. ORFEUS Echelle Spectra and O VI

O VI has been detected in absorption by gas in the halo of the Milky Way on the line of sight to the SMC star HD 5980 (Widmann et al. 1998). It also shows up at $v_{rad} \simeq 140$ km s^{-1}, the 'normal' SMC velocity. Furthermore, it is detected at $v_{rad} \simeq 300$ km s^{-1}, the velocity of exceptional gas noted before by de Boer & Savage (1980) and described in detail by Fitzpatrick & Savage (1983). Gas in this exceptional state (showing all species from Si II and C II through Si III to C IV, N V and O VI) most likely is produced by a supernova shock through gas in the foreground of HD 5980. This gas clearly deserves to be the subject of further investigation.

Acknowledgments. I like to thank the team at the Sternwarte where over the years so many studies of aspects of the Magellanic Clouds could be completed, H.J. Widmann and the ORFEUS team in Tübingen for the pleasant cooperation in the H_2 and O VI projects, and the DFG for the grant Bo779/23 to travel to this conference.

References

Aanestad, P.A. 1989, ApJ, 338, 162

Blondiau, M.J., Kerp, J., Mebold, U., & Klein, U. 1997, A&A, 323, 585

Bomans, D.J., Dennerl, K., & Kürster, M. 1994, A&A, 283, L 21

Bomans, D.J., de Boer, K.S., Koornneef, J., & Grebel, E.K. 1996, A&A, 313, 101

Clayton, G.C., Green, J., Wolff, M.J., et al. 1996, ApJ, 460, 313

de Boer, K.S., & Savage, B.D. 1980, ApJ, 238, 86

de Boer, K.S., Braun, J.M., Vallenari, A., & Mebold, U. 1998a, A&A, 329, L 49

de Boer, K.S., Richter, P., Bomans, D.J., Heithausen, A., & Koornneef, J. 1998b, A&A, 338, L 5

Dickey, J.M., Mebold, U., Marx, M., Amy, S., Haynes, R.F., & Wilson, W. 1994, A&A, 289, 357

Domgörgen, H., & Mathis, J.S. 1994, ApJ, 428, 647

Domgörgen, H., Bomans, D.J., & de Boer, K.S. 1995, A&A, 296, 523

Fitzpatrick, E.L., & Savage, B.D. 1983, ApJ, 267, 93

Jenkins, E.B., & Peimbert, A. 1997, ApJ, 477, 265

le Coarer, E., Rosado, M., Georgelin, Y., Viale, A., & Goldes, G. 1993, A&A, 280, 365

Marx-Zimmer, M., Herbstmeier, U., Dickey, J.M., Stavely-Smith, L., Mebold, U., & Zimmer, F. 1999, A&A, to be subm.

Mebold, U., Greisen, E.W., Wilson, W., Haynes, R.F., Herbstmeier, U., & Kalberla, P.W.M. 1991, A&A, 251, L 1

Mochizuki, K., Nakagawa, T., Doi, Y., et al. 1994, ApJ, 430, L 37

Osterbrock, D.E. 1988, *Astrophysics of Gaseous Nebulae and Active Galactic Nuclei*; Univ. Science Books, Mill Valley

Panagia, N., Gilmozzi, R., Macchetto, F., Adorf, H.-M., & Kirshner, R.P. 1991, ApJ, 380, L 23

Poglitsch, A., Krabbe, A., Madden, S., et al. 1995, ApJ, 454, 293

Richter, P., de Boer, K.S., Bomans, D.J., Heithausen, A., & Koornneef, J. 1998a, in: *The Magellanic Cloud and other dwarf galaxies*, (eds) T. Richtler & J.M. Braun, p. 189

Richter, P., Widmann, H., & de Boer, K.S., et al. 1998b, A&A, 338, L 9

Rosado, M., Laval, A., Le Coarer, E., Georgelin, Y.P., Amram, P., Marcelin, M., Goldes, G., & Gach, J.L. 1996, A&A, 308, 588

Spyromilio, J., Malin, D.F., Allen, D.A., Steer, C.J., & Couch, W.J. 1995, MNRAS, 274, 256

Spitzer, L., & Zweibel, E.G. 1974, ApJ, 191, L 127

Spitzer, L., Cochran, W.D., & Hirschfeld, A. 1974, ApJS, 28, 373

Vladilo, G., Molaro, P., Monai, S., et al. 1993, A&A, 274, 37

Wakker, B.P., Howk, J.C., Chu, Y.-H., Bomans, D., & Points, S.D., 1998, ApJ, 499, L 87

Widmann, H., de Boer, K.S., Richter, P., Kraemer, G., et al., 1998, A&A, 338, L 1

Xu, J., Crotts, A.P.S., & Kunkel, W.E., 1995, ApJ, 451, 806

Xu, J., & Crotts, A.P.S., 1999, ApJ, 511, 262

Discussion

Dennis Zaritsky: Are there many more lines of sight for which H_2 absorption could be detected with current instrumentation?

de Boer: In principle yes. There should be enough dense gas in front of the star to expect the molecules. But then the extinction should be small enough to not lose too much signal needed to record the spectrum. Whether ORFEUS will fly again is unclear. Hopefully FUSE, to be launched in 1999, will provide lots of opportunities.

Wolfgang Gieren: Can you estimate the depth of the LMC in the line of sight from the interstellar medium?

de Boer: Unfortunately not. What one can do, of course, is to find the sequence in depth of clouds and stars, but no, no depth in parsecs. However, taking the radial velocities of the gas clouds one can estimate how soon clouds might collide and by including probability estimates guess how many parsecs the separation in depth is.

Arlin Crotts: In response to the question regarding the lack of knowledge of the scale height of the gas in the LMC: the light echoes from SN 1987A show dust (and associated gas) spread over at least one kiloparsec in depth.

Daniel Wang: You did not mention any results from HST. Could you comment on the results, especially on absorption line studies?

de Boer: Several of those results will be presented by others. Yes, we detected CIV absorption to stars inside LMC 4 demonstrating the cavity is filled with hot gas (Bomans et al. 1996, A&A, 313, 101). And I am, of course, pleased that Wakker et al. (these proceedings) confirm (after 19 years of uncertainty; see de Boer 1984 in IAU Symp 108) the existence of coronal gas around the LMC.

Nolan Walborn: There is some current progress on interpreting the relative distribution of different emissions from 30 Doradus you showed, which will be presened by Monica Rubio this afternoon and myself tomorrow. The filaments west and northeast of R136 are interfaces between the outflows from the latter and dust clouds beyond, in which new star formation is being triggered. Hence, the filaments are bright in Brγ, while the CO and H_2 are concentrated beyond them, with respect to R136.

New Views of the Magellanic Clouds
IAU Symposium, Vol. 190, 1999
Y.-H. Chu, N.B. Suntzeff, J.E. Hesser, and D.A. Bohlender, eds.

The UM/CTIO Magellanic Cloud Emission-line Survey

R. Chris Smith[1,2]
Cerro Tololo Inter-American Observatory, NOAO, Casilla 603, La Serena, Chile

and the MCELS team[3]

Abstract. The Magellanic Clouds are unique in providing sites to study the interstellar medium (ISM) and its components at all scales. To promote the pursuit of such studies, we have begun the Magellanic Cloud Emission-line Survey (MCELS), a deep imaging survey of both of these nearby galaxies in the emission of Hα, [S II], and [O III]. The emission-line images will be used in detailed optical and multiwavelength studies of H II regions, supernova remnants, planetary nebulae, superbubbles, and supergiant shells. Together with parallel surveys at other wavelengths, this survey will provide the foundation upon which to build a deeper understanding of the ISM in the Clouds and other galaxies, from small scales (~1 pc) all the way up to global scales. We present a sample of recent and current work using the MCELS dataset.

While we can study some H II regions, supernova remnants (SNRs), planetary nebulae (PNe), and other nebular structures in detail in our own Galaxy, we lack the global perspective to study their general properties, especially in the optical where interstellar absorption prevents observations of all but the relatively local environment. Studies of distant galaxies provide the global view, but are limited in their ability to examine the detailed interactions on the scales of filaments or stars. The Large and Small Magellanic Clouds are unique in providing us with sites to study the interstellar medium (ISM) and its components at all scales. Their proximity and low foreground absorption make them the ideal laboratories both for studies of individual H II regions, SNRs, and superbubbles, and for investigations of global properties using samples of these objects.

To undertake such research, uniform datasets are needed to provide the basis for the identification, classification, and detailed study of the many components of the ISM. Until recently, optical surveys of the ISM in the Magellanic Clouds were limited to photographic studies performed in individual emission lines (e.g., Hα in Davies, Elliott, & Meaburn 1976, [O III] in Morgan et al.

---

[1] Also Dept. of Astronomy, Univ. of Michigan, Ann Arbor

[2] Dean B. McLaughlin Fellow

[3] MCELS team includes: J. Bregman, Y.-H. Chu, R. Ciardullo, G. Jacoby, R. Kennicutt, S. Oey, F. Winkler, & D. Zaritsky

1995). More recent surveys with modern CCD detectors have generally been limited either in area or in resolution (e.g., Kennicutt et al. 1995). With the relatively recent combination of a large format CCD detector with the Schmidt telescope at CTIO, we have initiated a deep multiple emission-line CCD survey of both the LMC and SMC, the UM/CTIO Magellanic Cloud Emission-line Survey (MCELS).

1. MCELS Details

Taking advantage of the wide field of view available on the UM/CTIO Curtis Schmidt telescope coupled to the sensitivity and large area of newer CCDs, we are imaging the central 8×8 deg. of the LMC and the central 3.5×4.5 deg. of the SMC, covering most of the gaseous extent of both galaxies. The 2048^2 CCD cameras used provide 3-4″ resolution with a field of view of more than 1 degree^2. A grid of multiple, overlapping fields has been planned to provide uniform coverage of the Clouds with multiple samples at each point.

In order to isolate the emission lines, we are using narrow-band interference filters of Hα (λ_c=6563, $\Delta\lambda$=30Å), [S II] (λ_c=6724, $\Delta\lambda$=50Å), and [O III] (λ_c=5007, $\Delta\lambda$=40Å). Images in two "continuum-band" filters (at λ_c=6850, $\Delta\lambda$=95Å and λ_c=5130, $\Delta\lambda$=155Å) are also obtained to allow subtraction of the stellar background, leaving only the emission-line contribution. We plan to reach a limiting surface brightness in all three emission lines of approximately 3×10^{-17} ergs cm^{-2} s^{-1} arcsec^{-2}, a surface brightness roughly equivalent to the diffuse Hα background measured in the LMC by Kennicutt et al. (1995).

2. Preliminary Survey Results

MCELS observations began in November 1995, and as of February 1998 we have completed almost three-quarters of the survey of both Clouds. Preliminary analysis of both survey and pre-survey emission-line data is already providing interesting and exciting results. Highlights of these include:

• **Newly Identified SNRs:** The sample of SNRs in both the LMC and SMC suffers from considerable incompleteness. This incompleteness has been highlighted by the ease with which we have discovered several previously missed SNRs in the LMC. Smith et al. (1994) identified two SNRs in a single $30' \times 30'$ field with the combination of optical emission-line and X-ray datasets (Figure 1). Since then, we have identified, or provided corroborating data for, five more SNRs (e.g., Chu et al. 1995), along with several more SNR candidates.

• **Multiwavelength Studies of Known SNRs:** Despite its incompleteness, the Magellanic Cloud sample of SNRs is still the best sample of remnants available for detailed study. Williams et al. have taken the lead on combining our MCELS emission-line data with both radio and X-ray datasets to examine several LMC SNRs in detail, including the colliding remnants DEM316 (Williams et al. 1997) and two examples of SNR blowouts, N11L and N86 (Williams et al. 1999). Tavarez et al. (1999) have also been pursuing such multiwavelength studies, including near-IR information into the analysis.

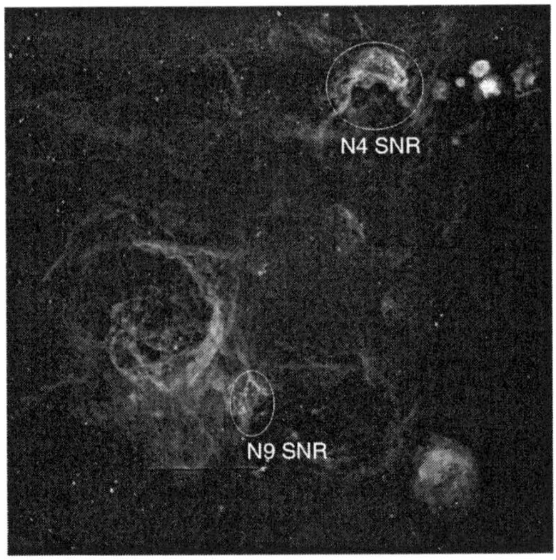

Figure 1. This continuum-subtracted Hα image of a $30' \times 30'$ region on the western edge of the LMC reveals the rich variety of structures which can be seen in the MCELS data, including the newly identified SNRs in the N4 and N9 HII complexes. The subtraction of stellar light allows even the faintest filaments to be traced out.

• **HII Complexes:** On the larger scale, our observations have provided input for several studies of HII complexes and superbubbles in the LMC. Mac Low et al. (1998) showed that the N11 superbubble is an intermediate case in the known range of X-ray overluminosity with respect to superbubble models, and Magnier et al. (1996) demonstrated that the filamentary structure to the south of the superbubble N44 (Figure 2) is most likely due to a blowout from the shell. Kim et al. (1998) added H I data to the analysis of N44, showing that the neutral component has as much as seven times the kinetic energy of the ionized component.

• **Large-scale Structures:** Points et al. (1999) have used MCELS data in their study of the supergiant shell LMC2. Combining the optical emission-line data with the X-ray dataset covering the region, they present a detailed case-study of LMC2. And at the largest scale, Winkler et al. (1999) have provided a global view of the ISM in the SMC, highlighting the variety of structure and detail which can be seen in the MCELS dataset.

3. Future

The work described above is just the beginning of our exploration of the ISM of the Magellanic Clouds using the MCELS observations. When complete, the survey will provide definitive samples of wind-blown bubbles, PNe, and SNRs, as well as calibrated multiple emission-line images of HII regions, superbubbles, and supergiant shells. More information about the MCELS project, as well as

Figure 2. The N44 superbubble in the light of Hα.

details about the datasets and their eventual public release, can be found at `http://www.astro.lsa.umich.edu/mcels`.

Acknowledgments. We gratefully acknowledge the support of CTIO and all of the assistance which has been provided in upgrading the Curtis Schmidt telescope. The MCELS is funded through the generous support of the Dean B. McLaughlin fund at the Univ. of Michigan and through NSF Grant #9540747.

References

Chu, Y.-H., et al. 1995, AJ, 109, 1729

Davies, R.D., Elliott, K.H., & Meaburn, J. 1976, MmRAS, 81, 89

Kennicutt, R.C., et al. 1995, AJ, 109, 594

Kim, S., et al. 1998, ApJ, 503, 729

Mac Low, M.-M., et al. 1998, ApJ, 493, 260

Magnier, E.A., et al. 1996, ApJ, 464, 829

Morgan, D.H., 1995, A&AS, 112, 445

Points, S., et al. 1999, ApJ, 518, 298

Smith, R.C., et al. 1994, AJ, 108, 1266

Tavarez, M., et al. 1999, in preparation

Williams, R., et al. 1997, ApJ, 480, 618

Williams, R., et al. 1999, ApJ, 514, 798

Winkler, P.F., et al. 1999, this volume

New Views of the Magellanic Clouds
IAU Symposium, Vol. 190, 1999
Y.-H. Chu, N.B. Suntzeff, J.E. Hesser, and D.A. Bohlender, eds.

10^6-10^7 K Gas in the Magellanic Clouds

S. L. Snowden

Universities Space Research Association, NASA/GSFC, Greenbelt, MD 20771, USA

Abstract. The Large Magellanic Cloud contains an extensive distribution of hot plasma and is one of the brightest extragalactic regions in the diffuse 0.5 – 2.0 keV X-ray sky. The plasma is not isothermal but increases in color temperature from west to east from $\sim 10^{6.6}$ K to $\sim 10^{6.9}$ K. The total flux from this plasma is $\sim 10^{38}$ ergs s^{-1}. The average emission measure is ~ 0.014 cm^{-6} pc, which if the emitting plasma is distributed uniformly throughout the LMC, implies a space density of $n_e \sim 0.002$ cm^{-3}. There is an apparent $^1/_4$ keV enhancement in the southwest of the LMC, which if associated with the LMC implies a considerable emission measure of $\sim 10^6$ K plasma. (The foreground column density of Galactic neutral hydrogen is $\sim 6 \times 10^{20}$ cm^{-2}, or several optical depths for $^1/_4$ keV emission.)

The Small Magellanic Cloud exhibits less diffuse X-ray emission in the 0.5 – 2.0 keV band than the LMC with a total flux of $\sim 4 \times 10^{36}$ ergs s^{-1}. The average emission measure of ~ 0.006 cm^{-6} pc also implies a space density of $n_e \sim 0.002$ cm^{-3}. The optical depth of Galactic HI for $^1/_4$ keV X-rays from the SMC is considerably lower than that for the LMC. However, while there is a significant variation in the $^1/_4$ keV band intensity over the SMC field, it is in general not particularly well correlated with anything associated with the SMC, although there is an enhancement which may be related to the leading edge of the SMC. Considerably more work is needed to unravel the origin of the structure, which may be associated instead with either emission or absorption variations in either the disk or halo of the Milky Way.

1. Introduction

The Magellanic Clouds (MCs) provide ideal subjects for the study of entire galactic systems. They are near enough so that a substantial fraction of their X-ray point and small-scale extended sources can be adequately resolved and separated from any background of diffuse origin. While diffuse X-ray emission (0.1 – 2.0 keV is the energy range of choice to observe the emission from 10^6 K to 10^7 K plasmas) has been previously observed in the Magellanic Clouds (e.g., Wang et al. 1991; Wang 1991) new maps of the Clouds are presented and analyzed here which include essentially all of the *ROSAT* PSPC observations.

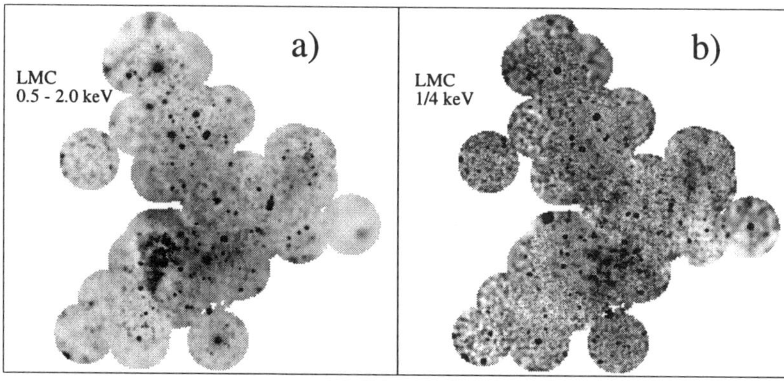

Figure 1. a) Hard band (0.5 – 2.0 keV) image of the LMC. Darker shading implies higher intensity with a dynamic range of 10^{-4} to 3×10^{-3} counts s^{-1} $arcmin^{-2}$. The data are square-root scaled. b) Soft band (0.1 – 0.284 keV) image of the LMC. The dynamic range is 3.5×10^{-4} to 9×10^{-4} counts s^{-1} $arcmin^{-2}$. The data are linearly scaled. The apparent circles of coverage are produced by individual observations and are $\sim 1.5°$ in diameter. The 30 Dor region and Bar are located in the southern half of the image.

2. Data and Data Reduction

The data presented here were obtained over the life time of the *ROSAT* (Trümper 1992) PSPC in a large number of mostly independent observations. The data were accessed through the *ROSAT* public archive of the HEASARC at NASA/GSFC. In order to create the large-area coverage of the fields, the observations ($\sim$ 140 for the LMC, $\sim$ 20 for the SMC) needed to be mosaicked together requiring careful analysis of non-cosmic background constituents and uniformity of treatment of the observations. The processing required three separate steps.

Step 1: Reduction of individual observations – Following the methods laid out in Snowden et al. (1994) and using the software[1] described in Snowden & Kuntz (1998), all pointings at targets in the LMC and SMC were analyzed. Times of anomalous backgrounds were excluded, the particle background, scattered solar X-ray background, and long-term enhancements (see Snowden et al. 1994; Snowden et al. 1995) were modeled and subtracted, and the effective exposure (including vignetting and detector artifacts) determined. Observed counts, modeled background counts, and effective exposure were cast into images in sky coordinates. Multiple pointings in the same target direction, from either the same or different observations, were merged into one set of maps.

Step 2: Determination of residual background offsets – Because of their temporal nature, the contamination from long-term enhancements mentioned

[1]The Extended Source Analysis Software, ESAS, package can be obtained from the NASA/GSFC *ROSAT* Guest Observer Facility through the legacy anonymous ftp account under ftp://legacy.gsfc.nasa.gov/rosat/software/fortran/sxrb.

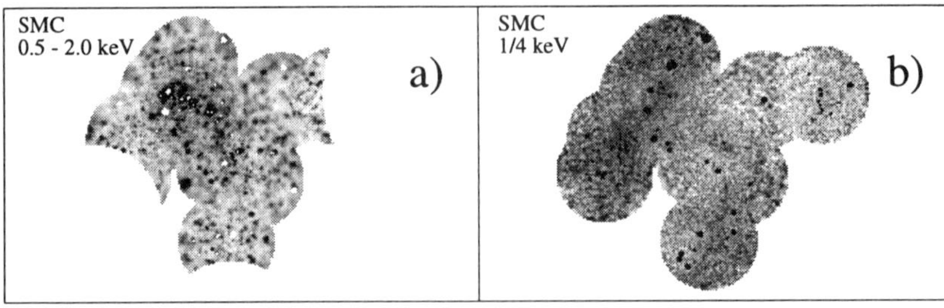

Figure 2. a) Hard band (0.5 – 2.0 keV) image of the SMC. Darker shading implies higher intensity with a dynamic range of 10^{-4} to 6×10^{-4} counts s^{-1} arcmin^{-2}. The data are linearly scaled and point sources have been removed for clarity of the diffuse emission. b) Soft band (0.1 – 0.284 keV) image of the SMC. The dynamic range is 7×10^{-4} to 1.4×10^{-3} counts s^{-1} arcmin^{-2}. The data are linearly scaled. The Bar runs from the southwest to the northeast through the image.

above in general can not be completely purged from individual observations. This can lead to a variable zero-level offset between observations. The zero-level offset, while perhaps not removed, can be at least made more uniform over the set of observations by analyzing the region of overlap between all adjacent pointings. For these observations, a set of simultaneous equations including all of the overlaps was set up and solved by single-value deconvolution to determine the best-fit offset values for each observation. Software to do this is included in the ESAS package and is discussed in Kuntz & Snowden (1998).

Step 3: Mosaicking of the observations – The final step was to mosaic the maps from the individual or combined observations into a common projection. This was done individually for the count, model background count, and exposure maps. The model offset counts from Step 2 were cast into sky coordinates at this time using normalized individual exposure maps as the probability distribution. When the model background and offset-count mosaics were subtracted from the count mosaic and divided by the exposure mosaic, with a proper scaling factor, the result was a count-rate image of the entire field normalized to the on-axis response of the PSPC.

3. Diffuse X-ray Emission from the Magellanic Clouds

Figures 1a,b and 2a,b show hard (0.5 – 2.0 keV) and soft (0.1 – 0.284 keV) band maps of the LMC and SMC. The LMC clearly exhibits extensive diffuse emission in the hard band while in the SMC it apparently exists but at a lower level and is confused with the emission from point sources. Higher angular resolution observations with the *ROSAT* HRI do not show evidence for additional point sources, implying that some of the unresolved emission originates in diffuse gas.

The MC maps in the soft band relate a different story. While the LMC $^{1}/_{4}$ keV map shows a region of significant enhancement in the southwest, the structure is completely unlike that seen at higher energies. The enhancement

Table 1. Diffuse emission parameters.

Cloud	Band	Temperature K	Luminosity ergs cm^{-2} s^{-1}	Region Diameter pc	n_e cm^{-3}
LMC	Hard	$10^{6.6} - 10^{6.9}$	1.1×10^{38}	3500	0.002
	Soft	$10^{6.0}$	1.1×10^{37}	1550	0.008
SMC	Hard	$10^{6.9}$	3.9×10^{36}	1660	0.001
	Soft	$10^{6.2}$	2.1×10^{35}	500	0.003

itself is unexpected as the column density of Galactic HI in the direction of the LMC is ~ 4 optical depths at $^1/_4$ keV. The structure in SMC in the soft band is also nearly completely different from that in the hard band. The structure is extensive enough in solid angle that it may not be related to the SMC at all but be associated with emission and/or absorption variations related to the Milky Way. There is a slight apparent $^1/_4$ keV enhancement along the bar of the SMC. It is this enhancement and the enhancement in the southwest of the LMC that are used to derive luminosity estimates.

A simple estimate can be made for the luminosity of the Magellanic Clouds in X-rays from diffuse emission. The hard and soft bands are first subdivided to provide additional spectral information. The excess emission (after point sources have been masked) over the average emission of the surrounding region is summed to provide a total intensity for each band. (The additional bands cover $0.47 - 1.29$ keV and $0.76 - 2.02$ keV for the hard band, $0.11 - 0.284$ keV and $0.14 - 0.284$ keV for the soft band, the limits are the 10% of peak area values in all cases.) The intensity ratio of the subdivided bands then provide an effective temperature for the broad-band emission assuming a thermal emission model (in this case, Raymond & Smith 1977 spectra, 1995 vintage, with Morrison & McCammon 1984 interstellar absorption). The total emission for each broad band can be scaled to a luminosity. Finally, using the angular extent of the emission enhancements and the assumption of a spherical emission region, the densities of the emitting regions can be determined. Table 1 shows the results. While there are a number of obvious caveats, such as the contribution of unresolved point sources to the diffuse emission and the possibility of additional absorption intrinsic to the Clouds, these results offer a reasonable order-of-magnitude estimate for the diffuse emission parameters.

References

Kuntz, K., & Snowden, S. L. 1998, NASA/GSFC, ROSAT GOF

Morrison, R., & McCammon, D. 1983, ApJ, 270, 119

Raymond, J. C., & Smith, B. W. 1977, ApJS, 35, 419

Snowden, S. L., & Kuntz, K. 1998, NASA/GSFC, ROSAT GOF

Snowden, S. L., et al. 1994, ApJ, 424, 714

Snowden, S. L., et al. 1995, ApJ, 454, 643

Trümper, J. 1992, QJRAS, 33, 165
Wang, Q. D. 1991, ApJ, 377, L85
Wang, Q. D., et al. 1991, ApJ, 374, 475

Discussion

You-Hua Chu: How do you determine the plasma temperatues without knowing the absorption column densities?

Snowden: I don't. I use the measured Milky Way column densities from existing surveys. However, I included no absorption which might be intrinsic to the Magellanic Clouds.

Despina Hatzidimitriou: In the $^1/_4$ Kev band map of the SMC shown, you mentioned that the observed enhancements are probably not associated with the SMC. However, I find it a bit disturbing that they happened to be approximately where the Magellanic Stream and the Magellanic Bridge are starting. Could you comment on that?

Snowden: I didn't say the $^1/_4$ emission enhancements in the SMC field are not associated with the SMC; I don't know the answer to that, and it would be very interesting if they are physically associated. There are alternative explanations which must be investigated. The enhancements are comparable in size and intensity to emission and absorption variations produced in the halo and disk of the Milky Way. Accurate maps of Galactic HI to separate absorption effects and inclusion of ROSAT all-sky survey data to obtain the larger-scale emission context are needed to address the issue.

Frank Winkler: In removing individual X-ray sources, you noted that you mask out not only point sources, but also individual extended sources. At what size scale do you draw the line between individual extended sources and the overall diffuse emission you have described?

Snowden: This is, of course, a subjective cut. Bright, relatively young SNRs are removed as they can easily be studied in the context of individual observations. Older, larger scale, and/or fainter superbubbles which extend beyond observation boundaries and need as much exposure as possible are included. The critical aspect is that when the analysis of the diffuse data is complete, it must be folded back in with the rest of the source populations to derive a global view of the Magellanic Clouds.

Tetsuo Hasegawa: How do the temperatures, luminosities, and densities you derive compare with those parameters for the Milky Way Galaxy?

Snowden: The temperatues of the diffuse plasmas in the Magellanic Clouds are quite consistent with those seen in the Milky Way. This may have a lot to do with regions of relative stability in the cooling curve of the diffuse plasma around 1-5$\times 10^6$ K. Densities of around 0.002 cm^{-3} are also quite reasonable. The $0.5 - 2.0$ kev luminosity of the Milky Way bulge is $\sim 2 \times 10^{39}$ ergs s^{-1}, so the LMC luminosity is reasonable given its relative size.

New Views of the Magellanic Clouds
IAU Symposium, Vol. 190, 1999
Y.-H. Chu, N.B. Suntzeff, J.E. Hesser, and D.A. Bohlender, eds.

Neutral Hydrogen in the Magellanic Clouds

L. Staveley-Smith

Australia Telescope National Facility, CSIRO, P.O. Box 76, Epping, NSW 1710, Australia

S. Kim

Astronomy Department, University of Illinois at Urbana-Champaign, Urbana, IL 61801, USA

S. Stanimirović

University of Western Sydney Nepean, P.O. Box 10, Kingswood, NSW 2747, Australia

Abstract. We review observations of neutral atomic hydrogen (HI) in the Magellanic Clouds (MCs). Being the nearest gas-rich neighbours of the Milky Way the MCs give us an excellent opportunity to study in detail the structure and evolution of the interstellar medium (ISM) and the effect of interactions between galaxies. HI in emission provides a probe of the structure and velocity field of the Clouds, allowing the study of their velocity dispersion, 3-D structure, and large-scale total-mass distribution. Recent data from Australia Telescope Compact Array surveys reveal a morphology (for both Clouds) which is heavily dominated by the effects of local star-formation, rotational shear, fragmentation, self-gravity and turbulence. The new data, which has a spatial resolution down to 10 pc, also allows the study of the distribution functions in velocity and mass for HI clouds. We discuss the morphology, dynamics and giant shell population of the LMC and SMC.

1. Introduction

In this paper, we concentrate on recent observations of HI in emission in the LMC and SMC. HI is an extremely important constituent of the ISM in both Magellanic Clouds. Emission in the $\lambda 21$ cm line is independent of temperature for typical conditions and, in the usual optically-thin regime, becomes a direct measure of column density (e.g. Kulkarni & Heiles 1988). The HI content of the Clouds is high – the HI mass-to-light ratios are 0.14 and 0.61 $M_{\odot}/L_{\odot}$ for the LMC and SMC, respectively (Staveley-Smith et al. 1998).

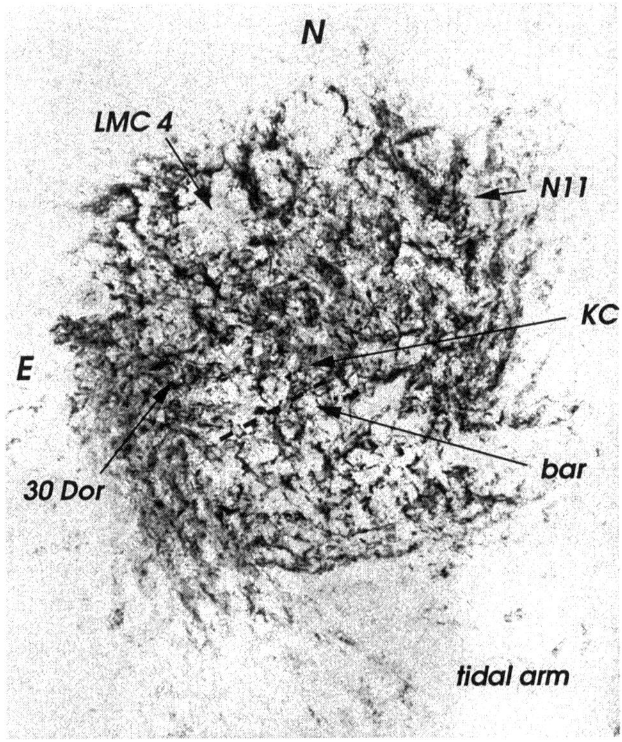

Figure 1. The peak-brightness-temperature image of the LMC from Kim et al. (1998). The HII regions N11 and 30 Doradus are marked, as are the positions of the optical bar and the kinematic center (KC). The tidal arm in the south stretches into the Bridge region between the LMC and SMC. This sub-image is $9^\circ\!.4 \times 10^\circ\!.7$ across.

For the LMC, the highest resolution study is the Australia Telescope Compact Array[1] (ATCA) mosaic by Kim et al. (1998). This provides an angular resolution of $1'\!.0$ and a velocity resolution of 1.6 km s^{-1} over the whole $10^\circ \times 12^\circ$ region encompassing the LMC. However, this study does not include large-scale features which can only be observed with a single-dish. For this, the observations of Luks & Rohlfs (1992) and Putman et al. (1999) with the Parkes telescope may be used. In a square-degree region around 30 Dor, Mebold et al. (1997) have obtained high-resolution ($30''$) ATCA data which they have combined with recent Parkes observations. For the SMC, Stanimirović et al. (1999) have combined recent Parkes observations with the ATCA mosaic of Staveley-Smith et al. (1997). This provides full spatial sampling with an angular resolution of $1'\!.6$

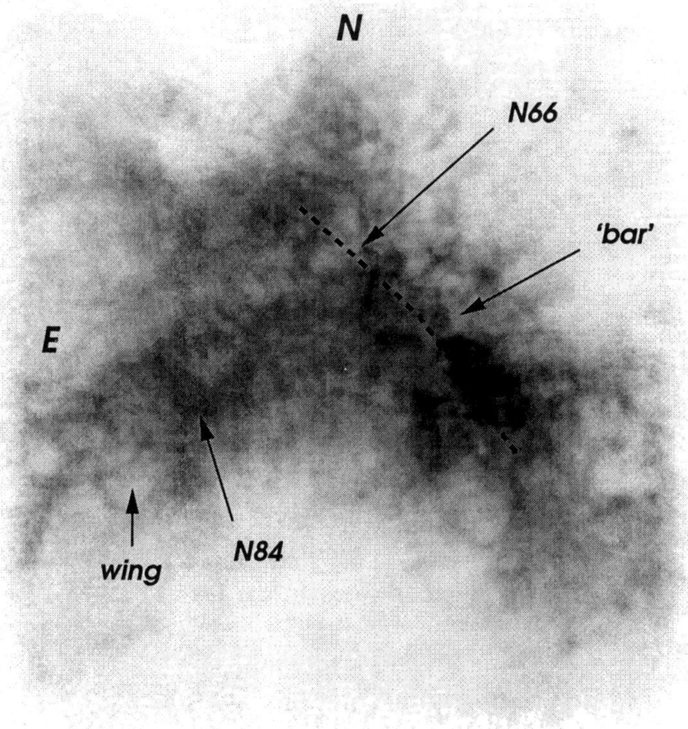

Figure 2. The column density image of the SMC from Stanimirović et al. (1999). The HII regions N66 and N84 are marked, as well as the positions of the optical bar and the eastern wing. The image is 4°.5 × 4°.5 across.

and a velocity resolution of 1.6 km s^{-1} over a region 4°.5 × 4°.5 encompassing the SMC.

2. HI Morphology and Dynamics

The distribution of HI in the LMC is shown in Figure 1. As noted by Kim et al. (1998), the galaxy is surprisingly symmetric and disk-like on the largest scales, compared with earlier optical studies. The outer, tenuous arms reveal its clockwise rotation. The well-known optical bar is not prominent in HI. However, a tidal arm to the south of the LMC is prominent, and appears to extend into the Bridge region connecting the LMC and SMC. The SMC, on the other hand, shows no large-scale symmetry (Figure 2), but is instead dominated by a central HI concentration associated with the optical 'bar', and an eastern wing

[1]The Australia Telescope is funded by the Commonwealth of Australia for operation as a National Facility managed by CSIRO.

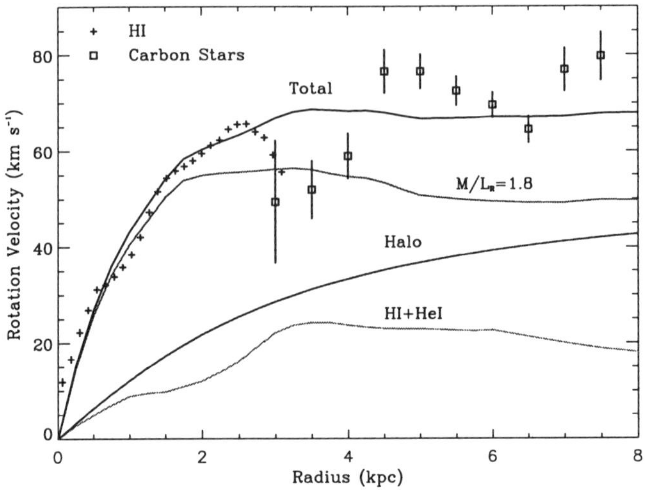

Figure 3. Rotation curve for the LMC from HI (+; Kim et al. 1998) and Carbon star (□; Kunkel et al. 1997) data. The model rotation curve (topmost solid line) is derived from a thick (scale height 600 pc, inclination 33°) stellar and HI/He disk, and a spherical isothermal halo.

(Stanimirović et al. 1999). On small scales, the ISM of both the LMC and SMC is dominated by the effects of stellar winds and SN, this being clearer in Figure 1 owing to the missing short spacings.

As Figure 3 shows, the rotation curve of the LMC can be decomposed into several significant components: a thick stellar disk, a thick gaseous disk and a dark halo. Within a radius of 4 kpc, Figure 4 shows that the respective masses are 1.4×10^9, 4×10^8 and 1.0×10^9 $M_\odot$.

The SMC also has a significant velocity gradient (Figure 5). If interpreted as simple, disk-like rotation, the inclination of the SMC would have to be $< 35°$ in order for the dynamical mass to exceed the obvious baryonic mass (HI, He, stars) of $\sim 1.3 \times 10^9$ $M_\odot$ within a 2 kpc radius. This would imply intrinsic rotation > 50 km s^{-1}. However it is unlikely that simple disk rotation model is relevant. Cepheid data (Caldwell & Coulson 1986) show that the average distance of the stellar population in the NE may be closer by $\sim$ 16 kpc compared with the SW. This, combined with modeling (Gardiner et al. 1994), shows that larger scale tidal forces are likely to be responsible for the current kinematics of the SMC.

3. The HI Shell Population

Recent (< 50 Myr) star-formation events leave indelible features on the ISM through the effects of stellar radiation, winds and supernova blast waves. In the HI data, these features reveal themselves most clearly as giant and supergiant shells. The physical properties of these features (mass, expansion velocity, radius, dynamical age, kinetic energy, number density, size distribution) can be measured and tested against models of star formation. The size distribution

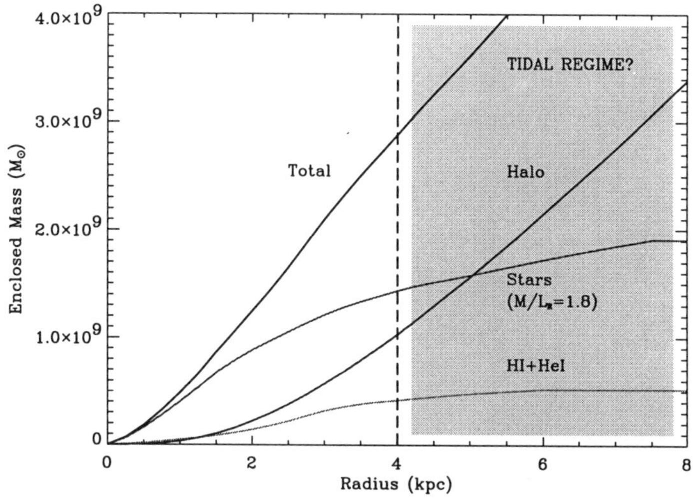

Figure 4. The enclosed masses of stars, gas and halo in the LMC. At 4 kpc radius, the enclosed mass is 3×10^9 $M_\odot$. Beyond 4 kpc, it is likely that tidal forces are important.

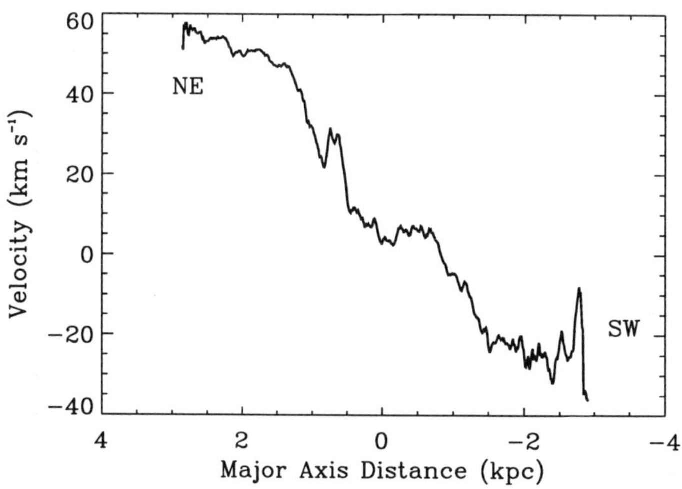

Figure 5. Radial velocities along the kinematic major axis of the SMC. Velocities have been corrected for Proper Motion and are in the Magellanocentric frame of Gardiner et al. (1994).

of the recently cataloged shells and candidate shells in the LMC and SMC is shown in Figure 6. For $R > 80$ pc, a limit dictated by resolution and completeness, the SMC has a shell-size distribution that can be approximated by a power law $N(\log R) \propto R^{-1.6}$. Using the standard model for adiabatically expanding shells (e.g., Weaver et al. 1977), Oey & Clarke (1997) predict a power law $N(\log R) \propto R^{-1.7\pm0.6}$, in good agreement. For the LMC, the size distribution

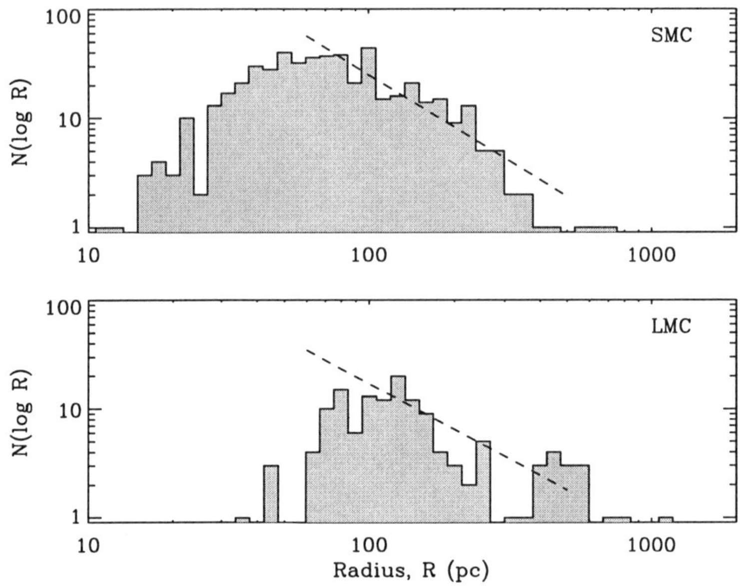

Figure 6. The distribution of shell radii for HI shells in the SMC (top) and HI shell candidates in the LMC (bottom). The dashed line represents a power law $N(\log R) \propto R^{-\alpha}$ with $\alpha = 1.6$ for the SMC and $\alpha = 1.4$ for the LMC.

for the supergiant and giant shell candidates (Kim et al. 1999) is also shown in Figure 6. There is an apparent excess of supergiant shells in the LMC, resulting in a shallower power law, $N(\log R) \propto R^{-1.4}$. However, the number of SMC shells (~ 500 cataloged) appears to greatly exceed the number of LMC shells (~ 130: Kim 1998, Kim et al. 1999). At a shell radius of 100 pc, the surface density of SMC shells is about 4.1 times that of LMC shells. This is surprising given that the specific star-formation rate of the LMC is probably higher than for the SMC and that, for a given age, shell radius is only a weak function of wind luminosity and ISM density. The surface density ratio of shells is comparable to the ratio of HI surface densities (~ 3.8, Staveley-Smith et al. 1998), implying that confusion and a more complex ISM may impede shell identification in the LMC. Another factor may be venting of LMC shells once their radius approaches the scale height of the disk. This causes more shells to stall, and become difficult to identify. Almost half the supergiant shell candidates noted by Kim et al. (1999) have uncertain expansion velocities, and may be stalled.

Acknowledgments. We thank J. Dickey, M. Dopita, K. Freeman, D. Hatzidimitriou, M. Kesteven, D. McConnell, R. Sault and S. Snowden who have greatly assisted us in this research.

References

Caldwell, J. A. R., & Coulson, I. M. 1986, MNRAS, 218, 223

Gardiner, L. T., Sawa, T., & Fujimoto, M. 1994, MNRAS, 266, 567

Kim, S. 1998, Ph.D. Thesis, Australian National University

Kim, S., Dopita, M. A., Staveley-Smith, L., & Bessell, M. 1999, ApJ, submitted

Kim, S., Staveley-Smith, L., Dopita, M. A., Freeman, K. C., Sault, R. J., Kesteven, M. J., & McConnell, D. 1998, ApJ, 503, 674

Kulkarni, S. R., & Heiles, C. 1988, in Galactic and Extragalactic Radio Astronomy, G. L. Verschuur & K. I. Kellermann, New York: Springer-Verlag, 95

Kunkel, W. E., Demers, S., Irwin, M. J., & Albert, L. 1997, ApJ, 488, L129

Luks, T., & Rohlfs, K. 1992, A&A, 263, 41

Mebold, U., Düsterberg, C., Dickey, J. M., Staveley-Smith, L., & Kalberla, P. 1997, ApJ, 490, L65

Oey, M. S., & Clarke, C. J. 1997, MNRAS, 289, 570

Putman, M. E., Gibson, B. K., & Staveley-Smith, L. 1999, this volume

Stanimirović, S., Staveley-Smith, L., Dickey, J. M., Sault, R. J., & Snowden, S. L. 1999, MNRAS, 302, 417

Staveley-Smith, L., Sault, R .J., Hatzidimitriou, D., Kesteven, M. J., & McConnell, D. 1997, MNRAS, 289, 225

Staveley-Smith, L., Kim, S., Putman, M., & Stanimirović, S. 1998, in Reviews in Modern Astronomy, R. E. Schielicke, Hamburg: Astronomische Gesellschaft, 117

Weaver, R., McCray, R., Castor, J., Shapiro, P., & Moore, R. 1977, ApJ, 218, 377

Discussion

Marc Kutner: When you do mosaicing how do you assure image fidelity on intermediate scales (i.e. structures that extend over a few fields)?

Stavely-Smith: The largest scale structures ($0\overset{\circ}{.}6$) are not present in the LMC data presented. On intermedidate scales ($0\overset{\circ}{.}3$ - $0\overset{\circ}{.}6$) fidelity is assured by Nyquist sampling of the field (primary beams spaced by half a FWHP beam). A maximum entropy procedure is used to combine information from neighboring fields.

Sally Oey: The relative number of HI shells compared to HII regions in the SMC is consistent with the relative life expectancies for shells vs. HII regions. However, the small number of shells in the LMC is startling, considering the higher star formation rate in the LMC. Why is there such a small number of shells in the LMC?

Stavely-Smith: I suspect the answer to this very interesting question is related to (A) the LMC having a thinner disk than the SMC which allows the easier venting of giant/supergiant shells; and (B) the LMC being much less HI-rich than the SMC. This may decrease the time required for the shells to dissipate into the general ISM.

David Graff: What are the large scale departures from the disk rotation solution in the LMC?

Stavely-Smith: The most significant departues are around the supergiant shells such as LMC4 and the 30 Dor regions. In the east, there is a spiral arm which is dynamically hot, perhaps a result of tidal interactives.

Daniel Wang: Any idea about the volume filling factor of HI gas?

Stavely-Smith: If we assume the HI shell interiors contain no HI, the volume filling factor of the HI gas is 0.5, or less in the SMC. If modelled as a fractal structure, a much smaller filling factor is suggested.

Nicholas Suntzeff: Stars in the LMC extend to ~8 kpc, while your maps and total HI gas extend only out to 4 kpc. Do you have total HI masses out to the extent of the stellar disk?

Staveley-Smith: The existing ATCA maps have no zero-spacing (single-dish) information, so I cannot say. The previous data of McGee and Milborn and Luks and Rohlfs appear to suggest that the HI mass beyond 4 kpc does not contribute much to the total HI mass. However, there appear to be calibration differences between their data sets.

New Views of the Magellanic Clouds
IAU Symposium, Vol. 190, 1999
Y.-H. Chu, N.B. Suntzeff, J.E. Hesser, and D.A. Bohlender, eds.

Interstellar Phases in the Magellanic Clouds

John M. Dickey

Univ. of Minnesota, 116 Church St. SE, Minneapolis, MN 55455, USA

Monika Marx-Zimmer, Christian Düsterberg, Ulrich Mebold

Radioastronomisches Instutut der Universität Bonn, Bonn, Germany

Snezana Stanimirović, Lister Staveley-Smith

Australia Telescope National Facility, Epping, NSW1710, Australia

H.A. Kobulnicky

Astronomy Dept., Univ. of California, Santa Cruz, CA 95064, USA

Abstract.
Surveys of λ21-cm absorption in the Magellanic System show that the cool phase of the HI is less abundant in the SMC than in the Milky Way, and may be so also in the LMC. The typical cool cloud temperature is colder than in the Milky Way, 30 to 40 K rather than 60 to 75 K. The lower abundance of cool phase HI can be traced to the lower heavy element abundances in the Magellanic environment. The cooler cloud temperatures are somewhat mysterious.

1. Background

The Magellanic Clouds contain interstellar matter showing the physical properties associated with all the various phases of the interstellar medium of the Milky Way. There is a hot (kinetic temperature $T_{kin} \simeq 10^5$ to 10^6 K) ionized medium traced by the soft X-ray emission and UV absorption (e.g. Snowden & Petre 1994; Wang et al. 1991; Bomans et al. 1996). There is abundant warm ionized medium ($T_{kin} \simeq 10^4$ K) both in classical HII regions and in diffuse ionized regions traced by Hα and many other emission lines (Kennicutt et al. 1995). There are clouds of molecular gas at temperatures from about 30 K to below 10 K, traced by emission from CO and other molecules (Israel et al. 1993; Rubio et al. 1993). Tracers of the molecular gas indicate that this phase is much less abundant in the Magellanic System than in the Milky Way, but this may be due to a lower ratio between CO emissivity and H_2 column density compared to that in the Galaxy (Maloney & Black 1988). In addition, the Magellanic Clouds contain abundant atomic hydrogen in both the warm ($T_{kin} \simeq 6000$ K) and cool ($T_{kin} \simeq 60$ K) neutral phases, similar to what is seen in the Milky Way (Milky Way reviews include Kulkarni & Heiles 1987, 1988, and Dickey & Lockman 1990). This paper concentrates on these latter two phases of the interstellar medium (ISM) of the Magellanic System.

To study the relative abundance and distribution of the warm and cool neutral gas requires observations of both emission and absorption spectra in the λ21-cm line. Though emission is relatively easy to detect, λ21-cm absorption toward compact background sources requires high angular resolution to separate from the emission. Thus it has only been possible to measure λ21-cm absorption through the Magellanic Clouds since the Compact Array of the Australia Telescope (ATCA) has become available. So all of our knowledge of the cool atomic phase of the ISM in the Clouds has come in the last five years or so.

The ultimate goal of HI absorption studies is to understand how the balance between heating and cooling of the HI in different environments sets the relative abundances of the warm and cool neutral media. The Magellanic Clouds are perfect as case studies in this subject, since they are so near, and yet so different from the Milky Way in their abundances of the elements critical for **cooling** (primarily C and O), and in their dust abundance which is critical for **heating**. By studying the distribution of temperatures in the HI we can determine how robust is the mixture of phases which we have in the solar neighborhood. If there are big differences in the phase mixture in the Magellanic environment, we must expect differences among galaxies, and with redshift as well. But if the mixture is similar to that in the Milky Way in spite of the differences in the environments, we may conclude that there is a universal mixture of HI phases which is somehow inescapable.

2. Surveys

The column density of HI is most easily traced by emission in the λ21-cm line, since the emission coefficient for this transition is proportional to density and independent of temperature. A survey of the brightness temperature of the line thus maps out the distribution of HI column density, $N_H \propto T_{em}$, with only small corrections needed for self-absorption. The optical depth, τ, is proportional to the column integral of density, n, divided by temperature. Thus cool gas shows itself prominently in absorption spectra, while the warm gas is almost undetectable in absorption, but the two phases both contribute to emission spectra in proportion to their masses. Thus comparing emission and absorption allows us to eliminate the density and measure the excitation temperature alone. The excitation temperature of the λ21-cm line is equal to T_{kin} since the spontaneous transition rate is extremely low.

Two kinds of absorption studies have been done with the ATCA : 1) surveys of ATCA absorption spectra toward many background sources, which have been paired with coarse resolution emission data taken in the same directions with the Parkes 64m telescope; and 2) a few full synthesis observations which compare the absorption with a high resolution map of the emission made with a combination of data from both Parkes and the ATCA. The former method is a quick way to measure the relative abundances of warm and cool gas in many directions, but to determine the temperature of the cool phase, T_{cool}, requires the latter approach. This is because a single Parkes emission spectrum (beamwidth 14′ or 200 pc at 50 kpc) and the absorption spectrum toward a compact background continuum source (typically 5″ in diameter) sample quite different volumes of gas. In order to sample nearly the same line of sight in emission as in absorption requires a small beam in the emission study, so as to get very close to the direction of the background source.

To analyse emission-absorption data taken in the quick survey mode we work with the velocity integrals of the spectra, assuming that the averages for the sample as a whole give representative information on the properties of the HI overall. Thus the mean excitation temperature,

$$< T_{spin} > \; = \; \frac{\sum \int T_{em}(v)\, dv}{\sum \int (1 - e^{-\tau})\, dv} \propto \frac{N_H}{\int \frac{n}{T_{kin}}\, ds}$$

measures the density weighted harmonic mean temperature of the gas, including both phases. (The sums are taken over all emission/absorption spectrum pairs in the survey, and the integral over ds is taken along the line of sight.) In the solar neighborhood we find a value of about 250 K for $< T_{spin} >$. To interpret this average we make the "two phase medium" assumption, i.e. that the warm gas is warm enough to make a negligable contribution to the absorption (in practice it must only be warmer than a few hundred K, though in general the warm phase is at least several thousand K). With this assumption, an observed value for $< T_{spin} >$ gives the relative fractions of HI in the warm and cool phases, f_{warm} and f_{cool}, if we know the value of T_{cool}. Thus :

$$f_{cool} \equiv \frac{N_{cool}}{N_{warm} + N_{cool}} = \frac{T_{cool}}{< T_{spin} >} = 1 - f_{warm}$$

As simple way to compare different environments we can assume the same value for T_{cool} and compute f_{cool} for each. For the Milky Way in the solar neighborhood we have some cases where T_{cool} is known to be 60±5 K, so this is a reasonable value to use for the comparison. Using this number gives $f_{cool} \simeq 25\%$ for the Milky Way.

Region	number of lines of sight	number with absorption	$< T_{sp} >$ (K)	f_c (%)	Reference
Halo & Stream	5	0	>600	<0.1	Mebold et al. (1991)
LMC	30	19	165	36	Dickey et al. (1994)
LMC	20	9	180	33	Marx-Zimmer et al. (1999)
Bridge & Halo	7	2	59	-	Kobulnicky and Dickey (1999)
30 Dor	5	5	30 - 40	→100?	Mebold et al. (1997)
SMC	32	11	465	13	Dickey et al. (1999)

Table 1. Magellanic Cloud λ21-cm surveys.

The table lists surveys of λ21-cm absorption in the Magellanic System. The overall result of all these studies is to show that the mixture of phases in the

HI **does** differ between the Magellanic Clouds and the Milky Way. Extensive observations of the LMC show that it contains **more** cool phase HI relative to warm phase than the Milky Way does (about 35% cool vs. 25% cool), though the warm phase still dominates. In the SMC the situation is reversed. There absorption is quite rare, indicating that no more than about 15% of the HI is in the cool phase. These numbers depend on the assumption that T_{cool} is the same in the Magellanic Clouds as in the Milky Way, if T_{cool} is lower there then f_{cool} would be lower in proportion.

We have some information on T_{cool} from two studies of the second type, listed as the last two lines on the table, which include full aperture synthesis of the emission (for the SMC taken from Stanimirović et al. 1999). These studies both show surprisingly cool temperatures, typically 30 to 40 K, for T_{cool}. Using this value would reduce the fraction of the HI needed in the cool phase to $f_{cool} \simeq 20\%$ and 8% for the LMC and the SMC, respectively. The question is, why should the kinetic temperature in the diffuse ISM clouds of the Magellanic System be so low, and why should that phase be so scarce in the SMC ?

3. Discussion

We can understand the balance between heating and cooling with reference to the theoretical study of Wolfire et al. (1995). They consider thermal equilibrium in an environment with low metallicity, which has the effect of raising the minimum pressure for which the cool phase can exist, and raising the maximum pressure for which the warm phase can exist. Thus decreasing the abundance of C and O makes it harder to have cool phase gas, and presumably would decrease the abundance of the cool phase relative to the warm phase. A similar result is obtained by increasing the UV radiation field. On the other hand, Wolfire et al. also consider decreasing the grain abundance, which reduces the heating rate. If both the metallicity and the dust-to-gas ratio decrease together, then the effect is similar to just reducing the metallicity, but less strong. Ultimately, heating by soft x-rays becomes more important than photoelectric heating, if the dust abundance is reduced far enough. In sum, the heating and cooling balance in an environment like that of the SMC can explain why the abundance of cool phase gas is decreased compared to the solar neighborhood.

Why the temperatures of the cool clouds are also lower than their Milky Way values is more of a puzzle. For this we may consider models of the equilibrium between molecule formation and destruction (e.g. Elmegreen 1993). With a much lower abundance of dust, and a higher intensity of UV which can photodissociate molecules, it may be that the molecular gas is driven deeper into the clouds, toward regions of higher protective column density, than in Milky Way clouds. Thus the HI could dominate over H_2 in some regions with quite high densities, $n \simeq 10^3$ cm^{-3}, perhaps. So in the Magellanic Clouds the HI absorption traces not only diffuse interstellar clouds, but dense, cold clouds as well. These clouds would be molecular if they were in the Milky Way, but in the Magellanic environment they remain predominantly atomic.

The presence of these cold, atomic clouds in the Magellanic System raises the question of whether a similar population exists in the Milky Way. A few examples of HI clouds with temperatures below 40 K have been noticed in studies of the Galactic ISM, but they are uncommon in the solar neighborhood. Surveys at low latitudes could reveal whether these "missing link" clouds, between the diffuse and molecular types, are common elsewhere in the Milky Way.

References

Bomans, D.J., de Boer, K.S., Koornneef, J., & Grebel, E.K. 1996, A&A, 313, 101.

Dickey, J.M., & Lockman, F.J. 1990, ARA&A, 28, 215.

Dickey, J.M., Mebold, U., Marx, M., Amy, S., Haynes, R.F., & Wilson, W. 1994, A&A, 289, 357.

Dickey, J.M., Mebold, U., Stanimirovic, S., & Staveley-Smith, L., 1999, in preparation.

Elmegreen, B.G. 1993, ApJ, 411, 170.

Israel, F.P., Johansson, L.E.B., Lequeux, J., Booth, R.S., Nyman, L.A., et al. 1993, A&A, 276, 25.

Kennicutt, R.C., Bresolin, F., Bomans, D.J., Bothun, G.D., & Thompson, I.B. 1995, AJ, 109, 594.

Kulkarni, S.R., & Heiles, C., 1987, in: *Interstellar Processes*, (eds.) D.J. Hollenbach & H.A.Thronson, (Dordrecht: Reidel), p. 87.

Kulkarni, S.R., & Heiles, C., 1988, in: *Galactic and Extragalactic Radio Astronomy*, 2nd ed., eds. G.L. Verschuur & K. Kellerman, (New York : Springer-Verlag), p. 95.

Kobulnicky, H.A., & Dickey, J.M. 1999, AJ, 117, 908

Maloney, P., & Black, J.H. 1988, ApJ, 325, 389.

Marx-Zimmer, M., Herbstmeier, U., Dickey, J.M., Staveley-Smith, L., Mebold, U., & Zimmer, F. 1999, A&A, in press.

Mebold, U., Greisen, E.W., Wilson, W., Haynes, R.F., Herbstmeier, U., & Kalberla, P.M.W. 1991, A&A, 251, L1.

Mebold, U., Dusterberg, C., Dickey, J.M., Staveley-Smith, L., Kalberla, P., Müller, H., & Osterberg, J. 1997, ApJ, 490, L65.

Rubio, M., Lequeux, J., Boulanger, F., Booth, R.S., Garay, G., et al. 1993, A&A, 271, 1.

Stanimirović, S., Staveley-Smith, L., Dickey, J.M., Sault, R.J., & Snowden, S.L. 1999, MNRAS. 302, 417

Snowden, S.L., & Petre, R. 1994, ApJ, 436, L123.

Wang, Q., Hamilton, T., Helfand, D.J., & Wu, X. 1991, ApJ, 374, 475.

Wolfire, M.G., Hollenbach, D., McKee, C.F., Tielens, A.G.G.M., & Bakes, E.L.O. 1995, ApJ, 443, 152.

Discussion

Dan Welty: You see evidence for a high percentage of cold gas in HI in the LMC, but heavy element abundances (from UV absorption lines) do not show evidence from "cold dust" depletions. Any comments/explanations?

Dickey - paraphrased: The bulk of the HI is still warm.

Unknown: How about the gas support in radial direction in the LMC, SMC and the Milky Way?

Dickey: I think the LMC is like the Milky Way in that it has a rotation curve, so the gas is supported by centrifugal force in the radial direction, not by a pressure gradient. For the SMC there is a velocity gradient, but it's not clear this is a rotating disk, so I don't know if we know the shape of the gravitational potential.

New Views of the Magellanic Clouds
IAU Symposium, Vol. 190, 1999
Y.-H. Chu, N.B. Suntzeff, J.E. Hesser, and D.A. Bohlender, eds.

New HI Features of the Magellanic System

M. E. Putman

Mount Stromlo and Siding Spring Observatories, ANU, ACT 2611, Australia

B. K. Gibson

CASA, University of Colorado, Boulder, CO 80309, USA

L. Staveley-Smith

Australia Telescope National Facility, CSIRO, NSW 2121, Australia

Abstract. The first results from the HI Parkes All-Sky Survey (HIPASS) provide a spectacular view of the global HI distribution in the vicinity of the Magellanic Clouds and the southern Milky Way. A 2400 square degree mosaic around the South Celestial Pole (SCP) reveals the existence of a narrow, continuous counter-stream which "leads" the direction of motion of the Clouds, i.e., opposite in direction to the Stream. This strongly supports the gravitational model for the Stream in which the leading and trailing streams are tidally torn from the body of the Magellanic Clouds. The data also reveal additional tidal features in the Bridge region which appear to emanate from the LMC, and a distinct spiral structure within the LMC itself.

1. Introduction and Observations

The gas distribution in the Magellanic Clouds, unlike the stellar component, immediately suggests that the Clouds are not independent entities, but part of a dynamically interacting system. Recent observations taken as part of the HI Parkes All-Sky Survey (HIPASS) reveal several exciting new features of this system. The single dish 21-cm line observations allow us to examine the large scale structure of the LMC and the overall spatial and velocity distribution of the SMC and Bridge gas, and resolve the long-standing controversy about the mechanism responsible for the formation of the Magellanic Stream.

HIPASS is a survey for extragalactic neutral hydrogen in the southern sky ($\delta \leq +2^{\circ}$), which has also proven useful for studies of the local gas distribution (i.e., -400 to $+400$ km s^{-1}). The survey is carried out by actively scanning the 64-m Parkes telescope, equipped with the 13 feed-horn multibeam receiver, across the sky in declination (Staveley-Smith 1997). The entire southern sky has already been scanned once with a spacing on the sky of $\sim 7'$ and a sensitivity of 20 mK (1σ). The complete survey will consist of five separate scans, resulting

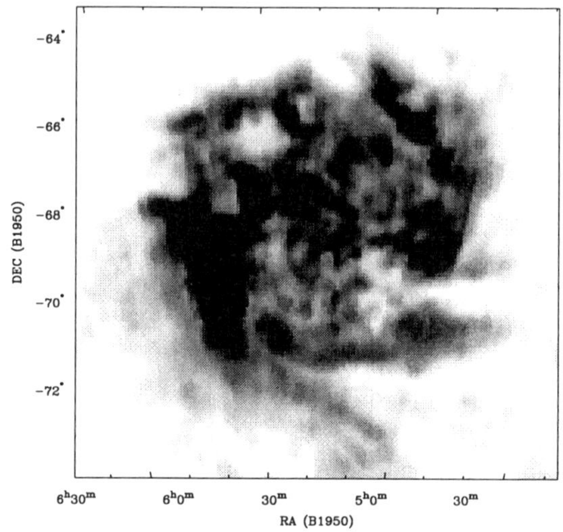

Figure 1. HIPASS map of the LMC. Note the distinct spiral arm in the south and the two stubs in the north.

in sub-Nyquist sampling and a resulting sensitivity of 9 mK (1σ). The channel spacing is 13.2 km s^{-1}; after hanning smoothing, the final resolution is 26.4 km s^{-1}. The survey's unbiased coverage of the southern sky and dense spatial sampling provide us with the ideal tools to reveal previously unidentified HI structure about the Magellanic Clouds.

2. Results

The data presented here are from the first scan of HIPASS data. Starting with the LMC (Figure 1), we immediately see the advantages of fully sampled single dish observations. Figure 1 highlights the distinct spiral structure of the LMC and also depicts the diffuse HI edges on all sides of the Cloud. This contradicts the suggestion that the gas is being compressed as the LMC moves through the halo of the Milky Way (e.g., de Boer et al. 1998); however, there is still a steep gradient in the east and the diffuse nature of the outer gas may be due to differential rotation.

The LMC appears to have 3 spiral "arms" (Figure 1) and one southern extension which looks like another thin spiral arm until its relationship to the Magellanic Bridge gas is examined. Figure 2 shows the LMC+Bridge+SMC as seen by HIPASS. In this picture the LMC's southern extension is clearly seen as a tidal feature which merges with gas from the SMC, forming the Bridge. This figure also shows the beginning of the Magellanic Stream which continues for another 100° to the north (Mathewson et al. 1974). The clumpy nature of the Stream is immediately apparent, even in this limited region.

In addition to the Magellanic Stream, Figure 2 also shows the beginning of a thin southern feature emanating from the SMC+Bridge region ($\ell, b = 292°, -32°$). Moving out to a 2400 square degree mosaic about the South Celes-

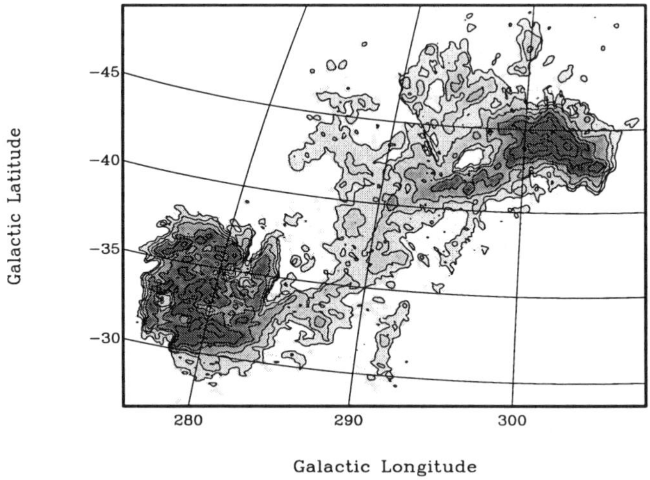

Figure 2. Brightness temperature map of the LMC+SMC+Bridge region including V_{LSR} = 85 - 400 km s^{-1}. Contours are a geometric progression from 1.3 - 83 K.

tial Pole (Figure 3), we see that this feature continues "towards" the Galactic Plane, on the leading side of the Clouds. This extension is at very high positive velocities and is clearly separate from Galactic emission (see Putman et al. 1998). The leading feature continues from the Clouds both spatially and in velocity, to $b \approx -8°$ and V_{LSR} = 350 km s^{-1}.

This leading extension is exactly the type of feature predicted to form during a tidal interaction between the Clouds and the Milky Way (e.g., Gardiner & Noguchi 1996). There are two principal interaction scenarios which attempt to explain the presence of the Magellanic Stream. One uses gravitational tides to pull out leading and trailing streams from the Clouds; the other employs ram-pressure stripping to rip the Stream out of the Clouds as they pass through an extended Galactic halo. The argument against tidal models (and for the ram-pressure models) is largely based on two observational features: no leading stream has been detected, and there are no stars observed in the Stream (stars should also be affected by tidal forces). The leading feature of Figure 3 removes the first discriminant against tidal models. The second issue, not finding stars in the Stream, is actually quite complex. It is common to find systems with strong HI tidal features and no optical counterpart (e.g., the M81 system – Yun et al. 1994) or HI streams offset from the optical streams (e.g., Smith et al. 1997). It may be that we have not looked in the right place for this optical counterpart to the Magellanic Stream, or, considering that the Stream is believed to be $\sim$ 1.5 Gyr old, we may not be looking for the right age of stars (Majewski et al. 1999). In addition, it is probable that the gas in the Clouds was originally more extended than the stars. Thus, the discovery of the Leading Arm is the strongest support in 25 years for a past tidal interaction between the Magellanic Clouds and the Milky Way.

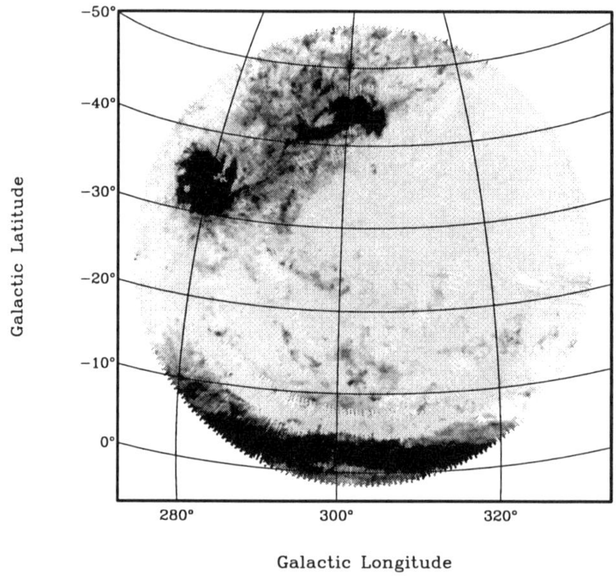

Figure 3. Brightness temperature map of the South Celestial Pole including $V_{LSR} = 85 - 400$ km s^{-1}.

There are several ways in which the tidal interaction scenario can be confirmed. HIPASS will continue to map the HI about the Magellanic System, allowing us to search for a continuation of the Leading Arm and other Magellanic-related features. A preliminary look at the data on the other side of the Galactic Plane suggests that the arm veers to the east and continues to $b \approx 30°$. We will also increase the sensitivity of our present data by adding more multibeam scans and increase our velocity resolution by observing these features with the narrow-band facility. Emission line observations and metallicity determinations will also prove useful for constraining the dominant ionization mechanism in the Leading Arm region and confirming the Magellanic origin.

Acknowledgments. We thank the multibeam working group for their help in the data collection and development of the instrumentation and software.

References

de Boer, K.S., Braun, J.M., Vallenari, A., & Mebold, U. 1998, A&A, 329, L49

Gardiner, L.T., & Noguchi M. 1996, MNRAS, 278, 191

Majewski, S., et al. 1999, this volume

Mathewson, D.S., Cleary, M.N., & Murray, J.D. 1974, ApJ, 190, 291

Putman, M.E., Gibson, B.K., Staveley-Smith, L., et al. 1998, Nature, 394, 752

Smith, B.J., Struck, C., & Pogge, R.W. 1997, 483, 754

Staveley-Smith, L. 1997, PASA, 14, 111

Yun, M.S., Ho, P.T.P., & Lo, K.Y. 1994, Nature, 372, 530

Discussion

Jim Hesser: Are any surveys underway to identify stellar populations associated with your "leading arm"?

Putman: Not presently, but Majewski et al. (this volume) have begun a search for stellar streamers about the clouds and has found a number of giant stars offset from the HI in the LMC. They have plans to search the Leading Arm region as well.

Lance Gardiner: Looking at the global view of HI in the Magellanic System, would you say that the Leading Arm feature originates from the inter-cloud region? The Gardiner and Noguchi model suggests that the feature emerges nearer the LMC.

Putman: It is difficult to say exactly. There appears to be a filament extending down from the SMC/Inter-Cloud regions into the arm, as well as a more tenuous filament emerging formt he bottom of the LMC. Higher velocity resolution observations should help to clarify this.

Klaas de Boer: Can you tell us what the column densities are in your leading arm? On the other side of the galaxy there are a lot of high velocity clouds belonging to the Milky Way halo. Can you exclude that the clouds you see in front of the LMC are plain normal galactic HVC's?

Putman: The maximum column density is approximately 6×10^{19} atoms cm^{-2} and there does not seem to be a systematic variation in column density along the arm. We do not yet know if high velocity clouds belong to the Milky Way halo. The Leading Arm is distinct from Milky Way gas in several ways. Its velocity ($\sim$310 km s^{-1}) is much higher than the velocity of the MW disk gas in this direction ($\leq$ 120 km s^{-1}). As seen in the channel maps, the arm is connected to the Magellanic Clouds and not to any Galactic emission. It also has a unique continuous structure for over 25°.

New Views of the Magellanic Clouds
IAU Symposium, Vol. 190, 1999
Y.-H. Chu, N.B. Suntzeff, J.E. Hesser, and D.A. Bohlender, eds.

Coronal C^{+3} in the LMC: Evidence for a Hot Halo

Bart Wakker

Dept. of Astronomy, Univ. of Wisconsin, Madison, WI 53706, USA

Abstract. Observations of interstellar C IV toward six stars in the Large Magellanic Cloud are described. These data were obtained with the *GHRS* on *HST* and with *IUE*. They provide the first unambiguous evidence that C^{+3} exists in the Large Magellanic Cloud away from regions where it could have been produced locally, as the six stars are of spectral type later than O8, as well as away from hot early-type stars or active regions. Significant differences between the velocities of the C^{+3} absorption and those of Hα emission near the stars shows that these ions are not co-spatial, implying that the C^{+3} originates in the halo of the LMC. It is shown that photo-ionization is unlikely to be the sole source of the C^{+3} and that the LMC halo thus contains hot gas.

1. Motivation

To understand the interstellar medium and how it is shaped by stars, it is necessary to characterize all of its phases at large and small scales. Studies of the LMC have the distinct advantage that a global view can be obtained, while small-scale structure can still be discerned. Other papers in these proceedings present new data on the small- and large-scale distribution of gas with temperatures of 10 K (seen in CO emission), 10^2–10^3 K (seen in H I emission and absorption), 10^4 K (seen in Hα emission) and 10^6 K (seen in X-ray emission).

This paper focuses on the gas at 10^5 K, as observed in absorption in UV resonance lines of highly-ionized atoms such as O^{+5}, N^{+4}, and C^{+3}. Since we can obtain velocity information, it is possible to assess the vertical distribution, unlike what is the case for the 10^6 K gas. The O VI and N V lines are only produced in hot gas. N V tends to have low optical depth, however. O VI cannot be observed by *GHRS*, *STIS* and *IUE*; it will be observed by *FUSE* in the near future. The lines of C IV and Si IV will be present in hot gas, but may also be produced in substantial amounts by photo-ionization (especially Si IV).

Observations of N V and C IV absorption in the Milky Way halo have been interpreted as evidence for some form of the galactic fountain (Savage et al. 1997). The 10^5 K gas in the LMC was studied by Chu et al. (1994), who concluded that it occurs in all superbubbles and supergiant shells, but that there was insufficient evidence to infer the presence of a hot halo. Such a halo is expected to be present if the hot gas inside bubbles and shells can break out and redistribute itself across the LMC. This provides a mechanism to homogenize the metal abundances in the ISM; without redistribution stars born at the same time in well-separated clusters might have large differences in metallicity.

2. Previous Observations

The first discussion of C IV absorption in the LMC halo was presented by de Boer & Savage (1980) and de Boer & Nash (1982). They had relevant data for two SMC stars and one LMC star (Sk−67 104 in N 51 D). All three were inside a bright H II region, and all showed interstellar absorption by highly-ionized atoms. A number of arguments were presented to conclude that some of the C IV absorption is related to a halo, including: (a) in other LMC and Milky Way stars in H II regions narrow lines of Si II* are found, but not in the three stars in question; (b) the C IV lines are broader than lines associated with the H II region; (c) the standard wind-blown bubble model (Weaver et al. 1977) predicts much lower column densities.

However, we now know that N 51 D also has strong X-ray emission, which is not expected for a wind-blown bubble but can be explained by off-center SNRs (Chu & Mac Low 1990); this also increases the expected strength of the high-ionization absorption. Further, Prévot et al. (1980) argued that a star in the SMC outside an H II region did not show C IV absorption, although Fitzpatrick (1984) found C IV in a deeper spectrum. Savage (1984) mentioned C IV absorption in the star Sk−67 147a, a B0.5 star outside an H II region, incapable of producing photo-ionized C^{+3}.

However, Chu et al. (1994) showed that C IV absorption is seen in all stars of type earlier than O7, all stars in superbubbles, and all stars in supergiant shells (such as Sk−67 147a in LMC 3). The only stars without interstellar C IV were Sk−67 05 and possibly Sk−67 111 and Sk−68 114, all of which were in the field. This showed that previous studies did not *conclusively* show that the LMC has hot halo gas, as in all cases it was possible to argue that the observed C IV absorption might be associated with more local phenomena.

3. Procedure to Detect Hot Halo Gas

One of the conclusions of Chu et al. (1994) was that to unambiguously detect hot gas in the LMC halo it is necessary to select stars with a low a-priori probability of having highly-ionized gas nearby. This excludes stars in superbubbles and supergiant shells, where supernovae may locally produce hot gas. It also excludes stars of type earlier than O7, which can photo-ionize C^{+2} (although not O^{+4} and N^{+3}). Stars later than B2 are also bad because of the increasing amount of stellar absorption lines.

To find probe stars fitting these criteria, the Sanduleak catalogue of LMC early-type stars and its updates were used for spectral types and magnitudes. For each star the spectral type gives an estimate of $T_{\rm eff}$. A Kurucz stellar atmosphere model for the temperature was then combined with the star's visual magnitude to predict the flux at 1550 Å and at 1035 Å. Requiring an S/N of >10 in the continuum, in practice about 100 stars are bright enough to observe halo C IV absorption, 170 to observe halo O VI.

Even if high-ionization absorption would be seen in such a halo probe, it is not necessarily true that this absorption originates in the LMC halo: it could be from the disk near the star. To settle this question we further need to observe the velocity of the Hα emission near the star, which should originate in the

LMC disk. If the velocities are the same, the warm and highly-ionized gas are probably co-spatial. Only if these velocities are different can we conclude that the C IV probably originates in a halo.

After showing that the highly-ionized gas occurs in the halo, it must be shown that photo-ionization is not the major producer. The models of Bregman & Harrington (1986) predict that the fraction of C in the form of C IV will be f(C IV)$\sim$0.1 for densities below $n_x \sim 10^{-3}\,\mathrm{cm}^{-3}$. So, assuming an exponential halo with scaleheight H, N(C IV) can be estimated as f(C IV) A(C) n_x H $\sim 3\times10^{13}$ H(kpc) cm^{-2}, where A(C) is the abundance of carbon ($\sim10^{-4}$ in the LMC). In the Milky Way, H is $\sim$4.5 kpc (Savage et al. 1997). Although there are no direct estimates, in the LMC H can be expected to be smaller than that, since the diameter of the ISM in the LMC is only about 5 kpc.

4. Results from GHRS

In *HST* cycle 5 we obtained C IV spectra for 5 stars, at about 15 km s^{-1} resolution, having a S/N ratio of $\sim$10. We further used the 4-m telescope at CTIO to observe the Hα emission near the stars. Figure 1 shows the resulting continuum-normalized spectra. The table below presents the measured column densities.

Table 1. Measured properties of C IV absorption.

Star	λ_0	Milky Way			LMC				
		v	W	N	v	W	N	v(Hα)	diff.
		km/s	km/s	cm^{-2}	km/s	km/s	cm^{-2}	km/s	km/s
Sk−67.05	1548	...	...	...	...	...	<3.0		
	1550	64	28	10.9	...	...	<2.1		
Sk−67.46	1548				272	80	9.1	289	−23
	1550	20	42	6.4	260	78	11.5		
Sk−67.76	1548	30	75	9.0	254	42	6.9	316	−59
	1550	25	45	8.7	259	54	6.7		
Sk−69.59	1548	37	75	12.1	256	59	7.1	281	−26
	1550	42	78	12.9	254	57	7.3		

Interstellar C IV absorption associated with both the Milky Way and the LMC is clearly detected in the spectra of Sk−67 46, Sk−67 76 and Sk−69 59. The flux of Sk−70 96 was much lower than expected, but C IV λ1550 is present. Stellar absorption hides C IVλ1548. The spectral type of Sk−69 110 was found to be B3Ia, rather than B1 as originally assumed; the interstellar lines are therefore hard to discern. A good non-detection is found for the LMC absorption toward Sk−67 05, the table gives the 5-σ detection limit.

The measurements lead to the following conclusions:

1. Our stars were selected to be away from regions where C^{+3} can be easily produced by LMC stars, supernova shocks, or stellar photo-ionization. Strong C IV absorption is still seen, however.

2. Since the velocity of Hα emission and C IV absorption near the star differ, not all of the C IV absorption originates in the LMC disk.

3. Photo-ionization is not the main producer of C^{+3} in the LMC.

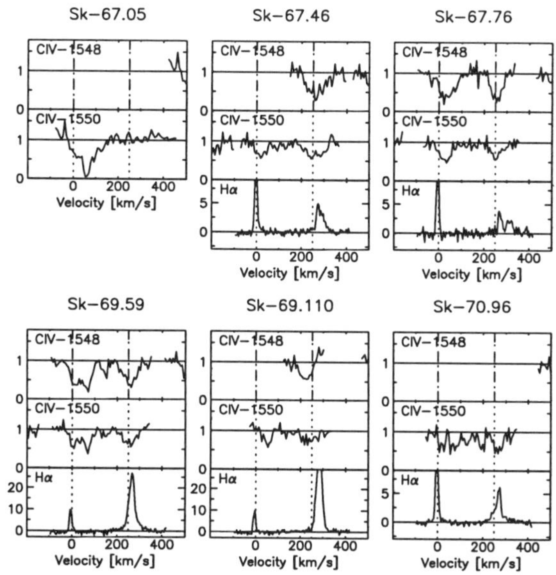

Figure 1. Continuum-normalized C IV and Hα spectra.

4. Combining 1), 2) and 3) we conclude that the LMC has a hot gaseous halo.

5. There are large variations in N(C IV): a factor 2 betweeen detections, down to zero toward Sk−67 05.

6. The column density range in the LMC is similar to that in the Milky Way: 9–16$\times10^{13}$cm^{-2} vs 8–14$\times10^{13}$cm^{-2}.

7. As the carbon abundance in the LMC is lower than that in the Milky Way, a higher percentage of carbon is in the form of C^{+3}, which can easily be understood in cooling flow models such as the galactic fountain.

A more complete discussion of this work is given by Wakker et al. (1998).

References

de Boer, K.S., & Nash, A.G. 1982, ApJ, 255, 447

de Boer, K.S., & Savage, B.D. 1980, ApJ, 238, 86

Bregman, J.N., & Harrington, J.P. 1986, ApJ, 309, 833

Chu, Y.-H, Wakker, B.P., Mac Low, M.-M., & García-Segura, G. 1994, AJ, 108, 1696

Chu, Y.-H., & Mac Low, M.-M. 1990, ApJ, 365, 510

Fitzpatrick, E.L. 1984, ApJ, 282, 436

Prévot, l., et al. 1980, A&A, 90, L13

Savage, B.D. 1984, in: *Future of UV Astronomy Based on Six Years of IUE Research*, NASA Conference Publication 2349, p3

Savage, B.D., Sembach, K.R., & Lu, L. 1997, AJ, 113, 2158

Wakker, B., Howk, J., Chu, Y.-H, Bomans, D., & Points, S. 1998, ApJ, 499, L87

Weaver, R., et al. 1977, ApJ, 218, 377

Discussion

Daniel Wang: In the field of the LMC, there must be some bright, nearby QSOs that could be used for FUV absorption studies. Are you aware of any such work?

Wakker: So far there are no known AGNs toward the LMC bright enough ($V < 15$) to observe at sufficiently high spectral resolution (15 kms^{-1}). For fainter AGN's impractically long integration times will be needed even if the resolution requirement is relaxed.

Nicholas Suntzeff: There is no evidence for a kinematical halo in the LMC. Does this mean that the "halo" of hot gas you are discussing has no stellar component?

Wakker: Yes, it is a gaseous halo, such as the MW has too, possibly formed by the cumulative effect of star formation that pushes gas out.

Robin Shelton: Can you constrain the filling factor of the CIV-rich gas?

Wakker: Not really, since we do not know the actual volume densities and path lengths.

Robin Shelton: Your last overhead specified that the CIV gas was 10^5 K. Is that from a constraint or can the gas be cooler but not yet recombined?

Wakker: Will be in the paper.

New Views of the Magellanic Clouds
IAU Symposium, Vol. 190, 1999
Y.-H. Chu, N.B. Suntzeff, J.E. Hesser, and D.A. Bohlender, eds.

A ^{12}CO Survey of the LMC with NANTEN

Y. Fukui, R. Abe, A. Hara, T. Hayakawa, S. Kato, A. Kawamura, A. Mizuno, N. Mizuno, H. Ogawa, T. Onishi, H. Saito, K. Tachihara, K. C. Xiao, N. Yamaguchi, and R. Yamaguchi

Department of Astrophysics, Nagoya University, Chikusa-ku, Nagoya, 464-8602, Japan

Y. Yonekura

Earth and Life Sciences, Osaka Prefecture University, Gakuencho 1-1, Sakai, Osaka 599-8531, Japan

M. Rubio

Departamento de Astronomia, Universidad de Chile, Casilla 36-D, Santiago, Chile

Abstract. We have made a ^{12}CO(J = 1–0) survey of the LMC with NANTEN. A sample of 55 giant molecular clouds has been identified and comparisons with stellar clusters, HII regions and SNRs are presented. The connection between the clouds and cluster formation is discussed.

1. Introduction

The Large Magellanic Cloud (LMC) classified as the barred sub-type of Hubble's irregular class is the nearest neighbor to our own. Studies of this galaxy have provided invaluable information for our understanding of the universe and galaxies in various aspects including evolution of stars and stellar clusters, owing to its proximity to the solar system ($D \sim 50$ kpc). Compared to the advanced evolutionary phase of stars, the process of star formation in the LMC has been only poorly understood, mainly due to the lack of comprehensive observations of giant molecular clouds where stars form. Previous observations of molecular gas in the LMC are either of low angular resolution or of small spatial coverage (Cohen et al. 1988; Israel et al. 1993; Kutner et al. 1997; Johansson et al. 1998). We have performed new observations of the LMC in the J = 1–0 rotational transition of interstellar carbon monoxide (CO) at 2.6 mm wavelength in order to reveal detailed molecular gas distribution at a linear resolution of $\sim$ 30 pc. These observations were made with NANTEN newly installed at Las Campanas Observatory, Chile in 1996 (Fig. 1) and should allow us to have a complete view of the giant molecular clouds in the LMC for the first time. Four related contributions included in these proceedings are complementary to this paper, describing more details of these CO observations and comparisons with other astronomical objects (Abe et al.; Mizuno et al.; Saito et al.; Yamaguchi et al.).

Figure 1. 4m millimeter-wave telescope, NANTEN, of Nagoya University located at Las Campanas Observatory, Chile.

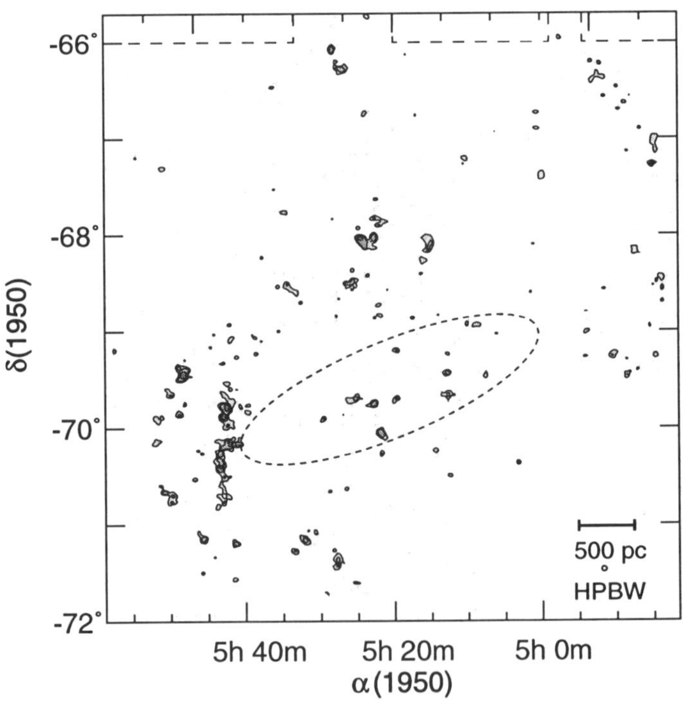

Figure 2. Velocity-integrated intensity of CO ($J = 1$–0) emission obtained with NANTEN. The lowest contour and the separations between contours are 3.0 K km s^{-1} for each. The dotted line shows the optical boundary of the Bar.

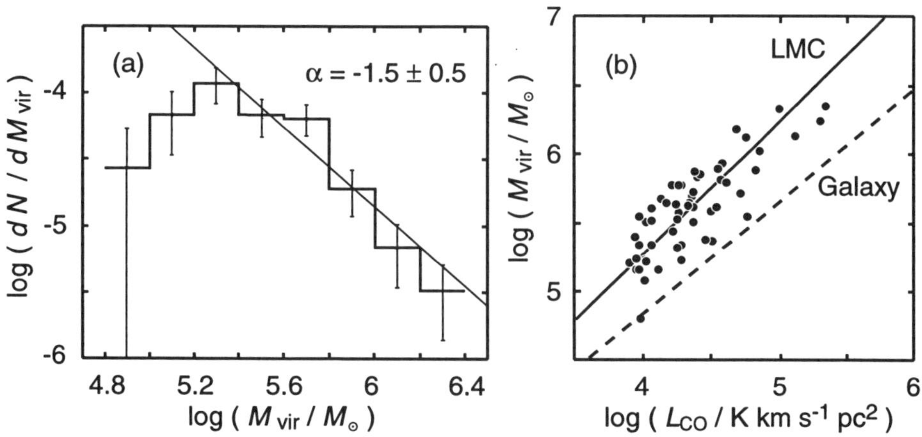

Figure 3. (a) Mass spectrum, $dN/dM_{\rm vir}$, of the 55 clouds. The error bars represent the $(dN)^{1/2}$ statistical errors. The straight line is the least-squares fit to the clouds with $M_{\rm vir} \geq 1.6 \times 10^5 M_\odot$ (by $[dN/dM_{\rm vir}] = 1.2 \times 10^4\ [M_{\rm vir}/M_\odot\ ]^{-1.5}$). (b) Virial mass, $M_{\rm vir}$, plotted against CO luminosity, $L_{\rm CO}$. The solid line represents the least-square fit and the dashed line indicates the same relation but derived for the Galactic molecular clouds (Solomon et al. 1987).

2. Properties of the Giant Molecular Clouds

Figure 2 shows the CO image obtained with NANTEN. This image consists of 32,800 spectra taken at every 2′, corresponding to $\sim$ 30 pc at a distance of 50 kpc, with a 2.′6 beam. The distribution is highly clumpy, extended over 36 square-degrees. Large CO complexes whose size is more than 100 pc are located at the southern part of 30 Dor and N44 regions. Smaller CO clouds are distributed over the observed area, with moderate concentration toward the Bar and toward several prominent HII regions. We note that an arclike semi-circular CO distribution whose diameter is $\sim$ 3 kpc is seen in the south-east boundary of the optical galaxy.

The number of the CO clouds in Figure 2 is more than 100. Among them, there are 55 CO clouds which are detected at more than three positions. The mass of these 55 clouds ranges from $\sim 10^5 M_\odot$ to $\sim 3 \times 10^6 M_\odot$, indicating that they are giant molecular clouds. In Figure 3a, we present the mass spectrum $dN/dM_{\rm vir}$ of the 55 clouds. It is fitted by a single power-law of an index of -1.5 in a cloud mass range 10^5–$10^6 M_\odot$, which is similar to that found in our Galaxy (Solomon et al. 1987). The lower boundary, $\simeq 10^5 M_\odot$, is the present detection limit in mass. Figure 3b shows the relation between the virial mass ($M_{\rm vir}$) and CO luminosity ($L_{\rm CO}$), for the 55 CO clouds. The CO luminosities in the LMC are weaker by a factor of $\sim$ 3 relative to those in the Galaxy. By using this factor, the total molecular cloud mass is estimated from the CO luminosities to be $\sim 4 \times 10^7 M_\odot$, 90% of which is included in the large 55 clouds whose H_2 column density is greater than 2×10^{21} cm^{-2}. According to the subsequent, more sensitive observations of selected regions, the cloud mass may increase up to $\sim 7 \times 10^7 M_\odot$ at the H_2 column density $\geq 1 \times 10^{21}$ cm^{-2} (see Mizuno et al. 1999 in these proceedings). We, therefore, estimate the molecular mass to be 4 – 7 $\times 10^7 M_\odot$, corresponding to 5 – 10% of the HI mass M(HI) $\sim 7 \times 10^8 M_\odot$ (McGee & Milton 1966).

The apparent weakness of CO relative to $M_{\rm vir}$ (Fig. 3b) is possibly caused by the lower metal abundance and by enhanced photodissociation due to relatively strong UV radiation fields under significantly smaller shielding. The conversion factor $X \equiv N(H_2) / I_{\rm CO}$, the ratio of H_2 column density to CO line intensity, varies strongly with such physical conditions in clouds. The conversion factor, X, for clouds in the LMC, determined by assuming that they are virialized, is 9×10^{20} cm^{-2}/(K km s^{-1}). This is smaller than the value from Cohen et al. (1988) obtained with the 1.2 m radio telescope, but about 3 times higher than the value for our Galaxy (Solomon et al. 1987).

3. Formation of Star Clusters

In order to study star formation in the LMC, detailed comparisons of the CO clouds with stellar clusters, HII regions, and SNRs have been made over the whole Cloud as shown in Fig. 4.

The following are the results of the comparison of CO clouds with *stellar clusters* and *OB associations* (hereafter, *clusters*). Out of the 55 CO clouds, 26 are associated with clusters. Using the age of the clusters estimated from their color indices, UBV (Bica et al. 1996), $\sim$ 90% of the clusters associated with the CO clouds are younger than 10 Myr (SWB0). The older clusters with $\tau > 10$ Myr (SWBII-VII) show little correlation with the CO clouds. The youngest clusters associated with massive CO clouds tend to be in a compact group of young stellar clusters, i.e., N159 and N44. These groups of clusters are located at or near the peak of the CO clouds, indicating that they have been just formed in massive CO clouds.

The youngest clusters show a significant correlation with the CO clouds; about $\sim$ 30% of them are associated with CO clouds. The remaining $\sim$ 70% of the youngest clusters tend to be scattered, away from the CO clouds. Though these clusters may be associated with clouds smaller than the present detection limit, this suggests that young stellar clusters can rapidly dissipate their surrounding CO clouds (see also Yamaguchi et al. in these proceedings.)

Comparisons of the 55 massive CO clouds with HII regions and with stellar clusters indicate the following:

1. 12 CO clouds show no sign of star formation; i.e., they are associated with no HII regions or stellar clusters.

2. 17 CO clouds are associated with small HII regions only, but with no stellar clusters.

3. 26 CO clouds are associated with stellar clusters and large HII regions, suggesting active, on-going formation of massive clusters (e.g., N44).

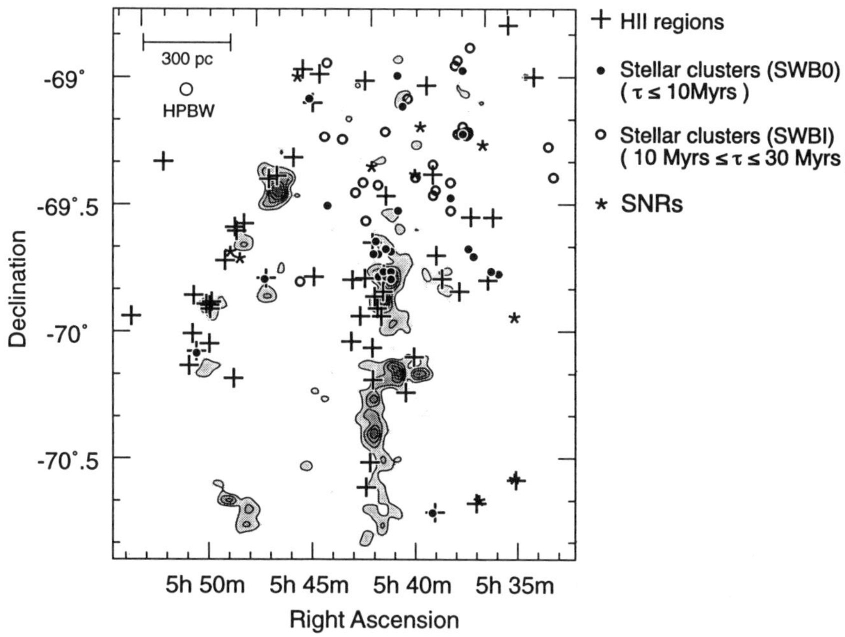

Figure 4. Distribution of the CO clouds, HII regions, stellar clusters and SNRs toward 30 Dor. The contours represent an intensity map of CO; the lowest contour and the separations are each 3.0 K km s^{-1} . The pluses indicate HII regions cataloged by Davies et al. (1976). The filled circles and squares are stellar clusters with $\tau \leq 10^6$ yr (SWB0) and 10^6 yr $\leq \tau \leq 3 \times 10^6$ yr (SWBI), respectively (Bica et al. 1996). The asterisks are SNRs (Mathewson et al. 1983).

On the other hand, there are several regions where stellar clusters (SWB0 and SWBI) are coexistent with large HII regions, but associated with only a small amount of CO clouds. Such an example is seen toward 30 Dor.

If we assume that the CO clouds in the LMC are being formed nearly steadily, the above comparisons may represent evolution of the CO clouds. The fact that about a half of the CO clouds are associated with the youngest clusters, SWB0, suggests that stellar clusters are activey formed over ~50% of a cloud's lifetime. This lifetime is roughly estimated to be ~ 3 Myr, since ~1/3 of the SWB0 clusters, which are younger than 10 Myr, are associated with CO clouds. By including the period prior to cluster formation, the typical lifetime of a CO cloud may be estimated as ~ 6 Myr the LMC. On the other hand, the absence of a massive CO cloud near the rest of the SWB0 clusters suggests that cloud dissipation is fairly rapid, possibly due to stellar UV photons (see Yamaguchi et al. in these proceedings).

To summarize, the complete sample of the giant molecular clouds in a single galaxy, the LMC, has allowed us to pursue how formation of stellar clusters takes place in them. The present results suggest that a giant molecular cloud forms massive stellar clusters fairly quickly, leading to very rapid dissipation of itself over a timescale of some 6 Myr.

Acknowledgments. The NANTEN project is based on the mutual agreement between Nagoya University and Carnegie Institution of Washington. We greatly appreciate the hospitality of all the staff member of the Las Campanas Observatory of the Carnegie Institution of Washington. We also proudly mention that this project has been supported by the contribution from many Japanese public donators and companies.

References

Abe, R., Hara, A., Hayakawa, T., Kato, S., Kawamura, A., Mizuno, A., Mizuno, N., Ogawa, H., Onishi, T., Saito, H., Tachihara, K., Xiao, K. C., Yamaguchi, N., Yamaguchi, R., Yonekura, Y., & Fukui, Y. 1999, this volume

Bica, E., Claria, J. J., Dottori, H., Santos, Jr. J. F. C., & Piatty, A. E. 1996, ApJS, 102, 57

Cohen, R. S., Dame, T. M., Garay, G., Montani, J., Rubio, M., & Thaddeus, P. 1988, ApJ, 331, L95

Davies, R. D., Elliott, K. H., & Meaburn, J. 1976, MmRAS, 81, 89

Israel, F. P., Johansson, L. E. B., Lequeux, J., Booth, R. S. , Nyman, L.-A., Crane, P., Rubio, M., de Graauw, Th., Kutner, M. L., Gredel, R., Boulanger, F., Garay, G., & Westerlund, B. 1993, A&A, 276, 25

Johansson, L. E. B., Greve, A., Booth, R. S., Boulanger, F., Garay, G., de Graauw, Th., Israel, F. P., Kutner, M. L., Lequeux, J., Murphy, D. C., Nyman, L.-A., & Rubio, M. 1998, A&A, 331, 857

Kutner, M. L., Rubio, M., Booth, R. S., Boulanger, F., de Graauw, Th., Garay, G., Israel, F. P., Johansson, L. E. B., Lequeux, J., & Nyman, L.-A. 1997, A&AS, 122, 255

Mathewson, D. S., Ford, V. L., Dopita, M. A., Tuohy, I. R., Long, K. S., & Helfand, D. J. 1983, ApJS, 51, 345

McGee, R. X., & Milton, J. A. 1966, Austr. J. Phys. 19, 343

Mizuno, N., Abe, R., Hara, A., Hayakawa, T., Kato, S., Kawamura, A., Mizuno, A., Ogawa, H., Onishi, T., Saito, H., Tachihara, K., Xiao, K. C., Yamaguchi, N., Yamaguchi, R., Yonekura, Y., & Fukui, Y. 1999, this volume

Saito, H., Abe, R., Hara, A., Hayakawa, T., Kato, S., Kawamura, A., Mizuno, A., Mizuno, N., Ogawa, H., Onishi, T., Tachihara, K., Xiao, K. C., Yamaguchi, N., Yamaguchi, R., Yonekura, Y., & Fukui, Y. 1999, this volume

Solomon, P. M., Rivolo, A. R., Barrett, J., & Yahil, A. 1987, ApJ, 319, 730

Yamaguchi, R., Abe, R., Hara, A., Hayakawa, T., Kato, S., Kawamura, A., Mizuno, A., Mizuno, N., Ogawa, H., Onishi, T., Saito, H., Tachihara, K., Xiao, K. C., Yamaguchi, N., Yonekura, Y., & Fukui, Y. 1999, this volume

New Views of the Magellanic Clouds
IAU Symposium, Vol. 190, 1999
Y.-H. Chu, N.B. Suntzeff, J.E. Hesser, and D.A. Bohlender, eds.

Molecular Gas in the Magellanic Clouds

Mónica Rubio

Departmento de Astronomía, Universidad de Chile, Casilla 36-D, Santiago, Chile

Abstract. The molecular gas content in the Magellanic Clouds has been studied, with different spatial coverage and resolution, through obervations of CO(1-0) line emission. In the LMC and the SMC the molecular gas is dominated by clouds whose properties are different from those of their Galactic counterparts. The relation between the intensity of CO emission and molecular hydrogen column density, or the conversion factor X, is different than that of molecular clouds in our Galaxy and depends on the ambient physical conditions. Studying the molecular gas through observations in the H_2 emission line may prove an alternative way to determine the molecular content associated with star forming regions in the Magellanic Clouds. In particular, results obtained towards 30 Doradus in the LMC are presented.

1. Introduction

The Large and Small Magellanic Cloud (LMC, SMC), the two nearest extragalactic systems of well known distance provide a unique opportunity to study physical processes in an interstellar medium different from that of our Galaxy. They have lower metallicities, higher gas to dust ratios and strong UV radiation field, and hence constitute the best target for a detailed study of the molecular gas content and physical properties, and the study of star formation processes associated with them.

In our Galaxy, extensive observations of the CO emission have shown that molecular gas (H_2) rather than atomic gas (HI) is the most active component of the interstellar medium. The molecular gas appears concentrated in massive (10^6 $M_\odot$) giant molecular clouds which are the sites of all known regions of massive and luminous stars (Combes 1991). Extragalactic CO observations have shown that molecular clouds within the disk of late type spirals appear to share the same characteristics as those of our Galaxy (Young & Scoville 1991).

A different situation is seen in irregular galaxies. The observations show a strong defficiency in CO emission compared to spirals of the same luminosity and star forming activity (Rubio et al. 1991 and references therein). Based on the critical assumption that the conversion factor used to derive molecular hydrogen masses from the observed CO emission is the same as in our Galaxy, these galaxies have been claimed to contain negligible amounts of molecular gas. On the other hand, irregular galaxies contain giant regions of ionized gas (HII regions) and luminous and massive stars (OB associations), whose presence suggests episodes of recent star formation. Therefore several questions are raised by the apparent contradiction between a high rate of star formation and the absence of molecular material.

Thus, a major goal of the study of the molecular gas content in the Magellanic Clouds has been to determine the relation between CO and H_2 under conditions significantly different from those found in our Galaxy. Another important aspect to be addressed, since stars form in molecular clouds, is the study of star formation processes in these low metallicities molecular clouds.

2. Millimeter-wave Studies of the Molecular Gas

Molecular gas in the Magellanic Clouds has been investigated from observations of the tracer molecule CO made with different spatial resolutions and coverages. The first complete CO(1-0) survey of the Magellanic Clouds was made using the Columbia Southern Telescope at Cerro Tololo with an $8'.8$ resolution (150 pc linear resolution) and covered an area of $6° \times 6°$ for the LMC (Cohen et al. 1988) and $3° \times 2°$ for the SMC (Rubio et al. 1991).

The ESO-SEST Key Programme: CO in the Magellanic Clouds (Israel et al. 1993; Rubio 1997) has produced a wealth of CO observations of the LMC and SMC, with an angular resolution of $45''$ at 115GHz ($\sim$ 10 pc linear resolution). The SEST telescope fully resolves the giant molecular clouds, allowing the determination of their sizes and critical parameters to obtain virial masses, which can be used in turn to calibrate the conversion factor between molecular hydrogen H_2 and CO emission.

A new complete CO(1-0) survey of the Large Magellanic Clouds has been recently undertaken with the 4m NANTEN radiotelescope at Las Campanas Observatory (Fukui et al. 1999). This survey, made with angular resolution of $2'.6$, ($\sim$ 30 pc linear resolution) covers the same $6° \times 6°$ area previously surveyed at $8'.8$ by the Columbia survey.

3. Results

The Columbia surveys showed the overall distribution of the molecular gas in both clouds. The CO emission appeared associated with HII and radio continuum regions and was detected towards the highest column densities of HI: N(HI)$> 10^{21}$ cm^{-2} in the LMC and N(HI)$> 10^{22}$ cm^{-2} in the SMC. In the LMC, about 40 complexes were identified with typical sizes of $\sim$ 200 pc and CO luminosities of $\sim 1 \times 10^5$ K km s^{-1} pc^{-2}. The CO complexes appeared to be underluminous in CO as compared to galactic GMC's with similar velocity line widths. Based on the observed $L_{CO} - \Delta v$ relationship, Cohen et al. (1988) and Rubio et al. (1991) suggested that the conversion factor to derive H_2 column densities from CO luminosities was a factor of 6 and 20 larger than the Galactic factor in the LMC and SMC, respectively. Using the derived factors, the total molecular mass of the LMC was determined to be $\sim$ 30% of the neutral hydrogen mass and in the SMC only $\sim$7%.

The NANTEN CO emission survey of the LMC confirmed the distribution of molecular gas as determined by Cohen et al. (1988). They disentagle the CO complexes in more than 100 molecular clouds with sizes ranging from 30 to

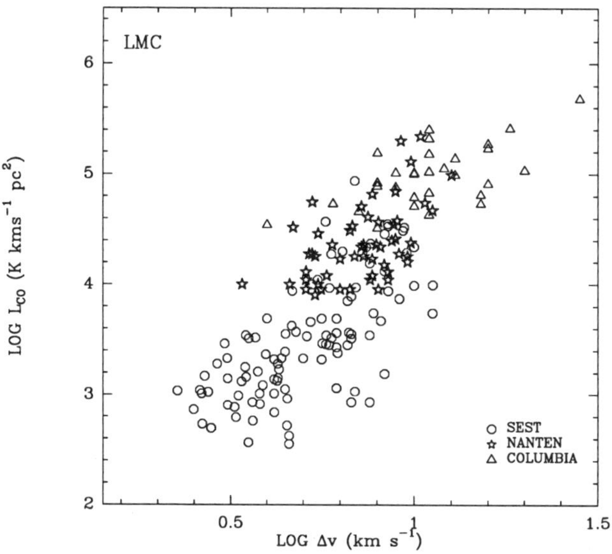

Figure 1. CO luminosity versus line width for molecular clouds in the LMC as defined by SEST(circles), NANTEN (stars) and Columbia (triangles) observations.

100 pc and virial masses exceeding 10^4 $M_\odot$ (Fukui et al. 1999 and references therein). Assuming a conversion factor of 2 to 4 times larger than the Galactic factor they obtain a total molecular hydrogen mass of 7×10^7 $M_\odot$ which is about 10% of the total HI mass.

The SEST observations have resolved the Columbia molecular complexes into clouds of sizes of 10 to 40 pc and virial masses of (0.3 to 5.5) $\times 10^5$ $M_\odot$. Determination of the physical properties of these molecular clouds is underway. In this regard, multiple-line observations in different transitions of the CO molecule as well as other molecules have been performed (e.g. Chin et al. 1998; Johansson et al. 1998).

Figure 1 shows the L_{CO} versus Δv realtionship for all the molecular clouds in the LMC as defined by the three sets of observations with different angular resolutions. These include about 100 molecular clouds from the SEST data (Rubio 1997; Kutner et al. 1998) and 55 clouds from the NANTEN survey. The SEST observtions appear concentrated in the lower-left part of the diagram while the Columbia observations are concentrated in the upper-right part. The NANTEN observations cover the range, both in L_{CO} and Δv in between the SEST (Johansson et al. 1998; Kutner et al. 1997; Rubio 1997) Columbia (Cohen et al. 1988).

Figure 2 shows the ratio X/X_{gal} as a funtion of cloud size, R, for SEST and NANTEN molecular clouds in the LMC, molecular clouds in the outer galaxy (May et al. 1997), and molecular clouds in the inner Galaxy (Bronfman et al., private communication), where $X_{gal} = 2.3 \times 10^{20}$ cm^{-2} (K km s^{-1})$^{-1}$. The X factor is defined as $X \equiv M(H_2)/L_{CO}$. To compute the X factor we used

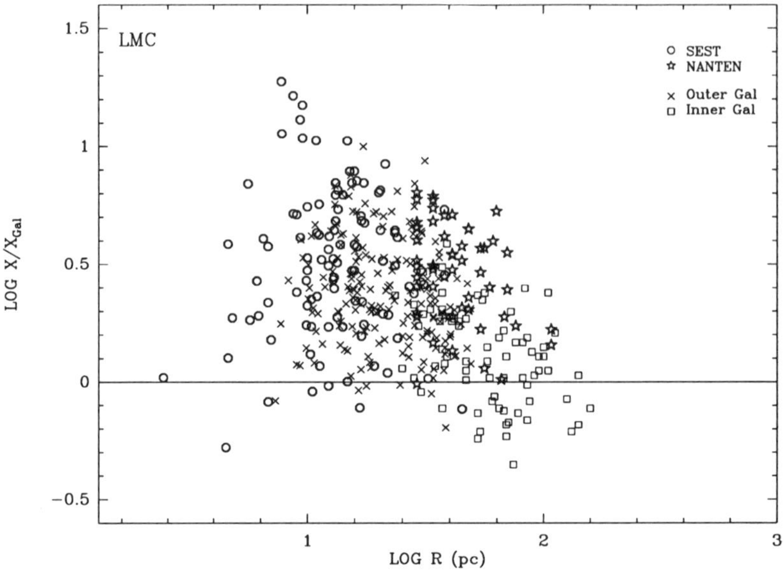

Figure 2. Ratio of the conversion factor X/X_{gal} for molecular clouds in the LMC as defined by SEST(circles) and NANTEN (stars), outer Galaxy (crosses), and inner Galaxy (rectangles) versus size.

as mass the virial mass, $M_{vir} = 190\ R\Delta v^2$. Although the points in this figure show a large spread, some trends are evident. Inner galaxy clouds tend to concentrate around the standard value for the Galaxy, X_{gal}, as determine by other methods (Strong et al. 1988), that is $X/X_{gal} \approx 1$. Outer galaxy clouds exhibit a larger spread, but with an average value of log $[X/X_{gal}]$=0.4, indicating that the conversion factor is about two times larger than the Galactic value (May et al. 1997).

The CO clouds in the LMC have similar sizes as those of outer galaxy clouds. The difference between the size of SEST and NANTEN clouds is due solely to the different angular resolution of the observations. The SEST and NANTEN observations show a clear difference in the X/X_{gal} value. Fukui et al. 1998 reports $X = (9 \pm 6) \times 10^{20}$ cm^{-2} $(K$ km s$^{-1})^{-1}$ about 2 to 4 times X_{gal}. SEST data indicates an average value of about 4 times larger than the standard Galactic value (Rubio 1997) although the spread in X values is about one order of magnitude. An explanation of this large spread requires a detailed study of the local ambient conditions of the molecular clouds over the different regions in the LMC.

Comparison of the CO emission within one Columbia beam to that observed by SEST within the same region, shows that the CO luminosity measured in one Columbia beam is equal to the sum of the luminosities of the smaller clouds observed by SEST in the corresponding area (Kutner at al. 1997). This result confirms the suggestion that the molecular gas in the LMC is dominated by clouds, with no diffuse CO emission in the intercloud medium. Further work comparing the properties of the CO clouds as derived from these three sets

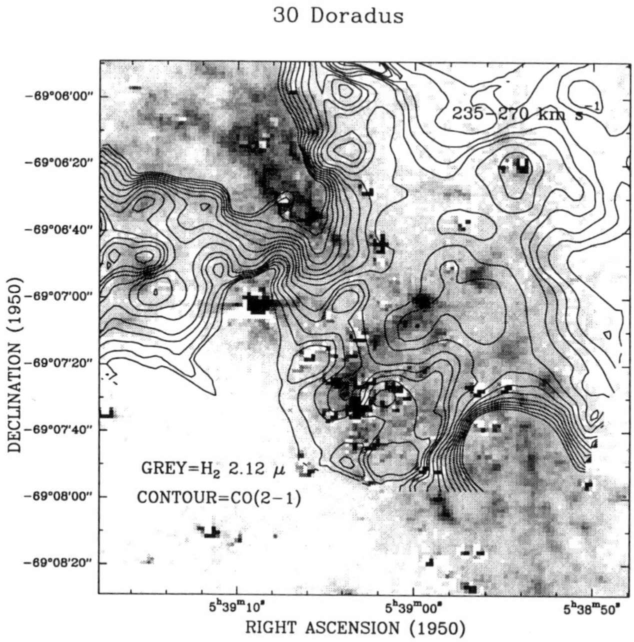

Figure 3. 2.12μ image of the 30 Doradus Nebula. Superimposed is the velocity integated CO(2-1) contour map.

of observations is envisaged as this comparison will be most valuable for the interpretation of extragalactic CO observations. This comparison should reveal the effects introduced in the determination of the properties of the molecular clouds when different spatial resolutions are used.

Our present knowledge of the molecular gas in the LMC indicates that the gas is dominated by clouds which are underluminous in CO and clumpy. The CO cloud properties are different from those of their galactic counterparts and this result is independent of the angular resolution used for the identification of the CO emission structures. The relation between the intensity of the CO emission and the molecular hydrogen density, the conversion factor X, is smaller than the factor derived for molecular clouds in our Galaxy. The X value is dependent on the local ambient conditions of the CO clouds. Clearly, more work is needed to determine how local ambient conditions, e.g., photodissociation, metallicity, etc, affect the properties of the CO clouds.

4. Near IR studies of the Molecular Gas in 30 Doradus

An important result of the studies of the molecular gas in the Magellanic Clouds is that CO molecules can only survive in the dense cores of the interstellar medium while they are photodissociated in the less dense medium. (Rubio et al. 1993; Lequeux et al. 1995). Therefore, the interclump medium may contain large amounts of H_2 as this molecule is more selfshielded against the UV field.

Near IR studies of star forming regions in the LMC has been made through deep imaging in the 2.12μ H_2 1-0 S(1) line and the 2.16μ Brγ emission line by Probst & Rubio (1997). The results obtained toward the giant HII region 30 Doradus are striking. The IR images show an intriguing molecular gas distribution. Figure 3 shows a 2.12μ image taken with CIRIM at the 1.5m telescope at Cerro Tololo of the central 4$'$.5$\times$4$'$.5 area of 30 Doradus. A chain of dense H_2 knots are seen to the north of the central cluster R136 and a reticulated structure is seen towards the northeast and southwest regions of the nebula. CO(1-0) emission had been reported by Johansson et al. (1998) from two molecular clouds to the northeast and southwest of the big arcs and filaments delineated by the ionized hydrogen gas of the nebula. The detection of the molecular hydrogen knots where no CO emission had been reported led us to perform deep CO(2-1) observations in the central area of the HII region. These observations revealed the presence of weak, clumpy CO emission in the area (Rubio, Garay, & Probst 1998). The velocity integrated CO(2-1) emission line contours have been superimposed on the 2.12μ image in Figure 3. The 2.12μ knots appear to be well correlated to the different CO(2-1) peaks seen in the contour map. This association is confirmed when the CO(2-1) emission is integrated in different velocity ranges. The H_2 knots and condensation are seen on the border or the peak of the CO(2-1) emission.

The detection of compact H_2 knots projected close to R136 seems to indicate that dense molecular stuctures can survive the strong winds and intense ionizing radiation produced by the luminous young stars of the compact R136 cluster.

Acknowledgments. We are grateful to the LOC and the IAU for their support to attend the meeting. This work has been funded through FONDECYT (CHILE) grant N° 1960926.

References

Combes, F. 1991, ARA&A, 29, 41

Cohen, R.S., Dame, T.M., Garay, G., at al. 1988, ApJ, 331, L95

Chin, Y. 1999, this volume

Fukui, Y., Abe, R., Hara, A., et al. 1999, this volume

Israel, F.P., Johansson, L.E.B., Lequeux, J., et al. 1993, A&A, 276, 25

Johansson, L.E.B., Greve, A., Booth, R.S., et al. 1998, A&A, 331, 857

Johansson, L.E.B., Heikkila, A., & Olofsson, H. 1999, this volume

Kutner, M.L., Rubio, M., Booth R.S., et al. 1997, A&AS, 122, 255

Kutner, M.L., Rubio, M., Chair, J., et al. 1999, In preparation

Lequeux, J., Le Bourlot, J., Pineau des Forets, G., et al. 1994, A&A, 292, 371

May, J, Alvarez, H., & Bronfman, L. 1997, A&A, 327, 325

Probst R., & Rubio, M. 1997, BAAS, vol 29, 834

Rubio, M. Garay, G., Montani, J., et al. 1991, ApJ, 368, 173

Rubio, M. Lequeux, J., & Boulanger, F. 1993, A&A, 271, 9

Rubio M., Garay, G., & Probst R. 1998, The Messenger, vol 93, p.34

Rubio, M. 1997, in IAU Symposium 170: *CO:Twenty-five years of Millimeter-wave Spectroscopy*, eds. W.B. Latter et al., Kluwer, p. 265.

Strong, A.W., Bloemen J.B.G.M., Dame, T.M., et al. 1988, A&A, 207, 1

Young, J.S., & Scolville, N. 1991, ARA&A, 29, 581

Discussion

Jan Palous: Is it possible, or how far it is possible, to use the virial masses of CO/H_2 clouds in regions of star formation and new clusters? The clouds are probably expanding and out of equilibrium. Is it not possible that a fraction of the difference in X factor is due to more non-equilibrium clusters in the LMC and the SMC than in the MW?

Rubio: In our galaxy, the virial mass determination gives similar results as the masses derived from the CO luminosity applying the galactic X factor. This result includes molecular clouds in the star forming regions, which may not be in virial equilibrium. The interesting result is that when we apply the galactic X factor to the CO luminosities measured in the Magellanic Clouds the derived masses are lower than those determined from the virial mass.

It would be an interesting exercise to see if there is a significant difference between the value of the X factor for molecular clouds associated with star forming regions in the galaxy and those that are not associated with star formation.

Since we cannot measure the H_2 directly we have to look for alternative ways to determine the total H_2 mass. Using the virial mass, at least, in our galaxy gives consistent results to other methods used.

Margarita Rosado: Monica, could you tell us which is the $X{=}H_2/CO$ fraction derived directly from your 2.12μ (H_2) and SEST (CO) observations in 30 Dor, and how does it compare with the previous estimates of X?

Rubio: The X factor derived from the CO luminosity and the virial determination of the mass of these knots is larger than the X factor derived for the ensemble of CO clouds mapped by SEST. It can be as high as 20 times the galactic conversion factor. To derive a mass from the 2.12μ flux we need to know the excitation mechanism for the H_2. We will have a spectroscopic run next semester to investigate this mechanism. Anyhow, the column density that we can derive from the 2.12μ H_2 emission will be of the hot gas and not the cold H_2 gas determined by the CO molecule. It will be an interesting result to compare with PDR's models that can give the fraction of warm to cold molecular gas.

Klaas de Boer: Since the "X-Factor" has been mentioned several times. I would like to comment on it. The X is defined as $N(H_2)/I(CO)$. However, the H_2 is not a measured parameter, it merely is derived from the far-infared luminosity by assuming a constant factor $L(IR)$ to $N(H_2)$. So all the discussion about variation in X has nothing to do with H_2, but only with variation of $L(CO)$. I would like to see the use of this X-factor be ended, but for historical reasons this will probably not happen.

New Views of the Magellanic Clouds
IAU Symposium, Vol. 190, 1999
Y.-H. Chu, N.B. Suntzeff, J.E. Hesser, and D.A. Bohlender, eds.

Supernova Remnants in the Magellanic Clouds

R. Petre

NASA Goddard Space Flight Center, Greenbelt, MD 20771, USA

Abstract. Over the past decade, substantial progress has been made in understanding the properties of Magellanic Cloud supernova remnants and their role in the ISM. Among the notable results are the "typing" of progenitors via the X-ray spectra of their remnants, the use of X-ray spectra to measure ISM abundances, and the discovery of remnants with unique properties. I summarize recent studies of MC SNRs, and describe how a refined understanding of the SNR population requires consideration of the unique attributes of each remnant.

1. Introduction

Supernova remnants (SNRs) play a key role in the evolution of the Magellanic Clouds, as they do in all galaxies. Their enormous explosive energy, the manifestation of the gravitational collapse of their progenitor stars, is one of the primary sources of kinetic energy within a galaxy. It helps to shape the interstellar medium (ISM) and possibly catalyze the birth of a new generation of stars. The nucleosynthesized material they produce and disperse seed subsequent generations of stars (and planets) with metals. Their shocks are responsible for accelerating the cosmic rays that permeate interstellar space.

The Magellanic Clouds have long been touted as "ideal" locations for studying SNRs, both individually and collectively, and their global effect on the ISM. Many of the difficulties associated with studying Galactic remnants are overcome: the remnants are at a uniform, well-established distance, and the interstellar extinction is far lower and more uniform than in the Galactic plane. The Clouds are near enough so that detailed studies of individual remnants can be performed, especially with the latest generation of ground and space-based instrumentation. The high degree of star formation activity guarantees the presence of many massive, rapidly evolving stars, hence, supernovae. And our external view of the Clouds allow us to more easily observe the consequences of interactions between SNRs and their surroundings. This idealistic view is tempered by reality. As we study MC remnants, we find that, like Galactic remnants, each has unique attributes. These must be taken into consideration before any credible general statement about MC SNRs can be made.

2. Demographics

There has been no "census" of MC SNRs since the work of Mathewson et al. (1983, 1984, 1985). They cataloged 32 LMC SNRs and 11 SMC SNRs. Based on this catalog and subsequent discoveries, the numbers of confirmed LMC and SMC SNRs stand at 38 and 23. The progenitor type of some is well established, while for others we can make a reasonable guess. The four LMC SNRs located in relatively gas free regions, whose optical emission consists exclusively of Hα filaments characteristic of nonradiative shocks, are thought to be remnants of Type Ia explosions. The X-ray spectra of two of these are dominated by line emission from Type Ia nucleosynthesis products, as is the spectrum of N103B. There are three O-rich remnants, N132D and N158A in the LMC and E0102-72 in the SMC, which most likely had massive progenitors. N158A also has a central pulsar, strengthening the massive progenitor case; a pulsar has also been recently discovered in N157B. Circumstantial evidence, i.e., proximity to H II regions or OB associations, indicates that at least 18 other LMC SNRs had massive progenitors (Chu & Kennicutt 1988; Smith et al. 1994). And of course, SN1987A was a Type II SN. In contrast, no systematic study of the environments of SMC SNRs has been carried out, so with the exception of E0102-72, none of the SMC SNRs have been "typed."

Table 1. Current Census of MC SNRs.

	Total	Type Ia	Type II/Ib/Ic	Unknown
LMC	38	5	24 (+1!)	8
SMC	23	0	1	22

In Table 2, we compare the morphological types of the SNRs in the Mathewson et al. catalog with those in the Galaxy. As in the Galaxy, the majority of the MC remnants are shell-like. Although the numbers are small, there is a dearth of known composite and mixed morphology remnants in the Clouds.

Table 2. Demographics of MC SNRs.

Type	LMC	SMC	Milky Way
Shell	25	8	148
Filled	2	1	9
Composite	2	1	30
Mixed Morphology	1	0	14
Undetermined	2	1	14
Central Compact Objects	3	0	34

3. Recent Discoveries - Individual Remnants

N103B: This remnant is the best example of why care must be taken before making generalizations about MC SNRs. Based on its interaction with an H

II region, it was thought to have a massive progenitor (Dickel & Milne 1995). The X-ray spectrum, however, shows clear evidence of Type Ia nucleosythesis products (Hughes et al. 1995). A recently acquired deep ASCA spectrum of N103B, reveals a strong Fe K line at ~6.5 keV (Hwang et al. 1999).

E0102-72: In constrast to the X-ray spectrum of N103B is that of the O-rich SMC remnant E 0102-72 (Hayashi et al. 1993). It is dominated by strong O and Ne lines, indicative of a massive progenitor. The O, Ne, and Mg lines are all produced in different ionization conditions, suggesting that the ejecta are stratified, unlike the turbulently-mixed Type II remnant Cas A. Radio observations reveal a central enhancement, which could be a plerion (Amy & Ball 1993).

N132D: HST narrow band images of this O-rich LMC SNR resolves spatial scales comparable to ground based images of Galactic SNRs (Morse et al. 1996). Ionization structure is discerned within shocked interstellar clouds and filaments, but not within the O-rich ejecta. The X-ray emission is closely correlated with shocked clouds, but not ejecta. The ionizing radiation from the main blast wave is not sufficient to generate the optical emission from the precursor outside the rim; photons produced in slower shocks moving through clouds around the outer rim can supply sufficient additional radiation.

N157B and N49: Two important discoveries have recently been made involving central compact objects within LMC SNRs. A pulsar has been found in N157B, whose 16 ms period is the fastest known for a pulsar associated with a SNR (Marshall et al. 1998). N49 has long been thought to house the soft γ-ray repeater SGR 0526-66, famous for its eruption on March 5, 1979. Danner et al. (1999) report variability from the X-ray counterpart of this object, which might have a periodicity of 95 days.

There are other notable results. The complex optical morphology of N63A can be explained as the result of the interaction between the shock front and ISM features of sizes between 0.01 and 5 pc (Chu et al. 1999). The complex structure of the Honeycomb SNR can be understood as an old shock front interacting with the dense, but porous, wall of a pre-existing cavity in the line of sight (Chu et al. 1995). The SMC remnant 0101-7226 has a X-ray to radio luminosity ratio at least a factor of three less than any other known SNR (Ye et al. 1995).

4. Recent Collective Studies

SNRs in OB Associations: Chu (1997) has performed a careful search for high velocity gas in 14 LMC SNRs positionally coincident with OB associations. She found that SNRs in H II regions all show classical SNR signatures (bright X-ray emission, nonthermal radio emission and enhanced [S II]/Hα ratio) as well as high-velocity gas, while some SNRs in superbubbles show only X-ray emission. These differences can be traced to the environments in which the remnants are expanding. Her results suggest that some SNRs might be missed if they are classified as such only if they have all three classical signatures.

X-ray Spectral Surveys: ASCA has produced other two key results. Spectroscopy of several evolved LMC remnants has been used to infer the ISM metal abundances (Hughes et al. 1998); the results are consistent with those of Russell & Dopita (1992). Petre et al. (1999) have searched for evidence of high energy cosmic ray acceleration in shell-like LMC remnants in the guise of nonthermal

hard X-ray tails. Several young SNRs might have nonthermal components comparable to that of the Galactic remnant SN1006, but none is found in the older remnants in the sample. This result places the first observational limit on the age to which SNRs efficiently accelerate cosmic rays.
Radio Survey: Dickel and co-workers (e.g., Dickel & Milne 1995) have used ATCA to measure the spectral index and produce polarization maps of approximately a dozen MC SNRs. They show that most of these appear morphologically similar to their Milky Way counterparts.

5. Conclusions

The abundance of data in all bands has provided an opportunity to study the MC SNRs at a level of detail comparable with many Galactic remnants. These studies have shown that the MCs possess the full range of analogs to Galactic SNRs, as well as some unique objects. All of the remnants studied show distinctive features. These studies also show that care must be taken when performing collective studies to account for distinctive properties, such as expansion into cavities and against H II regions and other large structures. Not taking these into account affects inferences about ages, energy, and progenitor type. Far less attention has been paid to SMC SNRs than their LMC counterparts. Now that a reasonably sized sample of candidates exists, this situation should be remedied.

References

Amy, S.W., & Ball, L. 1993, ApJ, 411, 761
Chu, Y.-H. 1997, AJ, 113, 1815
Chu, Y.-H., et al. 1995, AJ, 109, 1729
Chu, Y.-H., et al. 1999, this volume
Chu, Y.-H., & Kennicutt, R.C. 1988, AJ, 96, 1874
Danner, R., et al. 1999, A&A, in press
Dickel, J.R., & Milne, D.K. 1995, AJ, 109, 200
Hayashi, I., et al. 1993, PASJ, 46, L121
Hughes, J.P., et al. 1995, ApJ, 444, L81
Hughes, J.P., et al. 1998, ApJ, 505, 732
Hwang, U., et al. 1999, this volume
Marshall, F.E., et al. 1998, ApJ, 499, L179
Mathewson, D.A., et al. 1983, ApJS, 51, 345
Mathewson, D.A., et al. 1984, ApJS, 55, 189
Mathewson, D.A., et al. 1985, ApJS, 58, 197
Morse, J.A., et al. 1996, AJ, 112, 509
Petre, R., et al. 1999, ApJ, submitted
Russell, S., & Dopita, M.A. 1992, ApJ, 384, 508
Smith, R.C., et al. 1994, AJ, 108, 1266
Ye, T., et al. 1995, MNRAS, 275, 1218

New Views of the Magellanic Clouds
IAU Symposium, Vol. 190, 1999
Y.-H. Chu, N.B. Suntzeff, J.E. Hesser, and D.A. Bohlender, eds.

Superbubbles in the Magellanic Clouds

M. S. Oey

Institute of Astronomy, Madingley Road, Cambridge CB3 0HA, U.K.

Abstract. Superbubbles that result from the stellar winds and supernovae of OB associations probably play a fundamental role in the structure and energetics of the ISM in star-forming galaxies. Their influence may also dominate the relationship between the different interstellar gas phases. How do superbubbles form and evolve? How do they affect the local and global ISM? The Magellanic Clouds provide a superior opportunity to study this shell-forming activity, since both stellar content and gaseous structure can be examined in detail. Here, the results of recent studies of superbubbles in the Magellanic Clouds are reviewed.

1. Introduction

It is now well-established that the kinetic feedback from massive stars creates shell structures in the interstellar medium (ISM). Indeed, studies of these objects in the Magellanic Clouds provide the best empirical understanding of this phenomenon. These types of shells can be roughly classified into three categories (e.g., Chu 1996; Meaburn 1980), which are summarized in Table 1: "bubbles" and supernova remnants (SNRs), which result from stellar winds and supernovae (SNe) of individual massive stars; "superbubbles," which result from the action of a few, to hundreds of, OB star winds and SNe clustered in an OB association; and "supergiant shells," which have sizes of order ~ 1 kpc or more. If the supergiant shells are analogously created by massive star feedback, they must be associated with starburst phenomena having total wind and SN energies of $\gtrsim 10^{54}$ erg. This review will focus on recent (1990's) studies of superbubbles in the Magellanic Clouds, with also a brief look at supergiant shells.

Table 1. OB star shells.

Type	Parent population	$\log E/\mathrm{erg}$	$\log R/\mathrm{pc}$
Bubble, SNR	single O, WR	51	0 – 1
Superbubble	OB assoc.	52 – 53	1 – 2
Supergiant shell	starburst	≥ 54	2 – 3

How do these shells evolve? Depending on the dominant physics, there are three self-similar relations that are usually used to describe the shell evolution. In the case of a single point energy deposition, as would apply to a SN, we have the traditional Sedov (1959) blastwave:

$$R \propto (E/n)^{1/5}\, t^{2/5} \quad , \tag{1}$$

where R is the shell radius, E is the input mechanical energy, n is a uniform ambient density, and t is elapsed time. For a constant mechanical power L over an extended period, there is a standard, adiabatic evolution (Pikel'ner 1968) given by,

$$R \propto (L/n)^{1/5}\, t^{3/5} \quad , \tag{2}$$

This model has a double-shock structure, in which the inner shock, close to the energy source, heats the wind ejecta to temperatures of order 10^6 K, driving the shell growth by the thermal pressure behind the outer shock at the shell. If the hot interior is able to radiate away its energy, the inner shock will collapse upon the outer one, and the shell may assume a momentum-conserving model (Steigman et al. 1975):

$$R \propto (L/nv_\infty)^{1/4}\, t^{1/2} \quad , \tag{3}$$

where v_∞ is the wind terminal velocity. This model describes a growth rate intermediate between those of equations 1 and 2. There are other variations of these models; for example, an extensive review of astrophysical blastwaves is given by Ostriker & McKee (1988).

Assuming coeval star formation in OB associations, the growth of superbubbles quickly becomes dominated by SN activity, although most of those found in Hα are extremely young objects that are still wind-dominated, $\lesssim 5$ Myr old.

2. Evolution

Most empirical, recent studies of superbubbles attempt to test the applicability of the standard, adiabatic model. The LMC in particular, offers a fine selection of objects, which have been used to evaluate the shell dynamics and search for enclosed hot gas.

Recent dynamical studies (Oey & Massey 1995; Oey 1996; Mac Low *et al.* 1998; Oey & Smedley 1998) of individual LMC superbubbles have found a class of objects whose observed v/R is far too high for self-similar growth, in view of the stellar population present (v is the shell expansion velocity). Based on excess observed X-ray luminosities, Chu & Mac Low (1990) and Wang & Helfand (1991a) suggest that most of these objects suffered recent SNR impacts on the shell walls. Complex velocity structures in several objects (Rosado et al. 1990; Ambrocio-Cruz et al. 1997) are consistent with this interpretation. However, Oey & Smedley (1998) and Mac Low et al. (1998) demonstrate that blowout from a high-density into a low-density region can perfectly mimic the shell kinematics.

Although many optical superbubbles show this high v/R evidence of disruption, it is important to note that other objects have been found whose dynamics are fully consistent with self-similar growth (Oey 1996). However, all of the objects, along with the high v/R objects, appear to indicate a growth rate discrepancy from the standard model, where the assumed L/n is overestimated by up to an order of magnitude. It is unclear whether this is caused by an overestimate in stellar wind power, or underestimate in ambient n, or both.

The search for hot gas within the superbubbles is an important test of the standard model. Several studies (Chu & Mac Low 1990; Wang & Helfand 1991a; Magnier et al. 1996; Mac Low et al. 1998) have found objects with X-ray emission in excess of predictions. As mentioned above, these may be caused by SNR impacts. Reassuringly, there are other superbubbles that do not show X-ray enhancements (Chu et al. 1995), and detection limits for these remain consistent with the predicted X-ray emission. Furthermore, in all superbubbles that have been examined for C IV and Si IV absorption, these high-ionization species have been found (Chu et al. 1994), and can be attributed to an interface layer between the shell wall and hot interior.

3. Supergiant Shells

Supergiant shells (SGS's) are more poorly understood than superbubbles. It is possible that many of these do not originate from star formation events at all. Alternative mechanisms include impacts by high-velocity clouds (e.g., Tenorio-Tagle et al. 1987) and gamma-ray bursters (Efremov et al. 1998; Loeb & Perna 1998). At any rate, the evolution of SGS's is likely to differ substantially from that of superbubbles because the size scales are similar to those of galactic parameters like scale height and length.

The SGS's in the LMC have been catalogued by Meaburn (1980), and only two of these have been examined in any detail. The most well-known example is LMC-4, associated with the Constellation III stellar region. This region is a remarkably well-defined shell, $\sim$ 600 pc in radius, seen in both H I and as a ring of optical H II regions. There have been numerous studies of LMC-4 over the last two decades; the most recent work is by Domgoergen et al. (1995) and Bomans et al. (1996) studying its gaseous properties, and by Olsen et al. (1997) and Braun et al. (1997) on the stellar content. The other LMC SGS that has been examined is LMC-2, just east of 30 Dor. Its gas properties were studied by Wang & Helfand (1991b), Caulet & Newell (1996), and Points et al. (1999).

4. Superbubbles and the Global ISM

Superbubbles, as a direct manifestation of massive star feedback, are of critical importance to the global properties of the ISM. These include ISM structure and kinematics, relationship between the different gas phases, and interstellar processes like star formation.

To gain insight on the ISM structure and kinematics caused by superbubbles, it is possible to use equation 2 to derive a size and velocity distribution for superbubbles in the ISM. This also serves as an additional test of the standard, adiabatic evolution over the full lifetimes of the superbubbles. Oey & Clarke (1997) computed the size distribution $N(R)\,dR \propto R^{-\alpha}dR$ for the SMC, using the H II region luminosity function and assuming a constant creation rate for OB associations. The observed slope $\alpha = 2.7 \pm 0.6$ for H I superbubbles (Staveley-Smith et al. 1997) is in remarkable agreement with the predicted slope of 2.8 ± 0.4, suggesting that superbubble activity can fully explain the H I structure of the SMC. This can be compared to $N(R)\,dR$ for holes in a fractal ISM structure (Elmegreen 1997), which is being quantified for the SMC by Stanimirović et al. (1998). They find a volume fractal dimension $D = 2.5$, translating directly to $\alpha = 2.5$. It is therefore too difficult to distinguish between a superbubble and fractal ISM model, using only $N(R)\ dR$. Indeed, the similarity in observed $N(R)\ dR$ for both cases in the SMC, might be indicative of a physical relation between the two structural models.

It is also possible to test for superbubble structure in the ISM by examining the morphology of the H I holes, of which > 500 were catalogued as shells, based on elliptical shape (Staveley-Smith et al. 1997). In addition, the velocity distribution $N(v)\, dv$ computed by Oey & Clarke (1998) is also consistent with the observed shell velocities for the SMC. A superbubble velocity structure could be a dominant input to the kinematics and turbulence of the ISM.

As superbubbles expand and age, their surface brightness quickly diminishes. Hence, they could be an important component of the diffuse, warm, ionized medium (WIM). Hunter (1994) obtained emission-line spectra of superbubbles in the LMC, which do show that some of the line ratios are intermediate between those for classical H II regions and WIM detected in other galaxies. Kennicutt et al. (1995) examined the morphological structure of the WIM in the Magellanic Clouds, and also found a substantial contribution from faint superbubbles and SGS's.

Finally, it is widely thought that superbubbles can trigger renewed star formation in their shells of swept-up cool gas. Several candidates have been examined in the LMC. The clearest example is DEM 34 / N11 (Walborn & Parker 1992; Rosado et al. 1996), and additional candidates are DEM 152 / N44 (Oey & Massey 1995) and DEM 192 / N51 D (Oey & Smedley 1998). As mentioned above, the SGS LMC-4 has a long history of investigation as a triggered star formation candidate (e.g., Efremov 1999).

Acknowledgments. I am grateful for an IAU Travel Grant that made possible my participation in this meeting.

References

Ambrocio-Cruz, P., Laval, A., Marcelin, M., & Amram, P. 1997, A&A, 319, 973

Bomans, D. J., de Boer, K. S., Koornneef, J., & Grebel, E. K. 1996, A&A, 313, 101

Braun, J. M., Bomans, D. J., Will, J.-M., & de Boer, K. S. 1997, A&A, 328, 167

Caulet, A., & Newell, R. 1996, ApJ, 465, 205

Chu, Y.-H. 1996, in *The Interplay between Massive Star Formation, the ISM, and Galaxy Evolution,* eds. D. Kunth, B. Guiderdoni, M. Heydari-Malayeri, & T. X. Thuan, Gif-sur-Yvette: Editions Frontières, 201

Chu, Y.-H., & Mac Low, M.-M. 1990, ApJ, 365, 510

Chu, Y.-H., Chang, H.-W., Su, Y.-L., & Mac Low, M.-M. 1995, ApJ, 450, 157

Chu, Y.-H., Wakker, B., Mac Low, M.-M., García-Segura, G. 1994, AJ, 108, 1696

Domgoergen, H., Bomans, D. J., & de Boer, K. S. 1995, A&A, 296, 523

Efremov, Y. N. 1999, this volume

Efremov, Y. N., Elmegreen, B. G., & Hodge, P. W. 1998, ApJ, 501, L163

Elmegreen, B. G. 1997, ApJ, 477, 196

Hunter, D. A. 1994, AJ, 107, 565

Kennicutt, R. C., Bresolin, F., Bomans, D. J., Bothun, G. D., & Thompson, I. B. 1995, AJ, 109, 594

Loeb, A., & Perna, R. 1998, ApJ, 503, L35

Mac Low, M.-M., Chang, T. H., Chu, Y.-H., Points, S. D., Smith, R. C., & Wakker, B. P. 1998, ApJ, 493, 260

Magnier, E. A., Chu, Y.-H., Points, S. D., Hwang, U., & Smith, R. C. 1996, ApJ, 464, 829

Meaburn, J. 1980, MNRAS, 192, 365

Oey, M. S. 1996, ApJ, 467, 666

Oey, M. S., & Clarke, C. J. 1997, MNRAS, 289, 570

Oey, M. S., & Clarke, C. J. 1999, in *Interstellar Turbulence*, eds. J. Franco & A. Carraminana, Cambridge: Cambridge Univ. Press, in press

Oey, M. S., & Massey, P. 1995, ApJ, 452, 210

Oey, M. S., & Smedley, S. A. 1998, AJ, 116, 1263

Olsen, K. A. G., Hodge, P. W., Wilcots, E. M., & Pastwick, L. 1997, ApJ, 475, 545

Ostriker, J. P., & McKee, C. F. 1988, Rev. Mod. Phys, 60, 1

Pikel'ner, S. B. 1968, Astrophys. Lett., 2, 97

Points, S. D., et al. 1999, ApJ, 518, 298

Rosado, M., et al. 1990, A&A, 238, 315

Rosado, M., et al. 1996, A&A, 308, 588

Sedov, L. I. 1959, *Similarity and Dimensional Methods in Mechanics*, New York: Academic

Stanimirović, S., Staveley-Smith, L., Dickey, J. M., Sault, R. J., & Snowden, S. L. 1999, MNRAS, 302, 417

Staveley-Smith, L., Sault, R. J., Hatzidimitriou, D., Kesteven, M., & McConnell, D. 1997, MNRAS, 289, 225

Steigman, G., Strittmatter, P. A., & Williams, R. E. 1975, ApJ, 198, 575

Tenorio-Tagle, G., Franco, J., Bodenheimer, P., & Różyczka, M. 1987, A&A, 179, 219

Walborn, N. R., & Parker, J. W. 1992, ApJ, 399, L87

Wang, Q., & Helfand, D. J. 1991a, ApJ, 373, 497

Wang, Q., & Helfand, D. J. 1991b, ApJ, 379, 327

Discussion

Jan Palous: The expanding shells and supershells are rather stable to Jeans, Elmegreen, Vishniac instabilities. To get them unstable we need pre-existing clouds or shell-shell collision. (See my poster, this volume)

Oey: Thank you. Indeed, we do not yet understand the conditions for triggered star formation.

Marc Kutner: When you look at the large scale effects of superbubbles in the MCs, can you speculate on their effect on the MW?

Oey: In our paper (Oey and Clarke 1997), we examined exactly this question. We found that if we apply our method of using the observed HII LF to derive the porosity of the Galactic ISM, we obtain a fairly small value of ~0.3. However, when we use the observed SN rate for the galaxy instead of the HII LF, we obtain a porosity value that is larger by an order of magnitude. This discrepancy is not seen in the 4 other galaxies we studied. Hence, it suggests some sort of discrepancy between the observed HII LF and SN rate in the Milky Way.

Daniel Wang: Do you think that we can distinguish supershells from supergiant shells in a more physical way? Say, supershells are produced by individual OB associations, whereas supergiant shells by a combination of multiple OB associations?

Oey: You point out an inconsistency in the definitions of the shells. It's true that we define supergiant shells by virture of their "supergiant" status. At present, we do not understand the creation and evolution of these objects, and they may originate from different and/or varying mechanisms. Multiple generations of OB associations could indeed be one of these mechanisms. Until we understand the nature of the origins better, it would seem best to leave the supergiant shells as morphologically defined for now.

New Views of the Magellanic Clouds
IAU Symposium, Vol. 190, 1999
Y.-H. Chu, N.B. Suntzeff, J.E. Hesser, and D.A. Bohlender, eds.

Modeling the ISM: Molecular Gas, Ionizing Radiation, and Numerical Simulations

Joel N. Bregman

Dept. of Astronomy, Univ. of Michigan, Ann Arbor, MI 48109, USA

Abstract. Three different topics regarding the ISM in the Magellanic Clouds are discussed. First, we examine how the Magellanic Stream can be used as a tracer of the ionizing radiation leaking out of Galaxy and the Magellanic Clouds. We show that the radiation reaching the Magellanic Stream is less than 1% of the ionizing radiation produced by Galactic O and B stars. Since about 14% of the ionizing radiation from these stars is required to ionize the Reynolds layer, which is within 1 kpc of the disk, most of this radiation must be absorbed before reaching the Stream.

Second, we examine the reliability of using CO as a tracer of H_2 in regions of low or modest column densities (not giant molecular cloud complexes). For our Galaxy, the usual CO to H_2 conversion factor overlooks a considerable amount of H_2 and the evidence suggests that this may be true in the LMC as well. Finally, we present numerical hydrodynamical calculations of the interstellar medium in disk galaxies for a region of size 2 kpc along the plane and 15 kpc out of the plane. The simulations reveal a rich structure of low density hot regions separated by cold dense material, with the resulting position velocity diagrams being qualitatively similar to the recent HI studies of the LMC. A number of other aspects of these simulations are discussed also.

1. Introductory Comments

I would like to discuss a few different topics in regard to the interstellar medium in the Magellanic Cloud region, with an effort to avoid topics already discussed by others at this conference. The interstellar medium can be studied in order to understand its structure and properties, or it can be used as a tool to study some other phenomenon. I will consider both types of topics, dealing with the radiation field of the Milky Way, of the molecular gas content of the LMC, and of the structure of the HI gas in the LMC.

2. The Magellanic Stream and the Ionizing Radiation from the Milky Way

The Magellanic Stream passes over the south Galactic pole, so it is directly exposed to the radiation escaping from the Milky Way. The flux of ionizing radiation reaching the Stream is the amount escaping the Galactic disk, less the absorption of this radiation in the 50-80 kpc journey to the Stream. We know that significant amounts of ionizing radiation escape the Galaxy because in order to maintain the Reynolds layer, the warm ionized gas within about 1 kpc of the plane, 5-15% of the ionizing radiation from all OB stars is required. We will show that this amount of ionizing radiation never reaches the Magellanic Stream, becoming absorbed by an intervening medium.

In the Magellanic Stream, the ionizing radiation field should be composed of three components, the metagalactic radiation field, radiation from the Milky Way, and radiation from the SMC & LMC. For the ambient metagalactic ionizing radiation field, Kulkarni and Fall (1995) estimated the field based upon the "proximity effect" of absorption line systems, and they find that $J \approx 0.006 \times 10^{-21}$erg cm^{-2} s^{-1} Hz^{-1}sr^{-1}, or equivalently, $\phi \approx 3 \times 10^3$ photons cm^{-2} s^{-1}. The Galactic contribution can be estimated, the relevant quantity being the amount of ionizing radiation leaking out of the Galaxy into the halo. When $\approx$10% of the ionizing radiation from the disk escapes into the halo (e.g., Bregman & Harrington 1986; Reynolds 1991; Bland-Hawthorn & Maloney 1997), the radiation field in the lower halo would be $\phi \approx 2 \times 10^6$ photons cm^{-2} s^{-1}, orders of magnitude larger than the metagalactic radiation field. Therefore, the ionizing radiation field from the Galaxy should dominate the local environment to at least a distance of 200 kpc, as elegantly demonstrated by Bland-Hawthorn & Maloney (1997). At the distance of the Magellanic Stream, the radiation contribution from the Galaxy would be $0.6 - 1 \times 10^6$ photons cm^{-2} s^{-1}, without the contribution from the Magellanic Clouds. The contribution from the LMC & SMC upon the Stream is difficult to assess for two reasons. It is hard to estimate the amount of radiation escaping the neutral disk of the Clouds without having a tracer such as the "Reynolds" layer around the Galaxy. Also, the Stream may be close to the plane of the LMC & SMC disks, which is the worst orientation for the escape of photons. If ionizing radiation escapes from the LMC & SMC as easily as from the Galactic disk, their contribution to the radiation field will be comparable to that of the Galaxy (especially for the parts of the Stream closest to the LMC & SMC), so the estimated radiation field could be a factor of two higher than the above value in parts of the Stream.

Because the column density of the Stream is typically 5×10^{19} cm^{-2}, the HI is highly opaque to ionizing radiation ($\tau \approx 30$), so in a steady-state, the number of ionizing photons into the Stream must be balanced by the emerging recombination photons. This leads to a simple relationship between the ionizing radiation field and the $H\alpha$ radiation field:

$$\varepsilon = 450(\phi / 1 \times 10^6 \text{photons/cm}^2\text{/s})\ mR$$

So depending upon location in the Stream, the Galaxy should produce an intensity of about 300-500 mR. A study of $H\alpha$ emission from the Stream was conducted by Weiner and Williams (1996), who looked at the leading edge of the Stream subunits MS II, MS III, and MS IV. In addition, they looked at regions that were not on the leading edge. They found that the leading edge regions were always detected, at levels of $\sim$200 mR. However, they found that the regions not at the leading edges were either very weak or were undetected, with the strongest limit obtained toward MS IVA of $< 40\ mR$. They argue that if photoionization from an ambient radiation field were responsible for the emission, the clouds would be lit up nearly equally since the line intensity would be proportional to the radiation field, with some variation due to the geometry of the surfaces and the shadowing of one cloud by another. Since there is at least an order of magnitude variation in the emission, and the leading edges are always emitters, they suggest that the ionization at the leading edges is due to ram pressure heating of the neutral clouds through a hot halo of dilute gas. They present feasibility calculations in support of this interpretation.

We find that their upper limits are of particular interest since it places upper limits on the emission that is an order of magnitude lower than originally expected. This has the important implication that the ionizing radiation field from the Galaxy that reaches the Stream is an order of magnitude lower than expected, implying that most of the ionizing radiation is absorbed somewhere between 1 kpc and 50 kpc above the disk. An order of magnitude decrease corresponds to an optical depth in the ionizing continuum (near 14 eV) of 1.2, or a column density of neutral gas of $\sim 3\times10^{17}$ cm^{-2} (the opacity may be greater since several of the $H\alpha$ measurements are upper limits). Although it is hardly surprising to find this amount of neutral gas above the disk with a covering factor of unity, it means that very little Galactic ionizing radiation escapes to large distances. If this is generally true for spiral galaxies, then the contribution by spiral galaxies to the metagalactic radiation field can be neglected. Another implication of this result is that it becomes difficult if not impossible to use $H\alpha$ intensities to determine distances to HI clouds in the vicinity of the Galaxy and the Magellanic clouds. It had been hoped that if the radiation field were known a priori, $H\alpha$ intensities would vary only as the distance of the gas cloud, but in the "test" case (the Stream), the wrong answer would have been returned.

3. Missing Molecular Gas

If the ISM in the Magellanic Clouds bears certain similarities to the Galaxy, the amount of molecular gas may be underestimated by a substantial amount. To understand this situation, we need to look at the Milky Way, where the Copernicus observatory showed that for $N_{HI} > 5\times10^{20}$cm^{-2}, H_2 is very prevalent and is a significant fraction of the total neutral column (25% of the total $HI+H_2$ column; Savage et al. 1977). In a completely independent study, where molecular gas is identified through the absorption strength of the HCO^+ millimeter line, HCO^+ absorption is seen in 18/22 lines of sight below 39° (Lucas & Liszt 1996), which corresponds to column densities above 5×10^{20} cm^{-2}; this is consistent with the Copernicus result. In another independent study in yet another waveband, X-ray absorption toward extragalactic sight lines indicates substantial absorption when $N_{HI} > 5 \times 10^{20}$cm^{-2} (at latitudes below 40°). The only possible ISM component that could be responsible for the extra absorption is H_2, which would account for about 40% of the total neutral gas column (Arabadjis & Bregman 1999), consistent with the above results.

These three results paint a consistent picture of the H_2 content in the Solar circle, but it differs from the picture that one obtains from CO studies, which find that CO is not prevalent in the Solar neighborhood. Recently, there was a survey of CO at latitudes away from the plane detected little CO (Hartmann, Magnani, & Thaddeus 1998). They found a covering factor of 0.004-0.008 and employing the usual CO to H_2 conversion ratio, a mass surface density of 0.015-

0.035 $M_\odot$ pc^{-2}. The inferred H_2 mass is about 50-130 times less than the H_2 mass determined from the studies above.

The difference between these two determinations of the H_2 mass can be understood by noting that the interaction between molecules and the radiation field is different for CO and H_2 at these column densities. This interaction is discussed by van Dishoeck & Black (1988; and others), who show that for the ambient radiation field in the disk, H_2 can shield itself against photodestruction when $A(V) > 0.25$ mag, which corresponds to $N_H > 5 \times 10^{20} cm^{-2}$. However, CO is still easily destroyed under these conditions, which leads to the situation where CO is nearly absent while H_2 is still abundant. This can lead to values for the CO to H_2 conversion factor (X) that is orders of magnitude larger than the value derived for giant molecular clouds. Therefore, for the Galaxy, CO observations can lead to an underestimate of the molecular gas mass when using the typical value of X.

This raises the issue of whether the molecular gas content in the LMC (or SMC) has been underestimated by the CO observations, and there is a hint that this may be the case. An important tracer is the [CII] 157 μm line, which usually arises from warm regions at the surface of molecular clouds (in the Galaxy). It is easily detected in the LMC but it is not correlated with CO emission (Israel et al. 1996). The 157 μm line emission is correlated with IRAS emission, which probably traces the entire neutral gas column. One likely explanation for this set of observations is that the [CII] emission traces the molecular gas that contains H_2 but where the CO cannot survive.

Programs to study the H_2 content directly are possible by using the UV absorption lines, as Copernicus did for our Galaxy. In this case, one needs UV-bright targets in or behind the LMC and SMC and a telescope capable of working in the 900-1100 A region. ORPHEUS is one such telescope, and it has observed a few hot stars in the LMC & SMC and has detected H_2 in absorption (see de Boer, this conference). In the near future, FUSE will be launched and it can observe many sources in or behind the LMC & SMC, which will lead to a more detailed census of the prevalence of H_2 in these systems.

4. Models for the Structure of the Cold Gas in the Clouds

One of the landmark observations in the study of the ISM in the Clouds has been the high-quality and rather beautiful HI maps obtained by Stavely-Smith and collaborators (Kim et al. 1998). These observations provide one of the most powerful constraints on any model of the ISM and I will describe a global hydrodynamical model that can be compared to these data.

The model for the ISM that we have developed is meant to simulate the density, temperature, and velocity structure of the multiphase medium in the disk. In this model, the gas is treated as one fluid and the stars as a second fluid, the two fluids being coupled through star formation, mass loss, and heating of the gas by the stars (supernovae plus stellar winds; see Rosen & Bregman 1995, and Rosen, Plewa, & Bregman, in preparation). The part of the disk for which I will show you a simulation is 2 kpc wide and the gas is followed to ± 15 *kpc* from the disk, although nearly all the gas lies within a few kpc of the disk. The simulations are carried out in two spatial dimension (one in the plane and one perpendicular to the plane) and the results shown here are an improvement from previous models because they are at higher resolution (600×1200 cells).

A typical example of the density, temperature, and velocity distribution is shown in Figure 1, which reveals a highly structured ISM. This type of structure changes in detail throughout the run, but not in its general nature. The densest cool gas is contained within several hundred parsecs of the midplane, while the warm gas (3×10^3-$10^5 K$) and the hot gas ($> 10^5 K$) have greater scale heights. The interaction between the stars and gas pushes the gas into dense regions, often with the shape of filaments (shells or partial shells in three dimensions) that often surround large volumes of low density gas (either warm or hot; these are bubbles in three dimensions). There is a characteristic length scale to the low density regions of a few hundred parsecs, largely due to effects such as the star formation time in associations and the propagation time of star-forming regions.

Although supernovae preferentially occur in denser than average gas, because they usually occur near sites of star formation, those that occur in the low density regions can be extremely destructive, causing rapid heating of gas. In these simulations, bubbles often connect, become quite large, and "break out" of the disk. Chimney-like structures develop that are typically 0.5-1 kpc wide and extend a few kpc out of the disk, and there are examples of this in Figure 1. This hot gas rarely escapes the galaxy, eventually cools (often in the upward flow), and the cooled material, having lost its buoyancy, falls backward to the disk. The return of the cooled gas is not randomly distributed over the disk, but occurs in a number of organized downward flow regions. In this sense, this "galactic fountain" has an organization such as is seen in convection.

These properties of the simulations are extremely robust for the gas density and supernova heating rate found in the LMC and the Galaxy. The implication for the LMC is that it should possess the same gaseous components as the Galaxy: a hot halo, a thick ionized gas layer, in addition to the usual HI and molecular gas components. Failure to detect the hot halo and thick ionized layer would lead to a major revision in these models.

The evidence for halo gas around the LMC is tantalizing, but to this point, rests largely on a single study. Some of the first observations of hot LMC stars with the IUE showed the presence of high ionization ions, similar to those seen in the halo of the Milky Way. Although it was tempting to conclude that the LMC has a hot halo like the Milky Way (de Boer & Savage 1980), there was the concern that the high ionization ions were an effect local to the hot stars, being produced by their intense ionizing radiation field (Chu et. al. 1994). Recently however, absorption by high ionization ions were seen toward four stars believed to be too cool to have produced them by photoionization, giving us the strongest current evidence that the LMC might have a hot halo (Wakker et al. 1998).

The other usual tracers of hot gas are presently ambiguous regarding the LMC halo. X-ray emission is detected over regions of the galaxy (Snowden & Petre 1994), but this emission, which is often associated with superbubbles, could be confined to the plane, rather than tracing halo gas. Another potential tracer is the pulsar dispersion measure, which measures the column density of electrons (possibly from hot gas) along a line of sight toward a pulsar. The three pulsars projected upon the LMC (McConnell 1991) have dispersion measures 2-3

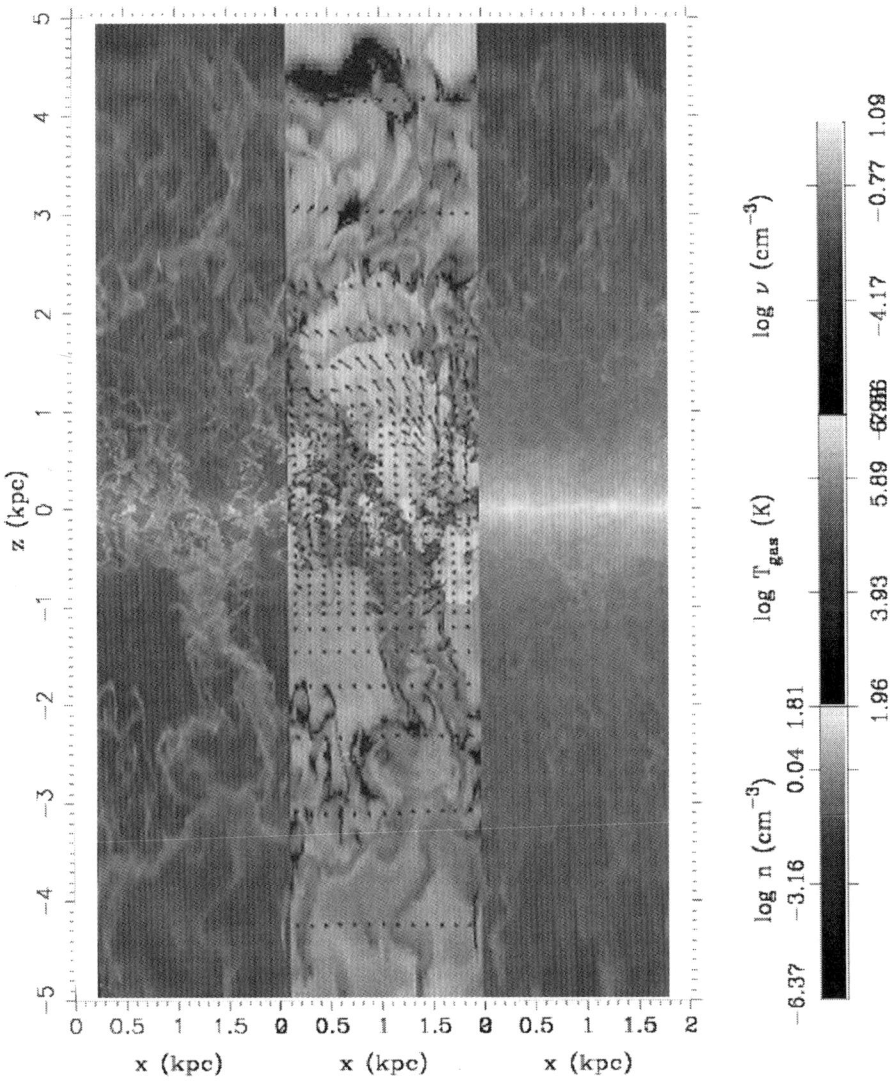

Figure 1. The gas density, gas temperature, and Population I stellar density in the simulation which is a two-dimensional cut through the midplane. The low density regions correspond to high temperature regions because the gas tries to approach pressure equilibrium. The stellar density is considerably smoother than the gas, which shows many examples of connected hot bubbles, cool clouds and shells, as well as a halo of material due to a galactic fountain.

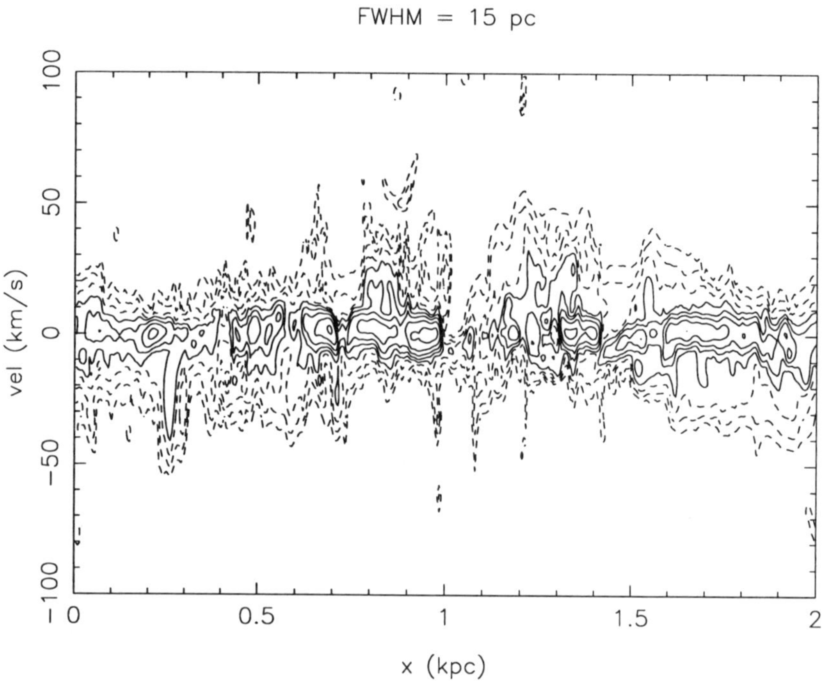

Figure 2. The position velocity diagram of the numerical hydrodynamic simulation for the cool material, smoothed to 15 pc resolution. The highly structured cool material bears a qualitative resemblance to the recent radio synthesis HI maps.

times that attributed to the Milky Way in that direction (Taylor and Cordes 1993), consistent with a model of hot halo gas around the LMC. However, it is impossible to determine whether the electron column lies in the LMC plane, halo, or an extended halo of the Milky Way. $H\alpha$ emission is generally present across the LMC (*e.g.*, Kim 1998), but it traces $10^4 K$ gas and it is more easily produced in the disk than the halo, judging from the Milky Way and other galaxies.

Another point of comparison between our simulations and the data is in the HI, and here the easiest comparison is for position-velocity diagrams. Our position-velocity diagrams (Figure 2) show a wealth of structure down to scales typical of the recent synthesis array survey. Although the minima of our $p - v$ diagrams may be deeper than the observed minima, the similarity in the nature of the distinct structures with observations presented elsewhere in this meeting is rather striking. The comparisons will become more meaningful when they can be made quantitatively and when the zero spacing data has be incorporated into the synthesis HI maps, and we hope to carry out work of that nature in the coming year.

References

Arabadjis, J.S., & Bregman, J.N. 1999, ApJ, 514, 607

Bland-Hawthorn, J., & Maloney, P.R. 1997, PASA, 14, 59.

Bregman, J.N., & Harrington, J.P. 1986, ApJ, 309, 833.

de Boer, K.S., & Savage, B.D. 1980, ApJ, 238, 86.

Chu, Y.-H., Wakker, B.P., Mac Low, M.-M., & García-Segura, G. 1994, AJ, 108, 1696.

Hartmann, D., Magnani, L., & Thaddeus, P. 1998, ApJ, 492, 205.

Israel, F.P., Maloney, P.R., Geis, N., Herrmann, F., Madden, S.C., Poglitsch, A., & Stacey, G.J. 1996, ApJ, 465, 738.

Kim, S., Staveley-Smith, L., Dopita, M.A., Freeman, K.C., Sault, R.J., Kesteven, M.J., & McConnell, D. 1998, ApJ, 503, 674.

Kulkarni, V.P., & Fall, S.M. 1995, ApJ, 453, 65.

Lucas, R., & Liszt, H. 1996, A&A, 307, 237.

McConnell, D., McCulloch, P.M., Hamilton, P.A., Ables, J.G., Hall, P.J., Jacka, C.E., & Hunt, A.J. 1991, MNRAS, 249, 654.

Reynolds, R.J. 1991, in: *The Interstellar Disk-Halo Connection in Galaxies*, ed. H. Bloemen (Kluwer: Dordrecht), p. 67.

Rosen, A., & Bregman, J.N. 1995, ApJ, 440, 634.

Rosen, A., Bregman, J.N., & Kelson, D.D. 1996, ApJ, 470, 839.

Savage, B.D. Bohlin, R.C., Drake, J.F., & Budich, W. 1977, ApJ, 216, 291.

Snowden, S.L., & Petre, R. 1994, ApJ, 436, L123.

Taylor, J.H., & Cordes, J.M. 1993, ApJ, 411, 674.

van Dishoeck, E.F., & Black, J.H. 1988, ApJ, 334, 771.

Wakker, B., Howk, J.C., Chu, Y-H., Bomans, D., & Points, S.D. 1998, ApJ, 499, L87.

Weiner, B.J., & Williams, T.B. 1996, AJ, 111, 1156.

As the smiles indicate on the faces of (left to right) Ed Olszewski, Karen Vanlandingham, Paul Scowen and Don Garnett, it is always easier to solve the problems of the Universe with a pint in one hand!

Part 2. Interstellar Medium

Section B. Poster Papers

Ray Haynes and Don Morton catch up on Australian news at the reception.

New Views of the Magellanic Clouds
IAU Symposium, Vol. 190, 1999
Y.-H. Chu, N.B. Suntzeff, J.E. Hesser, and D.A. Bohlender, eds.

Deep Wide-field Hα Images of the Magellanic Clouds

D.H. Morgan

Royal Observatory, Blackford Hill, Edinburgh EH9 3HJ, UK

Q.A. Parker

Anglo-Australian Observatory, Siding Spring Mountain, Coonabarabran, NSW 2357, Australia

S. Phillipps

Astrophysics Group, University of Bristol, Bristol, BS8 1TL, UK

Abstract. A new Hα survey of the Magellanic Clouds which is being carried out on fine-grained Tech-Pan emulsion with the UK 1.2m Schmidt Telescope will have the best combination of depth and resolution of any that cover such a wide area in and around the Magellanic Clouds. Preliminary results show that the films will provide identifications of new emission-line stars and nebulae.

1. Introduction

A new, large, single-element, Hα filter is being used with the UK 1.2m Schmidt Telescope for a survey of the Milky Way and Magellanic Clouds (Parker & Phillipps 1998) using fine-grained, panchromatic Tech-Pan emulsion on film as a detector. The standard exposure time is 180 minutes and the useful area covered is circular and 5.5 deg across (Parker & Bland-Hawthorn 1998). The survey of the Magellanic Clouds covers an area of ~650 sq deg using 40 field centres spaced 4 deg apart. The Hα exposures are accompanied by contemporaneous short R-band exposures (15mins) using Tech-Pan film and an OG590 filter.

As well as obtaining complete coverage of the Magellanic Cloud area, the UKST has an additional programme in which a set of ten 120 minute exposures is taken on the same centre and digitally co-added (stacked) to form a very deep image. The stacking programme has started on two centres, one in each Cloud.

2. Observations

The LMC field centre is $5^h\ 22^m$, $-69°$ (1950.0). To date, three good quality 120 minute exposures have been obtained. These films have been scanned using the SuperCOSMOS machine with a pixel size of 10 microns. The pixel maps were mapped on to the x-y coordinate grid of one chosen as master image, and were then added using a preliminary version of the stacking software which in its final form will register and add the data and filter extraneous noise features (Knox et al. 1998). Data from the new short red films are not yet available, but older measurements of an SR Survey plate on IIIa-F emulsion with the RG630 filter have been used instead for initial comparison with the Hα data.

3. Results

The limiting magnitude of each exposure is around R$\leq$ 19.5, based on identifications in the sequence by Will et al. (1997) in association LH47.

One purpose of obtaining these films was to search for point sources that emit in Hα (see Morgan 1998). Accordingly, the first test was to search a crowded 2000 × 2000 pixel area in the Bar on the master Hα film and the SR plate for objects visible only on the former. Three small, faint nebulae were found: two known planetary nebulae – J15 (Jacoby 1980) and MG28 (Morgan & Good 1992) – and one unknown nebula. The last of these cannot be seen in [OIII]λ5007 on deep UKST objective-prism plates and appears resolved in Hα with a diameter of ~5 arcsec. Its brightest pixel corresponds to ~20σ, whereas the two known planetary nebulae are ~50σ and 60σ respectively, σ being the rms of the sky background. The nebula J15 is one of the faintest known planetary nebulae and has an Hα flux of $1.9 \times 10^{-14}\,\mathrm{erg\,cm^{-2}\,s^{-1}}$ (Boroson & Liebert 1989).

A similar exercise was carried out around the bright ($m_v = 13.4$) Bep star BE189 (Bohannan & Epps 1974) which has an Hα equivalent width of ~112Å (Morgan et al. 1992). It too is easily identified.

These preliminary new images already have the best combination of depth and resolution of any that cover such a wide area in and around the Magellanic Clouds. They will be used to identify new Hα-emitting objects such as planetary nebulae (especially low-excitation objects), faint, diffuse nebulae and emission-line stars. The full stacking procedure will reduce the background noise significantly and will allow detection of even fainter nebulae.

Acknowledgments. The authors would like to thank the staffs of the UK Schmidt Telescope Unit (AAO) and the SuperCOSMOS Unit (ROE) for their help in producing the films and measurements.

References

Bohannann, B., & Epps, H.W. 1974, A&AS, 18, 47

Boroson, T.A., & Liebert, J. 1989, ApJ, 339, 844

Jacoby, G.H. 1980, ApJS, 42, 1

Knox, R.A., Hambly, N.C., Hawkins, M.R.S., & MacGillivray, H.T. 1998, MNRAS, 297, 839

Morgan, D.H. 1998, PASA, 15, 123

Morgan, D.H., Watson, F.G., & Parker, Q.A. 1992, A&AS, 93, 495

Parker, Q.A., & Bland-Hawthorn, J. 1998, PASA, 15, 33

Parker, Q.A., & Phillipps, S. 1998, PASA, 15, 28

Will, J.-M., Bomans, D.J., & Dieball, A. 1997, A&AS, 123, 455

New Views of the Magellanic Clouds
IAU Symposium, Vol. 190, 1999
Y.-H. Chu, N.B. Suntzeff, J.E. Hesser, and D.A. Bohlender, eds.

A Digital Emission-Line Survey of the Small Magellanic Cloud

P. Frank Winkler, Yasser Rathore

Middlebury College, Middlebury, VT 05753, USA

R. Chris Smith

University of Michigan, Ann Arbor, MI 48109, USA

Abstract. We present results on the SMC from the first full season of the Michigan/CTIO Magellanic Cloud Emission-Line Survey, being carried out from CTIO. Images are being obtained in Hα, [S II] λλ 6717, 6731, and [O III] λ 5007, plus red and green continuum bands for star subtraction. Data from the 1996-97 season have been assembled into large mosaic images which reveal the rich variety of nebulosity in the SMC in unprecedented detail. These are providing definitive samples of the active, occasionally violent, ISM on scales including superbubbles, wind-blown bubbles, supernova remnants, H II regions, and planetary nebulae.

The Michigan/CTIO Magellanic Cloud Emission-Line Survey (MCELS) will image virtually all of both the Clouds at the wavelengths of Hα, [S II], and [O III]. (See the paper by R.C. Smith in this volume for further details on the MCELS.) During the 1996-97 season at CTIO, a total of 34 overlapping 1°1 square fields in the SMC were observed from the 0.6/0.9m Curtis Schmidt and 2K STIS CCD. Conditions were excellent; 14 of the 18 separate nights of observations were photometric.

The survey fields cover virtually the entire SMC, but for this presentation we concentrate on the 2°7 × 2°1 central region. The data for this region have been assembled into mosaic images on a standard coordinate system at a scale of 2″0 pixel^{-1}. The continuum images have been scaled and subtracted from those in the emission lines, in order to remove most of the stellar contribution and reveal faint nebulosity. In Figure 1 we show the continuum-subtracted Hα mosaic. The rms noise level corresponds to 2.5×10^{-17} ergs cm^{-2} s^{-1} arcsec^{-2}; diffuse features down to almost this level are readily apparent. The sensitivity is comparable in the [S II] and [O III] bands.

Through comparison of the emission in different bands we can classify many of the emission features; e.g., discrete structures with [S II]/Hα flux ratio > 0.4 are probably supernova remnants, while marginally resolved objects with high [O III]/Hα and weak [S II] are almost certainly planetary nebulae. All but one of the previously cataloged SNRs are apparent, and numerous new ones are seen as well.

Acknowledgments. This work has been supported in part by the NSF through grants AST-9618465 and AST-9530747, with additional support from the W.M. Keck Foundation through the Keck Northeast Astronomy Consortium.

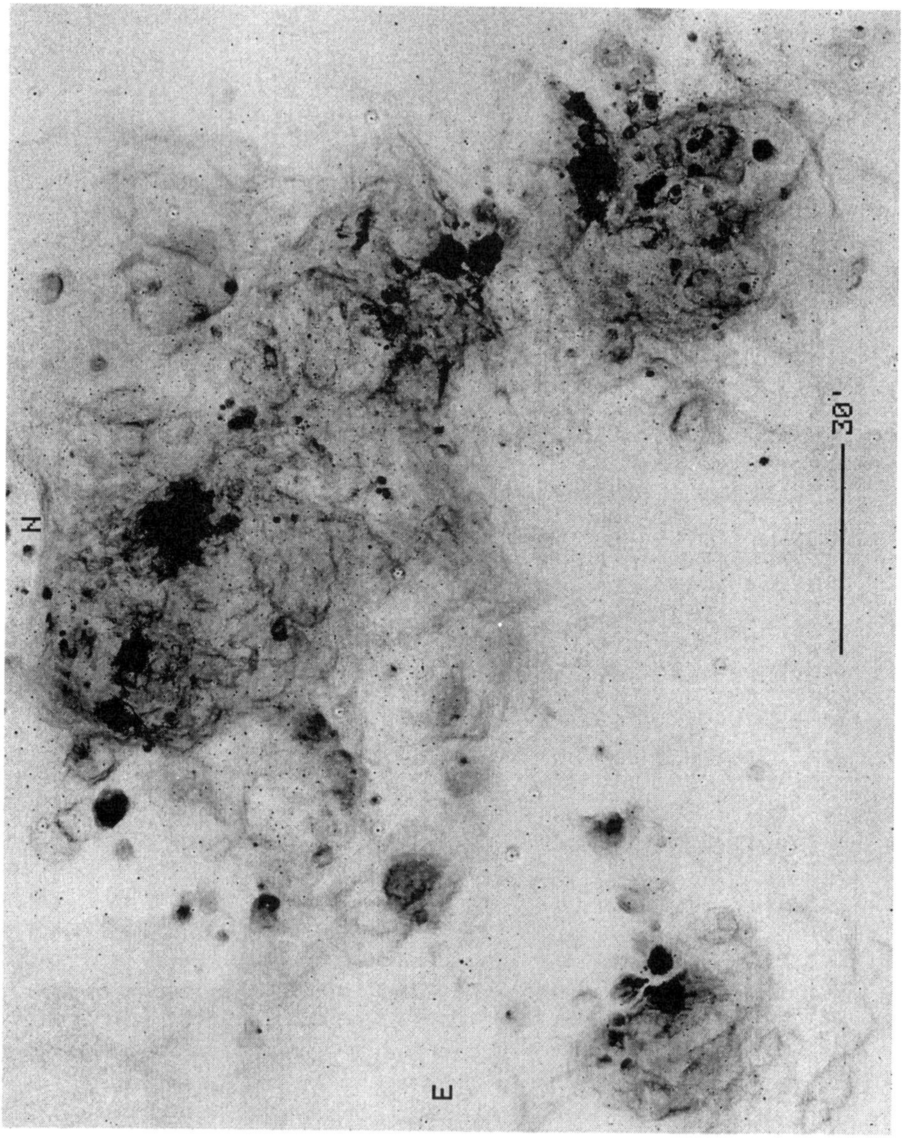

Figure 1. This continuum-subtracted Hα mosaic image of the central region of the SMC reveals a rich variety of structures on a wide range of scales. This image covers a $2^{\circ}\!.7 \times 2^{\circ}\!.1$ region at a scale of $2''\!.0$ pixel^{-1}; the full survey covers 4 times this area—essentially the entire SMC.

New Views of the Magellanic Clouds
IAU Symposium, Vol. 190, 1999
Y.-H. Chu, N.B. Suntzeff, J.E. Hesser, and D.A. Bohlender, eds.

A Wide-Angle Hα Image of the LMC

John E. Gaustad

Department of Physics and Astronomy, Swarthmore College, Swarthmore, PA 19081, USA

Wayne Rosing

Las Cumbres Obs., 18716 Favre Ridge Road, Los Gatos, CA 95033, USA

Peter R. McCullough

Astronomy Department, University of Illinois, Urbana, IL 61801, USA

Dave Van Buren

IPAC, California Institute of Technology, Pasadena, CA 91125, USA

1. The Survey

We are conducting a wide-angle Hα survey of the southern sky at CTIO using a robotic CCD camera. The survey consists of 283 fields covering the sky from $\delta = -90°$ to $\delta = +10°$, with the same centers as those in the IRAS Sky Survey Atlas. As of July 1, 1998, it was about 45% complete. When all the images are obtained and fully processed, the survey will be made available to the scientific community on the web and on CD-ROM.

2. The Robotic Camera

Our CCD camera, from SpectraSource, Inc., contains a 1024×1024 TI chip with 12-μm pixels. The sky is imaged onto the chip with a 50-mm f.l. Canon lens operated at f/1.6, yielding a field of view of 12.6°×12.6° and a scale of 45″ pixel^{-1}. A filter wheel mounted in front of the lens contains an Hα filter of 3-nm bandwidth and a dual-band notch filter which excludes Hα but transmits two 6-nm bands of continuum radiation on either side of Hα. The camera sits in a 10-ft dome (from Technical Innovations, Inc.) on a concrete pad near the 1-m telescope building. Because of its small size compared to the other telescopes at CTIO, our instrument has earned the nickname of El Enano (The Dwarf).

We can inquire of the status of the robot via email, and make changes in the program or observing schedule via ftp. But we do not operate the camera in real time–it is a true robot. The robot has its own weather station and will not attempt observations if humidity or wind conditions exceed safe limits, but as an extra precaution it also asks permission (via email) of the 4-m telescope operator before opening the dome. The operator can also close the dome via email command. The only other human intervention, during normal operations, is for weekly tape-changing and preventive maintenance.

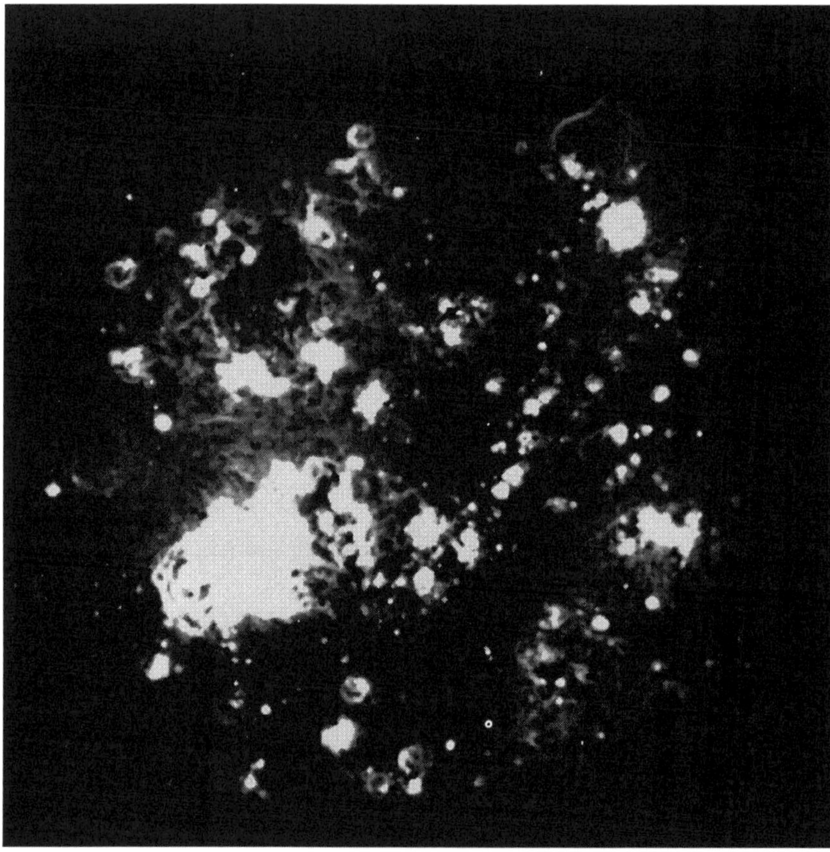

Figure 1. An Hα image of the LMC.

3. An Image of the LMC

An Hα image of the LMC is shown in Figure 1. It was made by combining ten 20-minute exposures taken through the Hα filter (total exposure time 200 minutes) and then subtracting twelve 5-minute exposures taken through the dual-band continuum filter. The size of the field is about $10°\times10°$.

Copies of this and other images of the LMC in FITS-format can be obtained via anonymous ftp from the site puck.swarthmore.edu in the directory /pub/LMCImages.

Further information about the survey, including the current stage of completion, is available at www.astronomy.swarthmore.edu or by writing the project director at jgausta1@swarthmore.edu

Acknowledgments. This project has been supported by grants from the National Science Foundation, Las Cumbres Obs., Dudley Obs., the Fund for Astrophysical Resedarch, the American Astron. Soc., and Swarthmore College.

New Views of the Magellanic Clouds
IAU Symposium, Vol. 190, 1999
Y.-H. Chu, N.B. Suntzeff, J.E. Hesser, and D.A. Bohlender, eds.

An H I Aperture Synthesis Mosaic Survey of the Large Magellanic Cloud

Sungeun Kim[1]

MSSSO, ANU, Weston Creek PO, Canberra, ACT 2611, Australia

L. Staveley-Smith and R. J. Sault

ATNF, CSIRO, P.O. Box 76, Epping, NSW 2121, Australia

M. A. Dopita and K.C. Freeman

MSSSO, ANU, Weston Creek PO, Canberra, ACT 2611, Australia

M. J. Kesteven and D. McConnell

ATNF, CSIRO, P.O. Box 76, Epping, NSW 2121, Australia

We present the results of an H I aperture synthesis mosaic of the Large Magellanic Cloud (LMC), made by combining data from 1344 separate pointing centers using the Australia Telescope Compact Array (ATCA). The resolution of the mosaiced image is 1′ (15 pc, using a distance to the LMC of 50 kpc).

The details of this survey and analysis of the results are given in Kim et al. (1998). The H I map was compared with an Hα image taken with a focal reducing camera and covering the same region of sky as our H I survey (Kim et al. 1999). The H I disk appears to be remarkably symmetric, and to have a well organized and orderly, if somewhat complex, rotational field (Kim et al. 1998). The mass of the disk component of the LMC is 2.5×10^9 M$_\odot$, and the upper limit to all mass within a radius of 4 kpc is $\sim 3.5 \times 10^9\ M_\odot$.

The structure of the neutral atomic ISM in the LMC is dominated by H I filaments combined with numerous H I shells and holes (Figure 1). The H I shell candidates chosen from visual inspection result in the division of the supergiant and giant shell classes on the basis of our estimate of the H I scale-height in the disk of the LMC (Kim et al. 1999). The expansion velocity increases from about 15 km s^{-1} for the smallest up to about 20–35 km s^{-1} for the giant shells, but the supergiant shells shows no trend with radius (Kim 1998).

We further classify H I shells into 5 different types, based on the comparision of the H I with their associated Hα emission (Kim 1998). In general, there is a weak correlation between the H I shells and the ionized gas traced out by the H II regions or H II shells and a number of the smaller shells have been found on the rim of supergiant shells (Kim et al. 1999).

---

[1] *Present Address*: Dept. of Astronomy, Univ. of Illinois, Urbana, IL61801, USA

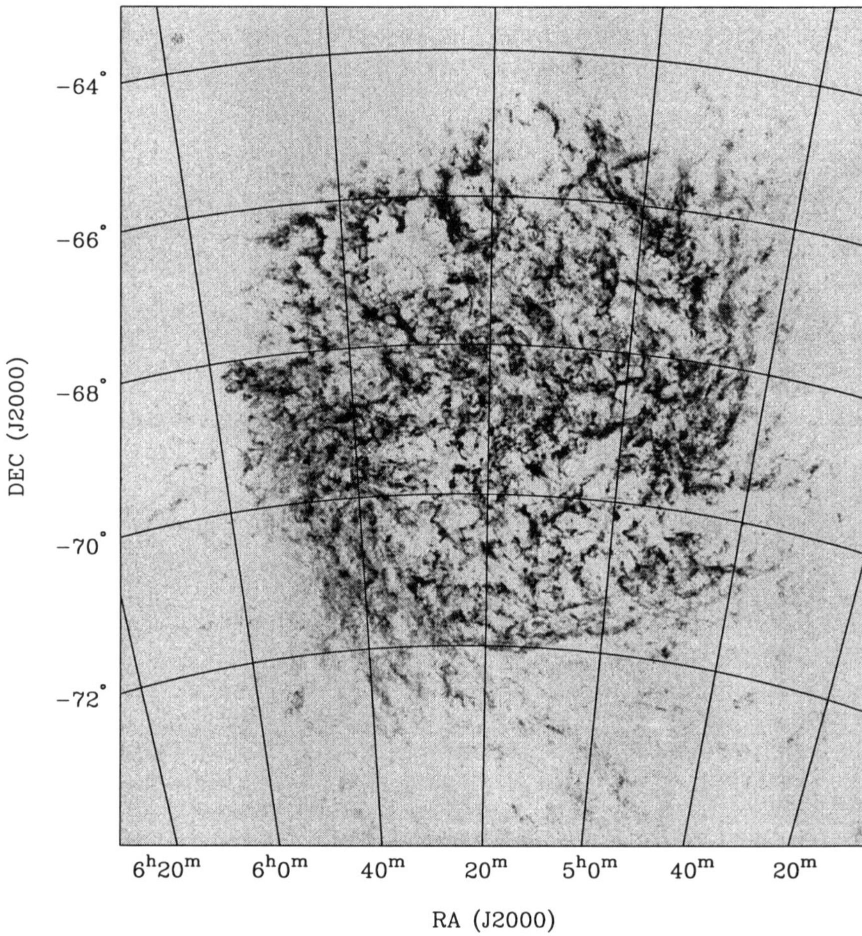

Figure 1. The H I peak brightness temperature map of the LMC.

References

Kim, S., Staveley-Smith, L., Dopita, M.A., Freeman, K.C., Sault, R.J., Kesteven, M.J., & McConnell, D. 1998, ApJ, 503, 674.

Kim, S. et al. 1999, in preparation.

Kim, S. 1998, PhD thesis, Australian National University.

New Views of the Magellanic Clouds
IAU Symposium, Vol. 190, 1999
Y.-H. Chu, N.B. Suntzeff, J.E. Hesser, and D.A. Bohlender, eds.

Fractal Structure in the Interstellar Medium of the Small Magellanic Cloud?

Snezana Stanimirović

University of Western Sydney Nepean, P.O. Box 10, Kingswood, NSW 2747, Australia

L. Staveley-Smith, R.J. Sault

Australia Telescope National Facility, CSIRO, P.O. Box 76, Epping, NSW 2121, Australia

J.M. Dickey

University of Minnesota, 116 Church St. SE, Minneapolis, MN 55455, USA

S.L. Snowden

Code 662, NASA/Goddard Space Flight Center, Greenbelt, MD 20771, USA

Abstract. We use the spatial power spectrum for a statistical study of HI in the SMC in order to probe the intrinsic topology of the ISM. A remarkable power-law fit to the spatial power spectrum was found, supporting an alternative ISM model having a fractal nature resulting from interstellar turbulence.

1. 2-D HI Spatial Power Spectrum

We have recently combined an ATCA HI mosaic of the SMC with the 'zero-spacing' Parkes data to obtain images sensitive to all spatial scales between 30 pc and 4 kpc (Stanimirović et al. 1999). This allows us to undertake a statistical study of HI in the SMC using the spatial power spectrum (e.g., Crovisier & Dickey 1983; Green 1993).

The 2-D spatial power spectrum of the whole SMC was derived by Fourier transforming images for each velocity channel and measuring the average value of the square of the modulus of the transform, $< \Re^2 + \Im^2 >$, in annuli of equal width in $\log \sqrt{u^2 + v^2}$ (u and v are measured in wavelengths, λ). We derive the following results:

- For all velocity channels, over the whole linear size range, the power spectra are remarkably well fitted by a power law (see Fig. 1), $P(k) \propto k^{\gamma}$ (k is in units of λ), with $< \gamma >= -3.04 \pm 0.02$.
- The power law slope (γ) is very similar to the ones derived previously for our own Galaxy (Crovisier & Dickey 1983; Green 1993), suggesting

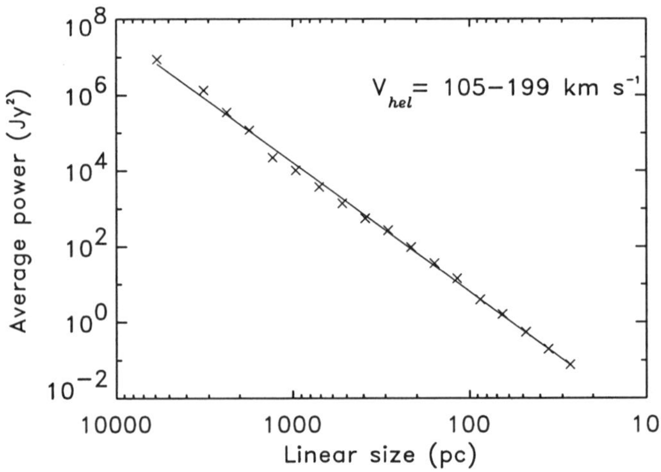

Figure 1. An average 2-D spatial power spectrum of the SMC. A least-squares fit is overlaid.

that similar processes are responsible for structure formation in both the Galaxy and the SMC.

- $P(k) \propto k^{\gamma}$ shows that the IS turbulence may play a significant role in the ISM of the SMC. However, in a compressible, self-gravitating environment, the situation is much more complex than the Kolmogorov's one which applies to incompressible fluids.

2. The ISM as a Fractal?

The observed HI power spectrum in the SMC could be a manifestation of a fractal structure created by the IS turbulence (Falconer 1997). The projected fractal dimension is $D_p = 1.5$, and the volume fractal dimension is $D = 2.5$. D_p is very similar to the ones obtained for molecular clouds, suggesting that fractal structure is pervasive in the ISM. This alternative fractal model of the ISM consists of a hierarchy of HI clouds, clustered and clumped into dense interstellar regions, while the bulk of the ISM has low density.

Acknowledgments. S. S. acknowledges supports by: a UWS Nepean travel grant, an ASA travel grant and an IAU 190 travel grant.

References

Crovisier, J., & Dickey, J.M. 1983, A&A, 122, 282

Falconer, K. 1997, in: *Fractal Geometry*, J. Wiley & Sons Ltd: Baffine Lane, England

Green, D.A. 1993, MNRAS, 262, 327

Stanimirović, S., Staveley-Smith, L., Dickey, J.M., Sault, R.J., & Snowden, S.L. 1999, MNRAS, 302, 417

Discussion

Sally Oey: I just want to suggest that the fractal dimension and turbulence may result from energy input from supernovae and stellar winds, rather than in opposition to this model.

Stanimirovic: Supernovae and stellar winds certainly contribute a lot to the IS turbulence, especially on intermedidate spatial scales, but they are not enough to explain the hierarchy of clouds going up to 4 kpc (the physical size of the SMC).

New Views of the Magellanic Clouds
IAU Symposium, Vol. 190, 1999
Y.-H. Chu, N.B. Suntzeff, J.E. Hesser, and D.A. Bohlender, eds.

How to Interpret the Bubble-Dominated ISM in the LMC?

Jan Palouš, Soňa Ehlerová

Astronomical Institute, Academy of Sciences of the Czech Republic, Boční II, 141 31 Prague 4, Czech Republic

Abstract. 3D computer simulations are used to discuss shapes, expansion velocties, masses and fragmentation of expanding shells in the LMC.

1. Model of the LMC

A three component model of the mass distribution in the LMC composed of the interstellar medium, stars and dark matter is proposed by Efremov et al. (1999). We assume that the large-scale shape of the gravitational potential is formed by stars and dark matter. The model reproduces the rotation curve of the LMC as it is given by Kim et al. (1998), the total mass within 8 kpc from the LMC center is $8 \times 10^9 M_\odot$, 60 % is in stars and 40 % is in the dark matter. The interstellar medium is distributed in a disk, which is partially supported by the centrifugal force and partially balanced with the pressure gradients. Such distribution has been used by Tomisaka & Ikeuchi (1988) and others for models of large-scale bipolar flows from nuclear starbursts in dwarf galaxies. For the gas turbulent velocity 8 km s^{-1} we get $2.2 \times 10^8 M_\odot$ for the total gas mass within 4 kpc, and about 375 pc for the thickness of the gas layer between $1 < r < 4$ kpc.

2. Shapes, Velocities and Masses of Expanding Shells in the LMC

The expansion of gas layers due to energy inputs from OB associations is described using a 3D computer code, which has been developed by Palouš (1990) and Ehlerová et al. (1997). The results of simulations with the average or increased densities are given in Table 1 and Fig. 1. Conditions for instability and fragmentation of expanding and decelerating shells have been discussed by Ehlerová et al. (1997). The fragmentation integral I_f is shown as a function of time in Fig. 1 (right panel). For values $I_f \sim 1$ the fragments are well developed, so that star forming clouds may form.

3. How to Trigger Star Formation?

Computer simulations of expanding shells in the LMC give sizes, expansion velocities and masses in agreement to the observed values. However, our analysis shows that, with the average gas density of the LMC, they are rather stable. Assuming that the gas density has been increased to higher values, probably

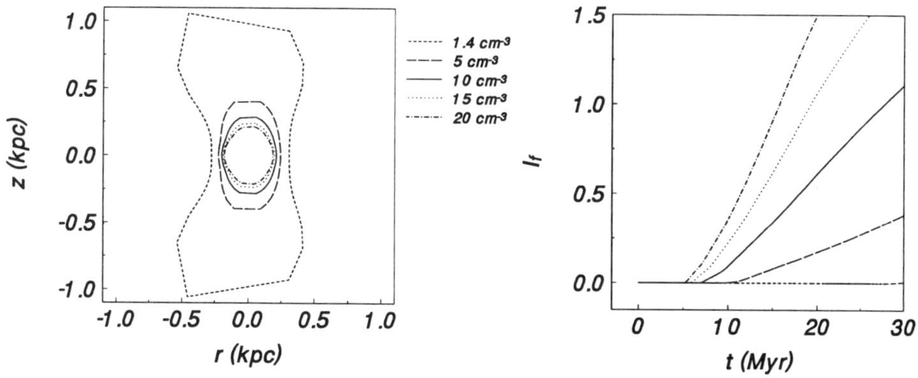

Figure 1. Shells after 15 Myr of expansion: $r_{sh} - z_{sh}$ cuts (left panel), and the value of the fragmentation integral I_f as a function of time (right panel). The values of the ambient density n_0 are indicated.

due to the compression by another nearby supershell, the expanding shell may become instable, fragment, and form the next generation stars.

	5 Myr				10 Myr				15 Myr			
n_0	r_{sh}	z_{sh}	v_{exp}	m_{sh}	r_{sh}	z_{sh}	v_{exp}	m_{sh}	r_{sh}	z_{sh}	v_{exp}	m_{sh}
1.4	157	190	23.3	0.7	225	410	40.7	1.9	278	1062	113.4	2.7
5	122	135	15.5	1.2	176	236	15.6	3.5	225	397	27.3	5.0
10	106	114	12.6	1.6	154	188	11.6	4.4	201	279	14.8	5.9
15	97	104	11.2	1.9	143	165	10.1	4.5	189	234	11.4	5.8
20	91	97	10.3	2.0	136	151	9.21	4.4	182	208	9.6	5.4

Table 1. Five models at three expansion times: 5, 10 and 15 Myr. The expansion starts with the total energy input $E_0 = 5 \times 10^{52}$ erg in the medium of the volume density n_0. The resulting typical radius r_{sh}, the z extent of the shell z_{sh}, both in pc, the typical expansion velocity v_{exp} in km s^{-1} and the total mass in the shell m_{sh} in $10^6 M_\odot$ are given.

Acknowledgements The authors gratefully acknowledge financial support by the Grant Agency of the Czech Republic under grant No. 205/97/0699 and support by the grant project of the Academy of Sciences of the Czech Republic No. K1-003-601/4.

References

Efremov, Yu. N., Ehlerová, S., & Palouš, J. 1999, A&A, submitted

Ehlerová, S., Palouš, J., Theis, Ch., & Hensler, G., 1997, A&A, 328, 121

Kim, S., et al. 1999, ApJS, submitted

Palouš, J. 1990, in The Interstellar Disk-Halo Connection in Galaxies, ed. H. Bloemen, Sterrewacht Leiden, p. 101

Tomisaka, K., & Ikeuchi, S., 1988, ApJ, 330, 695

New Views of the Magellanic Clouds
IAU Symposium, Vol. 190, 1999
Y.-H. Chu, N.B. Suntzeff, J.E. Hesser, and D.A. Bohlender, eds.

Narrow Band HI System for the Parkes Telescope Multibeam Package

Raymond Haynes, Lister Staveley-Smith

Australia Telescope National Facility, CSIRO, PO Box 76, Epping 1710, NSW, Australia

Ulrich Mebold, Peter Kalberla

RAIUB, University of Bonn, Bonn, Germany

Keith Jones

Physics Department, University of Queensland, Australia

Graeme White, Paul Jones, Miroslav Filipovic

University of Western Sydney, Nepean, Australia

John Dickey

Astronomy Dept., Univ. of Minnesota, Minneapolis, MN 55455, USA

Anne Green

School of Physics, University of Sydney, Sydney, Australia

Abstract. A narrow-band translator and filter system has been constructed for the Parkes Multibeam receiver. This brief note outlines the characteristics of the system and indicates the range of new projects that will be possible in studies of the Magellanic Clouds.

1. Introduction

In a proposal funded by the ATNF, the University of Queensland and the University of Bonn, the ATNF has now developed a narrow-band back-end system for the Parkes Multi-Beam system.

The project has provided a narrow-band back-end filter and correlation-system able to use the inner 7-beams of the multi-beam system for observations of spectral lines over 2048 channels with a total bandwidth across each spectrum of either 4 or 8 MHz. This will provide a spectral resolution of down to 0.4 km s^{-1} for each of the two polarization channels on the inner 7-beams of the multi-beam system.

This makes an important addition to the multi-beam system which can then be used to obtain "zero-spacing" data at Parkes on the multi-beam system to complement Australia Telescope Compact Array (ATCA) data or to map larger areas of the sky in HI with the observing speed-advantage obtained by multi-beaming the Parkes Telescope.

2. System Characteristics

The new Parkes narrow-band HI system will provide:

1. 7 beams on the sky;
2. 14.′4 beam-width on each beam;
3. T_s on the sky on each beam ∼24K
4. 2 orthogonal polarizations per beam;
5. 14 IF channels of 4 or 8 MHz bandwidth;
6. 2048 channels of spectral resolution;
7. velocity coverage is −800 to 800 km s^{-1} (in 8 MHz bandwidth);
8. velocity resolution down to 0.4 km s^{-1} per channel.

3. Projects

Initial projects being undertaken with the facility by the team will include: a study of the stray-radiation problems at 21-cm wavelength with the Parkes Telescope and the multi-beam system preparatory to using the facility for mapping of HI in the Galaxy; a project to map the HI in a test region along the Galactic Plane; and a project to survey the extended HI in the LMC region at Parkes.

In the future other feasible projects with the system will include: fully sampled surveys of the Magellanic System (including the the LMC and SMC, Magellanic Bridge and Stream); gas flow into the Magellanic Stream from the Magellanic Bridge; searches for HI gas in the neighborhood of the discrete HI clouds in the Magellanic Stream; projects related to the interaction of Magellanic Cloud gas and HI in the halo of the Galaxy[1]; high-velocity HI cloud[1] studies; surveys of HI in the Galactic plane[1]; rotation curves of larger galaxies discovered in the HIPASS and ZOA surveys; simultaneous HI and continuum polarization studies at 1420 MHz; HI observations of newly discovered galaxies in the Zone of Avoidance survey[1]; HI observations of galaxies found in the HIPASS multi-beam survey; and "zero-spacing" data acquisition to complement HI studies on the ATCA.

Acknowledgments. We would like to thank the ATNF's electronics, receiver and computing staff for their support in developing the narrow-band system for the Parkes multi-beam system.

---

[1] this project requires a good understanding of stray-radiation effects at the Parkes Telescope

New Views of the Magellanic Clouds
IAU Symposium, Vol. 190, 1999
Y.-H. Chu, N.B. Suntzeff, J.E. Hesser, and D.A. Bohlender, eds.

ISO-[CII]-Investigation of Cool HI Clouds in the Large Magellanic Cloud

M. Marx-Zimmer and F. Zimmer

Radioastronomisches Institut der Universität Bonn, Auf dem Hügel 71, 53121 Bonn, Germany

U. Herbstmeier

Max-Planck-Institut für Astronomie, Königstuhl 17, 69117 Heidelberg, Germany

J. M. Dickey

Department of Astronomy, University of Minnesota, Minneapolis, Minnesota 55455, USA

1. Introduction

Despite a strong UV radiation field, the Large Magellanic Cloud (LMC) shows a relatively large abundance of cool HI gas. Neither CO- nor [CII]-lines have been detected in most of these regions in previous surveys. The energy balance of these cool clouds, of which some are located in warm surroundings, is still an open question. The improved resolution and sensitivity of the ISO telescope compared to previous measurements offers a unique opportunity to study the heating and cooling of these clouds in the LMC. Here we present first results of an investigation of the dominant cooling line, [CII] (158 μm), toward cool HI clouds.

2. Observations

We searched for [CII] line emission at 158 μm toward 13 lines of sight showing cool HI gas using the Long-Wavelength Spectrometer (LWS, Clegg et al. 1996) on board the Infrared Space Observatory[1] (ISO, Kessler et al. 1996) with an angular resolution of $1\overset{'}{.}65$. The sources have been selected from the HI absorption line surveys of Dickey et al. (1994) and Marx-Zimmer et al. (1998a). The typical integration time was 9 minutes on each source, which provides an rms of 0.8 Jy. The velocity resolution is 260 km s^{-1}.

---

[1]Based on observations with ISO, an ESA project with instruments funded by ESA Member States (especially the PI countries: France, Germany, the Netherlands and the United Kingdom) with the participation of ISAS and NASA.

3. Results

Our ISO-[CII]-observations toward cool atomic clouds reveal emission of ionized carbon toward 6 out of the 13 lines of sight. [CII] emission has been detected in the surroundings of 30 Doradus, in direction of the giant molecular cloud south of 30 Doradus, near LMC4 and toward the eastern steep HI boundary. Three of these lines of sight do not show [CII] emission in the previous survey of Mochizuki et al. (1994) due to a smaller beam filling factor (12′ resolution).

There is no clear correlation of [CII] emission with the spin temperature or the optical depth, τ_{HI}, of the cool HI, although the probability of finding [CII] emission in direction of cool HI gas increases with τ_{HI}.

We have compared the ISO-[CII]-data with our SEST-CO-survey of cool HI clouds in the LMC (Marx-Zimmer et al. 1998b). No CO emission has been detected in direction of the [CII] emitting gas. CO seems to be completely dissociated in these regions. In contrast, one HI cloud in the outer region of the LMC shows CO but no [CII] emission, indicating a low strength of the UV radiation field in the far south-west of the LMC.

The large amount of cool HI gas in the LMC and the non-detection of the dominant cooling lines CO and [CII] in most directions of the cool HI suggest that the low temperature of the atomic phase is not the result of a high cooling rate. This indicates a low heating rate of gas in many directions of the LMC despite a high UV radiation field. One possibility for a lower heating rate of gas in the LMC compared to the Milky Way might be a reduced number of small dust grains, since photoelectric heating from small grains is expected to dominate the heating of the cool neutral phase (Wolfire et al. 1995).

Acknowledgments. M. Marx-Zimmer thanks for financial support through the DLR (Deutsches Zentrum für Luft- und Raumfahrt e.V.) grant 50OR9615.

References

Clegg, P.E., Ade, P.A.R., Armand, C., et al. 1996, A&A, 315, L38

Cohen, R.S., Dame, M.T., Garay, G., et al. 1988, ApJ, 331, L95

Dickey, J.M., Mebold, U., Marx, M., Amy, S., Haynes, R.F., & Wilson, W. 1994, A&A, 289, 357

Kessler, M.F., Steinz, J.A., Anderegg, M.E., et al. 1996, A&A, 315, L27

Marx-Zimmer, M., Herbstmeier, U., Zimmer, F., Dickey, J.M., Staveley-Smith, L., & Mebold, U. 1998a, Proceedings of the Workshop *The Magellanic Clouds and other Dwarfs Galaxies*, p. 273

Marx-Zimmer, M., Zimmer, F., Herbstmeier, U., Chin, Y.-N., Dickey, J.M., & Mebold, U. 1998b, Proceedings of the Workshop *The Magellanic Clouds and other Dwarfs Galaxies*, p. 275

Mochizuki, K., Nakagawa, T., Doi, Y., Yui, Y.Y., Okuda, H., Shibai, H., Yui, M., Nishimura, T., & Low, F.J. 1994, ApJ, 430, L37

Wolfire, M.G., Hollenbach, D., McKee, Ch.F., Tielens, A.G.G.M., & Bakes, E.L.O. 1995, ApJ, 443, 152

New Views of the Magellanic Clouds
IAU Symposium, Vol. 190, 1999
Y.-H. Chu, N.B. Suntzeff, J.E. Hesser, and D.A. Bohlender, eds.

The Cool Atomic Gas in the Large Magellanic Cloud

M. Marx-Zimmer and F. Zimmer

Radioastronomisches Institut der Universität Bonn, Auf dem Hügel 71, 53121 Bonn, Germany

U. Herbstmeier

Max-Planck-Institut für Astronomie, Königstuhl 17, 69117 Heidelberg, Germany

J. M. Dickey

Department of Astronomy, University of Minnesota, Minneapolis, Minnesota 55455, USA

L. Staveley-Smith

Australia Telescope National Facility, PO Box 76, Epping, NSW 2121, Australia

1. Introduction

Studying the cool atomic phase of the interstellar medium is of special significance as cool atomic clouds can become the raw material for star formation and so determine the evolution of the whole galaxy. The cool atomic interstellar medium of the Large Magellanic Cloud (LMC) seems to be quite different from that in the Milky Way. In three 21 cm absorption line surveys using the Australia Telescope Compact Array (ATCA)* the physical properties of the cool atomic hydrogen in the LMC and the halo of the Magellanic Clouds have been studied. Here we present the results of the third HI absorption line survey. A detailed investigation of the cool HI has been done toward the supergiant shell LMC4, the surroundings of 30 Doradus and in the direction of the eastern steep HI boundary. The data have been compared with survey 2 (Dickey et al. 1994) to probe the cool gas fraction for these different regions of the LMC and to study the differences of the cool atomic phase of the LMC and that of the Milky Way.

2. Observations

The 21 cm absorption line survey has been done toward 20 compact continuum sources, which have been selected from our ATCA snapshot survey at 1.4 GHz (Marx et al. 1997). The sources show peak flux densities between 21 mJy and 80 mJy. For the spectral line observations we used the single 6 km configuration

*The Australia Telescope is funded by the Commonwealth of Australia for operation as a National Facility managed by CSIRO.

(6D) of the ATCA. The resulting synthesized beamwidth is $\sim 7''$, the velocity resolution is $\Delta v = 1.65$ km s^{-1}. We have integrated two to seven hours on each source depending on the continuum flux density. This provides an optical depth sensitivity σ_τ between 0.1 and 0.25.

3. Results

We have identified 20 absorption features toward nine of the 20 lines of sight. The 30 Doradus complex shows an unusually large amount of cool HI, about half of the atomic neutral hydrogen. No other galaxies are known to possess regions with such high fractions of cool HI gas compared to the warm. The cool gas in the vicinity of 30 Doradus shows a complex dynamic structure even beyond the optical part of this giant star forming region. Whereas all lines of sight toward the 30 Doradus complex show cool HI clouds, only about half of the lines of sight toward LMC4 reveal HI absorption features. Cool HI clouds have been even less frequently detected in direction of the eastern steep HI boundary. The number of detected cool HI clouds and their properties, such as the optical depth or the absorption integral, suggest a higher cooling rate near LMC4 and 30 Doradus compared to reference positions, caused by an enhanced density near shock fronts. We do not find a statistically significant enhancement of cool clouds toward the eastern boundary of the LMC, in spite of an expected compression zone due to the motion of the galaxy through the halo of the Milky Way (Mathewson & Ford 1984). But the detected cool HI clouds at the leading edge differ from atomic clouds at reference positions by higher values of the optical depth and the absorption integral, which seems to indicate a higher cooling rate of gas behind a large shock on the east side.

Comparing the cool atomic phase of the LMC with that of the Milky Way we find: HI clouds in the LMC are either unusually cool ($< T_c > \approx 30$ K) compared to the cool phase in the Milky Way ($< T_c > \approx 60$ K, Kalberla et al. 1985), or the cool atomic phase of the interstellar medium is more abundant in the LMC ($f_c = 35\%$ for $T_c = 60$ K) relative to the warm neutral medium than in our Galaxy ($f_c = 24\%$ for $T_c = 60$ K). Even if we exclude lines of sight toward the 30 Doradus complex and toward LMC4 we find, assuming $T_c = 60$ K, a similar mixture of warm and cool interstellar phases compared to that in the Milky Way, despite the completely different radiation field, heavy element abundance and dust-to-gas ratio in the LMC.

Acknowledgments. M. Marx-Zimmer thanks for financial support through the DLR (Deutsches Zentrum für Luft- und Raumfahrt e.V.) grant 50OR9615.

References

Dickey, J.M., Mebold, U., Marx, M., Amy, S., Haynes, R.F., & Wilson, W. 1994, A&A, 289, 357

Kalberla, P.M.W., Schwarz, U.J., & Goss, W.M. 1985, A&A, 144, 27

Marx, M., Dickey, J., & Mebold, U. 1997, A&AS, 126, 325

Mathewson, D.S., & Ford, V.L. 1984, IAU Symp. 108, 125

New Views of the Magellanic Clouds
IAU Symposium, Vol. 190, 1999
Y.-H. Chu, N.B. Suntzeff, J.E. Hesser, and D.A. Bohlender, eds.

Molecular Cloud Structure in the Magellanic Clouds: CO-to-H_2 Conversion Factor

Soojong Pak

Astronomy Department, Seoul National University, Kwanak-Gu, Seoul 151-742, Republic of Korea

D. T. Jaffe

Astronomy Department, University of Texas at Austin, Austin, TX 78712, USA

1. Introduction

In the Milky Way, the ^{12}CO $J = 1 \rightarrow 0$ line traces the molecular gas content. The conversion factor, X^{GAL}, between the H_2 column density, $N(H_2)$, and the velocity integrated intensity of CO, $I(CO)$, has been measured via the virial theorem or via γ-ray emission (Solomon et al. 1987; Bloemen et al. 1986; Digel et al. 1997; and references therein).

How can we apply X^{GAL} to other galaxies where the metallicities are different from that of our Galaxy? The metallicity dependence of the conversion factor has been an issue. Cohen et al. (1988), Wilson (1995), and Arimoto, Sofue, & Tsujimoto (1996) argued that the value of X increases as the metallicity of the individual galaxy decreases. In contrast, Taylor, Kobulnicky, & Skillman (1996) showed that some low abundance galaxies have lower X.

2. The Cloud Size Versus Metallicity

In our previous work (Pak et al. 1998), we examined the chemical structure of neutral clouds in low-metallicity environments by comparing the observed emission lines (i.e., far-IR, ^{12}CO $J = 1 \rightarrow 0$, [CII] 158 μm, and H_2 $(1,0)S(1)$) from photo-dissociation regions (PDRs) and the simulated results from numerical codes (i.e., a PDR code and a radiative transfer code). The column density of the cloud increases as the metallicity decreases. Figure 1 shows that the typical cloud size in the LMC is 4 times bigger than that of the Galaxy. Our result also agrees with the photoionization-regulated star formation theory of McKee (1989).

3. The CO-to-H_2 Conversion Factor in the Magellanic Clouds

Neglecting the micro-turbulent width of the CO line, the CO-to-H_2 conversion factor, X, can be expressed by the radius of the neutral cloud, R_{cloud}:

$$X = \frac{N(H_2)}{I(CO)} \propto \frac{M_{cloud}}{L(CO)} \propto \frac{R_{cloud}^3}{R_{cloud}^2} \propto R_{cloud} \ ,$$

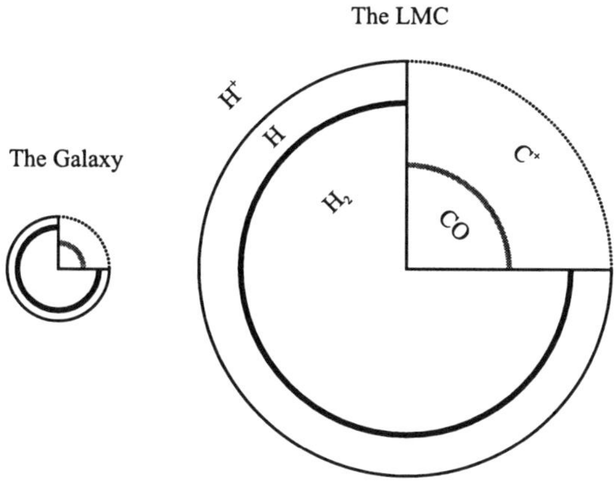

Figure 1. Schematic of chemical structures and sizes of typical clouds in the Galaxy and the LMC. The size of the star-forming clouds is inversely proportional to the metallicity of the galaxy. We assumed that the far-UV field and the gas density are same in both clouds.

where M_{cloud} is the mass of the neutral cloud, and $L(CO)$ is the luminosity of the CO line emission from the cloud. From our previous results (Pak et al. 1998), we deduced that $R_{cloud} \propto Z^{-1}$, where Z is the metallicity. Therefore, the conversion factor is inversely proportional to the metallicity:

$$X \propto Z^{-1} ,$$

which agrees with the results of Wilson (1995) and Arimoto, Sofue, & Tsujimoto (1996).

References

Arimoto, N., Sofue, Y., & Tsujimoto, T. 1996, PASJ, 48, 275

Bloemen, J. B. G. M., et al. 1986, A&A, 154, 25

Cohen, R. S., Dame, T. M., Garay, G., Montani, J., Rubio, M., & Thaddeus, P. 1988, ApJ, 331, L95

Digel, S., et al. 1997, in: *CO: Twenty-Five Years of Millimeter-Wave Spectroscopy*, IAU Symp. 170, (eds.) W. B. Latter, S. J. E. Radford, P. R. Jewell, J. G. Mangum, & J. Bally (Kluwer Academic Pub.), p. 22

McKee, C. F. 1989, ApJ, 345, 782

Pak, S., Jaffe, D. T., van Dishoeck, E. F., Johansson, L. E. B., & Booth, R. S. 1998, ApJ, 498, 735

Solomon, P. M., Rivolo, A. R., Barrett, J., & Yahil, A. 1987, ApJ, 319, 730

Taylor, C., Kobulnicky, H., & Skillman, E. 1996, BAAS, 189, 122.01

Wilson, C. D. 1995, ApJ, 448, L97

New Views of the Magellanic Clouds
IAU Symposium, Vol. 190, 1999
Y.-H. Chu, N.B. Suntzeff, J.E. Hesser, and D.A. Bohlender, eds.

Molecular Line Observations in the Magellanic Clouds

Lars E.B. Johansson, Arto Heikkilä

Onsala Space Observatory, S-439 92 Onsala, Sweden

Hans Olofsson

Stockholm Observatory, S-133 36 Saltsjöbaden, Sweden

Abstract. We have observed spectral line emission from a sample of clouds in the Magellanic Clouds to investigate the effects of metallicity and FUV radiation on the physical and chemical properties of the interstellar medium. The clouds were identified by CO surveys and selected to cover a wide range of environments. We present molecular line data for five clouds in the LMC and one cloud in the SMC.

1. Introduction

The Magellanic Clouds provide us with excellent laboratories for studies of the effects of metallicity variations on the chemical and physical properties of the interstellar medium. Here we present observational results from the Swedish-ESO Submillimetre Telescope (SEST) which allows studies of individual molecular clouds at scales smaller than 5 pc, i.e. well below the sizes of the largest molecular cloud complexes in the Galaxy. We present molecular line data for a sample of clouds, identified by CO surveys. The LMC sample (30Dor-10, 30Dor-27, N160, N159W, N159S; see Johansson et al. 1998) is located in an area including the 30 Doradus nebula and its southern HII regions N160 and N159. One cloud in the SMC bar, N27 (also denoted LIRS 49), is included in the sample.

2. CO in the Magellanic Clouds

The CO surveys (see e.g. Johansson et al. 1998 and references therein) have shown that the bulk of the CO emission originates in clouds or cloud complexes of sizes no larger than about 20 pc, surrounded by large areas apparently devoid of CO gas. Thus, the surface filling factor of the CO emission is small on size-scales of a few 100 pc. Low CO filling factors, relative to the Galaxy, seem to apply to the clouds as well. This is indicated by radiative transfer calculations using CO isotopomers observed in the three lowest rotational transitions - typical volume filling factors of the CO gas are less than 10^{-1}. These properties suggest that photon-dominated regions occupy a larger fraction of the ISM in the Magellanic Clouds than in the Galaxy, a result of the reduced dust abundances and, accordingly, a deeper penetration of the UV radiation into the ISM.

3. Molecular Abundances

The number of detected interstellar molecules is at present 16 (26 including isotopomers) in the LMC and 10 (14) in the SMC. The most recent detections are those of hydrogen sulphide (H_2S) and methyl acetylene (CH_3CCH) in N159W in the LMC and $H^{13}CO^+$ in N27 in the SMC (Heikkilä et al. 1999). In addition, there are a few tentative detections, e.g. $C^{34}S$ in the SMC (Chin et al. 1998) and H_2CS in the LMC (Heikkilä et al. 1999).

To derive fractional abundances we have estimated the H_2 contents using the virial mass concept. Typical abundances are 5×10^{-10} and 1×10^{-10} in the LMC and the SMC, respectively. After the CO isotopomers, the ethynyl radical (C_2H) is the most abundant trace molecule observed.

Relative to Galactic clouds, the derived abundances in N159W are typically 5 to 20 times lower. Two of the clouds, N27 in the SMC and 30Dor-10 in the central parts of the 30 Doradus nebula show significantly lower abundances than the rest of the sample; on the average a factor of 7 relative to those in N159W. In N27, the most likely explanation is the lower metallicity in the SMC while the underabundances in 30Dor-10 probably are due to higher photodissociation rates caused by the intense FUV radiation field in this area.

4. Physical and Chemical Properties of the Gas

The kinetic temperatures derived mirror the environment of the clouds: regions associated with vigorous star formation activity are warmer. A comparison between number densities and average densities, calculated from the virial masses, indicates that the clouds have a very clumpy structure. The number density and kinetic temperature estimates are generally based on an excitation analysis of CO, CS, SO, HCO^+ and H_2CO.

Relative to the Galaxy, the Magellanic Clouds seem to be characterised by a higher degree of photon-driven chemistry. This is indicated by the relatively high abundances of C_2H, but also as abundance variations of C_2H and HCO^+ with the environment. The formation of both molecules have reaction routes involving C^+. Our data suggest that more complex molecules do not show significantly smaller abundances than simpler species in the ISM of the Magellanic Clouds. For example, the methanol abundance in N159W is similar to that of less complex molecules.

References

Chin, Y.-N., Henkel, C., Millar, T.J., et al. 1998, A&A, 330, 901

Heikkilä, A., Johansson, L.E.B., & Olofsson, H. 1999, A&A, 344, 817

Johansson, L.E.B., Greve, A., Booth, R.S., et al. 1998, A&A, 331, 857

New Views of the Magellanic Clouds
IAU Symposium, Vol. 190, 1999
Y.-H. Chu, N.B. Suntzeff, J.E. Hesser, and D.A. Bohlender, eds.

Molecular Clouds in the Magellanic System

Yi-nan Chin
Academia Sinica Institute of Astronomy & Astrophysics, PO Box 1-87 Nankang, 11529 Taipei, Taiwan

Abstract. Several molecular cores in the Magellanic Clouds have been investigated. While previous works only concentrated on "warm" clouds, our study provides the first systematic study to cover different environments. Preliminary results are presented with brief discussions.

1. Introduction

Temperature, density, metallicity, and radiation field are important parameters to characterize the physical and chemical state of molecular clouds. Observationally, it is possible to vary the first two parameters (temperature and density) within our Milky Way by observing clouds in different locations. The metallicity, however, does not change drastically in the plane of the Milky Way. As two of the closest galaxies, the Magellanic Clouds provide metallicities which are factors of 3 and 10 lower (Westerlund 1991). If we treat our Galaxy as a "chemically" evolved system, the Magellanic System are without doubt still in an early stage of "chemical" evolution, with elemental abundances that may resemble those characterizing larger galaxies at high redshifts. In addition, the radiation field is stronger than in the solar neighborhood. As a consequence of the low metallicities and strong UV radiation field, the Magellanic Clouds are characterized by low dust-to-gas mass ratios. They are thus a "laboratory" where we can study molecular clouds with varied boundary conditions and it is easy to foresee that detailed observations will have a great impact on our general knowledge of astrochemistry and astrophysics of interstellar clouds.

2. Observations

The observations were carried out during several periods between 1996 and 1998 using the 15-m Swedish-ESO Submillimetre Telescope (SEST) at La Silla, Chile. A prominent molecular cloud in each of the Magellanic Clouds has been observed. Nevertheless, this work as well as the previous study (e.g., Johansson et al. 1994; Chin et al. 1997, 1998) have so far concentrated only on molecular cores associated with prominent H II regions. This does not sample the entire range of physical and chemical properties. Extending our studies to dark clouds would introduce a second molecular environment and would make a systematic comparison between the Magellanic system and our own Galaxy more meaningful. Recently, a hot interstellar region with enhanced X-ray emission in the LMC has been detected by *ROSAT* and showed a large X-ray shadow to its western side.

No H II emission was seen toward this shadow. This result encourages us to look for dense molecular cores in this area. Diffuse CO gas with some clumps has been detected and one of these quiescent molecular cores has been investigated in detail.

3. Discussion

In Table 1 some selected line intensity ratios of observed molecules are listed toward two molecular cores associated with prominent H II regions (N113 in the LMC and LIRS36 in the SMC), as well as a quiescent molecular core (DC 1) for the first time found in the LMC. Previous studies toward various Galactic molecular cores and extragalactic centers have also been presented in the table for comparison. There are several factors which can affect the molecular abundance. For instance, the under-abundance of nitrogen-bearing molecules characterizes an early phase of "chemical" evolution.

Table 1. Selected line intensity ratios of observed molecules †

Source	$\frac{HCO^+}{HCN}$	$\frac{HCN}{HNC}$	$\frac{CN}{HCN}$	$\frac{HC_2}{HCN}$	$\frac{CS}{HCN}$	$\frac{CS}{CN}$
N113	1.51	2.56	1.16	0.51	0.85	0.73
LIRS 36	3.39	2.87	< 0.27	2.25	2.07	> 7.5
DC 1	1.68	2.18	0.38	0.89	1.44	3.77
N159 ‡	2.21	2.97	0.48	—	0.84	1.75
S 138	0.67	3.75	—	—	—	—
M17SW	0.16	1.9	$\ll 1$	—	—	—
IRC+10216	—	8.24	1.24	—	0.45	0.36
IC 443	1.2	7.0	1.7	—	0.85	0.5
W49N	1.0	6.0	0.5	> 25.0	1.0	2.0
NGC 253	0.81	1.3	1.5	0.20	0.77	0.51
IC 342	0.51	3.3	0.79	—	0.25	0.32
M 82	2.1	2.0	1.4	0.46	0.95	0.68
NGC 4945	0.93	2.03	1.48	0.45	0.52	0.35

† All transitions are in the 3 mm range, i.e., the $J = 1 - 0$ transitions for the listed molecules, except $J = 2 - 1$ for CS.
‡ Another molecular cloud near 30 Dor in the LMC, observed by Johansson et al. 1994.

References

Chin, Y.-N., Henkel, C., Whiteoak, J.B., et al. 1997, A&A, 317, 548

Chin, Y.-N., et al. 1998, A&A, 330, 901

Johansson, L.E.B., Olofsson, H., Hjalmarson, Å., Gredel, R., & Black, J.H. 1994, A&A, 291, 89

Westerlund, B.E. 1990, A&AR, 2, 29

New Views of the Magellanic Clouds
IAU Symposium, Vol. 190, 1999
Y.-H. Chu, N.B. Suntzeff, J.E. Hesser, and D.A. Bohlender, eds.

Molecular Clouds in the LMC Observed with NANTEN I. Physical Properties of the Clouds

N. Mizuno, R. Abe, A. Hara, T. Hayakawa, S. Kato, A. Kawamura, A. Mizuno, H. Ogawa, T. Onishi, H. Saito, K. Tachihara, K. C. Xiao, N. Yamaguchi, R. Yamaguchi, and Y. Fukui

Department of Astrophysics, Nagoya University, Chikusa-ku, Nagoya, 464-8602, Japan

Y. Yonekura

Earth and Life Sciences, Osaka Prefecture University, Gakuencho 1-1, Sakai, Osaka 599-8531, Japan

Abstract. Fully sampled ^{12}CO(J=1–0) observations of the whole extent of the LMC have been made with a linear resolution of $\sim$ 30 pc at a detection limit of $N(\mathrm{H_2}) = 2 \times 10^{21}$ cm^{-2}. In addition, several selected regions have been mapped with higher sensitivity corresponding to a detection limit of 1×10^{21} cm^{-2}. Based on these results, a new estimate of the molecular mass in the LMC is presented.

1. Observations

A ^{12}CO(J=1–0) survey of the LMC was made in the period between 1997 February and November, using the 4 meter millimeter-wave telescope NANTEN of Nagoya University located at the Las Campanas Observatory, Chile. The 4 K cooled SIS mixer receiver provided a typical single sideband (SSB) system temperature of 210 K toward the zenith including the atmosphere. The spectrometer was an acousto-optical spectrometer (AOS) which covers a velocity range of 100 km s^{-1} and has a velocity resolution of 0.1 km s^{-1}. All the observations were performed using the position-switching technique. The integration time at each position was about 1 minute and the rms noise fluctuations were around 0.35 K at a velocity resolution of 0.1 km s^{-1}.

Subsequently, to reveal the distribution of faint molecular gas, additional more sensitive observations were made from 1998 April through June, at the rms noise level of 0.16 K, in the same line at the same angular resolution for several selected regions including 30 Dor, the southern large CO complex, N11, N44, and the Bar.

2. Molecular Cloud Properties

A fully sampled CO map of the LMC and the general properties of giant molecular clouds are presented by Fukui et al. (1998). These observations have been made at a grid spacing of 2′ ($\sim$ 30 pc) with a 2.′6 beam. In the observed area

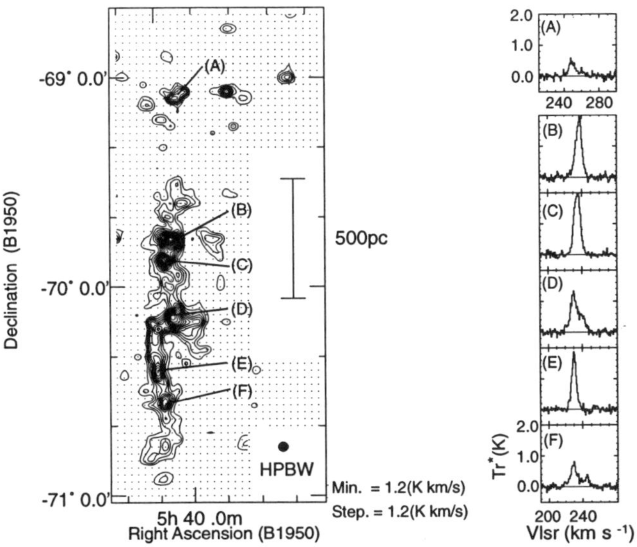

Figure 1. A contour map of the integrated ^{12}CO(J=1–0) emission in 30 Doradus and the southern large CO complex. The spectra of six major peak positions of clouds are also shown.

we identified 55 molecular clouds which are detected at more than 3 observed points having a ^{12}CO integrated intensity greater than the 4σ noise level (2.4 K km s^{-1}). The CO clouds have radii ranging from 30 to 100 pc, line widths between 3.5 and 13 km s^{-1}, CO luminosities between 8×10^3 and 2×10^5 K km s^{-1} pc^2, and virial masses ranging from 6×10^4 to $2\times10^6 M_\odot$.

In Figure 1, we present the map as well as several sample spectra from the deep observations. The peak intensities are $T_r^* = 2$ K in Figure 1, while the typical ^{12}CO spectrum in the LMC has a peak intensity of $T_r^* \lesssim 1$ K.

According to these deep observations, the masses of the large CO clouds, which were detected at more than 3 observed points in the first observations, increase by 120%, and the masses of the smaller CO clouds increase by 160%. We also detected some new, small clouds whose mass range from 10^4 to 10^5 $M_\odot$. From these results, we estimate that the total molecular mass may increase up to $\sim 7\times10^7$ $M_\odot$, about twice as large as our previous estimate, $\sim 4\times10^7$ $M_\odot$, if the deep observations were made over the whole galaxy.

We also identified clouds at the 3σ and the 6σ noise level, in order to test how much the conversion factor, X, depends upon the cloud definition. We find that the X factor is independent of cloud definition for a sample of 20 GMCs.

Acknowledgments. We appreciate the hospitality of all the staff members of the Las Campanas Observatory of the Carnegie Institution of Washington.

References

Fukui et al. 1999, this volume

New Views of the Magellanic Clouds
IAU Symposium, Vol. 190, 1999
Y.-H. Chu, N.B. Suntzeff, J.E. Hesser, and D.A. Bohlender, eds.

Dust and Gas in the Small Magellanic Cloud

Tj.R. Bontekoe[1]

Bontekoe Data Consultancy, Herengracht 2-B, 2312 LD Leiden, The Netherlands

D.J.M. Kester

Space Research Organisation Netherlands, P.O. Box 800, 9700 AV Groningen, The Netherlands

S. Stanimirović[2]

University of Western Sidney Nepean, P.O. Box 10, Kingswood, NSW 2747, Australia

L. Staveley–Smith

Australia Telescope National Facility, CSIRO, P.O. Box 76, Epping, NSW 2121, Australia

J.M. van der Hulst

Kapteyn Astronomical Institute, P.O. Box 800, 9700 AV Groningen, The Netherlands

High resolution IRAS images of the SMC

Here we present four IRAS high resolution images of the Small Magellanic Cloud, as reconstructed by the HIRAS program using the Pyramid Maximum Entropy method (Bontekoe et al. 1994). (HIRAS is not to be confused with HiRes (Aumann et al. 1990).) The images show much more detail than in Schwering & Israel (1990). Dust temperatures range from 25 – 41 K (Stanimirović et al. in prep).

References

Aumann, H.H., Fowler, J.W., & Melnyk, M. 1990, AJ, 99, 1674

Bontekoe Tj.R., Koper E., & Kester D.J.M. 1994, A&A, 284, 1037

Schwering, P.B.W. & Israel, F.P. 1990, Atlas and catalogue of infrared sources in the Magellanic clouds, Kluwer Academic Publishers.

[1] Astronomical Institute "Anton Pannekoek", Unversity of Amsterdam, Kruislaan 403, 1098 SJ Amsterdam, The Netherlands

[2] Australia Telescope National Facility, CSIRO, P.O. Box 76, Epping, NSW 2121, Australia

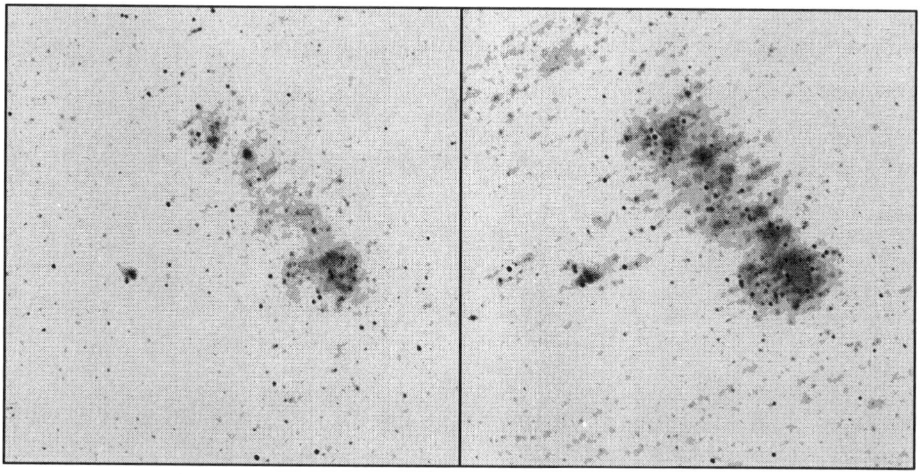

Figure 1. IRAS high resolution images of SMC, $12\mu m$ (left) and $25\mu m$. Spatial resolution of both images is about $1'$. Grey scale is identical for both images (0.3 0.4 0.5 0.6 0.8 1 2 5 10 15 20 MJy/Sr).

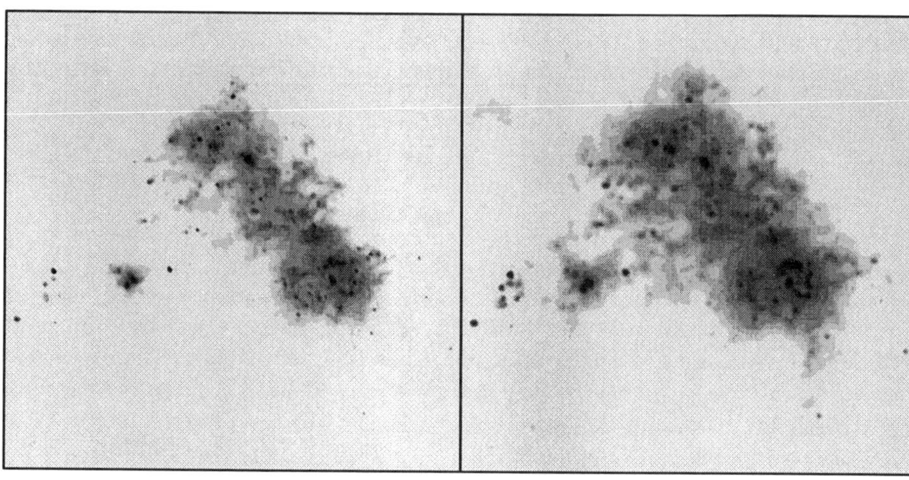

Figure 2. IRAS high resolution images of SMC, $60\mu m$ (left) and $100\mu m$. Spatial resolutions are about $1'$ and $1.7'$ resp. Grey scale is identical for both images (2 3 4 5 7.5 10 15 20 30 40 50 75 100 150 200 MJy/Sr).

New Views of the Magellanic Clouds
IAU Symposium, Vol. 190, 1999
Y.-H. Chu, N.B. Suntzeff, J.E. Hesser, and D.A. Bohlender, eds.

CO Emission toward HI Absorption Sources in the Large Magellanic Cloud

M. Marx-Zimmer and F. Zimmer

Radioastronomisches Institut der Universität Bonn, Auf dem Hügel 71, 53121 Bonn, Germany

U. Herbstmeier

Max-Planck-Institut für Astronomie, Königstuhl 17, 69117 Heidelberg, Germany

J. M. Dickey

Department of Astronomy, University of Minnesota, Minneapolis, Minnesota 55455, USA

1. Introduction

In the Large Magellanic Cloud (LMC) a large number of cool HI clouds have been detected with temperatures much lower than those found for atomic clouds in the Milky Way (Dickey et al. 1994; Mebold et al. 1997; Marx-Zimmer et al. 1998). Apparently, the population of cool HI clouds reaches kinetic gas temperatures down to as low as 10 or 20 K. These clouds may play an important role in the formation of stars in the LMC. We studied the association between the cool atomic gas and molecular gas in the LMC by ^{12}CO(1-0) line observations in directions of cool HI clouds using the 15-m Swedish-ESO Submillimetre Telescope (SEST[1]).

2. Observations

We used the SEST to survey ^{12}CO(1-0) line emission toward 25 lines of sight showing cool HI (57 HI absorption features) with spin temperatures down to 4 K. The sources have been selected from the HI absorption line surveys of Dickey et al. (1994) and Marx-Zimmer et al. (1998). We also studied the surroundings of 13 of these lines of sight by a four-point map around the central position with $40''$ spacing. The ^{12}CO(1-0) observations were carried out in a dual beam-switching mode (switching frequency 6 Hz) with a beam throw of $11'37''$ in azimuth and in a frequency-switching mode with a frequency offset of 15 MHz (40 km s^{-1}). At the frequency of 115 GHz of the J=1-0 transition of ^{12}CO the SEST beamwidth is $45''$, corresponding to a linear resolution of about 10 pc in the LMC. The velocity resolution is 0.11 km s^{-1}. Typical integration times were 20 min on the source, which provides an rms of about 0.06 K.

[1] Based on observations with the Swedish-ESO Submillimetre Telescope (SEST) at the European Southern Observatory (ESO), La Silla, Chile.

3. Results

Out of a total sample of 25 lines of sight showing cool atomic hydrogen CO emission has been detected toward 8 lines of sight with T_{MB} between 0.3 K and 8 K. Five of these lines of sight showing CO emission are in direction of the 30 Doradus complex. The other clouds have been detected toward the bar of the LMC, toward the far southwest, and in direction of LMC4. Four of these clouds have not been detected in the CO survey of Cohen et al. (1988) due to a smaller beam filling factor ($8'.8$ resolution). We mapped five of the detected CO clouds and found cloud sizes between about $110'' \times 80''$ (26 pc × 19 pc) and $600'' \times 180''$ (144 pc × 43 pc). The cool HI gas is mostly located toward the rim of the molecular clouds.

There is no clear correlation of CO emission with the HI spin temperature or the optical depth, τ_{HI}, although the probability of finding a CO-HI association increases with τ_{HI} and with the absorption integral. CO emission always occurs near but does not necessarily coincide with an HI absorption feature. The velocity differences $Dv = v_{HI} - v_{CO}$, where v_{HI} is the HI feature closest in velocity to the CO feature, are between 1 and 6 km s^{-1}. In general the value is smaller than the FWHM of the HI feature detected. Turbulent gas motion near 30 Doradus and LMC4 can cause this velocity shift of the CO gas with respect to the HI gas.

Although LMC clouds seem to have lower temperatures than atomic clouds in the Milky Way, the fraction of cool HI clouds showing CO emission is the same as in the Milky Way, where about 19% of the HI absorption components show CO emission (Despois & Baudry 1985). Whereas in the Milky Way the transition from atomic to molecular gas (H_2, CO), leaving little HI, seems to take place at temperatures below about 30 to 40 K, the non-detection of CO towards unusually cool HI clouds in the LMC might indicate that here the gas phase transition is determined by a different temperature-density combination. Another explanation of our result is that the conversion factor from CO emission to H_2 column density is lower in the LMC than in the Milky Way. The unusually cool HI clouds may be the key in understanding the high star formation activity in the LMC.

Acknowledgments. M. Marx-Zimmer thanks the financial support through the DLR (Deutsches Zentrum für Luft- und Raumfahrt e.V.) grant 50OR9615.

References

Cohen, R.S., Dame, M.T., Garay, G., et al. 1988, ApJ, 331, L95

Despois D., & Baudry, A. 1985, A&A, 148, 83

Dickey, J.M., Mebold, U., Marx, M., Amy, S., Haynes, R.F., & Wilson, W. 1994, A&A, 289, 357

Marx-Zimmer, M., Herbstmeier, U., Zimmer, F., Dickey, J.M., Staveley-Smith, L., & Mebold, U. 1998, Proceedings of the Workshop *The Magellanic Clouds and other Dwarf Galaxies*, p. 275

Mebold, U., Düsterberg, C., Dickey, J.M., Staveley-Smith, L., & Kalberla, P. 1997, A&A, 490, 65L

New Views of the Magellanic Clouds
IAU Symposium, Vol. 190, 1999
Y.-H. Chu, N.B. Suntzeff, J.E. Hesser, and D.A. Bohlender, eds.

Detection of H_2 in UV Absorption in the LMC

P. Richter and K.S. de Boer

Sternwarte, University of Bonn, Germany

D. J. Bomans

Astronomisches Institut, University of Bochum, Germany

A. Heithausen

Radioastronomisches Institut, University of Bonn, Germany

J. Koornneef

Kapteyn Institute & SRON, University of Groningen, Netherlands

Abstract. Based on measurements with the ORFEUS far UV echelle spectrograph, we present the first detection of H_2 in absorption in the Large Magellanic Cloud (LMC) in the spectrum of the star LH 10:3120.

1. Introduction

Molecular hydrogen (H_2) is the most abundant molecule in the interstellar medium and it therefore plays an important role in the physics and chemistry of the interstellar gas. The cool gas component of H_2 can only be studied through UV absorption lines near 1000 Å. Its presence in the Milky Way has been demonstrated with the *Copernicus* satellite in the early seventies (Spitzer et al. 1974). Owing to the limited sensitivity of former UV satellites, the measurement of H_2 in the Magellanic Clouds via UV spectroscopy had not been possible until the ORFEUS space shuttle mission in 1996.

Studies of cool H_2 in the Magellanic Clouds are of importance because of the lower metal content of these galaxies compared to the Milky Way and the different gas to dust ratios.

2. Data

The LMC star LH 10:3120 is located in the association LH 10 in the H II region N11B in the northwest corner of the LMC. It has a spectral type O5.5Vf with $V = 12.80$ and $E(B-V) = 0.17$ (Parker et al. 1992). LH 10:3120 was observed with a total observing time of 6000 s with the ORFEUS echelle spectrograph during the ORFEUS II mission in Nov./Dec. 1996. A wavelet algorithm has been used to smooth the data and to suppress the noise in the spectra.

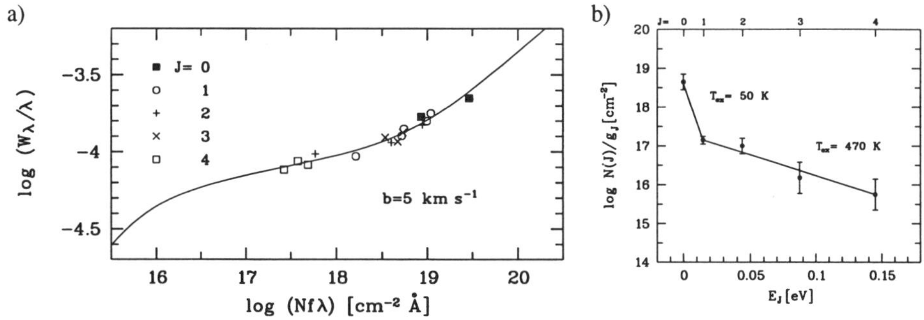

Figure 1. a) Empirical curve of growth for H_2 in the LMC gas towards LH 10:3120; b) Excitation temperatures for $J \leq 2$ and $2 \leq J \leq 4$

3. Results

All in all, we found 17 blend-free H_2 absorption lines from the lowest 5 rotational states near LMC velocities at 270 km s^{-1} for further analysis. For each rotational state curves of growth had been used to determine the individual column densities. From this analysis we derive individual column densities between 10^{17} and 10^{19} cm^{-2} and a total H_2 column density of $N_{J\leq 4} = 6.6 \times 10^{18}$ cm^{-2} for the gas in the LMC. Moreover, we derived equivalent excitation temperatures for the H_2 gas by fitting the population of the rotational states with theoretical Boltzmann distributions. The lowest two rotational states can be fitted with $T_{ex} \leq 50$ K, indicating a low kinetic temperature of the gas. Such low gas temperatures for the LMC also have been found in LMC H I 21 cm observations (Marx-Zimmer et al. 1999). For $2 \leq J \leq 4$ we derive a value of $T_{ex} \simeq 470$ K, similar to the temperatures found in the Milky Way for higher rotational states. This temperature is not the kinetic temperature of the H_2 gas, but gives the population of the excitation states due to the effect of UV pumping (Spitzer & Zweibel 1974). Our results represent not only the first qualitative detection of H_2 absorption profiles in LMC gas, but also show that the properties of the H_2 gas in the LMC along this individual line of sight are very similar to those found in the Milky Way. The full data analysis can be found in de Boer et al. (1998).

In closing we note that these findings represent, together with the detection of H_2 in the SMC (Richter et al. 1998), the first studies of cool molecular hydrogen in the Magellanic Clouds.

References

de Boer, K.S., Richter, P., Bomans, D.J., et al. 1998, A&A, 338, L5

Marx-Zimmer, M., Herbstmeier, U., Dickey, J.M., et al. 1999, A&A, submitted

Parker, J.W., Garmany, C.D., Massey, P., & Walborn, N.R. 1992, AJ, 103, 1205

Richter, P., Widmann, H., de Boer, K.S., et al. 1998, A&A, 338, L9

Spitzer, L., & Zweibel, E.G. 1974, ApJ, 191, L127

New Views of the Magellanic Clouds
IAU Symposium, Vol. 190, 1999
Y.-H. Chu, N.B. Suntzeff, J.E. Hesser, and D.A. Bohlender, eds.

Photodissociation Regions in the Large Magellanic Cloud and IC 10

Alberto D. Bolatto and James M. Jackson

Dept. of Astronomy, Boston University, Boston, MA 02215, USA

Christine D. Wilson

Department of Physics and Astronomy, McMaster University, Hamilton, Ontario L8S 4M1, Canada

Xiaolei Zhang

Harvard-Smith. Center for Astrophysics, Cambridge, MA 02138, USA

Both theory and observation suggest that metallicity profoundly alters the properties and structure of the Photodissociation Regions (PDRs; e.g., Madden et al. 1997). Several factors contribute to the differences between high and low metallicity systems: altered gas phase and grain surface chemistry due to the low Si, C and S elemental abundances, and diminished dust shielding because of the low dust-to-gas ratio. Since there is less dust shielding, UV photons penetrate more deeply into the molecular clouds leaving H_2 unaffected but photodissociating most other molecules everywhere except in the most opaque clumps. Thus, a low-metallicity system contains large regions where hydrogen remains molecular but the usual tracers of molecular gas like CO are photodissociated.

The Large Magellanic Cloud (LMC) and IC 10 are two irregular galaxies with low metalicities, and therefore provide excellent sites to study the behavior of PDRs in low-metalicity environments. We have measured the $^3P_1 \rightarrow ^3P_0$ fine structure transition of neutral carbon ([C I], $\lambda = 609\ \mu$m), as well as other tracers of dust and molecular gas. These measurements were performed using the Antarctic Submillimeter Telescope and Remote Observatory (AST/RO) at the South Pole station, and the James Clerk Maxwell Telescope (JMCT) at Mauna Kea, Hawaii. Neutral carbon originates in the PDRs, where CO is split into its atomic components by UV radiation. [C I] emission is thus thought to trace this transition occurring at a visual extinction of a few ($A_v \sim 1 - 3$). The results of our study are summarized in Fig. 1 for the LMC and Fig. 2 for IC 10.

Acknowledgments. This research was supported in part by CARA (NSF OPP 89-20223).

References

Israel, F. P. , et al. 1996, ApJ, 465, 738

Madden, S.C., et al. 1997, ApJ, 483, 200

Wilson, C.D., & Reid, I.N. 1991, ApJ, 366, L11

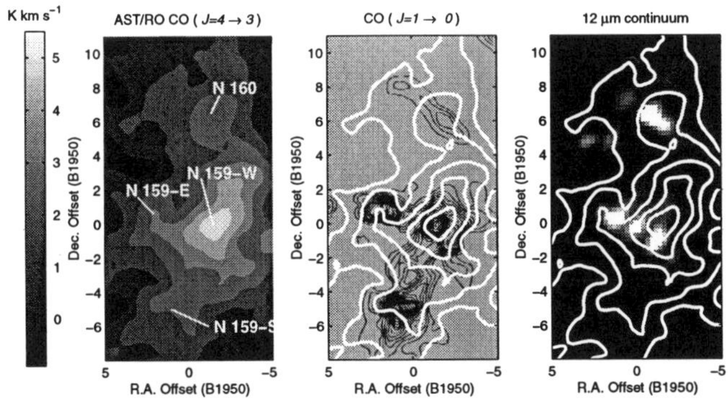

Figure 1. The molecular cloud/ HII region complex N 159/N 160 in the Large Magellanic Cloud. This complex is part of a increasingly young progression of star-forming regions south of 30 Doradus. (*Left*) AST/RO's CO ($J = 4 \rightarrow 3$) map (resolution ≈ 2 arcmin, preliminary calibration). (*Center*) Comparison with published SEST CO ($J = 1 \rightarrow 0$) observations (resolution ≈ 1 arcmin; Israel et al. 1996). (*Right*) Comparison with HIRES IRAS 12 μm showing the star formation activity in the area. The CO ($4 \rightarrow 3$)/CO ($1 \rightarrow 0$) ratio is affected by both the density and the temperature of the molecular gas and is similar for N160 and N159-W, the two active star forming clouds, while N159-S appears to harbor very little warm or dense gas.

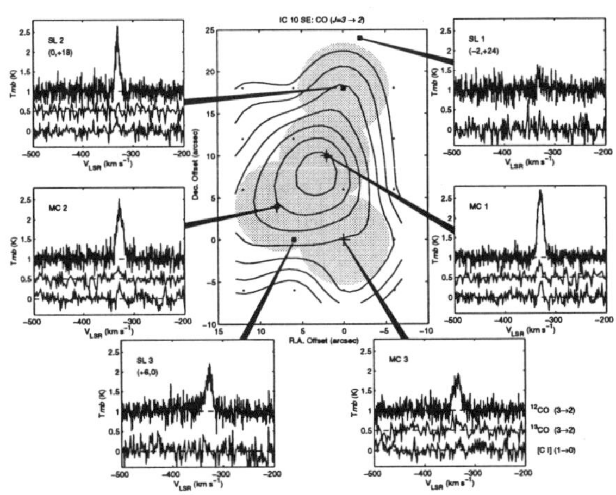

Figure 2. [C I] and CO ($J = 3 \rightarrow 2$) in the low metallicity, dwarf irregular galaxy IC 10. The contour map corresponds to the ^{12}CO integrated intensity. The dots show the actual CO pointings that constitute the map. The crosses are placed at the positions of CO clumps identified interferometrically (Wilson & Reid 1991). The squares show the placement of our [C I] observations, while the circles illustrate the size of the JCMT [C I] beam. Our six [C I] spectra (bottom) are shown here together with the corresponding ^{12}CO ($J = 3 \rightarrow 2$) (top) and ^{13}CO ($J = 3 \rightarrow 2$) (middle, scaled up by a factor of 2).

New Views of the Magellanic Clouds
IAU Symposium, Vol. 190, 1999
Y.-H. Chu, N.B. Suntzeff, J.E. Hesser, and D.A. Bohlender, eds.

Molecular Clouds in the LMC Observed with NANTEN: III. Comparison with HII Regions

R. Abe, A. Hara, T. Hayakawa, S. Kato, A. Kawamura, A. Mizuno, N. Mizuno, H. Ogawa, T. Onishi, H. Saito, K. Tachihara, K. C. Xiao, N. Yamaguchi, R. Yamaguchi, and Y. Fukui

Department of Astrophysics, Nagoya University, Chikusa-ku, Nagoya 464-8602, Japan

Y. Yonekura

Earth and Life Sciences, Osaka Prefecture University, Gakuencho 1-1, Sakai, Osaka 599-8531, Japan

Abstract. We have made ^{12}CO(J=1–0) observations in the LMC with NANTEN, and compared the detected giant molecular clouds (GMCs) with HII regions and stellar clusters. It is found that $\sim$ 80% of the GMCs are associated with HII regions. The results of comparisons of the GMCs with the HII regions and the stellar clusters are presented.

1. Introduction

A comparison of molecular clouds with HII regions is important for studying the process of massive-star and stellar-cluster formation. The LMC is the best target to carry out such a study in external galaxies because it is the nearest galaxy.

In the LMC, a few hundred HII regions are catalogued by Davies et al. (1976, DEM catalog) and Kennicutt & Hodge (1986, KH catalog). The diameters range from $\sim$ 10 pc to $\sim$ 400 pc and the luminosities from $\sim 10^{36}$ erg s^{-1} to $\sim 10^{40}$ erg s^{-1}. We used these two catalogs and the stellar clusters younger than 3×10^7 yr (Bica et al. 1996) to compare with the GMCs.

We have made a survey for molecular clouds in the ^{12}CO(J=1–0) in the LMC, using the NANTEN 4-meter millimeter wave telescope at the Las Campanas Observatory in Chile. About 100 GMCs are newly detected, and the physical parameters are determined for 55 of them. The general properties of the GMCs are reported by Fukui et al. (1999) and Mizuno et al. (1999) in these proceedings. In this contribution, we will present the results of the detailed comparisons of the GMCs with the HII regions and the stellar clusters.

2. Results and Discussion

1. Association between the GMCs and the HII regions is studied. The main results are as follows; $\sim$ 80% of the GMCs (43/55) are associated with HII regions. $\sim$ 50% of the GMCs (26/55) are associated both with HII regions

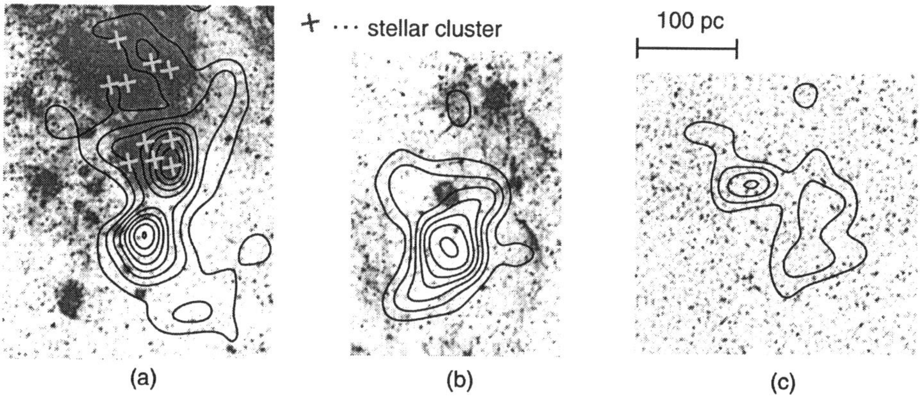

Figure 1. GMCs associated with (a) HII regions and stellar clusters (b) only HII regions (c) neither HII region nor stellar cluster.

and with stellar clusters (see Figure 1a), ~ 30% (17/55) only with HII regions (Figure 1b), ~ 20% (12/55) neither with an HII region nor with a stellar cluster (Figure 1c). There is no GMC associated only with a stellar cluster.

2. We studied the difference in the physical parameters, such as mass and line width etc., between the GMCs that are associated HII regions and/or stellar clusters and those that are not. It is found there is no significant difference in the physical parameters between the GMCs with HII regions and/or stellar clusters and those without star formation. This indicates the GMCs associated neither with HII regions nor stellar clusters have a possibility of forming massive stars or stellar clusters in the future.

3. HII regions associated with GMCs are studied. It is found the HII regions associated with stellar clusters are larger and more luminous than those not associated with clusters.

Acknowledgments. We greatly appreciate the hospitality of all the staff members of the Las Campanas Observatory of the Carnegie Institution of Washington.

References

Bica, E., Claria, J. J., Dottori, H, Santos, J. F. C. Jr., & Piatti, A. E. 1996, ApJS., 102, 57

Davies, R. D., Elliott, K. H., & Meaburn, J. 1976, MmRAS, 81,89

Fukui, Y. et al. 1999, this volume

Kennicutt, R. C., & Hodge, P. W. 1986, ApJ, 306, 130

Mizuno, N., et al. 1999, this volume

New Views of the Magellanic Clouds
IAU Symposium, Vol. 190, 1999
Y.-H. Chu, N.B. Suntzeff, J.E. Hesser, and D.A. Bohlender, eds.

The Compact HII Regions N11A and N88A in the LMC and SMC

G. Testor

Observatoire de Meudon, 92195 Meudon Cedex, France

C. S. Rola

Institute of Astronomy, University of Cambridge, United Kingdom

A. B. Whiting

Physics Department, U.S. Naval Academy, Annapolis, MD 21402, USA

Abstract. Several spectra have been obtained in the 3600–10000 Å range, for the HII regions N11A and N88A in the LMC and SMC, respectively. The spectral type of their exciting star was determined and from the emission-line intensities we have derived the gas electron density and temperature, and computed the chemical abundances of He, O, N, Ne, S, and Ar. These abundances are then compared with the ones found for other HII regions in the Magellanic Clouds.

1. Introduction and Observations

The compact HII regions N88A and N11A, with $FWHM_{H\beta} \sim 2''.4$ and $5''$ respectively (Heydari & Testor 1985; Kurt et al. 1995), are extremely bright and their emission line strengths enable accurate derivation of their physical conditions. A series of long- and short-exposure long-slit visible and near infrared spectra with a resolution of 1.3 - 6.4 Å were obtained during several runs at the ESO 1.5-m telescope. We determined the spectral types of the stars by comparing our spectra with those of the standards in Walborn & Fitzpatrick's atlas (1990). The nebular and stellar profiles of the HeIλ4471 line can be distinguished in our highest-resolution spectra. Physical parameters of the nebulae were determined with a five-level atom code, using the latest atomic data. The emission-line intensities were corrected for reddening using the Seaton law (Seaton, 1979). Three electron temperatures T_e, associated with different ionization regions in the nebulae, were computed from the line ratios [OIII] $\lambda\lambda$4363/4959+5007 (T_e^{O3}), [NII] $\lambda\lambda$5755/6548+6584 (T_e^{N2}) and [OII] $\lambda\lambda$3727/7319+7330 (T_e^{O2}). The electron density n_e was computed from the [SII]$\lambda\lambda$6717/6731 intensity ratio. Due to a low S/N in the [NII]λ5755 line, only T_e^{O3} and T_e^{O2} were used to calculate the corresponding emissivities.

2. Results

From the Lyman continuum flux (N_{Lyc}), N88A and N11A should be excited by an O5 and an O6 star, respectively (Wilcots 1994; Heydari & Testor 1985). However, a single source and multiple ionizing sources cannot be distinguished with the N_{Lyc} alone. From the averaged spectra we clearly see the HeII λ4541 and strong HeII λ4686 absorption lines, indicating that each nebula is ionized by a central and single star of spectral type earlier than an O5 V for N88A and O6 V for N11A. The mean T_e determined from the three line ratios (see above) are

- for N11A, $T_e^{O2} = 12350$ K, $T_e^{O3} = 9210$ K, $T_e^{N2} = 11075$ K
- for N88A, $T_e^{O2} = 15570$ K, $T_e^{O3} = 13450$ K, $T_e^{N2} = 15960$ K.

The mean n_e, derived for N11A is 370 cm^{-3} and 1110 cm^{-3} for N88A. The comparison with averaged abundances derived from other regions of the SMC and LMC shows slight variations but they are generally within the errors (Table 1). However, N88A presents a systematic slight O enrichment of the gas. A more detailed analysis of these objects will be submitted soon to MNRAS.

Table 1. Element abundances in N11A and N88A in the form (12+log[N(X)/N(H)]), derived from 2″ and 4″ slit width spectra.

Name	O	N	Ne	Ar	S	He	C(Hβ)	slit ″
N11A-91	8.40	7.24	7.52	6.41	5.81	10.94	0.24	2
N11A-97	8.44	7.19	7.52	6.29	6.81	10.93	0.19	4
LMC^R	8.35± 0.06	7.14± 0.18	7.61± 0.05	6.29± 0.25		10.95		
LMC^D	8.43± 0.08	6.97± 0.10	7.64± 0.10	6.20± 0.06		10.92		
N88A-91a	8.11	6.75	7.27	5.92	5.32	10.93	0.34	2
N88A-91b	8.12		7.23	5.81	5.31	10.91	0.56	2
N88A-91c	8.10	6.50	7.23	5.77	5.27	10.92	0.62	2
N88A-97	8.08	6.67	7.22	5.84	6.17	10.92	0.28	4
SMC^R	8.03± 0.10	6.63± 0.20	7.27± 0.20	5.81±0.08		10.91		
SMC^D	8.02± 0.08	6.46± 0.12	7.22± 0.12	5.78±0.12		10.92		
N88A^G	8.09± 0.04	6.60± 0.34						

^{D} Dufour et al. (1982), ^{G} Garnett et al. (1995), ^{R} Russell et al. (1990)

References

Dufour, R.J., Shields, G.A., & Talbot, R.J. 1982, ApJ, 339, 872

Garnett, D., Skillman, E., Dufour, R., Peimbert, M., Torres-Peimbert, S., Terlevich, R., Terlevich, E., & Shields, G. 1995, ApJ, 443, 43

Heydari-Malayeri, M., & Testor, G. 1985, A&A, 144, 98

Kurt, C., Dufour, R., Garnett, D., Skillman, E., Mathis, J., Peimbert, M., & Torres-Peimbert, S. 1995, Rev.Mex.AA 3, 223

Russell, S.D., & Dopita, M.A. 1990, ApJS, 74, 93

Seaton, M. J. 1979, MNRAS, 73P

Testor, G., & Pakull, M. 1985, A&A, 145, 170

Walborn, N., & Fitzpatrick, E. 1990, PASP, 102, 379

Wilcots, E.M. 1994, AJ, 107, 1338

New Views of the Magellanic Clouds
IAU Symposium, Vol. 190, 1999
Y.-H. Chu, N.B. Suntzeff, J.E. Hesser, and D.A. Bohlender, eds.

Formation of Ring Nebulae around Massive Stars in LMC H II regions

Kerstin Weis[1]

ITA, Univ. Heidelberg, Tiergartenstr. 15, 69121 Heidelberg, Germany
Astronomy Department, University of Illinois, Urbana, IL 61801, USA

Wolfgang J. Duschl

ITA, Univ. Heidelberg, Tiergartenstr. 15, 69121 Heidelberg, Germany
MPI für Radioastronomie, Auf dem Hügel 69, 53121 Bonn, Germany

Abstract. Massive stars have strong stellar winds and consequently a high mass loss during their lifetimes. Therefore they can form ring nebulae by stellar winds sweeping up the ambient medium in the main sequence phase or through wind-wind interaction or eruptions in the evolved state. We present preliminary results of a search for single bubbles and ring-nebulae around massive stars in the Large Magellanic Cloud (LMC).

1. Introduction

Massive stars with ZAMS masses above $\sim 25\,\mathrm{M}_\odot$ during most of their lifetimes lose mass through stellar winds of at least 10^{-6} $\mathrm{M}_\odot$ yr^{-1}. Main-sequence O stars, for instance, exhibit wind velocities of 1500-3000 kms^{-1} and shed half of their mass into the interstellar medium within less than 10^7 yrs. The exact value strongly depends on the ZAMS mass and the stars' metallicities ($Z \downarrow \Rightarrow \dot{M} \downarrow$). Evolving further, massive stars pass the red supergiant phase or, if they are more massive than $\mathrm{M}_{\mathrm{ZAMS}} > 50\,\mathrm{M}_\odot$, enter an unstable phase as *Luminous Blue Variables* (LBVs). If massive main-sequence stars form ring nebulae these are called *interstellar bubbles* since most of the bubbles' material is swept up ISM, while for evolved stars part of the ring nebula's material could be stellar material lost via slow winds during the red supergiant phase or via outbursts during the LBV phase. The ring nebulae around evolved stars are therefore called *circumstellar bubbles* or *Luminous Blue Variable nebula* (LBVN), depending on their origin.

The Large Magellanic Cloud reveals different large-scale bubbles formed by the combined actions of stellar winds and supernovae in OB associations, like superbubbles and supergiant shells. Surprisingly only a small number of bubbles around *single* OB stars is known. The same paucity of OB star bubbles is observed in the Milky Way. To differentiate whether this is a real physical

---

[1] Visiting Astronomer, Cerro Tololo Inter-American Observatory. CTIO is operated by AURA, Inc. under cooperative agreement with the National Science Foundation.

effect or an observational bias we intend to improve the statistics by performing a comprehensive search for small single star bubbles in the LMC.

2. The Survey in the Magellanic Clouds

To search for bubbles and ring nebulae we used the CTIO 0.9 m telescope and took images in H_α as well as in the broad band B and R filters. The analysis was concentrated on fields which are expected to include a large number of massive stars (like OB stars in H II regions). The core of 30 Dor was avoided as were the supergiant shells since in these environments we expect that bubbles may be confused with effects from groups of neighboring stars or could be undetectable due to a low-density, hot-gas environment. The identifications of ring nebulae was performed by visually inspecting our H_α images. The 28 fields used cover a total of 1.2°×1.2° ($\approx$ 1 kpc × 1 kpc using a distance modulus of $18^{m}.5$). The seeing was better than $1''.4$ in all our images, therefore we can safely detect structures around stars down to 0.5 pc. Our images reach down to a limiting surface brightness of about 1×10^{-17} ergs cm^{-2} sec^{-1} arcsec^{-2}.

3. Discussions and First Results

In the data set described above we identified 36 ring-nebulae/bubble candidates around single stars. All of them show a spherical geometry. As selection criterion we used a maximum diameter of the bubbles of 15 pc. This allowed us to differentiate between bubbles created by a single star and larger structures, which are probably created collectively from a number of massive stars. Nevertheless kinematics for all candidates is needed to differentiate classical Strömgren spheres from expanding bubbles.

A statistical analysis shows an increased number of bubbles with diameters of 2-3 pc clearly indicating bubbles intrinsic to single stars. Typical dynamical ages are of the order of 10^5 yrs if they are interstellar bubbles. If the ring nebulae are circumstellar, ages cannot be estimated, since we cannot decide from the current data whether wind-wind interaction or a single eruption in the LBV phase created the nebulae. With sizes as small as a few pc these bubbles are comparable to typical LBVNs in the LMC like R 127 or S 119. A second accumulation of diameters around 6-7 pc indicates bubbles around more evolved stars with nebulae formed through wind-wind interactions or older interstellar bubbles.

We conclude that with our survey we are able to identify bubbles around reasonably isolated massive stars in the LMC. The sizes for the bubbles we find range from 1-15 pc with the upper limit given by our selection criterion. If they are interstellar bubbles, we pick up interactions between very young stars and the surrounding ISM. In case of the stars being very massive, their ring nebulae could be circumstellar. They then are candidates for new LBVN or wind-wind interaction nebulae in the evolved states of the stars. To differentiate between these possibilities, chemical abundances and kinematic analyses are needed.

New Views of the Magellanic Clouds
IAU Symposium, Vol. 190, 1999
Y.-H. Chu, N.B. Suntzeff, J.E. Hesser, and D.A. Bohlender, eds.

Birth of Supernova Remnant 1987A

George Sonneborn, C. S. J. Pun

NASA's Goddard Space Flight Center, Code 681, Greenbelt, MD 20771, USA

Peter Garnavich, Robert Kirshner

Harvard/Smithsonian Center for Astrophysics, Cambridge, MA 02139, USA

Abstract. Recent observations of SN 1987A made with HST/STIS show that the shock interaction of supernova debris with the innermost edge of the circumstellar ring has begun.

1. Observations and Discussion

The discovery of circumstellar (CS) gas close to SN 1987A quickly led to the realization that the SN debris would soon collide with the CS gas (Fransson et al. 1989). Increasing radio and X-ray emission from SN 1987A since 1991 indicate that some type of interaction was taking place with gas inside the inner CS ring. Recent observations of SN 1987A with the Space Telescope Imaging Spectrograph (STIS) on HST have measured the velocity and location of the collision between high-velocity ejecta ($v \sim 15,000$ km s^{-1}) and CS gas (Sonneborn et al. 1998). The high-velocity Lyα and Hα emission is associated with the reverse shock of this collision (Michael et al. 1998).

STIS (Sonneborn et al. 1998) and WFPC2 (Garnavich et al. 1997) data also reveal the presence of a second interaction occuring in a highly localized region. This so-called "hot spot" is associated with the forward shock encountering a region of dense gas on the inner-most edge of the inner ring. The spatial relationship between the two shock signatures are illustrated schematically in Figure 1. The observations indicate that the long-awaited collision of the SN debris and the inner circumstellar ring has begun and should increase in brightness many fold across the electromagnetic spectrum in the coming years.

References

Fransson, C., et al. 1989, ApJ, 336, 429

Garnavich, P., et al. 1997, IAU Circular No. 6710

Michael, E., et al. 1998, ApJ, 492, L143

Sonneborn, G., et al. 1998, ApJ, 492, L139

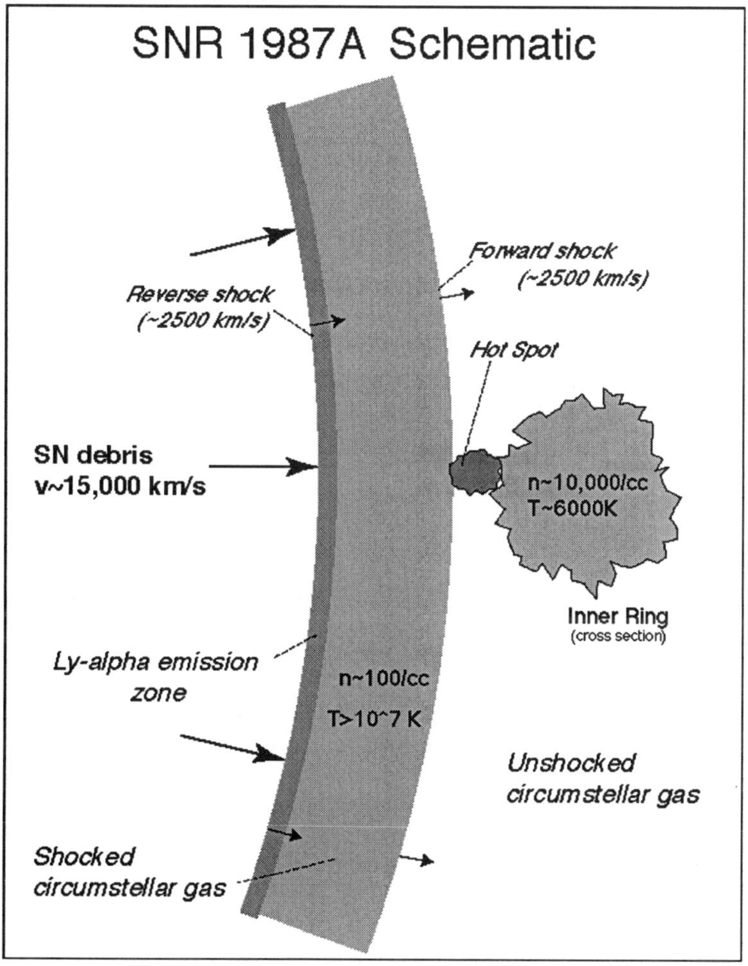

Figure 1. Schematic diagram of the shock interaction around SN 1987A. The Ly-α emission originates at the reverse shock where low-density, high-velocity ($\sim 15,000$ km s^{-1}) neutral debris from the SN explosion collides with circumstellar gas. This CS gas lies interior to the well-known CS ring, has a low density ($n_e \sim 100$ cm^{-3}), and originated as mass lost from the progenitor after the CS ring formed. This collision is driving a shock into the CS gas at ~ 2500 km s^{-1}, raising the temperature in the shocked gas above 10^7 K, and is the source for the radio and X-ray emission. The "hot spot" is the result of the forward shock from this collision encountering the innermost protrusion of the much denser circumstellar ring ($n_e > 10^4$ cm^{-3}).

Discussion

Nolan Walborn: Can you explain why Hα is stronger than [NII] in the brightened knot, while the opposite is true throughout the ring?

Sonneborn: The physical conditions in the ring, producing the line emission is one of pure recombination. The gas in the ring was ionized by the brief (few hours) burst of ionizing radiation from the break out of the SN shock on 23 Feb 1987. The gas has been recombining ever since. In contrast, the hot spot is very different, a high velocity shock colliding with what we believe to be dense gas. However, the density and temperature of the shocked gas are not known today. It is likely that the [N II] lines are not the dominant ionization stages and are probably formed under non-equilibrium conditions. Therefore, the [N II]/Hα line ratios could be very different from that in the ring, even for the same N/H abundance ratio. Future observations and shock models should produce a physical interpretation of the hot spot.

New Views of the Magellanic Clouds
IAU Symposium, Vol. 190, 1999
Y.-H. Chu, N.B. Suntzeff, J.E. Hesser, and D.A. Bohlender, eds.

Radio Properties of Magellanic Cloud SNRs

John R. Dickel

Astronomy Department, University of Illinois, Urbana, IL 61801, USA

Abstract.
Supernova remnants in the Magellanic Clouds may have somewhat lower surface brightnesses at radio wavelengths than their counterparts in the Milky Way but otherwise they appear to be very similar. Balmer dominated SNRs have a lower radio luminosity than other shell remnants.

1. Introduction

Because of their common distance, the supernova remnants (SNRs) in the Magellanic Clouds allow us to establish some statistical properties more accurately than possible for Milky Way remnants. We can also see if there are any major differences between remnants in the different galactic environments.

2. Luminosities

Some properties can be seen in Figure 1a. There is a weak correlation between the x-ray and radio luminosities which becomes stronger if the four Balmer-dominated remnants in the MCs, indicated by the dots, are excluded. Apparently the relativistic particles responsible for the radio synchrotron emission are not readily accelerated in the shocked neutral medium which is collisionally excited to produce Balmer line emission. The underluminosity of these Balmer-dominated SNRs in the radio is not related to the supernova type as the remnant of another young type Ia explosion, N103B, (Hughes et al. 1995) has a normal radio luminosity.

The 1 GHz luminosities of MC SNRs cover a range of about a factor of 1000, similar to that of Milky Way remnants. MC SNRs tend to be larger at the same luminosity, however. For example N132D, the brightest radio SNR in the LMC has a luminosity of 2 x 10^{25} erg sec^{-1} Hz^{-1}, about half the luminosity of Cas A but it is over 5 times larger and so has a mean surface brightness of about 1/50 that of Cas A. The average ISM may be less dense in the MCs than in the Milky Way so remnants expand to a larger size before the radio emission is generated.

Without greater knowledge of the spectral cutoffs, it is not possible to accurately determine the integrated radio synchrotron luminosities but they are approximately 1/100 of the integrated thermal x-ray luminosities.

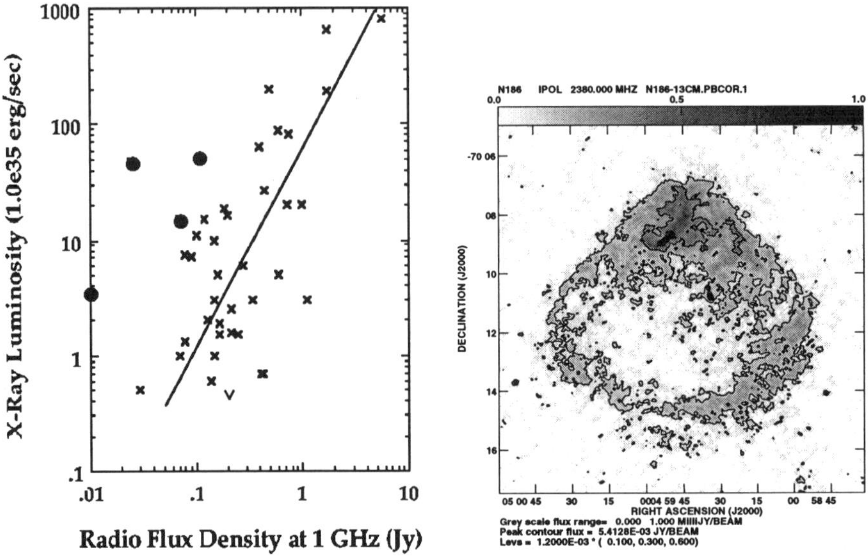

Figure 1. (a) Luminosity Comparison of MC SNRs. The fitted line has a slope of 1.8. (b) 13-cm radio image of N186 with a resolution of 6 arcsec.

3. Structure

N186, shown in Figure 1b, is 125 pc in diameter, one of the largest SNRs known. It may represent a class of object which has hit the wall of a large cavity and has only recently turned on.

Many MC SNRs show strong brightness gradients from one side to the other. These are probably associated with density gradients in the surrounding material. Good examples include N49, N23:SNR, and N103B (Dickel & Milne 1995, 1998). The general phenomenon is present in both radio and x-ray images but the radio gradient is smaller than the x-ray one. Point-by-point brightness comparisons in the two wavelength bands show no significant correlation.

Acknowledgments. The study of Magellanic Cloud SNRs using the Australia Telescope Compact Array has been a joint program with Doug Milne. We have had many valuable discussions.

References

Dickel, J. R., & Milne, D. K. 1995, AJ, 109, 200

Dickel, J. R., & Milne, D. K. 1998, AJ, 115, 1057

Hughes, J, P., Hayashi, I., Helfand, D., Hwang, U., Itoh, M., Kirshner, R., Koyama, K., Markert, T., Tsunemi, H., & Woo, J. 1995, ApJ, 444, L81

New Views of the Magellanic Clouds
IAU Symposium, Vol. 190, 1999
Y.-H. Chu, N.B. Suntzeff, J.E. Hesser, and D.A. Bohlender, eds.

X-ray Spectroscopy of the Supernova Remnant N103B

U. Hwang, R. Petre, and E. Gotthelf

NASA Goddard Space Flight Center, Greenbelt, MD 20771, USA

J. Hughes

Rutgers University, Piscataway, NJ 08855, USA

J. Keohane

NASA Goddard Space Flight Center, Greenbelt, MD 20771, USA

Abstract. We report and interpret a new, 100 ks X-ray observation of the supernova remnant N103B obtained by the ASCA Observatory.

1. Introduction

The LMC supernova remnant N103B has provided some surprises. It is located in a star-forming environment, and is interacting with an HII region (Chu & Kennicutt 1988), but its X-ray spectrum shows that its element abundances have the classic signature of a Type Ia progenitor (Hughes et al. 1995). N103B's interaction with its environment has also left strong imprints. In the radio and in X-rays, its morphology is compressed toward the HII region, and its brightness enhanced (Dickel & Milne 1995; Tavarez et al. 1997).

2. Observations

We obtained a new, deep, 100 ks observation of N103B with the ASCA X-ray Observatory. The ASCA instruments provide good spectral resolution over the energy range 0.5–10 keV. The best ASCA spectrum, obtained by SIS0, is shown in Figure 1. The Fe K line blend near 6.5 keV was undetected in previous X-ray observations, including an earlier ASCA observation (Hughes et al. 1995; Hayashi 1997). The model shown is a Sedov model for shock temperature $kT =$ 0.70 keV and ionization parameter (electron density times remnant age) $nt =$ 1.4×10^{11} cm^{-3} s. This model provides a satisfactory qualitative fit to the ASCA spectrum, including the Fe K line, and is consistent with the average temperature and ionization age measured for the remnant from emission line ratio diagnostics. The interpretation of the Fe K emission, however, is subject to checking the accuracy of our models. With the Sedov models of Hamilton, Sarazin & Chevalier (1983), which uses older atomic physics, the Fe K blend is severely underpredicted by the models, although other spectral features are adequately modeled.

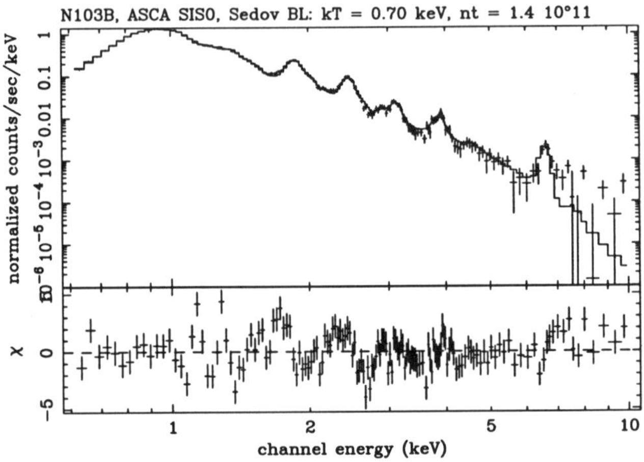

Figure 1. ASCA SIS0 observation of N103B, subtracted for background, and with the best-fit Sedov model folded through the instrument response and overlaid.

3. Discussion

The best-fit Sedov model provides reasonable consistency in the implied physical parameters. The dynamical Sedov age agrees with the ionization age to within a factor of two. A swept-up mass of $\sim 1\ M_\odot$ is comparable to the ejecta mass for a Type Ia remnant, and indicates that N103B is beginning its transition to the Sedov phase. The explosion energy is a factor of 15 lower than the canonical 10^{51} ergs, but this may be due to deceleration of the remnant by the HII region. Interaction with the HII region can also explain why this young, ejecta-dominated remnant is well-described by a Sedov model. The side of the remnant undergoing the interaction encounters higher densities and will thus both dominate the X-ray flux and evolve more quickly to the Sedov stage. This idea is supported by the observation that the optical filaments, located off-center toward the HII region, have characteristic ISM abundances (Danziger & Leibowitz 1985; Tavarez et al. 1997).

Acknowledgments. We are grateful to K. Borkowski and J. Lyerly for providing their Sedov models, and to K. Gendreau for providing the deep blank sky data used as background.

References

Chu, Y.H., & Kennicutt, R. C. 1988, AJ, 96, 1874
Dickel, J. R., & Milne, D. K. 1995, AJ, 109, 200
Hamilton, A. J. S., Sarazin, C. L., & Chevalier, R. A. 1983, ApJS, 51, 115
Hayashi, I. 1997, PhD Thesis, University of Kyoto
Hughes, J. P., et al. 1995, ApJ, 444, L81
Mathewson, D. S., et al. 1983, ApJS, 51, 345
Tavarez, M., Smith, R. C., & Petre, R. 1997, BAAS, 191, 4003

New Views of the Magellanic Clouds
IAU Symposium, Vol. 190, 1999
Y.-H. Chu, N.B. Suntzeff, J.E. Hesser, and D.A. Bohlender, eds.

N63A: A Supernova Remnant in a Cloudy Medium

You-Hua Chu, Adeline Caulet, John Dickel, Sean Points, Rosa Williams
Astronomy Department, University of Illinois, Urbana, IL 61801, USA

Lorena Arias-Montaño, Margarita Rosado
UNAM, Apdo Postal 70-264,04510 México DF, México

Patricia Ambrocio-Cruz, Annie Laval
Observatoire de Marseille, F-13248 Marseille Cedex 04, France

Dominik J. Bomans
Astronomisches Institut, Ruhr-Univ. Bochum, 44780 Bochum, Germany

Abstract. Our recently obtained HST WFPC2 images of N63A reveal shocked cloudlets in the SNR interior. The study of N63A helps us understand the structure and evolution of SNRs in a cloudy medium.

1. N63A - A Puzzling SNR in an OB Association

N63A was one of the first three supernova remnants (SNRs) discovered in the LMC based on its strong, nonthermal radio emission. Its identification as a SNR was subsequently confirmed by its bright X-ray emission and high [S II]/Hα ratios. As shown by the X-ray contours in Fig. 1a, N63A is located in the northeastern quadrant of the H II region N63 associated with the OB association LH 83. The progenitor of the supernova was likely a member of LH 83 (van den Bergh & Dufour 1980). The currently most luminous star is an O7 V((f)) star with a mass of $\sim$40 $M_\odot$ (Oey 1996). Therefore, the progenitor of N63A's supernova was probably more massive than 40 $M_\odot$ and its main-sequence spectral type earlier than O7.

The spatial extent of N63A's X-ray emission is similar to that of its radio emission, $\sim 70''$, and defines the size of the SNR. Within the SNR's boundary, ground-based optical images show three bright nebular knots. Spectroscopic observations show that the two eastern knots have high [S II]/Hα ratios, indicating shock excitation, and that the western knot has a spectrum consistent with photoionization (Levenson et al. 1995). Besides these bright knots, no other optical features can be recognized as belonging to the SNR. The two [S II]-bright knots together extend over $\sim 23''$, much smaller than the SNR's size. The absence of optical emission from the rest of the remnant cannot be caused by foreground obscuration, as A_v is generally <0.5 mag for stars in LH 83 (Oey 1996). *Why is N63A's optical emission region so much smaller than the SNR's extent?*

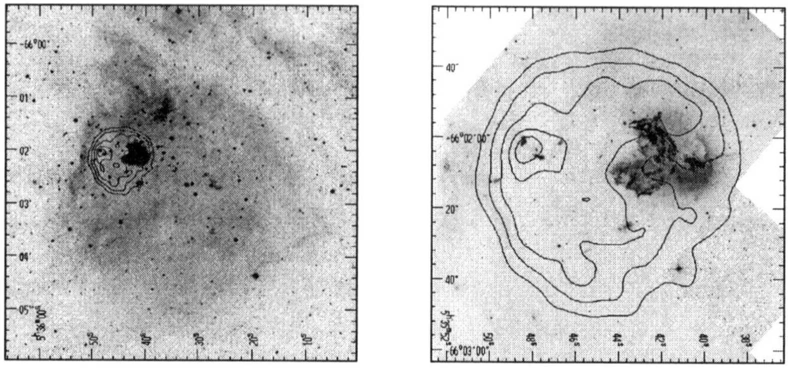

Figure 1. (a) Hα image of N63 overlaid by X-ray contours. (b) HST WFPC2 Hα image of N63A overlaid by X-ray contours.

2. HST Images Reveal Shocked Clouds in N63

The Hubble Space Telescope (HST) WFPC2 images of N63A in the Hα and [S II] lines have revealed shocked clouds with sizes ranging from 5 pc down to 0.1 pc (Fig. 1b). N63A is clearly in a cloudy medium. The three bright nebular knots identified in ground-based images are well resolved in the WFPC2 images. The two eastern knots show distinct filamentary structures indicative of a compression by SNR shocks, while the western knot shows only diffuse ionized gas and dust clouds typical for H II regions. The two eastern knots might belong to the same cloud; their combined size is ~5 pc, comparable to the Orion Nebula.

Within the SNR boundary, numerous [S II]-bright cloudlets are detected. The high [S II]/Hα ratios indicate that these cloudlets have been shocked and are located in the SNR interior. Some cloudlets show a diffuse morphology with radial density gradients, suggestive of isotropic evaporation and implying that the magnetic field is either very weak or tangled. Some cloudlets contain multiple cores, each ~0.07 pc across. The rms densities of the cloudlets range from 150 to 700 H cm^{-3}, and the masses range from 0.01 to ~1 M$_\odot$.

ROSAT and ASCA observations of N63A were used to derive a SNR interior T$_e$ of ~6$\times$10^6 K, N$_e$ of ~5 cm^{-3}, and a mass of ~160 M$_\odot$. In such a hot medium, the evaporation timescale for the embedded cloudlets is a few $\times$10^3 yr, comparable to the dynamic age of N63A. It is conceivable that a significant amount of cloud mass has already been injected into the hot SNR interior through cloud evaporation.

References

Levenson, N. A., Kirshner, R. P., Blair, W. P., & Winkler, P. F. 1995, PASP, 110, 739

Lucke, P. B., & Hodge, P. W. 1970, AJ, 75, 131

Oey, M. S. 1996, ApJ, 465, 231

van den Bergh, S., & Dufour, R. J. 1980, PASP, 92, 32

New Views of the Magellanic Clouds
IAU Symposium, Vol. 190, 1999
Y.-H. Chu, N.B. Suntzeff, J.E. Hesser, and D.A. Bohlender, eds.

Supernova Remnants in the Large Magellanic Cloud: A Multiwavelength Study of Energetics and Environments

Rosa Murphy Williams, You-Hua Chu, and John R. Dickel

Astronomy Department, University of Illinois, Urbana, IL 61801, USA

R. Chris Smith

Cerro Tololo Inter-American Observatory, La Serena, Chile

Abstract.
We analyze optical, radio and X-ray data from supernova remnants (SNRs) in the Large Magellanic Cloud (LMC) in order to make a systematic study of a full sample at a known distance. By this multiwavelength examination of SNRs from various progenitors and environments, we will build a comprehensive picture of the evolution of SNR structure and energetics.

1. Introduction

The Large Magellanic Cloud (LMC), provides an ideal setting for the study of supernova remnants (SNRs) at a common distance and with little interstellar absorption. We analyze optical, radio and X-ray data from these remnants, in order to make a systematic study of the full LMC sample. These data are being used to investigate such questions as the partitioning among thermal, kinetic, possibly magnetic, and relativistic-particle energies; the pressures from different components within a remnant; the density and clumpiness of the gas inside and immediately outside the remnants; and the effects of local interstellar gas on SNR development.

2. Available Datasets

X-ray images and spectra are available through archival and proprietary data from the *ROSAT* and *ASCA* satellite instruments. X-ray images show the distribution of million-degree gas within an SNR, providing information on shock structure for comparison with details of the local environment, as well as the possible presence of internal cloudlets. X-ray spectra address questions of internal temperatures, densities, hot gas (thermal) pressures, and energetics.

Long-slit echelle spectra for most LMC SNRs have been obtained through observations with the 4m telescope at Cerro Tololo Inter-American Observatory (CTIO). Echelle spectra provide velocity information via the Doppler shifts of the Hα line emission. These spectra give expansion and shock velocities for an SNR, which can be used in the derivation of ages and kinetic energies.

A variety of optical images have been made available by the Magellanic Cloud Emission-Line Survey, using the Curtis Schmidt telescope at CTIO. Additional images taken with the CTIO 1.5m and 4m telescopes are also available. Optical images show the distribution of cooler gas in the shell of the SNR, with different ions being produced by different shock conditions. Hα, [S II], and [O III] images can be combined for a detailed picture. The surface brightness of the Hα emisson, combined with estimates of the shell geometry, can give approximate figures for particle density; from this, the shell mass, kinetic energy, and cool gas (thermal) pressure can be derived.

Approximately half of the known LMC SNRs have been observed at radio wavelengths using the Australia Telescope Compact Array (ATCA). Radio images show the location of synchrotron emission, which is generated by the motion of relativistic electrons around magnetic field lines compressed behind the shock front. From the spectral index of the radio observations, we can derive the energy and pressure of magnetic fields within a remnant.

3. The Big Picture

By analyzing the data for each SNR at different wavelengths, we can build a comprehensive picture of its structure and energetics. Combining the data with observations of the SNR's immediate surroundings provides clues to the effects of local conditions on remnant development, as with a break-out into a local cavity, or the sweeping up of cloudlets into the SNR interior. Because the LMC contains a large sample of SNRs with different initial conditions, in a wide variety of environments, we compare the remnants to one another in order to determine the relative influence of various factors on SNR evolution, and, in return, the effects of the SNRs on their surroundings. Examples of such analyses have been published for the colliding remnants DEM L 316 (Fig. 1a; Williams et al. 1997) and two SNRs, N 11 L and N 86, which show breakout structures (Fig. 1b; Williams et al. 1999). We hope by these means to gain a clearer picture of the role SNRs play in the structure and evolution of the interstellar medium.

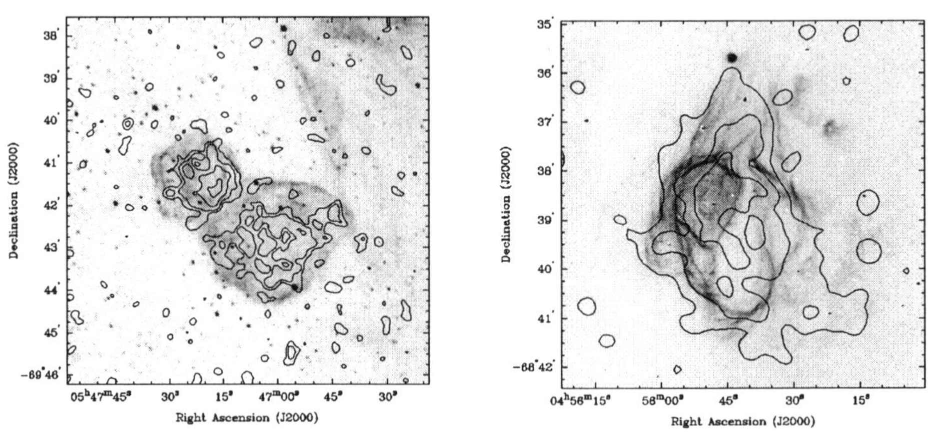

Figure 1. Hα images of (a) DEM L 316 (b) N 86 with X-ray contours.

References

Williams, R. M., et al. 1997, ApJ, 480, 618
Williams, R. M., et al. 1999, ApJ, 514, 798

Discussion

Nolan Walborn: Is there velocity information which supports the interpretation at the "double SNR" as two separate expansions, as opposed to a single bipolar one? The latter seems more probable from the morphology to me. The two lobes have similar, mirror image shapes, and the larger one has a lower surface brightness in X-rays. It must be very improbable to get such a configuration from two unrelated events. I suggest that a model with a very dense equatorial disk should be considered.

Williams: In fact, echelle velocity spectra show DEM L 316 as two separately expanding shells. In addition, were DEM L 316 the result of a bipolar outflow, the two shells must either be precisely perpendicular to the line of sight - an unlikely coincidence - or else show different radial velocities at the extreme ends. Instead, both converge to the object's 245 kms^{-1} heliocentric velocity. Futher evidence for interaction is provided by the coincidence of the X-ray brightening of one shell with the outer [O III] edge of the other, as well as an observed change in the radio polarization between the shells. Finally, a bipolar expansion would require the dense equatorial disk which you mentioned; we have looked for evidence of such material in H I maps provided by Stavely-Smith and Kim, but find no local increase at the juncture between the shells. A full treatment of these arguments is in Williams et al., ApJ, 480, 618 (1997).

New Views of the Magellanic Clouds
IAU Symposium, Vol. 190, 1999
Y.-H. Chu, N.B. Suntzeff, J.E. Hesser, and D.A. Bohlender, eds.

Molecular Clouds in the LMC Observed with NANTEN: IV. Comparison with SNRs

H. Saito, R. Abe, A. Hara, T. Hayakawa, S. Kato, A. Kawamura, A. Mizuno, N. Mizuno, H. Ogawa, T. Onishi, K. Tachihara, K. C. Xiao, N. Yamaguchi, R. Yamaguchi, and Y. Fukui

Department of Astrophysics, Nagoya University, Chikusa-ku, Nagoya 464-8602, Japan

Y. Yonekura

Earth and Life Sciences, Osaka Prefecture University, Gakuencho 1-1, Sakai, Osaka 599-8531, Japan

Abstract. We have made $^{12}CO(J{=}1\text{–}0)$ observations of the LMC with NANTEN. We report the results of a comparison between CO clouds and SNRs in the LMC. Among the 35 known SNRs, only 10 are possibly associated with CO clouds. These 10 CO clouds and SNRs deserve follow-up studies for possible interactions. We present overlays of CO clouds on the optical images of some of these SNRs.

1. Introduction

The interaction of supernova remnants (SNRs) with dense molecular clouds is an important subject in astrophysics. Molecular clouds swept up by supernova blast waves provide an important laboratory in which the physical processes associated with shocks can be studied. The LMC is well suited for the study of SNR and molecular cloud interactions. Thirty-five SNRs have been identified in the LMC from X-ray, optical, and radio continuum observations (Mathewson et al. 1983, 1984, 1985; Smith et al. 1994).

We have made a survey for giant molecular clouds in the LMC in the $^{12}CO(J{=}1\text{–}0)$ emission, using the NANTEN 4-meter millimeter wave telescope at Las Campanas Observatory in Chile. The survey covered 6×6 square degrees with a $2'$ grid spacing. We have identified $\sim$100 distinct clouds. The results of these observations are presented by Fukui et al. (1999). In this contribution, we report the results of a comparison between the CO clouds and the SNRs.

2. Comparison and Results

The sizes of the cataloged SNRs range from a few pc to $\sim$100 pc, and their ages are 10^2–10^6 yrs. We carried out a comparison between the 35 SNRs and the 100 giant molecular clouds. We define: (1) "SNR with CO cloud" – SNRs associated with molecular clouds; the SNR is superposed on a CO cloud or in the immediate neighborhood of a CO cloud; see Fig. 1a; and (2) "SNR near CO

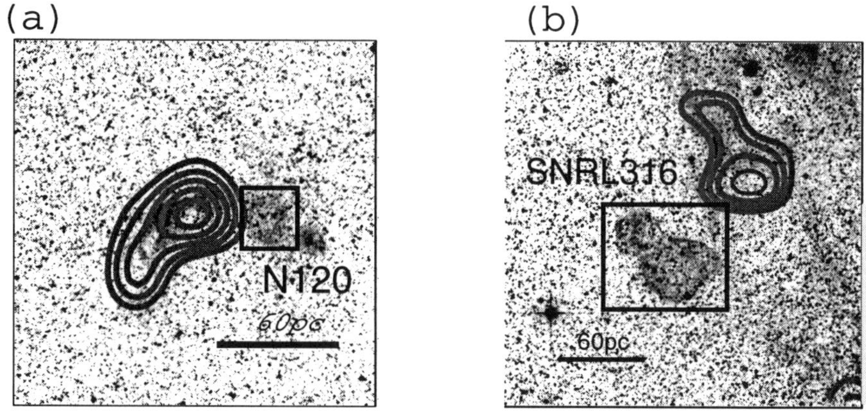

Figure 1. Contour maps of velocity-integrated $^{12}CO(J=1\text{–}0)$ emission. The optical images are taken from the DSS. SNRs are in a black square. (a) "SNR with CO cloud". (b) "SNR near CO cloud".

cloud" – SNRs located near molecular clouds; the distance between the SNR and a CO cloud is $\leq$30 pc; see Fig. 1b. Out of the 35 known SNRs, only 5 are associated with CO clouds. These CO clouds have virial masses of $\sim$1–3 $\times 10^5$ $M_\odot$. Another 5 SNRs are located near CO clouds. These CO clouds are more massive, $\sim 8\times10^5$ $M_\odot$. In both cases, the CO clouds associated with SNRs tend to be of intermediate masses among the present sample (M $\sim 10^4$–10^6 $M_\odot$), suggesting that giant molecular clouds are significantly dissipated prior to a supernova explosion.

Acknowledgments. We greatly appreciate the hospitality of all the staff members of the Las Campanas Observatory of the Carnegie Institution of Washington.

References

Fukui, Y., et al. 1999, this volume

Mathewson, D. S., Ford, V. L., Dopita, M. A., Tuohy, I. R., Long, K. S., & Helfand, D. J. 1983, ApJS, 51, 345

Mathewson, D. S., Ford, V. L., Dopita, M. A., Tuohy, I. R., Mills, B. Y., & Turtle, A. J. 1984, ApJS, 55, 189

Mathewson, D. S., Ford, V. L., Tuohy, I. R., Mills, B. Y., Turtle, A. J., & Helfand, D. J. 1985, ApJS, 58, 197

Smith, R. C., Chu, Y.-H., Mac Low, M.-M., Oey, S. M., & Klein, U. 1994, AJ, 108, 1266

New Views of the Magellanic Clouds
IAU Symposium, Vol. 190, 1999
Y.-H. Chu, N.B. Suntzeff, J.E. Hesser, and D.A. Bohlender, eds.

Molecular Clouds in the LMC Observed with NANTEN II. Comparison with Stellar Clusters

R. Yamaguchi, R. Abe, A. Hara, T. Hayakawa, S. Kato, A. Kawamura, A. Mizuno, N. Mizuno, H. Ogawa, T. Onishi, H. Saito, K. Tachihara, K. C. Xiao, N. Yamaguchi, and Y. Fukui

Department of Astrophysics, Nagoya University, Chikusa-ku, Nagoya 464-8602, Japan

Y. Yonekura

Earth and Life Sciences, Osaka Prefecture University, Gakuencho 1-1, Sakai, Osaka 599-8531, Japan

Abstract. We have made $^{12}CO(J{=}1\text{–}0)$ observations of the LMC with the NANTEN millimeter-wave telescope and identified about 100 distinct giant molecular clouds (GMCs). A detailed comparison of the GMCs with stellar clusters and a UV image is discussed.

1. Introduction

The LMC possesses "populous clusters" whose properties are intermediate between the Galactic open and globular clusters (e.g., Hodge 1961). Since some of the populous clusters are younger than a few 10 Myr (e.g., Searle, Wilkinson, & Bagnuolo 1980; van den Bergh 1981), their parental molecular clouds may still be observable at present. Previous millimeter-wave observations of the LMC were either of low angular resolution or of limited spatial coverage, hence were not appropriate for comparisons between molecular clouds and stellar clusters. We have made a new $^{12}CO(J{=}1\text{–}0)$ survey of the LMC with the NANTEN millimeter-wave telescope of Nagoya University at the Las Campanas Observatory, Chile. The details of the observations and the results are presented by Fukui et al. (1999) and Mizuno et al. (1999). In this contribution, we present the results of a comparison between 55 newly identified GMCs and stellar clusters, and discuss the formation of clusters in the LMC.

2. Results and Discussion

We used 388 stellar clusters and 114 OB associations compiled in the literature (Bica et al. 1996). They are classified into 9 types according to their ages (SWB0–VII), estimated from their color indices, $B - V$ and $U - B$.

A comparison of the GMCs with the clusters indicates that 26 GMCs, out of the 55, are associated with clusters. The youngest clusters with $\tau \leq 10$ Myr are at/near the peaks of the GMCs and often found in groups, while the older clusters (SWBI–VII) are away from the GMCs (See Fig. 1) and scattered

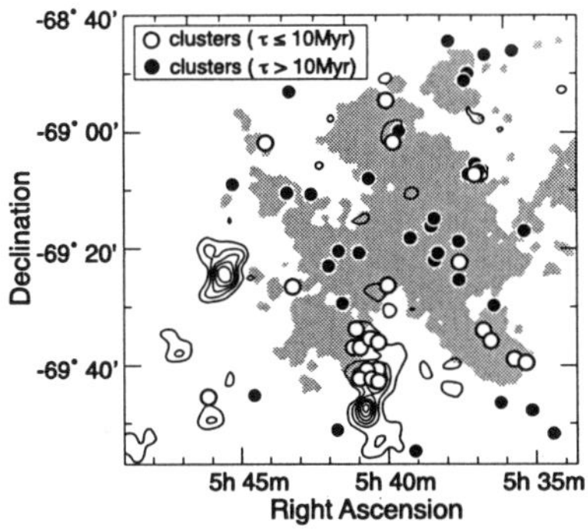

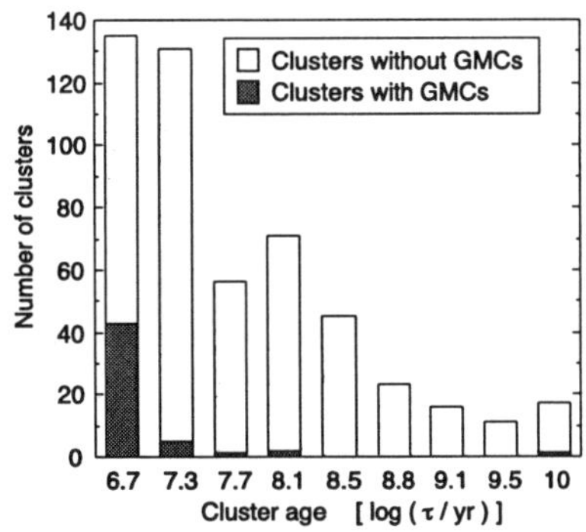

Figure 1. (*left*) Distribution of GMCs and clusters in the 30 Doradus region. Hatches are strong UV-emitting fields (Simith et al. 1987).

Figure 2. (*right*) Spatial correlation between GMCs and clusters for different ages. The ages correspond to the median of each SWB type.

spatially. Fifty-two clusters are associated with GMCs and $\sim$ 90% of them are younger than 10 Myr. Fig. 2 shows the spatial correlation between the clusters and the GMCs for each age. About 30% of the youngest clusters are associated with the GMCs. The rest may be associated with smaller clouds, whose mass is below the present detection limit (M $\sim 10^5$ $M_\odot$). The older clusters (SWBI–VII) show little/no correlation with the GMCs. The apparent association of the very old clusters with $\tau > 30$ Myr (SWBII–VII) with GMCs may be caused by chance superpositions along the same line of sight.

Next, we compared the distribution of the youngest clusters with a 1500 Å UV image (Smith et al. 1987). As shown in Fig. 1, the youngest clusters without GMCs are in strong UV radiation fields, and the GMCs tend to lie outside. Only small clouds (M $\sim$ a few 10^4 $M_\odot$) are in the field. These features are also found in most of the cluster forming regions, i.e., the N44 and the "Bar" regions, but not in the supergiant shells. This suggests that GMCs are rapidly dissipated and/or dissociated by UV photons from the clusters.

In summary, we suggest that a large fraction of the GMCs actively form clusters, and may be dissipated in $\leq$ 10 Myr by the UV fluxes from the clusters.

Acknowledgments. We appreciate the hospitality of all the staff members of the Las Campanas Observatory of the Carnegie Institution of Washington.

References

Bica, E., et al. 1996, ApJS, 102, 57

Fukui, Y., et al. 1999, this volume

Hodge, P.W. 1961, ApJ, 133, 413

Mizuno, N., et al. 1999, this volume

Searle, L., Wilkinson, A., & Bagnuolo, W.G. 1980, ApJ, 239, 803

Smith, A.M., Cornett, R.H., & Hill, R.S. 1987, ApJ, 320, 609

van den Bergh, S. 1981, A&AS, 46, 79

New Views of the Magellanic Clouds
IAU Symposium, Vol. 190, 1999
Y.-H. Chu, N.B. Suntzeff, J.E. Hesser, and D.A. Bohlender, eds.

Kinematic Study of the Northwestern Part of the Bar of the LMC

P. Ambrocio-Cruz, A. Laval, M. Marcelin, and P. Amram

Observatoire de Marseille. 2, Place Le Verrier, F-13248 Marseille Cedex 04, France

Abstract. We have been conducting a Fabry-Perot kinematic survey of ionized gas in the Magellanic Clouds in the Hα and [O III]λ5007 lines, using a 36-cm telescope at the European Southern Observatory. This poster presents our kinematic study of HII regions located at the northwestern parts of the LMC bar.

1. The Survey

An Hα and [OIII]λ5007 survey of the Magellanic Clouds is being carried out at the European Southern Observatory in La Silla, Chile (Le Coarer et al. 1993 for the SMC; Laval et al. 1987 for the LMC). The instrument consists of a 36-cm telescope with a focal reducer, a scanning Fabry-Perot interferometer, a photon counting camera, and a series of interference filters. We obtain images with a $38' \times 38'$ field of view, a $9''$ angular resolution, and a spectral sampling steps of 16 and 7 km s^{-1} for the Hα and [OIII] lines, respectively. The free spectral range is 377 km s^{-1} for Hα and 288 km s^{-1} for [OIII].

2. The Nebulae

In this poster we report on the sample of nebulae located on the northwest part of the LMC bar: DEM L117, DEM L115, DEM L107, DEM L99, DEM L98, DEM L97, DEM L89, DEM L76, N105, N103, N100, N33, N30, and N23. Table 1 shows the main characteristics of these nebulae. The linear diameters are derived by adopting a distance of 50 kpc to the LMC (Feast 1991). Columns 3 and 4 give the systemic heliocentric velocities measured in Hα and [OIII]. The last column indicates whether the line profiles show splittings.

3. Results

Five of the surveyed objects show splittings in their line profiles. By calculating the energy injected into the gas we have concluded that the splittings in DEM L84 is caused by a supernova explosion (Ambrocio-Cruz et al. 1997). The peculiar velocity field of DEM L86 and its morphology indicate that it contains four bubbles and two bright compact HII regions, almost coeval, embedded inside a large, faint nebular shell. This structure is formed by a combination of

Table 1. Main Characteristics of the Nebulae

Name of nebula	Diameter (pc)	$V_{H\alpha}$ (km s^{-1})	$V_{[OIII]}$ (km s^{-1})	Splitting
DEM L121 (N33)	24	302	–	no
DEM L117	22	306	–	no
DEM L115	26	305	–	no
DEM L107	126	304	–	no
DEM L106 (N30)	74	290, 310	–	yes
DEM L105 (N30)	130	285, 305	–	yes
DEM L99	61	298	–	no
DEM L98	68	305	–	no
DEM L97	65	303	–	no
DEM L89	131	293	–	no
DEM L86 (N105)	112	255	255	yes
DEM L84 (N103)	153	252	263	yes
DEM L79 (N100)	48	245	–	no
DEM L76	153	264	–	no
DEM L74 (N23)	39	259, 300	–	yes
DEM L73 (N23)	114	265	–	no
DEM L70 (N23)	81	265	–	no
DEM L66 (N23)	72	274	276	no

stellar winds and "champagne" outflows in a medium with a negative density gradient.

The line profiles observed in DEM L74 can be fitted by the sum of two Gaussians. The brighter component at 300 km s^{-1} is seen over the whole nebula and the fainter one at 259 km s^{-1} is seen only on the northwestern side. The intensity of this low-velocity component increases toward DEM L73 (located at northwest of DEM L74 and with a velocity of 265 km s^{-1}). Thus, these splittings cannot be interpreted as caused by expansive motions. This low-velocity component is most likely part of DEM L73 spilt over DEM L74.

The line profiles observed in DEM L105 and 106 can be fitted by the sum of two Gaussians. The first velocity component has a heliocentric velocity of 305-310 km s^{-1} and the second component has a velocity of 285-290 km s^{-1}. Since the second velocity component is best detected at the bright parts along the edge of the N30 bubble, the line split cannot be interpreted as caused by expansive motions. This 285 km s^{-1} component is probably associated with a low-velocity HI component; it is possible that some UV flux of the blue supergiant stars of N30 escapes and contributes to the ionization.

Acknowledgments. This work is supported by a fellowship from DGAPA-UNAM.

References

Ambrocio-Cruz, P., Laval, A., et al. 1997, A&A, 319, 973

Ambrocio-Cruz, P., Laval, A., et al. 1998, A&A, 339, 173

Feast, M. W. 1991, In: *The Magellanic Clouds*, IAU Symp. 148, p. 1

Laval A., Boulesteix, J., et al. 1987, A&A, 175, 199

Le Coarer, E., Rosado, M., et al. 1993, A&A, 280, 365

New Views of the Magellanic Clouds
IAU Symposium, Vol. 190, 1999
Y.-H. Chu, N.B. Suntzeff, J.E. Hesser, and D.A. Bohlender, eds.

The Absorption Line Systems of the Supershell LMC2 in the Large Magellanic Cloud

Adeline Caulet

Astronomy Department, University of Illinois, Urbana, IL 61801, USA

Abstract. The interstellar medium of LMC2, a well studied supershell in the Large Magellanic Cloud, has been probed in UV and optical absorption lines. The data allow to derive the kinematics, abundances and depletions of gas clouds in this supershell. The relative gas-phase abundances of observed elements with respect to sulphur are useful to determine the origins of the supershell absorption-line clouds.

1. Introduction

Supershells (or supergiant shells, SGS) which are large-scale structures in galaxies have a multi-phase ISM. They are made of gas clouds with a large range of velocities, densities and temperatures. It is interesting to determine whether these clouds originate in a cold or warm disk or in a hot halo, so that we may understand the role of supershells in galaxies. Because supershells represent large reservoirs of energy, and can reach large dimensions in and above galactic disks, they can play an important role in galactic structure, e.g., in the formation and evolution of galactic halos, via the interaction of multiple supernovae explosions with the ISM.

2. Observations

To probe the multi-phase ISM of the well-studied supershell LMC2 in the Large Magellanic Cloud, we have analyzed the high resolution optical (Caulet & Newell 1996) and UV (Fig. 1) spectra of 7 OB supergiant stars in the field of LMC2. Line-profile fits of the absorption lines give the velocities, column densities, and Doppler widths of the absorbing clouds along the stellar lines-of-sight. For the spectral ranges observed, the ions C I (λ1260, λ1560), and Ti II (λ3384) arise in cold and warm neutral clouds, and provide information about cool clouds kinematically associated with HI structures (such as the flocculent HI seen in ATCA maps by Kim et al. 1998), the compressed gas behind the outward moving shocks, and cool clouds embedded in the hot gas inside the SGS. The ionized medium comes from both photoionization of the SGSs and post-shock relaxation of the gas: Mg II (λ1240), S II (λ1250, λ1253, λ1259), Si II (λ1260), and Ca II (λ3934). The photo-ionized gas and coronal gas are traced by the C IV $\lambda\lambda$1548, 1550 (T $\sim 10^5$ K) lines. N V $\lambda\lambda$1238, 1240 (T $\sim 2\times10^5$ K) lines trace shocks, and collisionally excited gas as N^{+4} may arise at the interfaces of cloud–hot gas in the SGS cavities. Many cold, warm and hot clouds are detected over the whole velocity range between the Galactic velocities ($v \leq 50$ km^{-1}) and the LMC (200-360 km^{-1}) velocities. While C IV is generally observed, no N V is detected down to a 3σ detection limit of 10 mÅ. The ions Si II* (λ1264), C I (λ1260), Mg II (λ1240) are identified in some clouds.

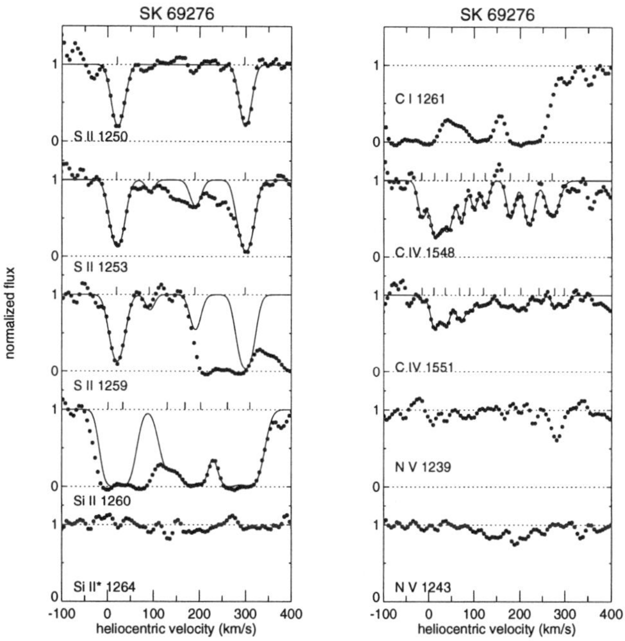

Figure 1. UV absorption lines toward SK-69276: no C I, Si II*, N V

Because it is not possible to determine directly distances for these clouds within the LMC, we examined first their relative positions by comparing their velocities with those of structures observed in emission lines, and neutral gas, and in absorption lines for other published lines-of-sight in the vicinity of LMC2 and the 30 Doradus region (see Paper I). Then, for each cloud, we plotted the gas-phase abundances of elements (relative to the "undepleted" sulpher) against their condensation temperatures to determine if these clouds are representative of disk or halo clouds (see Savage & Sembach 1996; Welty et al. 1997). The completed study will be reported in a refereed paper.

Acknowledgments. The author thanks You-Hua Chu and the Astronomy Deparment of the University of Illinois for their hospitality during her visit and the financial support to attend this symposium.

References

Caulet, A., & Newell, R. 1996, ApJ, 465, 205

Kim, S., et al. 1998, ApJ, 503, 674

Savage, B.D., & Sembach, K.R. 1996, ARA&A, 34, 279

Welty, D.E., et al. 1997, ApJ, 489, 672

New Views of the Magellanic Clouds
IAU Symposium, Vol. 190, 1999
Y.-H. Chu, N.B. Suntzeff, J.E. Hesser, and D.A. Bohlender, eds.

The Kinematic Structure of the Supergiant Shell LMC 2

Sean Points, You-Hua Chu, Sungeun Kim, Robert Gruendl
Astronomy Dept., Univ. of Illinois, Urbana, IL 61801, USA

R. Chris Smith
Dept. of Astronomy, Univ. of Michigan, Ann Arbor, MI 48109, USA

Steve Snowden
NASA/Goddard Space Flight Center, Greenbelt, MD 20771, USA

Wolfgang Brandner
JPL/IPAC, Pasadena, CA 91125, USA

1. Introduction

LMC 2 was one of the first supergiant shells in the Large Magellanic Cloud (LMC) to be identified, based on the detection of long, curved Hα filaments extending over 900 pc (Goudis & Meaburn 1978). LMC 2 is located to the east of the active star formation region south of 30 Doradus. LMC 2 is a spectacular supergiant shell in the LMC, having the most coherent filamentary structure and the highest X-ray surface brightness. As shown in Fig. 1, the diffuse X-ray emission from LMC 2 is bounded by the Hα filaments in the north and east. A bright X-ray arc is seen in the southwest quadrant, extending from N 158 and N 159.

The kinematic structure of LMC 2 has been investigated by various groups with conflicting results. Caulet et al. (1982) studied the kinematics of the 10^4 K ionized gas and concluded that LMC 2 was a coherently expanding shell-like structure. Meaburn et al. (1987) obtained profiles of the H I 21-cm emission line toward LMC 2. They detected no evidence that LMC 2 was coherently expanding. We have obtained high-resolution echelle spectra of the Hα line and Austrailia Telescope Compact Array (ATCA) aperture synthesis maps of the H I 21-cm line emission of LMC2 to study its dynamics.

2. Kinematic Structure of LMC 2

Our echelle spectra along three cuts toward LMC 2 (designated E I, E II, & E III in Fig. 1) reveal two major velocity components of the ionized gas at $V_{hel} \sim 250$ and 275 km s^{-1}. The velocity structure of these components across the echelle cuts does not show a velocity variation conforming to that expected in an expanding hemisphere (Points et al. 1999).

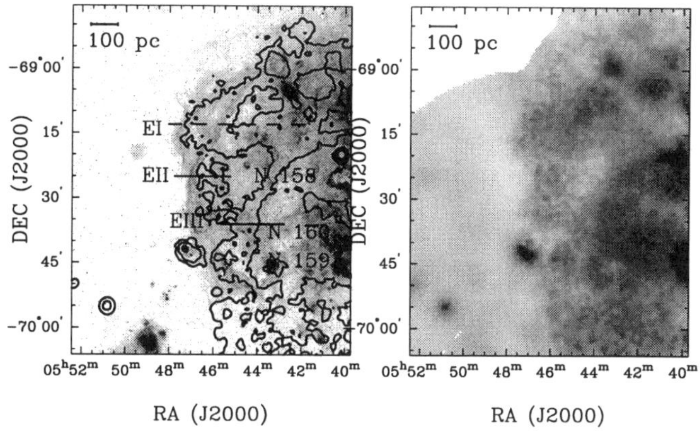

Figure 1. **Left:** Hα image of LMC 2 overlaid with *ROSAT* PSPC contours. **Right:** Grayscale *ROSAT* PSPC mosaic of LMC 2.

The ATCA aperture synthesis maps of the H I 21-cm line show the neutral hydrogen toward LMC 2 to be distributed in the form of long filaments and discrete clouds. The H I filaments show clear correspondence with the filaments seen in the Hα image. The H I filaments at the periphery of LMC 2 have velocities of 250, 275, and 300 km s^{-1}. If LMC 2 were a coherently expanding shell, we expect that the velocity extremes would be observed toward the interior of the shell, not the edge. Thus, the H I channel maps provide further evidence that LMC 2 is not coherently expanding.

3. Conclusions

We have investigated the kinematic structure of the supergiant shell LMC 2 using high-resolution echelle spectra of the Hα line and ATCA aperture synthesis maps of the H I 21-cm line. High-velocity gas is detected toward X-ray bright regions. In the northern region, this high-velocity gas is indicative of localized heating by a supernova remnant. Along the western boundary, the high-velocity gas may be associated with an outflow of material from the adjacent star-forming region. However, we detect no evidence of a global expansion pattern.

References

Caulet, A., Deharveng, L., Georgelin, Y. M., & Georgelin, Y. P. 1982, A&A, 110, 185

Goudis, C., & Meaburn, J. 1978, A&A, 68, 189

Meaburn, J., Marston, A. P., McGee, R. X., & Newton, L. M. 1987, MNRAS, 225, 591

Points, S. D., Chu, Y.-H., Kim, S., Smith, R. C., Snowden, S. L., Brandner, W., & Gruendl, R. A. 1999, ApJ, 518, 298

Wang, Q. D., & Helfand, D. J. 1991, ApJ, 379, 327

New Views of the Magellanic Clouds
IAU Symposium, Vol. 190, 1999
Y.-H. Chu, N.B. Suntzeff, J.E. Hesser, and D.A. Bohlender, eds.

The Physics of 10^6 K Gas in LMC Supergiant Shells

D.J. Bomans

Astronomisches Institut, Ruhr-Universität Bochum, Germany

K. Dennerl

MPI für extraterrestrische Physik, Garching, Germany

Abstract. We present first results of our study of the interplay between the 10^6 K, 10^4 K, 10^2 K gas phases in the supergiant shell LMC 4.

1. Introduction

Supergiant shells are the largest interstellar structures in galaxies. The Magellanic Clouds harbour more than ten supergiant shells like LMC 1, LMC 4, and SMC 1 (Meaburn 1980), which appear as complete shells of Hα filaments with some normal HII regions distributed along the rim. LMC 4 is the largest of the supergiant shells, with a diameter larger than 1.4 kpc. LMC 4 also contains Shapley Constellation III, a very large OB association (e.g., Braunsfurth & Feitzinger 1983). The most promising formation mechanism of Hα supergiant shells calls for the combined effect of fast winds of massive stars and SNe to create the expanding bubble (e.g. Mac Low et al. 1989). We have shown earlier, that the cavity of LMC 4 contains 10^6 K hot gas (Bomans et al. 1993) and that 10^5 K gas exists on boundary layers between hot and cold gas (Bomans et al. 1996). Supergiant shells are therefore exquisite laboratories for studying the interplay of the cold, warm, and hot gas phases of the interstellar medium.

2. Data

The presented Hα image is a mosaic of five fields taken with the Curtis Schmidt at CTIO. Exposure time of each field is 30 minutes. The IRAS 100μ image is produced by the SKYVIEW facility using the IRAS Sky survey (Wheelock et al. 1991). The X-ray image is mosaicked from our ROSAT PSPC raster pointings of LMC 4 and all available archived PSPC pointings in the LMC 4 region (about 200). We excluded the regions around point sources to minimize the contamination of diffuse emission, and transformed the remaining photons into a common reference frame. The mosaicking software produces spectra of defined regions by collecting the corresponding photons from all observations and correcting them individually for vignetting.

3. Large Scale Structure

The Hα image shows the known 1.4 kpc ring of LMC 4. This very deep image reveals the complexity of the filaments, making it sometimes difficult to define the exact boundary of LMC 4. On large scales Hα filaments are located just inside the peaks of the neutral gas ridge. We overlaid the X-ray medium band image with contours of the IRAS 100μ emission. A ring of enhanced X-ray emission follows the outline of the neutral gas hole, with the X-ray surface brightness decreasing toward the center. The center is still above the surface brightness of the extragalactic background, as measured outside the LMC. This structure may imply that we simply see the edge-brightening of a shell of gas with a very low density (and therefore low surface brightness) in the central region. This is exactly what theory predicts for a superbubble (Weaver et al. 1978). The higher X-ray surface brightness along the rim is partly caused by the mass evaporation across the conduction zone between the dense cold shell and the hot interior.

4. Temperature Structure of the Hot Gas

The medium to soft band ratio map shows a harder emission in the high surface brightness regions near the shell walls. This is consistent with recent supernovae from Shapley Constellation III hitting the shell walls and producing hotter and relatively dense plasma. Such a scenario was suggested for the superbubble N51D by Chu & Mac Low (1990). The spot of lowest X-ray surface brightness coincides with the minimum of neutral gas column density. Here, the HI 21cm line also shows high velocity components receding relative to the rest velocity of the LMC (Domgoergen et al. 1995). The low X-ray surface brightness supports that in this region LMC 4 has started to break out into the halo of the LMC.

Acknowledgments. We are grateful to W.J. Duschl and K. Weis for helpful discussions on the layout and optimization of this poster.

References

Bomans, D.J., Dennerl, K, & Kürster, M. 1993, A&A, 283, L21

Bomans, D.J., de Boer, K.S., Koornneef, J., & Grebel, E.K. 1996, A&A, 313, 101

Braunsfurth, E., & Feitzinger, J.V. 1983, A&A, 123, 113

Chu, Y.-H., & Mac Low, M.-M. 1990, ApJ, 365, 510

Domgoergen, H., Bomans, D.J., & de Boer, K.S. 1995, A&A, 296, 523

Mac Low, M.-M., McCray, R., & Norman, M.L. 1989, ApJ, 337, 141

Meaburn, J. 1980, MNRAS, 192, 365

Weaver, R., McCray, R., Castor, J., Shapiro P., & Moore R. 1978, ApJ, 218, 377

Wheelock et al. 1991, IRAS Sky Survey Atlas Explanatory Supplement

New Views of the Magellanic Clouds
IAU Symposium, Vol. 190, 1999
Y.-H. Chu, N.B. Suntzeff, J.E. Hesser, and D.A. Bohlender, eds.

The IRAS Vela Shell: a "Super Shell" in the Making?

Jayadev Rajagopal and G. Srinivasan

Raman Research Institute, Bangalore-560 080, India.

Abstract.
The IRAS Vela shell was first seen as a ring of infrared emission in the IRAS maps of the Gum-Vela region coincident with the Gum Nebula. We have studied the kinematics of the molecular material (^{12}CO) associated with the IRAS point sources and Southern Dark Clouds seen in this region and established the presence of a giant expanding shell of molecular gas. The estimated mass, size and expansion velocity of the shell lead us to conclude that it is the remnant of a GMC, in the process of disintegration through the influence of a central OB association. We conjecture that the IRAS Vela shell is a nearby example of a "super shell" in its early stages of evolution and as such is an ideal opportunity to study these objects now widely seen both in our Galaxy and in new observations of the Magellanic Clouds.

1. Introduction

The IRAS-Vela shell is a ring-like structure discovered from the IRAS Sky Survey Atlas in the 25, 60 and 100 μm maps. It is a large feature seen in the longitude range $l^{\rm II} = 245°$ to $275°$ and up to 15° latitude below the Galactic plane. The shell also shows up as an enhancement in Hα emission and is seen projected against a large HII region, the Gum Nebula. A detailed study of the region has been carried out by Sahu (1992). Based on the fact that the kinematics of the shell is quite different from that of the Gum Nebula, she concluded that it was a separate entity. The shell is also clearly seen in the IRAS Point Source Catalogue (IPSC). We carried out a survey of molecular gas in the shell. We also examined the kinematics of this gas and estimated its mass in an attempt to understand the nature of the shell.

2. Observations and Results

The observations were carried out from the 10.4 m millimetre wave telescope at the Raman Research Institute in the 115 GHz line from the J= 1$\rightarrow$ 0 transition of the ^{12}CO molecule. We looked for molecular gas in the direction of the infrared point sources seen in the IPSC. To maximize the chances of detections we selected sources which were probable Young Stellar Object (YSO) candidates. We observed in the direction of about 100 of the ~3750 IPSC sources in the region of the shell which satisfied the YSO criteria. The spectral resolution was ~ 0.26 km s^{-1} and the rms noise level was ~ 0.2 K antenna temperature.

The observations yielded detections for 42 of the 100 or so sources. This confirms the presence of molecular gas associated with the shell. The distribution of this molecular gas is clumpy and also hints at a shell-like structure.

3. Kinematics

Previous studies of ionized gas in the region had indicated that the shell could be expanding. The only available study of molecular gas in the vicinity of the shell was restricted to the ~ 35 or so cometary globules in the Gum Nebula (Sridharan 1992). These dark clouds with head-tail morphology also show systematic expansion from a common center at about 12 kms^{-1}. We analysed the radial velocities of our detections to check for expansion. We also extended this study to all the Southern Dark Clouds (SDCs) in the region. The radial velocities of these were made available to us from a survey of SDCs by Otrupcek *et al.* (1995). Our analysis clearly shows that the entire shell of molecular gas including the large number of SDCs in the region is expanding about a common center at ~ 13 km s^{-1}.

4. Nature of the IRAS-Vela Shell

Our study reveals the presence of a very large shell (angular diameter $\sim 12.5°$) of molecular gas associated with IRAS-Vela shell. If we assume that the typical molecular clump in the shell has a mass equal to a large cometary globule, the total mass of molecular gas in the shell is of the order of 10^5 $M_{\odot}$. We arrive at a similar estimate as well by assuming a typical star formation efficiency and counting the number of YSOs in the shell. This mass is comparable to that of a GMC. The shell is symmetrically placed around the OB association Vela OB2. Distance estimates to Vela OB2 place it at 450 pc, which is also the estimated distance of the shell. The clumpy nature of the shell, its expansion, the presence of the cometary globules etc. all lead us to confirm the conjecture that the shell is the remnant of the GMC within which Vela OB2 was born. Under the influence of stellar winds and supernovae from the association, the gas is being dissociated and swept outwards. Although at present there is no conclusive evidence for neutral gas in the shell, we estimate that as the shell continues to sweep up gas, it will be visible as a "supershell" in HI. Such shells are prevalent in our Galaxy and recent HI surveys of the Magellanic Clouds reveal a large number of shells in them. The IRAS-Vela shell could be a nearby young supershell and as such provides an ideal subject to study the early evolution of these objects.

References

Sahu, M.S. 1992, Ph.D. Thesis, University of Groningen

Sridharan, T.K. 1992, JAp&A, 13, 217

Otrupcek, R.E., Hartley, M., Jing-Sheng, W. 1995, unpublished (private communication, Otrupcek)

New Views of the Magellanic Clouds
IAU Symposium, Vol. 190, 1999
Y.-H. Chu, N.B. Suntzeff, J.E. Hesser, and D.A. Bohlender, eds.

The ISM near SN 1987A: Kinematics and 3-D Structure

Arlin P.S. Crotts and Jun Xu

Dept. of Astronomy, Columbia University, New York, NY 10027, USA

Abstract. We present two views of the region around SN 1987A: a three-dimensional map of structure revealed by light echoes, and a multicomponent velocity map at 10 km s^{-1} resolution in [N II], correlating the various superbubble structures found in both, with implications for their age and dynamics, as well as the history of the SN progenitor

The Large Magellanic Cloud (LMC) exhibits gas shells from ~100 pc to 1000 pc across. A new tool for unfolding the three-dimensional (3D) configuration of these structures seen to overlap in projection is the light echo from SN 1987A. We (Xu, Crotts, & Kunkel 1995) discuss the large scale interstellar medium (ISM) distribution in the SN foreground, and construct a map of these complicated structures from the light echoes, identifying dust sheets from 310 to 3240 pc in front of the SN. Additionally, we report results from a map of the velocity structure in the same field, revealed by long-slit echelle spectroscopy in [N II] (Xu & Crotts 1999). This information is tied to the same features seen in the echoes, revealing the kinematics and dynamics of the bubbles.

Significantly, N157C is not just the small shell of Hα filaments north-east of SN1987A. Instead, echoes from Hα structure extending south-west of the SN form a sphere with center of curvature on OB association LH90, at N157C's center, so N157C extends to ~220 pc in diameter. X-ray maps (Wang & Helfand 1991a) reveal a bubble ~120 pc across, while N157C seen in 21-cm emission extends over $\gtrsim$200 pc (Kim et al. 1998), nearly coextensive with the echo. The dust coincides with the H I, but sits outside the X-ray-emitting region.

Two additional bubbles are indicated along the sightline to the SN. Two echoes sit ~1 kpc in front of SN 1987A. These and N157C may form opposite sides of a giant bubble 600 pc across. Dense clouds about 100 pc in front of SN1987A bound a large region evacuated around SN 1987A, and therefore may contain the bubble blown by the SN and its companions. We also notice components at 269 and 301 km s^{-1} within $\sim 20''$ of SN1987A. These structures are probably due to the red supergiant wind, and are treated in a separate paper.

High resolution (10 km s^{-1}) [N II] spectra, sampled every 13 arcsec in a $6'$ square around SN1987A were obtained on the CTIO 4m/Echelle (Xu & Crotts 1999). Three components, V_{hel} = 265, 277 and 285 km s^{-1}, were identified as N157C, having the same morphology as the echoes corresponding to the shell centered on LH90. Wang and Helfand (1991b) reported X-ray emission in N157C, corresponding to very hot gas; therefore, N157C may be in the energy-conservation stage. The radius of this superbubble is found to expand at 10 km s^{-1}, with a lifetime of 6×10^6 years and a total energy of 3×10^{51} ergs, determined according to superbubble theory (McCray & Kafatos 1987): $R_S =$

97 $pc\ (N_\star E_{51}/n_0)^{1/5} t_7^{3/5}$ (Eqn. 1), and $V_S = 5.7\ km\ s^{-1}\ (N_\star E_{51}/n_0)^{1/5} t_7^{-2/5}$ (Eqn. 2). Dividing one equation by the other, $t_7 = (R_S V_S)/[(5.7\ km\ s^{-1})(97\ pc)]$ (Eqn. 3), where n_0 is the gas atomic density, the age $t_7 = t/10^7$ yr, $R_S = 100$ pc, and $V_S = 10$ km s^{-1}. We estimate N157C to have $t = 6$ Myr, still very young.

The morphology of $V_{hel} = 235$ km s^{-1} component corresponds to the echo 1 kpc in front of the SN, the near side of the 600 pc cavity. With nothing between this structure and N157C, this indicates a bubble expanding at ~40 km s^{-1} relative to the centroid of N157C. We suspect that this 600 pc bubble has blown out of the LMC disk: not only is its wall fragmented, but Wang and Helfand (1991b) report that this region has soft X-ray emission much higher than the background but much lower than other young superbubbles in 30 Doradus. Also, 600 pc is one of the largest superbubbles ever observed in the LMC, and approaches the limit on the superbubble size imposed by LMC disk thickness. McCray and Kafatos (1987) suggest that superbubbles breaking out of the galaxy, distort by gravitational instability, collapse to form giant clouds, and ignite propagating star formation. LH90 may have formed in the collapse of the 600 pc bubble.

If this superbubble blew out in its snow-plow stage, prior to this it was described in terms of the time (t_c) and expansion radius (R_c) beyond which cooling becomes important in the interior e.g. post-adiabatically, according to McCray and Kafatos: $R(t) \approx R_c(t/t_c)^{1/4}$, for $R_c = 50\ pc\ \zeta^{-0.9}(N_\star E_{51})^{0.4}\ n_0^{-0.6}$, and $t_c = 4\times10^6\ \zeta^{-1.5}(N_\star E_{51})^{0.3}\ n_0^{-0.7}$, where $\zeta \approx 0.3$ is metallicity, $N_\star$ is number of stars $\geq 7M_\odot$, $E_{51} = E/10^{51} ergs$, and $n_0 \approx 0.35$ is gas atomic density. Let's guess $N_\star = 50$ (like N157C) and $E_{51} = 1$, hence $t_c = 160$ Myr, and $R_c \approx 1.3$ kpc, much bigger than 300 pc. Therefore, this bubble was probably hot when it blew out, implying an age at blowout of ~10 Myr from eq. 3, or 15 Myr using eq. 1. This appears older than N157C, and may have created LH90.

Two components, $V_{hel} = 255$ and 245 km s^{-1} are identified as the inner major echo ~130 pc in front of SN1987A. Two components, 300 and 313 km s^{-1}, may be the far side of a bubble in which SN1987A exploded. They do not correspond to any known absorption lines or echoes (which show up mainly in front), hence probably lie behind the SN. If these components form the far side of a bubble surrounding SN1987A, its radius expands at 30 km s^{-1}. If we adopt a 90 pc radius, the bubble's age is ~2 Myr (from eq. 1).

We find that the time it took the SN1987A progenitor to move to the current location 300 pc behind N157C (at $\gtrsim 20$ km s^{-1}) is comparable to the age of N157C as well as that of the progenitor itself ($\sim 10^7$ y). This suggests that perhaps LH90 and the progenitor were created in the same previous superbubble collapse. Suggestively, the age the 600 pc giant bubble is consistent with this.

References

Kim, S., et al. 1998, ApJ, 503, 674

McCray, R., & Kafatos, M. 1987, ApJ, 317, 190

Wang, Q., & Helfand, D.J. 1991a, ApJ, 370, 541

Wang, Q., & Helfand, D.J. 1991b, ApJ, 373, 497

Xu, J., & Crotts, A.P.S. 1999, ApJ, 511, 262

Xu, J., Crotts, A.P.S., & Kunkel, W.E. 1995, ApJ, 451, 806 (Erratum: 463, 391)

Discussion

David Graff: What is the density of the different clumps?

Crotts: Most of the extinction is in N157C. There is greater extinction in patches of the cloud 1 kpc in front of the SN, but this feature is very non-homogeneous. The wall 100 pc in front of the SN is a few times thinner than N175C. The total amount of material corresponds to about 2×10^{21} cm^{-2} in H I.

New Views of the Magellanic Clouds
IAU Symposium, Vol. 190, 1999
Y.-H. Chu, N.B. Suntzeff, J.E. Hesser, and D.A. Bohlender, eds.

A Radio-Continuum Study of the Magellanic Clouds

M. Filipović

Max-Planck-Institut für extraterrestrische Physik, Garching, Germany
University of Western Sydney, Nepean, NSW 2747, Australia
Australia Telescope National Facility, CSIRO, NSW 2121, Australia

L. Staveley-Smith and R. Haynes

Australia Telescope National Facility, CSIRO, NSW 2121, Australia

G. White and P. Jones

University of Western Sydney, Nepean, NSW 2747, Australia

Abstract. We review all radio-continuum surveys of the Magellanic Clouds (MCs) and also present latest results from our new mosaic surveys.

1. Introduction

Radio-continuum investigations of the MCs started with the detection by Mills & Little (1953). Since that time, numerous radio surveys have been undertaken. The low-resolution observations helped to determine the large-scale structure and the total radio spectrum of the MCs (Haynes et al. 1991) whereas high-resolution studies concentrated on discrete sources in the Clouds.

2. Past Radio-Continuum Surveys

The first high-frequency radio-continuum detection of discrete sources in the MCs was made by Mathewson & Healey (1964). During the 1970s, McGee et al. (1972; 1976) made detailed observations of the MCs at various frequencies with the Parkes radio telescope and established catalogues of discrete radio sources. These observations became part of the PKSCAT-90 catalogue. Clarke et al. (1976) observed the MCs with the Molonglo radio telescope at 408 MHz and produced the MC4 catalogue of 227 sources towards the LMC and 75 towards the SMC. Milne et al. (1980) observed a number of interesting LMC sources at 14.7 GHz to estimate their radio spectral index. Mills et al. (1984) and Ye (1988) observed the MCs at 843 MHz with the Molonglo Synthesis Telescope (MOST). Following the first radio classification of sources in the MCs based on studies of the spectral index (McGee & Newton 1972), Milne et al. (1980) and Loiseau et al. (1987) provided the first detailed analysis of the spectra of some 40 LMC and 28 SMC sources, most of which were SNRs.

3. New Parkes Radio Surveys

In 1986, an international collaboration was launched using the Parkes 64-m telescope to survey the MCs at several radio frequencies with moderate angular resolution using a full polarization system. This project has produced several publications reporting the surveys (Haynes et al. 1991), the FIR/radio correlation (Xu et al. 1992), magnetic fields in the MCs (Klein et al. 1993), radio source catalogues at six frequencies (Filipović et al. 1995, 1997), the radio/X-ray discrete source correlation (Filipović et al. 1998a), the radio spectral index catalogue of sources in the MCs (Filipović et al. 1998b) and the radio/IR discrete source correlation (Filipović et al. 1998c). A total of 483 sources towards the LMC and 224 towards the SMC have been detected. Most of the MC's sources have been classified in one of three groups: SNRs, H II regions or background sources. In total, 209 discrete radio sources in the LMC and the 37 sources in the SMC are classified to be intrinsic.

4. The ATCA Observations of the MCs

The Australia Telescope Compact Array (ATCA) mosaic observations of the SMC were taken in October 1992. The baseline of 375 m was used at frequencies of 1.4 and 2.3 GHz with corresponding angular resolutions of $\sim 90''$ and $45''$. The total of 320 pointings covered area of ~20 square degrees (Staveley-Smith et al. 1997). The LMC mosaic observations are underway and more details can be found in Kim et al. (1998). The SMC image at 2.3 GHz is available and "corrected" for missing short-spacing by adding Parkes survey data from corresponding frequency. Over seven hundred sources were found in this survey. As a continuation of Parkes radio-continuum investigations and as an addition to the ATCA mosaic surveys, we observed most of intrinsic sources using ATCA at λ=6 and 3 cm with the 375 m configuration. These observations together with the ATCA mosaic surveys of both Clouds at 20/13 cm and the MOST survey at 843 MHz will give us wide frequency coverage of these objects with sufficient resolution to resolve them. The aim was to image a large number of H II regions and SNR in the snapshot mode. Using this method we observed, so far, all SNRs and SNR candidates in both Clouds (~80). A general SNR-like morphology was clearly detected in all observed SNR objects.

References

Clarke, J.N., Little, A.G., & Mills, B.Y. 1976, Aust. J. Phys. Suppl., 40, 1

Filipović, M.D., Haynes, R.F., White, G.L., et al. 1995, A&AS, 111, 311

Filipović, M.D., Jones, P.A., White, G.L., et al. 1997, A&AS, 121, 321

Filipović, M.D., Pietsch, W., Haynes, R.F., et al. 1998a, A&AS, 127, 119

Filipović, M.D., Haynes, R.F., White, G.L., &Jones, P.A. 1998b, A&AS, 130, 421

Filipović, M.D., Haynes, R.F., White, G.L., & Jones, P.A. 1998c, A&AS, 130, 441

Haynes, R.F., Klein, U., Wayte, S.R., et al. 1991, A&A, 252, 475

Kim, S., Staveley-Smith, L., Dopita, M.A., et al. 1998, ApJ, 503, 674

Klein, U., Haynes, R.F., Wielebinski, R., & Meinert, D. 1993, A&A, 271, 402

Loiseau, N., Klein, U., Greybe, A., Wielebinski, R., & Haynes R. F. 1987, A&A, 178, 62

McGee, R.X., Brooks, J.W., & Batchelor, R.A. 1972, Aust. J. Phys., 25, 581

McGee, R.X., & Newton, M.L. 1972, Aust. J. Phys., 25, 619

McGee, R.X., Newton, L.M., & Butler, P.W. 1976, Aust. J. Phys., 29, 329

Mills, B.Y., & Little, A.G. 1953, Aust J. Phys., 6, 272

Mills, B.Y., Turtle, A.J., Little, A.G., & Durdin, M.J. 1984, Aust. J. Phys., 37, 321

Milne, D.K., Caswell, J.L., & Haynes, R.F. 1980, MNRAS, 191, 469

Staveley-Smith, L., Sault, R.J., Hatzidimitriou, D., et al. 1997, MNRAS, 289, 225

Xu, C., Klein, U., Meinert, D., Wielebinski, R., & Haynes, R.F. 1992, A&A, 257, 47

Ye, T. 1988, PhD Thesis, Sydney University

New Views of the Magellanic Clouds
IAU Symposium, Vol. 190, 1999
Y.-H. Chu, N.B. Suntzeff, J.E. Hesser, and D.A. Bohlender, eds.

Kinematics of the Ionized Gas in the Irregular Galaxies IC 10 and NGC 4449

M. Rosado, A. Bullejos, M. Valdez, L. Georgiev and C. Lacey

Instituto de astronomía, UNAM. Apartado Postal 70-264, CP04510, México, D.F., México

J. Borissova

Institute of Astronomy, Bulgarian Academy of Sciences, 72 Tsaridradsko chaussee, BG1784 Sofia, Bulgaria

C. Esteban

Instituto Astrofísico de Canarias, Tenerife, Canarias, España

Abstract. We have searched for supernova remnants (SNRs) in the irregular galaxies IC 10 and NGC 4449 by means of Fabry-Perot interferometry in Hα and [SII] lines, complemented by radio continuum observations. We find several new SNR candidates in each galaxy in addition to the known ones. Although these galaxies have several common properties, the distributions of SNRs in each galaxy is different. This shows that the star formation mechanisms in these galaxies act in different ways, in favor of the formation of a large nebular HII complex in IC 10, similar to 30 Dor in the Large Magellanic Cloud. We are studying the conditions over which these different star formation histories take place.

1. Introduction

We have selected two irregular galaxies of the Magellanic type in order to identify the supernova remnants (SNRs): IC 10 and NGC 4449. These galaxies share several properties. Here we show the results of our optical observations (Hα and [SII] lines) using a scanning Fabry-Perot Interferometer PUMA (Rosado et al. 1995). We compare the results with radio continuum observations.

2. Results

IC 10 – The supernova (SN) explosions seem to be concentrated only to the south of the giant HII complex HL111 (Hodge & Lee 1990), forming a non-thermal superbubble (Yang & Skillman 1993). The HII complex HL111 appears to be similar to 30 Dor in the LMC but in a later stage of evolution. Figure 1 (left) shows a PUMA velocity channel image in [SII]. The circle marks the position and extent of the non-thermal radio superbubble reported by Yang & Skillman (1993). This non-thermal superbubble has an optical counterpart

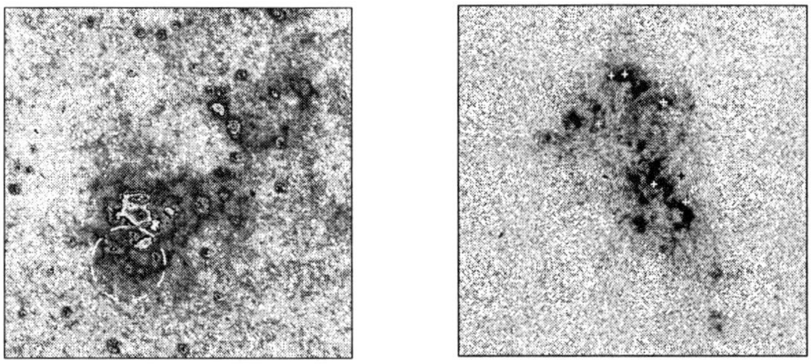

Figure 1. [SII] velocity channel images of IC 10 (left) and NGC 4449 (right). The white circle marks the non-thermal radio superbubble and the crosses mark the non-thermal radio sources.

revealed as a series of nebulosities and a curved filament running to the west of the radio superbubble over larger dimensions. Also, several of the HII regions catalogued by Hodge & Lee (1990) are seen interconnected by faint filamentary features, forming larger shells.

NGC 4449 – Recent VLA radio continuum observations of NGC 4449 detect at least 7 non-thermal sources, including a known SNR and the nucleus of this galaxy (Lacey et al. 1998, and references therein). Figure 1 (right) shows a PUMA [SII] velocity channel image of this galaxy. We have also marked the non-thermal radio sources found by Lacey et al. (1998). All, but one, of the radio sources are embedded in optical nebulosities. We are studying the kinematics of these regions to search for optical counterparts of these non-thermal sources. Also, we have detected networks of filaments extending radially from the bar of NGC 4449. As in the case of IC 10, several of the HII regions (catalogued by Sabbadin & Bianchini 1979) are interconnected by filaments, forming larger structures.

3. Discussion

We note that the distributions of SNRs are different in these galaxies. In IC 10 the SNRs seem to be concentrated near its largest HII complex, whereas in NGC 4449 the SNRs are distributed in a more uniform way. On the other hand, while in NGC 4449 it seems that the SN explosions are widely separated, in IC 10 the SNe combine to form larger structures. This shows that despite the several similar properties, the star formation histories of these galaxies are different.

References

Hodge, P., & Lee, M.G. 1990, PASP, 102, 26

Lacey, C.K., et al. 1998, BAAS, 192, 5310

Rosado, M., et al. 1995, RevMexAA Conf. Ser., 3, 268

Sabbadin, F., & Bianchini, A. 1979, PASP, 91, 280

Yang, H., & Skillman, E.D. 1993, AJ, 106, 1448

Despina Hatzidimitriou appears to provoke a skeptical reaction from Carme Gallart and Jay Gallagher during the reception.

Part 3. Massive Stars and 30 Doradus

Section A. Invited Reviews

Renewing acquaintances at the reception. (Top) Jon Holtzman and John Hutchings, (bottom) Alistair Walker and Bruce Bohannan.

New Views of the Magellanic Clouds
IAU Symposium, Vol. 190, 1999
Y.-H. Chu, N.B. Suntzeff, J.E. Hesser, and D.A. Bohlender, eds.

Massive Stars in the MCs: What They Tell Us about the IMF, Stellar Evolution, and Upper Mass "Cutoffs"

Philip Massey

National Optical Astronomy Observatories, P.O. 26732, Tucson, AZ 85726-6732, USA

Abstract.
Studies over the past decade have shown that the initial mass function (IMF) is the same for massive stars born in the OB associations of the LMC and SMC as in the associations of the Milky Way: the slope of the IMF is essentially Salpeter ($\Gamma \sim -1.3$), despite the factor of 4 difference in metallicity between these systems, and despite a factor of several hundred in stellar density between the sparsest and richest OB associations. However, there does appear to be a number of massive stars that are born in relative isolation, and the IMF of this mixed-age, *field* population is quite different than that of OB associations, with $\Gamma \sim -4$ in all three galaxies. The distribution of stars in the HR diagram is in excellent agreement with the Geneva group's evolutionary models for stars with masses $> 25\mathcal{M}_{\odot}$, with no "main-sequence widening problem" left to be solved. The massive stars born in clusters are formed quite coevally ($\Delta\tau < 1 - 2$Myr), which allows us to use the "turn-off masses" to determine what mass objects become Wolf-Rayet stars of various types, and new results are briefly described. For the LMC, WNEs come from a wide range of masses, WCs come only from the highest mass stars, and Ofpe/WN9 "slash" stars come from lower mass OBs. Recent work on the R136a cluster (described in Hunter's review talk) suggest that there is no such thing as an upper mass cutoff to the IMF, at least not one that has been found observationally: for the youngest clusters (2 Myr and younger), the mass of the highest mass star present is simply dependent upon how populous the cluster is; i.e., the IMF is truncated by statistics, not physics.

1. Introduction

When I was a postdoc at the DAO, Dr. van den Bergh would corner me after each trip and ask me what new, interesting thing I had learned. I remember coming back from Boulder once, circa 1981, and telling him about a result that Katy Garmany and Peter Conti had just obtained for massive stars in the Milky Way: it appeared that the initial mass function (IMF) showed a gradient with galactocentric distance, with proportionally more of the highest mass stars being formed in the inner regions of our Galaxy (Garmany, Conti, & Chiosi 1982). Sidney wrinkled his brow, and said that this really would be quite surprising if true, as one would expect that it would be harder to form higher mass stars where the metallicity is higher. This remark got me thinking about the Magellanic Clouds (MCs), and what fun one could have there in determining massive star IMFs, and then comparing them to those of the higher-metallicity Milky Way (MW). Little did he know he was launching a 17 year study! (He was also right about the result being wrong, for reasons I'll explain below.)

Why study massive stars in the MCs? The MCs have low reddening and are nearby, and hence complete samples of massive stars can be identified, albeit with some effort. The low metallicities serves as a useful contrast to Milky Way studies, as it is reasonable to wonder if the IMF slope or upper-mass cut-off does depend upon metallicity z: if radiation pressure on grains were the dominant effect in the mass of the most massive star we see, then the upper mass cutoff should go as $1/\sqrt{z}$ (Shields & Tinsley 1976). If that were the case, then, we might expect that the most massive star we see in the SMC would be about a factor of 2 higher in mass than in the Milky Way, and the most massive stars we see in the LMC should be a factor of 1.5 times more massive than in the MW ($\log(O/H)$=8.13, 8.37, and 8.70 in the SMC, LMC, and MW, respectively, according to Russell & Dopita 1990 and Esteban & Peimbert 1995). Secondly, we know that massive star evolution *should* depend upon z, as the mass loss rates scale as $\sqrt{z}$, and thus studies of massive stars in the MCs provide a critical test of evolutionary models.

There are two major complications in studying massive stars in the MCs, both of which are discussed extensively elsewhere (Massey et al. 1995a, 1995b; Massey 1998a, 1998b) and which I'll mention only briefly here: (a) In a *mixed-age* population the *visually* brightest stars will not be the most *bolometrically* luminous or massive. The result of this is that luminosities will not tell you *anything* about the underlying IMF. (See Fig. 1 of Massey 1998a.) (b) The colors of hot stars are degenerate with effective temperature ($T_{\rm eff}$), and hence with the bolometric corrections (BCs). (See Fig. 3 of Massey 1998a, and discussion in Massey 1998b.) But it is only through knowledge of the BC that we can get any insight into the total luminosity (and hence mass) of an O-type star, with the BC changing by about 1.5 mag, or a factor of 2 in mass (eq. 4 of Massey 1998a). However, spectroscopy allows us to address these concerns easily, as there is good $T_{\rm eff}$ resolution with spectral type for the hottest stars— at least until O3 is reached ($T_{\rm eff} \sim 50,000°$), beyond which the classification is degenerate.

How much does this really matter? Massey et al. (1995b) derived an IMF slope of $\Gamma = -1.4 \pm 0.2$ for the OB association LH58 in the LMC. Hill, Madore, & Freedman (1994) had analyzed the same cluster, and had found a slope of $\Gamma = -2.5 \pm 0.3$ based upon only photometry. We demonstrated that if we ignored our own spectroscopy—simply used our own photometry, and the transformations of Hill et al.—that the IMF slope we derived would indeed steepen from -1.4 to -2.0! The difference is in fact simply due to the degeneracy for the hottest and massive stars, and this serves as a cautionary tale to those wanting to derive IMFs from luminosity functions even in a coeval association (e.g., Fig. 2 of Massey 1998a).

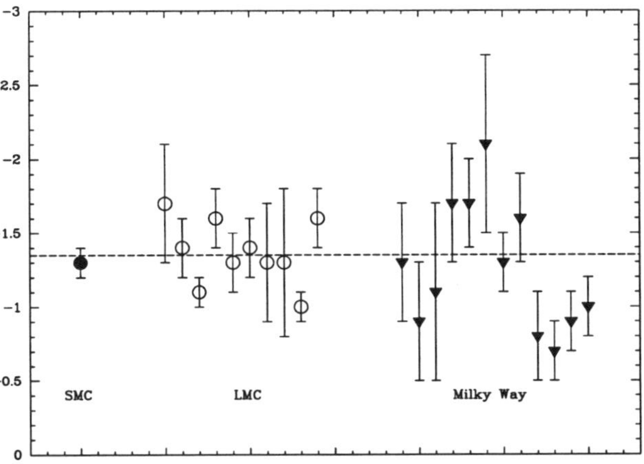

Figure 1. The IMF slope Γ for 23 OB associations in the SMC, LMC, and MW. We have indicated the value of a Salpeter $\Gamma = -1.35$ IMF.

2. Results: Massive Stars in OB Associations in the SMC, LMC, and MW

My colleagues and I have now determined the IMFs for 23 OB associations in the SMC, LMC, and Milky Way (Massey et al. 1995a,b; Massey & Hunter 1998). I review how this is done and some of the associated errors in Massey (1998a). Here, I will just make the two points: (1) If a region is strictly coeval, then the slope of the present day mass function (PDMF) *is* the slope of the IMF, with the only effect of age to deplete the upper-most mass bin. (2) The "absolute" value of the IMF slope is considerably less certain than the *relative* values: we make no allowance for binaries, but implicitly assume that the binary frequency and mass ratios are the same in these systems. One must analyze the data with the same set of models (at the appropriate metallicity) for intercomparisons to be meaningful. If you take values for the IMF slopes from many different sources, analyzed in different ways, you will of course convince yourself that the IMF slope is anything *but* constant—as Scalo (1998) does in his recent review.

However, as Figure 1 shows, the observational evidence suggests that the IMF slope *is* constant, at least for stars with masses $> 10\mathcal{M}_\odot$ formed in OB associations in the SMC, LMC, and MW—despite the factor of 4 change in metallicity, and despite a factor of several hundred difference in stellar density (going from the least populous cluster to that of R136a)!

We have also learned that massive stars that form in clusters do so very coevally, with an age spread $< 1 - 2$ Myr. In the case NGC 6611 in the MW, Hillenbrand et al. (1993) conclude that their data are consistent with all the massive stars "having been born on a particular Tuesday." Interesting, the intermediate mass stars continued forming even after the high mass stars had formed, contrary to the conventional wisdom. DeGioia-Eastwood et al. (1999) find the same for the MW associations Tr 14/16. However, in R136, star formation shut down once its extreme population of very massive stars formed (Massey & Hunter 1998), although some star formation continues in the general 30 Dor region to this day (Walborn & Blades 1987; see also Walborn & Barba in these proceedings).

3. Results: Massive Stars in the Field

For a mixed-age population, the PDMF will be related to the IMF via the main-sequence life-time, assuming that the star-formation rate has been relatively constant over the past several million years in the region in question. Massey et al. (1995b) analyzed the *field* population of the LMC and SMC, starting with the Rousseau et al. (1978) catalog updated by Fitzpatrick & Garmany (1990) for the LMC, and the Azzopardi & Vigneau (1975, 1982) catalogs for the SMC. We found that only about half the stars in these catalogs were actually in OB associations. (The fraction "half" is primarily an indication of how sparse our knowledge had been of the individual members of OB associations; only a handful of the massive O stars for which we obtained spectra in our association work were listed in these catalogs.) But, if these stars weren't in OB associations, where were they? Even a cursory examination of the Sanduleak (1969) atlas shows that many of these stars are well isolated—this is not a case of there being numerous OB associations that had been missed by Lucke & Hodge (1970).

Rather, it seems likely that these stars formed near to their current location from more modest star-forming events than those that produced large clusters. After all, we know from the mere existence of R136a that star-formation occurs on a variety of scales. Why should we be surprised that there is a "low end" to star formation that does, stochastically, produce the occasional high-mass star? (See the study of small Galactic H II regions by Hunter & Massey 1990.)

The lifetimes of these stars are short, and during that time these stars could not have simply wandered far from parent OB associations. For instance, a few of these "field" stars are O3 stars, which have an age of < 1 Myr. In that time, not even a runaway star can wander very far: at 50 km s^{-1} such a star would travel 50 pc in 1 Myr. Massey et al. (1995b) found a substantial population of massive stars using the "extreme" criterion of 300 pcs distance to the nearest OB association (see their Figs. 4 and 5).

Rousseau et al. (1978) were well aware of the completeness problems inherent in a V-limited catalog, that their sample would over-represent modest mass B-supergiants relative to stars of higher bolometric luminosities but higher temperatures. In order to correct for selective incompleteness at high temperatures, Massey et al. (1995a) obtained CCD photometry and spectroscopy of several regions, and found they could reliably correct the statistics down to 25 $\mathcal{M}_\odot$.

The rather surprising results of this study are that the IMF slope of these *field* stars is remarkably steep: $\Gamma \sim -4$. The same result is found for MW field stars, and the relative distribution of field and association stars seem to be responsible for the original Garmany et al. (1983) IMF "gradient" that started these inquiries! Massey et al. (1995b) also found that, although they are proportionately rarer, stars as massive as those found in associations are found in the field. Fig. 4 of Massey (1998a) compares the environments of four LMC O3 stars, two in the field, and two in associations.

4. Results: Stellar Evolution

The MCs really *are* the "laboratories" for probing stellar evolution, as many of us have claimed in our proposals. Let me touch briefly on two results. First, in our study of the field population of the LMC and SMC it was possible to provide a very critical test of the Geneva evolutionary models: how well do they match the observations? If the isochrones are "right" then we would expect to find the same number of stars between 1 and 2 Myrs as between 2 and 3 Myrs at a given mass. And, indeed this is just what we find! Thus there can not be very strong contamination by "blue loop" SN1987A precursors (at least not above 25 $\mathcal{M}_\odot$!), and there is no "main-sequence widening problem" to be solved. The difference between our results, and those of Fitzpatrick & Garmany (1990), was that we were able to correct the statistics for the hotter stars that would be selectively excluded in any V-limited Rousseau et al. (1978) sample (Massey et al. 1995b).

Secondly, it is possible to use the coeval associations of the SMC, LMC, and MW to test how the later stages of massive star evolution depend upon metallicity. My summer student Liz Waterhouse is working on a project with me currently in which we are measuring the "turn-off" masses in clusters which contain Wolf-Rayet (WR) stars of various types. We will shortly be able to look for metallicity dependent effects by comparing the results for the SMC, LMC, and MW; for now, let me just mention our very preliminary results for WRs in the LMC. We find that early-type WN stars (WN3-5) come from a wide variety of stellar progenitors (40-100 $\mathcal{M}_\odot$). The WC stars in the LMC are found only in OB associations with large turn-off masses ($> 70\mathcal{M}_\odot$), consistent with the expectations of the "Conti scenario"only the highest luminosity stars will evolve to WC type at lower metallicities due to the decreased mass-loss rates. And, we find that the so-called "slash" stars (Ofpe/WN9) are found preferentially in clusters with very *low* turn-off masses: 20-40 $\mathcal{M}_\odot$. The latter is in accord with the recent findings of St-Louis et al. (1998).

5. Is There Such a Thing as an Upper-Mass Cut-off?

In studying the OB associations in the SMC, LMC, and MW, we had found that mass of the highest mass star was very similar in all three galaxies ($\sim 90-100\mathcal{M}_\odot$), despite the predictions that there should be a 2$\times$ difference due to metallicity. Thus, radiation pressure acting on grains was not what was limiting the highest mass we saw. I continued to puzzle about this until we obtained data on R136a, the central "supercluster" of 30 Dor (Massey & Hunter 1998). Deidre Hunter discusses these results in this symposium; here let me emphasize that the upper mass cut-off that we had been observing was not physical in natural, but rather statistical. R136a is young enough so that its most massive stars have not yet died, and we see stars that are conservatively estimated to be $> 150\mathcal{M}_\odot$. But in fact for R136a this is just where the (completely normal!) IMF peters out—where statistically you would expect just one star, because it is so rich and populous. Looking back over the data for the more modest OB associations, we find that the upper mass "cut-offs" were simply what you would expect given the relative richness (and age). Doubtless there is a physical limit to how massive a star actually can form, but so far we have yet to encounter this limit in nature.—instead, the "limits" we've been up against have been statistical rather than physical.

Sidney Wolff had once suggested as much to me, but it had remained untested until the R136a data. So, you can see that two people named Sidney have played important roles here!

Acknowledgments. The work I've described here has been done with excellent collaborators, including summer students. Let me explicitly acknowledge the roles played by Peter Conti, Kathy Eastwood, Katy Garmany, and Sally Oey, and the REU students K. E. Johnson, J. Johnson, C. Lang, and E. Waterhouse. The R136a work was done with my partner Deidre Hunter, whose knowledge and insights have helped shaped my thinking on these matters.

References

Azzopardi, M., & Vigneau, J. 1975, A&AS, 22, 285

Azzopardi, M., & Vigneau, J. 1982, A&AS, 50, 291

DeGioia-Eastwood, K., Throop, H., Walker, G., & Cudworth, K. M. 1999, AJ, submitted

Esteban, D., & Peimbert, M. 1995, Rev. Mex. A&A, Ser. Conf. 3, 133

Fitzpatrick, E. L., & Garmany, C. D. 1990, ApJ, 363, 119

Garmany, C. D., Conti, P. S., & Chiosi, C. 1982, ApJ, 263, 777

Hill, R. J., Madore, B. F., & Freedman, W. L. 1994, ApJ, 429, 204

Hillenbrand, L. A., Massey, P., Strom, S. E., & Merrill, K. M. 1993, AJ, 106, 1906

Hunter, D A., & Massey, P. 1990, AJ, 99, 846

Lucke, P. B., & Hodge, P W. 1970, AJ, 75, 171

Massey, P. 1998a, in *The Stellar Initial Mass Function*, ed. by G. Gilmore & D. Howell (San Francisco, ASP), 17

Massey, P. 1998b, in *Stellar Astrophysics for the Local Group*, ed. A. Aparicio, A. Herrero, & F. Sanchez (Cambridge, Cambridge Univ. Press), 95

Massey, P., & Hunter, D. A. 1998, ApJ, 493, 180

Massey, P., Johnson, K. E., & DeGioia-Eastwood, K. 1995a, ApJ, 454, 151

Massey, P., Lang, C. C., DeGioia-Eastwood, K., & Garmany, C. D. 1995b, ApJ, 438, 188

Rousseau, J., Martin, N., Prevot, L., Rebeirot, E., Robin, A., & Brunet, J. P. 1978, A&AS, 31, 243

Russell, S. C., & Dopita, M. A. 1990, ApJS, 74, 93

Sanduleak, N. 1969, Contr. Cerro Tololo Inter-American Obs., 89

Scalo, J. 1998, in *The Stellar Initial Mass Function*, ed. G. Gilmore & D. Howell (San Francisco, ASP), 201

Shields, G. A., & Tinsley, B. M. 1976, ApJ, 203, 66

St-Louis, N., Moffat, A. F. J., Turbide, L., & Bertrand, J.-F. 1998, in *Boulder-Munich II: Properties of Hot, Luminous Stars*, ed. I. D. Howarth (San Francisco, ASP), 326

Walborn, N. R., & Blades, J. C. 1987, ApJL, 323, 65

Discussion

You-Hua Chu: An O3 star born in the field without any interstellar gas left implies a 100% star formation efficiency. This is surprising. It may be possible that the O3 stars were formed by mergers.

Massey: There are about 100 stars with masses > 25 $M_\odot$ known in the LMC and SMC more than 200 pc from the nearest OB association. Did all of these form by mergers? If so, it's even more remarkable that their distribution in the HRD matches the evolutionary models so well. Why assume 100% efficiency? You can disrupt a 1000 $M_\odot$ cloud pretty quickly via stellar winds and the emission measure drops down past our threshold.

After all - there are entire OB associations without much nebulosity - does the lack of nebulosity mean all the hot stars in these formed by mergers? Or is the ISM just lumpy and stellar winds good at blowing away gas?

Jon Holtzman: How much would you have to relax the assumption of steady-state star formation to get the field IMF to agree with the association IMF?

Massey: It would require the SFR to have been substantially higher 6 Myr ago - and by the same ratio in the Milky Way, LMC, and SMC.

Nolan Walborn: This was a very nice review containing important insights on the relevant issues. However, I have several comments. (1) The steeper IMF slope in the field means that there are proportionately fewer of the most massive stars there. Certainly far more than half of the most massive stars are in associations. (2) Some isolated objects may well be mergers or other peculiar binary evolution phenomena. Sk−67°22 is a case in point: it is a very peculiar 03 star, with a main-sequence luminosity but a broad, WR-like HeIIλ4686 emission feature. (3) Some massive associations are coeval, but others clearly have subgroups of different ages, which may be sequential in some cases. (4) The M_V - spectral type relation may not be "great" but for 100 stars in 30 Dor, I find the average difference between derived and calibration M_V's to be 0.05 mag; and for 70 of them, the individual deviations are within ±0.6 mag, or they are subluminous dwarfs or superluminous supergiants which are reasonable phenomena for future investigation of spectroscopic luminosity criteria.

Massey: Given the incompleteness both of massive stars in associations AND "in the field" I would hesitate to agree with your statement that "certainly far more than half" of the OB stars are in associations. Certainly the current data (Massey et al. 1995, ApJ, 438, 188) don't support your assertion, but I agree that the incompleteness factor for unstudied OB associations is large, but so what? The "low mass" stars we're talking about are 25 $M_\odot$? What I'm saying is that how these stars form "in the field" appears to be different when they form in small groups, or isolation.

I completely agree with you that some catalogued OB associations are coeval, and others are not. For instance, no one could rationally agree that the entire 30 Dor region is coeval or a single OB association. Lucke & Hodge (LH), afterall, catalogued several different associations in the 30 Dor region. So far I've found that most, but not all, of the LH associations are in fact coeval groupings - really a remarkable piece of work that has held up well for 25 years now.

Hans Zinnecker: You showed us several cases of isolated "field" O-stars. With no lower-mass stars around them, the slope of the IMF there would be infinite? How then did you get a slope for the field stars IMF of $\Gamma = -4$? Please explain.

Massey: The IMF is really just the probability of forming a star of a given mass. So for stars outside of coeval group, you can determine an IMF as long as you're careful to account for the star formation history. In the case of the field massive stars, I assumed steady-state that averaged over the entire LMC or SMC that the SFR has not varied significantly over the past 10 Myr. (If you want the relevant equations, see my write up in Gerry Gilmore's IMF Conference last year, but if we couldn't properly talk about the "IMF of the field" a lot of people would be out of work.)

Giuseppe Bono: Stars more massive than 80 $M_\odot$ are dynamically unstable (Pappoloziu 1979). Do you think that this is a statistical or a physical effect?

Massey: Perhaps this is a "statistical effect" in the sense that if you took a large number of theoreticians and had them produce a number they were comfortable with, you would probably not get a delta function, but rather a wide distribution of answers. Seriously, we see stars in the R 136 cluster which appear to be - by conservative assumptions - up at around 150 $M_\odot$. But no massive stars (of $>$ 50 $M_\odot$) are "stable" (whatever that means) - they all have strong outflows, with mass-loss rates of 10^{-5} to 10^{-4} $M_\odot$ yr^{-1}. These mass estimates require that the Geneva evolutionary tracks are more or less right, and we only have observational confirmation of the mass luminosity up to about 30 $M_\odot$ (Burkholder, Massey & Niemela 1997 ApJ).

New Views of the Magellanic Clouds
IAU Symposium, Vol. 190, 1999
Y.-H. Chu, N.B. Suntzeff, J.E. Hesser, and D.A. Bohlender, eds.

Massive Binary Stars in the Magellanic Clouds

V. S. Niemela[1,2,3] and N. I. Morrell[2,4]

La Plata Observatory, Paseo del Bosque s/n, (1900) La Plata, Argentina

Abstract. We discuss the present status of knowledge for massive binary stars in the Magellanic Clouds. New data for known binaries, and radial velocity orbits for new binaries are reported.

1. Introduction

Many stars in the Magellanic Clouds are observed to have luminosities corresponding to stars more massive than 100 $M_{\odot}$, when compared with the theoretical mass-luminosity relation or evolutionary tracks. However, the empirical mass-luminosity relation is unknown in the Magellanic Clouds, and in our Galaxy it is well known only up to about 25 $M_{\odot}$. Therefore, the masses derived from stellar luminosities in the Magellanic Clouds lack the much needed empirical verification. Studies of binary stars are necessary for estimates of empirical stellar masses.

This is a progress report about our ongoing observing program of searches and studies of massive binary stars with OB and Wolf-Rayet (WR) type components in the Magellanic Clouds.

2. Observations

Our observations have been obtained:

1. between 1980 and 1990, at CTIO, Chile, we used the Cassegrain IT spectrograph on the Yale 1-m telescope. The detector consisted of photographic plates with fine grain emulsion until 1985, and an image tube with a CCD after 1985. These spectra cover the wavelength range $\lambda\lambda$3700-4800Å.

---

[1]Visiting Astronomer, Cerro Tololo Inter-American Observatory (CTIO), operated by AURA, Inc. under cooperative agreement with the National Science Foundation

[2]Visiting Astronomer, CASLEO, operated under agreement between CONICET, SeCyT, and the Universities of La Plata, Córdoba and San Juan, Argentina

[3]Member of Carrera del Investigador, CIC, Prov. Buenos Aires, Argentina

[4]Member of Carrera del Investigador, CONICET, Argentina

2. after 1992 at Complejo Astronómico El Leoncito (CASLEO), San Juan, Argentina, with the Cassegrain and echelle spectrographs on the 2.15m telescope. A CCD was used as the detector.

The photographic observations were digitized with a Grant engine at La Plata. The digital data were processed and measured for determination of radial velocities using IRAF routines.

3. Results and Discussion

In what follows we will present spectral data, radial velocities, and velocity curve fits of some massive binaries both in the LMC and SMC, as results of our continuing observing program of OB and WR type stars.

3.1. Massive Binaries in the Large Magellanic Cloud

Sk–67°105 The first double-lined O type spectroscopic binary discovered in another galaxy than our own, was the O4f+O6 type system Sk–67°105 in the LMC (Niemela & Morrell 1986). This binary was subsequently also found to show light variations (Haefner et al. 1994). We have obtained further spectra of Sk–67°105, which improve the binary period, and confirm the previous orbital elements and spectral types of the components. However, due to the short binary period (3.3 days), and early spectral types of the components, this system probably has considerable interaction effects. Therefore it may not be particularly suitable for a comparison with evolutionary tracks of single stars. Figure 1 illustrates the spectrum of this binary.

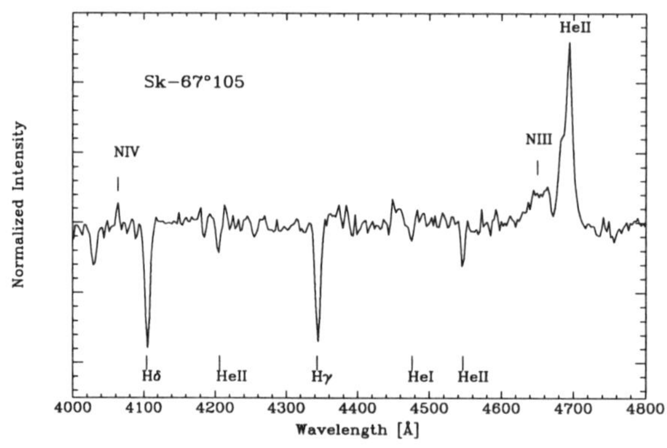

Figure 1. A blue spectrum of Sk–67°105, obtained at CASLEO.

HV 2241 An orbit based on the radial velocities of the eclipsing binary HV 2241 has been published by Niemela & Bassino (1994), who found spectral types of O7V and O8III: for the binary components, and minimum masses of 34 $M_{\odot}$ and 17 $M_{\odot}$, respectively. This binary has been the subject of a recent CCD light curve study by Prichart et al. (1998), who derived an orbital inclination of 85°.

Our new higher S/N CCD spectra agree with the previous radial velocity orbit, but we find a slightly different spectral type for the secondary. Figure 2

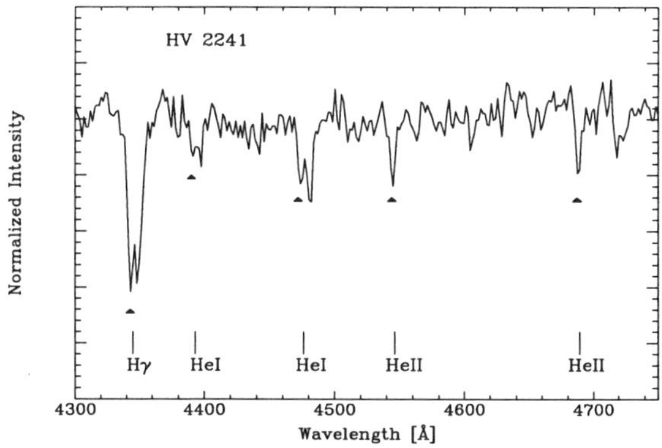

Figure 2. A CCD spectrum of HV 2241 obtained at CASLEO, showing double lines. The spectral lines corresponding to the primary are indicated by filled triangles.

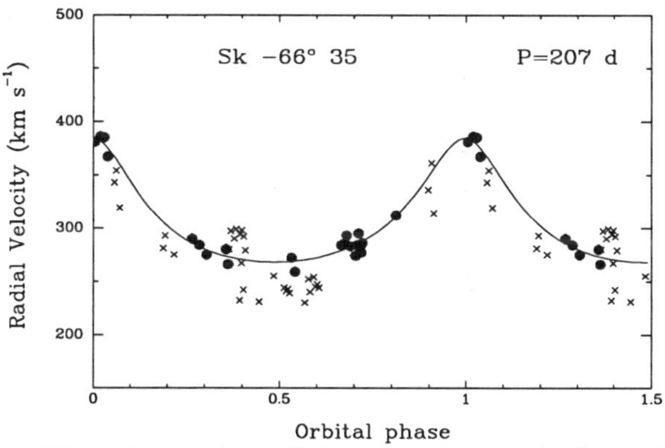

Figure 3. The observed radial velocities and derived orbit of Sk−66°35. Filled circles represent high resolution echelle spectra, and crosses lower resolution Cassegrain spectra.

shows a blue spectrum of HV 2241 obtained during the orbital phase of maximum separation of the spectral lines. This spectrum clearly indicates that the primary component is of spectral type O7V, but the spectral type of the secondary appears somewhat later than estimated previously, since the He II absorption lines of this component are very faint. The secondary probably is of spectral type O9-B0III. HV 2241 is a potentially very interesting binary system for a high accuracy study of stellar masses in the LMC.

Sk−66°35 This bright blue supergiant has been classified as BC1Ia by Fitzpatrick (1991). From observations obtained between 1985 and 1998 we have found Sk−66°35 to be a single-lined binary system with a period of 207 days. A preliminary radial velocity curve is shown in Figure 3.

Sk−71°34 A preliminary orbit for this O+WN type binary has been reported by Niemela (1991). A slightly different, but still preliminary, radial velocity orbit including our new data is shown in Figure 4.

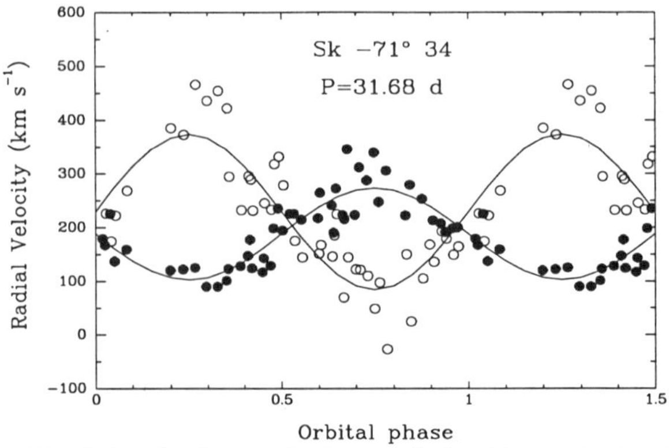

Figure 4. Radial velocity variations of the NV emission line of the WN component (open circles) and the absorption lines of the O type component (filled circles) in the spectrum of Sk−71°34.

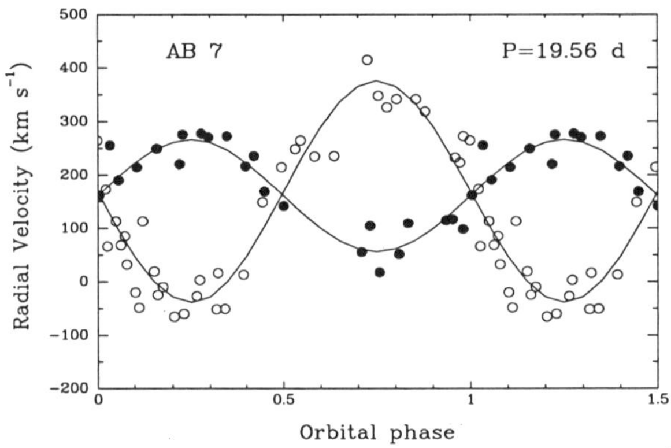

Figure 5. Radial velocity variations of the HeII λ 4686Å emission (open circles) and of the absorption lines of H and He (filled circles) in the spectrum of AB7 phased with the period of 19.56 days. The lines represent the orbital solution from Table 1.

3.2. Massive Binaries in the Small Magellanic Cloud

SMC/AB7 The star number 336a in the catalog of SMC members (Azzopardi & Vigneau 1975) is a WN type star also known as SMC/AB7 (Azzopardi & Breysacher 1979). Moffat (1988) classified this star as WN3:+O7: and found the radial velocity to be variable.

The blue optical spectrum of AB7 is dominated by the broad HeIIλ 4686Å emission line of the WN component. We have combined our radial velocities of this line with the data published by Moffat (1988), and find the most probable period of the radial velocity variations to be 19.56 days. Figure 5 shows the radial velocity orbits of both the absorption and emission lines phased with this period. The orbital elements, listed in Table 1, indicate that AB7 is a very massive binary, certainly deserving further study.

HD 5980 This eclipsing binary is the brightest WR star in the SMC. Surprisingly in 1994 HD 5980 began a LBV-like eruption (Barbá et al. 1995) transforming its spectral type from WN3 to WN6 to WN11. After the outburst, only one spectrum, of approximate type WN6, has been visible and its radial velocity variations follow those of the component which formerly was considered to have a WN3 type spectrum. An intriguing question remains: is the behaviour of HD 5980 an exception or the rule for massive binaries?

A recent discussion of the behaviour of HD 5980 can be found in e.g. Niemela et al. (1999).

4. Conclusions

In Table 1 we have summarized the most relevant information for the binaries discussed in the present paper. However, despite of years of observations, many of the parameters in this Table, especially the data for the longer period binaries, are still preliminary values. Information for the orbital inclinations is only available for the first two stars in Table 1. We note in particular that the mass ratios of O+O and O+WR binaries appear to be quite similar.

Table 1. Orbital Parameters for Massive Binaries in the Magellanic Clouds.

Star	Period (days)	Spectrum	e	K_1 km s^{-1}	K_2 km s^{-1}	$M_1 sin^3 i$ $M_\odot$	$M_2 sin^3 i$ $M_\odot$
Sk-67°105	3.301	O4f+O6:	0.0	224	355	41	26
HV 2241	4.342	O7+B0III:	0.0	162	322	34	17
Sk-66°35	207	BC1Ia	0.34	58	..	..	..
Sk-71°34	31.68	O7+WN3	0.0:	85	145	25	15
AB7	19.56	O6+WN3:	0.0:	105	207	41	21
HD 5980[1]	19.26	WN3:+WN4:	0.28	128	226	50	28

[1]The values for HD 5980 are for data before the LBV eruption (cf Niemela et al. 1997)

Acknowledgments. We are grateful to the directors and staff of CTIO and CASLEO for the use of their facilities over many years of observations. This observing program is a collaboration between astronomers at La Plata Observatory, including R. Barbá, G. Bosch, P. Ostrov, S. Giménez. We thank F. Bareilles and R. Gamen for "helping hands" with the figures.

References

Azzopardi, M., & Breysacher, J. 1979, A&A, 75, 120

Azzopardi, M., & Vigneau, J. 1975, A&AS, 22, 285

Barbá, R., Niemela, V. S., Baume, G., & Vazquez, R. A. 1995, ApJ, 446, L23

Fitzpatrick, E. 1991, PASP, 103, 1123

Haefner, R., Simon, K. P., & Fiedler, A. 1994, A&A, 288, L9

Moffat, A. F. J. 1988, ApJ, 330, 766

Niemela, V. S. 1991, in *Wof-Rayet Stars and Interrelations with Other Massive Stars in Galaxies*, K. A. van der Hucht & B. Hidayat, Dordrecht: Kluwer, 201

Niemela, V. S., Barbá, R. H., Morrell, N. I., & Corti, M. 1997, ASP.Conf Ser., 120, 222

Niemela, V. S., Morrell, N. I., & Barbá, R. H. 1999, New Astronomy, in press

Niemela, V. S., & Bassino, L. P. 1994, ApJ, 437, 332

Niemela, V. S., & Morrell, N. I. 1986, ApJ, 310, 715

Pritchard, J. D., Tobin, W., Clark, M., & Guinan, E.F. 1998, MNRAS, 297, 278

Discussion

Phil Massey: In our PSF photometry of NGC 346 (Massey, Parker, & Garmany 1989, AJ, 98, 1305) we found that HD 5980 was better fit by two stars rather than one, with a separation of a few tenths of an arcsec. Now, I've never completely believed this result - PSF fitting is not the technique of choice for separating unresolved doubles - but I wonder if the absorption lines in HD 5980 might not be due to a distant third companion. I suppose that a couple of WFPC observations obtained a decade or so apart could establish or refute the existence of such a companion.

Walborn: There is no optical companion to HD 5980 in a WFPC2 image down to 3 mag fainter. Of course, the resolution of 0.1 arcsec corresponds to 7000 AU at 70 kpc, so there is plenty of room for unresolved companions.

New Views of the Magellanic Clouds
IAU Symposium, Vol. 190, 1999
Y.-H. Chu, N.B. Suntzeff, J.E. Hesser, and D.A. Bohlender, eds.

ROSAT Survey of X-ray Sources in the Magellanic Clouds

F. Haberl, W. Pietsch and M. Filipovic[1]

Max-Planck-Institut für extraterrestrische Physik, Postfach 1603, D-85740 Garching, Germany

Abstract. More than 250 pointed observations of the Magellanic Clouds were performed with the ROSAT PSPC during its nominal operation phase. From an analysis of the images we obtained a list of about 750 point-like sources in the LMC and more than 500 sources in the SMC region. In combination with the ROSAT all-sky survey this large amount of data offers the unique opportunity to study the detailed physical properties of individual X-ray sources as well as the global statistical properties of different types of X-ray sources of a small galaxy as a whole. We report on the first results obtained from cross-correlations of our PSPC source catalogues with catalogues from other wavelength bands. Proper selection criteria, based on comparisons with known sources, permits the identification of new candidates for various source classes in the Magellanic Clouds such as supersoft sources, X-ray binaries and supernova remnants, and makes it possible to discriminate them from foreground stars and background objects.

1. Introduction

The Magellanic Clouds are among the very few galaxies sufficiently close to allow the major fraction of their X-ray sources to be resolved by imaging instruments. The first soft X-ray surveys performed with the Einstein satellite (LMC: Long et al. 1981, Wang et al. 1991; SMC: Wang & Wu 1992) revealed more than 100 discrete X-ray sources in the LMC and about 70 in the SMC regions. Due to the vicinity to the south ecliptic pole the LMC was covered with high sensitivity during the ROSAT (Trümper 1983) all-sky survey. This revealed more than 500 point(-like) sources and large-scale diffuse emission in a $13.5° \times 13.5°$ field centered on the LMC (Pietsch & Kahabka 1993). ROSAT pointed observations using the PSPC detector (Pfeffermann et al. 1986) have covered large parts of the Magellanic Clouds (LMC: Snowden & Petre 1994; SMC: Kahabka & Pietsch 1996). Here we present the first results of cross-correlations of X-ray source catalogues for the LMC and SMC with catalogues from other wavelength bands with the aim to derive classification schemes for the new, unidentified sources.

---

[1] Visiting from University of Western Sidney, Nepean, P.O. Box 10, Kingswood, NSW 2121, Australia

2. ROSAT PSPC catalogues of LMC and SMC

More than 250 pointed observations of the Magellanic Clouds were performed with the ROSAT PSPC during the nominal operation of the detector between June 1990 and September 1994. From the analysis of all individual PSPC images we obtained a list of about 750 point-like sources in a $10° \times 10°$ field of the sky centered on the LMC. A preliminary analysis of a $6° \times 6°$ field including the SMC yields more than 500 X-ray sources. Our catalogues include various properties of the X-ray sources (and their associated uncertainties) such as celestial coordinates, count rates in different energy bands, and hardness ratios which are useful as indicators for the spectral shape and source extent. From multiple detections of a source in different observations only the one with the best position (smallest error) entered the catalogues which were visually screened to remove spurious detections like e.g. multiple sources in diffuse emission regions or false detections near the support structure of the detector entrance window. The catalogues will be published separately.

3. First results from correlations with other catalogues

We cross-correlated the LMC PSPC catalogue with the SIMBAD database which contains catalogues from various wavelength bands from radio to optical and X-rays. For the current work we only use positional coincidence within $60''$ as identification criterion. So far we found 137 "identifications" for PSPC sources in the LMC catalogue. The majority originate from galactic foreground stars ($\sim 45\%$) and supernova remnants (SNR) in the LMC ($\sim 28\%$).

Seventeen X-ray binaries (XRBs) including five new candidates from a variability study of the PSPC sources (Haberl & Pietsch 1999, hereafter HP99) are found which are mostly high mass X-ray binaries (HMXBs). HP99 also suggest 2 new candidates for supersoft sources (SSSs) increasing the number of these systems to 9. The fraction of background objects such as AGNs and galaxies ($\sim 9\%$) is expected to increase after careful identification work since SIMBAD is biased towards stellar identifications.

In Fig. 1 the hardness ratios of LMC sources are shown, which corresponds to a colour-colour diagram. The identified sources are marked with different symbols, excluding the new, not yet confirmed candidates from HP99. Only sources with sufficiently small errors in the hardness ratios are shown. Most of the foreground stars cluster around HR1 and HR2 of 0 and X-ray point sources near this location are strong candidates for foreground stars. XRBs, SNRs and some AGNs are hard ROSAT sources and cover partly overlapping regions in the hardness ratio diagram. Additional information is required for their classification.

Fig. 2 presents source extent and its likelihood. The class of SNRs is separated from the majority of sources. However, finding point sources with extent likelihood up to ~ 50 shows that a source extent can only be treated as secure for a likelihood above that value. Point sources with very high statistics appear extended (lower left part) when deviations of the real point-spread-function to the assumed Gaussian shape become highly significant. Sources detected in the outer part of the detector are found to have artificial extent, which increases

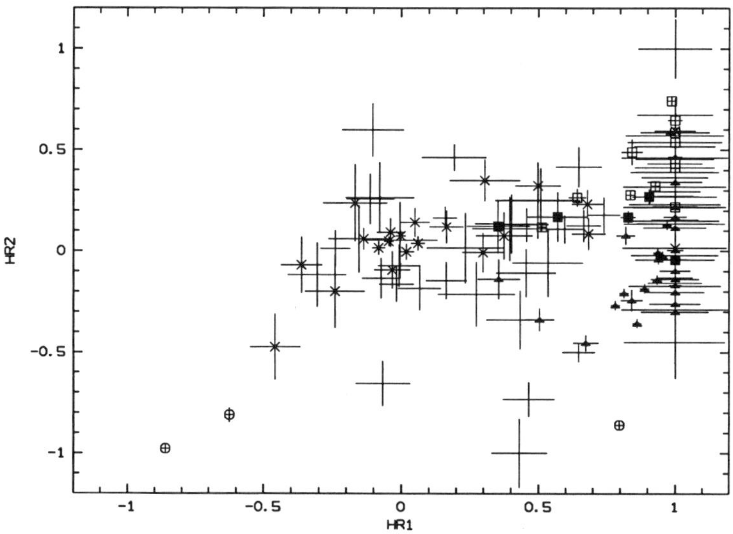

Figure 1. The location of different source classes in the hardness ratio diagram. Softest sources are found in the lower left part of the figure while hardest sources have HR1 and HR2 near +1. Plotted are only LMC sources that have 1σ errors on HR1 and HR2 < 0.2. Identified sources are marked with different symbols: X-ray binaries - open squares, SNRs - filled triangles, foreground stars - $\times$, SSS - o, and AGN - filled squares.

with off-axis angle and are not shown in Fig. 2. One new source with very likely extent is found. The hardness ratios indicate a somewhat harder spectrum than the average SNR but are still compatible with an SNR classification. The position of the source is 6.7″ from an entry in the Hubble GSC which is also flagged as extended. Follow-up studies are required to distinguish between an SNR or a possible background object (galaxy cluster?).

4. Discussion

Correlations of our ROSAT PSPC source catalogues of the Magellanic Clouds with SIMBAD yield first identifications with known sources from various classes like XRBs, SNRs, SSSs, AGNs and foreground stars. These samples permit a definition of selection criteria for the classification of previously unknown X-ray sources. Using these criteria together with time variability studies of the PSPC sources in the LMC, HP99 find new candidates for X-ray binaries, SSSs and foreground stars.

The spatial distribution of foreground stars in the LMC region is uniform across the observed field. A significant concentration of evolutionary young HMXBs is found in the area of the supergiant shell LMC4 (HP99) while only one known HMXB is located in the bar region. In contrast supersoft sources are

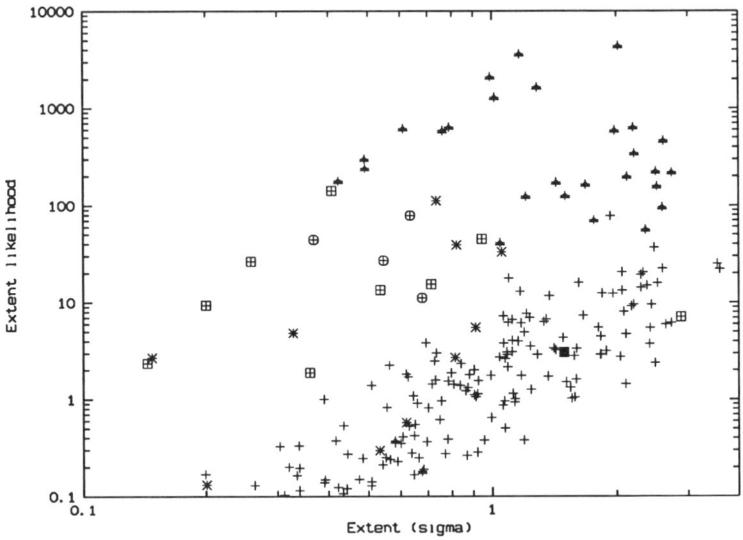

Figure 2. Source extent and extent-likelihood for LMC sources with off-axis angles less than 18′. Identified sources are marked as in Fig. 1. The extent is given in units of 15″.

so far only detected around the bar of the LMC. Due to the higher absorption the bar itself hides SSSs (as it is observed in M31, Supper et al. 1997). This is consistent with the age structure of the LMC derived from stellar formation history studies (Grebel & Brandner 1999).

Acknowledgments. The ROSAT project is supported by the German Bundesministerium für Bildung, Wissenschaft, Forschung und Technologie (BMBF/DLR) and the Max-Planck-Gesellschaft. This research made extensive use of the SIMBAD database operated at CDS, Strasbourg, France.

References

Grebel, E.K., & Brandner, W. 1999, this volume

Haberl, F., & Pietsch, W. 1999, A&A, 344, 521 (HP99)

Kahabka, P., & Pietsch, W. 1996, A&A, 312, 919

Long, K.S., Helfand, D.J., & Grabelsky, D.A. 1981, ApJ, 248, 925

Pfeffermann, E., Briel, U.G., Hippmann, H., et al. 1986, Proc. SPIE 733, 519

Pietsch, W., & Kahabka, P. 1993, in *Lecture Notes in Physics 416: New Aspects of Magellanic Cloud Research*, eds. B. Baschek, G. Klare, J. Lequeux, 59

Snowden, S.L., & Petre, R. 1994, ApJ, 436, L123

Supper, R., Hasinger, G., Pietsch, W., et al. 1997, A&A, 317, 328

Wang, Q., Hamilton, T., Helfand, D.J., & Wu, X. 1991, ApJ, 374, 475

Wang, Q., & Wu, X. 1992, ApJS, 78, 391

Discussion

Tetsuo Hasegawa: Can the lack of binary sources in the region near SN 1987A be due to confusion by diffuse emission?

Haberl: Very unlikely. The diffuse emission only increases the background. The hard X-ray binaries should easily be detected even if embedded inside the gas (HMXBs are observed throughout our own Galaxy).

Yuri Efremov: If I properly understood that there is a concentration of X-ray binaries just in the area of the supershell LMC4?

Haberl: Yes.

Daniel Wang: Could you elaborate a bit more on how you distinguish X-ray binaries from background AGNs? A comment on background QSOs suitable for absorption studies: from our optical identifications of X-ray sources for other nearby galaxy fields, we find that typically a couple of QSOs brighter than V $\sim$ 18 m are present. So you may expect about a dozen of such objects in the entire field of the LMC.

Haberl: Hardness ratios near +1 and high variability favour the selection of X-ray binaries.

Dominik Bomans: How many of the X-ray binaries are actually inside the supergiant shell LMC4 and how many are outside the Hα/HI boundary?

Haberl: From the 9 high mass X-ray binaries (including 4 new candidates in the LMC4 area) 3 are outside the Hα/HI boundary.

Hans Zinnecker: How many new quasars has ROSAT found behind the LMC/SMC? Log N - Log S statistics would predict lots of them (Hasinger et al. 1993, AA 275, 1).

Haberl: We so far have only done correlations with existing catalogues of various kinds (e.g., SIMBAD). This yields 11 positional coincidences from our LMC catalogue with background objects. No attempt was made to investigate these in more detail, nor have we looked for further candidates due to the mentioned difficulties to select them.

New Views of the Magellanic Clouds
IAU Symposium, Vol. 190, 1999
Y.-H. Chu, N.B. Suntzeff, J.E. Hesser, and D.A. Bohlender, eds.

Massive Star Evolution

N. Langer

Institut für Physik, Universität Potsdam
Postfach 601553, D–14415 Potsdam, Germany

A. Heger[1]

Max-Planck-Institut für Astrophysik
Postfach 1523, D–85740 Garching, Germany

Abstract. The evolution of massive stars is far from being fully understood, as we outline by pointing to a number of open problems related to massive stars in the Magellanic Clouds. We argue that rotation may be a key ingredient in the physics of massive stars. We report on recent results obtained including rotation, and their relevance to these remaining questions.

1. Introduction: Current Problems in Massive Star Evolution

The quantitative theory of stellar structure is more than 100 years old (Emden 1907). Our understanding of the stellar interior has improved dramatically since then, especially so since the 1950's when it became possible to construct detailed stellar models with the help of computers. Although there are still many fundamental questions left in the field of low- and intermediate mass stars, we can say that their evolutionary path — i.e., their route from the main sequence through the red giant and asymptotic giant branch to the white dwarf stage — has been known in detail for several decades (see, for example, Iben 1974). In addition, the principles of the evolution of massive stars' cores through the major nuclear fusion stages from hydrogen to silicon burning have also been known for a similar length of time (Weaver et al. 1978).

However, even today our understanding of the evolution of the observable properties of massive stars ($M_{\rm ZAMS} \gtrsim 8\ {\rm M}_\odot, \log L/{\rm L}_\odot \gtrsim 4$) is still rudimentary, as demonstrated by the following six problems related to massive stars in the Magellanic Clouds, a list which is far from complete.

1. Where is the post-main sequence gap in the distribution of massive stars in the HR diagram of the LMC (Blaha & Humphreys 1989; Fitzpatrick & Garmany 1990)? It is predicted by *all* stellar evolution calculations (Fig. 9 in Langer 1991) but there is no trace of it in the data. Can this gap be filled by secondary stars in massive close binary systems (Fig.1)?

[1]present address: UCO/Lick Observatory, UC Santa Cruz, CA 95064, USA

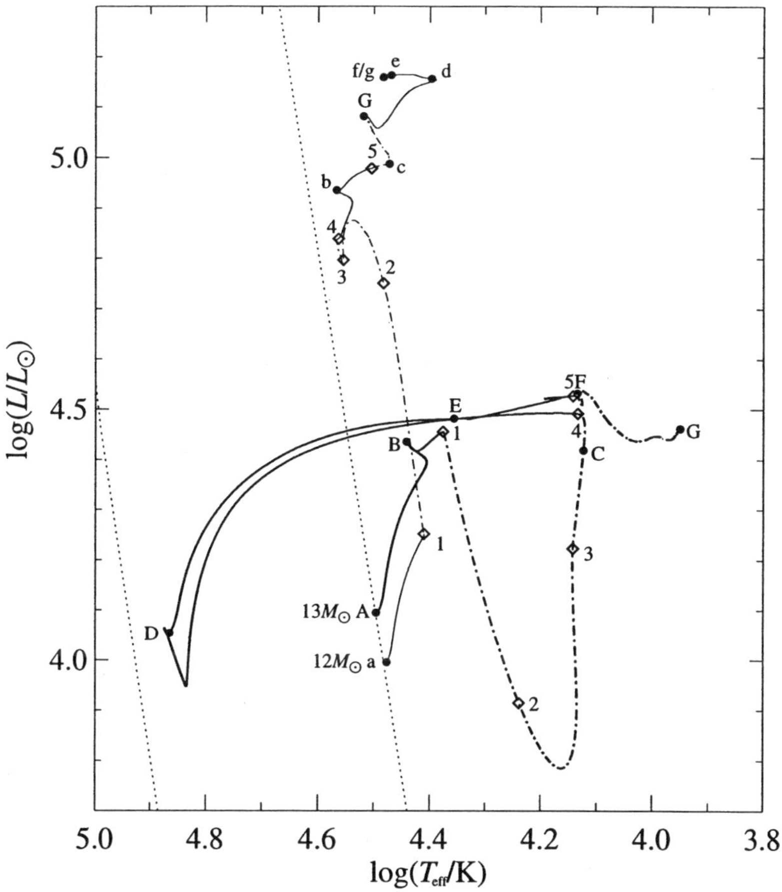

Figure 1. Evolutionary tracks in the HR diagram of the components of a 13+12 $M_\odot$ case B close binary system with a metallicity of $Z_\odot/4$ and an initial period of 3.1 days. The path of the primary component (initial mass 13 $M_\odot$) is marked by the thick line and upper case letters, that of the secondary by the thin line and lower case letters. Mass transfer stages correspond to the dot-dashed parts of the lines. The thin dotted lines designate the zero age main sequence and the location of pure helium stars (helium main sequence). The letters designate beginning and end of nuclear burning stages, i.e., core hydrogen burning (a/A – b/B), core helium burning (c/C – d/D), core carbon burning (e/E – f/F), and neon ignition (g/G). Numbers designate mass transfer events for both stars. 1: beginning of Case B mass transfer, 2: maximum of mass transfer rate, 3: start of slow phase of Case B mass transfer, 4: end of Case B mass transfer, 5: start of Case BB mass transfer. The final masses of the primary and secondary are 1.6 $M_\odot$ and 22 $M_\odot$, respectively. Note that this kind of binary evolution may be relevant to problems 1) and 3) mentioned in Section 1. See Braun & Langer (1999) for more details.

2. The observed trend of the blue-to-red supergiant ratio as function of the metallicity is opposite to the one predicted by any stellar evolution models (Langer & Maeder 1995).

3. When in 1987 a *blue* supergiant exploded as a supernova in the Large Magellanic Cloud (Arnett et al. 1989) it was a surprise to the community. The reason for the blue color of the progenitor star is still not unambiguously explained (see Langer et al. 1989, Woosley et al. 1998, and see Fig. 1).

4. What is the origin of the B[e] supergiants primarily found in the Magellanic Clouds (Gummersbach et al. 1995), i.e., how can evolved supergiants have disks or disk winds around them (Langer & Heger 1998)?

5. What is the evolutionary status of the WNL/Ofpe stars which again are primarily found in the Magellanic Clouds (Pasquali et al. 1997; Crowther & Bohannan 1997)?

6. Does the upper stellar mass limit depend on metallicity? While extremely massive stars ($\sim 200\,\mathrm{M}_\odot$) are found in the LMC (Kudritzki et al. 1996), recent investigations of stars near the Galactic Center (Figer et al. 1998) indicate the presence of very massive stars also in high metallicity environments.

The uncertainty in massive star evolution is dominated by the lack of understanding of two kinds of processes: the physics of mass loss from the stellar surface, and the physics of internal mixing processes (Langer 1994). In the present paper, we want to outline that both processes, despite acting in quite different locations in massive stars, may be strongly influenced by a process which has mostly been neglected in past studies of massive star evolution: rotation.

2. Massive Star Evolution with Rotation

Massive stars are rapid rotators. Equatorial rotation velocities span the range $v_{\mathrm{rot}} = 100 - 400\,\mathrm{km\,s^{-1}}$, with B stars rotating closest to their break-up speed v_{crit}, and values of $v_{\mathrm{rot}}/v_{\mathrm{crit}} = \omega_{\mathrm{rot}}/\omega_{\mathrm{crit}} = 0.5$ frequently observed for O stars (e.g., Howarth et al. 1997). During the last decade, many observations have revealed unusual surface abundances that may require additional internal mixing (beyond that of plain convection) for their explanation (Langer et al. 1998, and references therein).

Fig. 2 gives an overview on the evolution of some surface properties of massive star models which include mixing due to rotationally induced instabilities and the effect of the centrifugal force. For the most massive models with the fastest rotation, one can even obtain a transition to Wolf-Rayet (WNL) type abundances (see also Langer 1992), which may be relevant for understanding the WNL/Ofpe stars (Pasquali et al. 1997). Furthermore, for a distribution of initial rotation rates, one obtains, at a given initial mass, a spread in main sequence luminosities and effective temperatures (Fig. 3). This might reduce the problem of the missing post-main sequence gap mentioned above, but seems not to solve it completely.

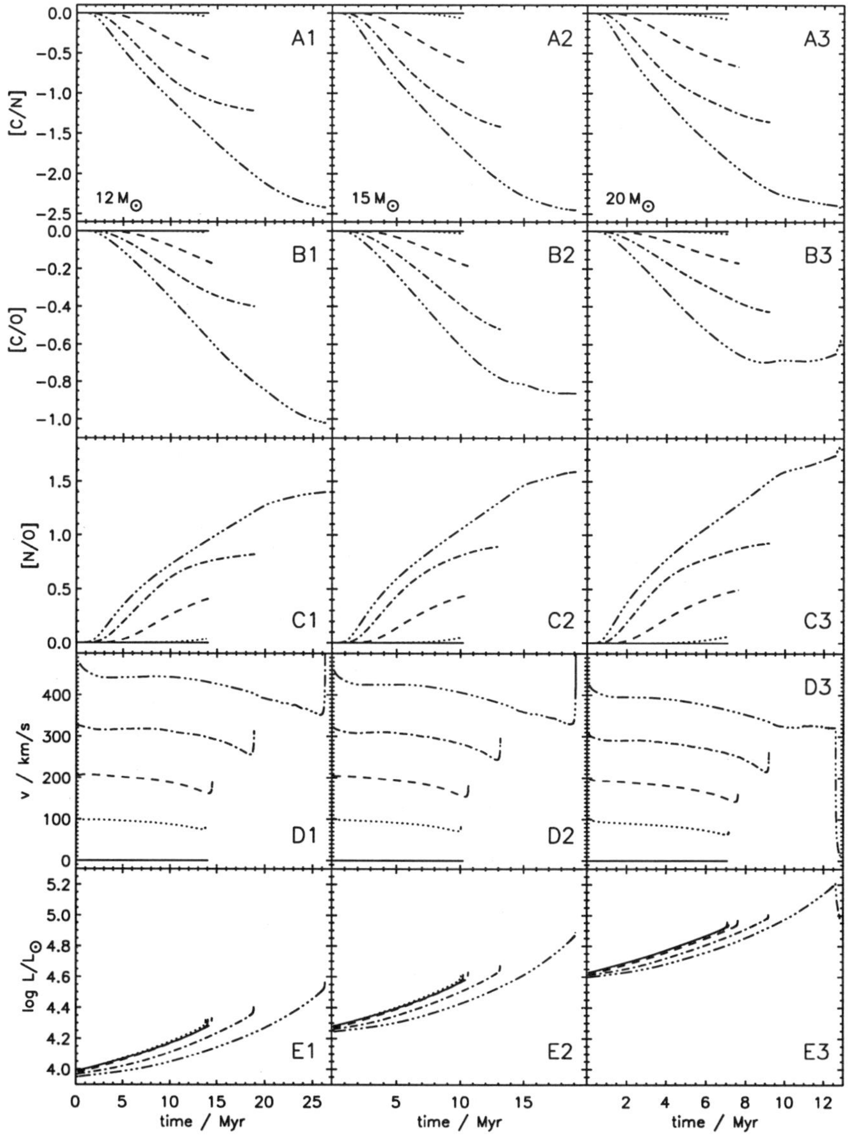

Figure 2. Evolution of surface properties of stellar models with rotation for initial masses of $12\,M_\odot$ (left panel), $15\,M_\odot$ (middle panel), and $20\,M_\odot$ (right panel) and various initial rotation rates (no rotation: solid lines, about 100, 200, 300, 450 $km\,s^{-1}$: dotted, dashed, dash-dotted and dash-triple-dotted lines, respectively) as function of time throughout core hydrogen burning. Shown are the the logarithmic increment of surface abundance ratios with respect to the solar ratio, the equatorial rotational velocity, and the stellar luminosity (see Heger 1998, and Heger et al. 1999 for more details).

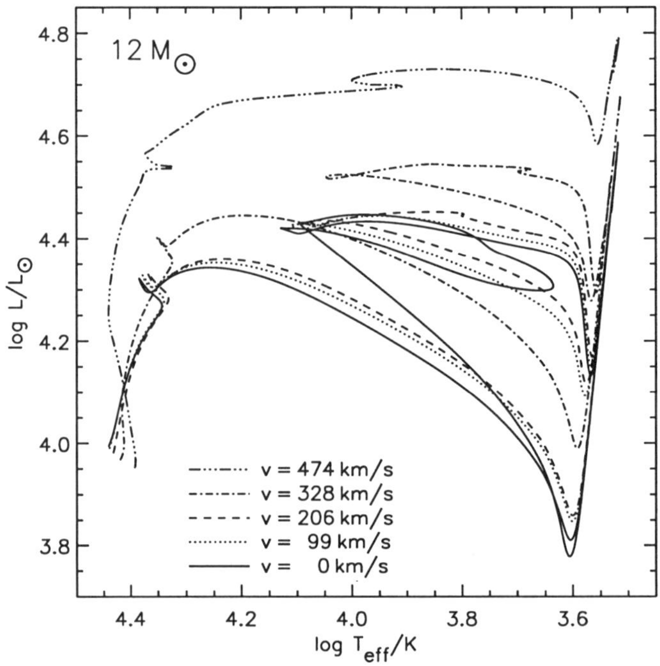

Figure 3. Evolutionary tracks in the HR diagram from the zero age main sequence to the pre-supernova position, for $12\,M_\odot$ stars with initial equatorial rotational velocities of $0\,km\,s^{-1}$, $99\,km\,s^{-1}$ (dotted line), $206\,km\,s^{-1}$ (dashed line), $328\,km\,s^{-1}$ (dash-dotted line), and $474\,km\,s^{-1}$ (dash-triple-dotted line). See Heger (1998) and Heger et al. (1999) for more details.

Rotation also increases the stellar mass loss rate (Langer 1998), which appears to be required by observations (Meynet et al. 1994). The anisotropy of the wind of rotating massive stars (Bjorkman & Cassinelli 1993) may also be related to the B[e] phenomenon (Langer & Heger 1998), to the bipolarity of LBV nebulae (Langer et al. 1999a), and perhaps to the rings around SN 1987A (Heger & Langer 1998).

In summary, rotation has the potential to assess most of the problems mentioned in Section 1 (see also Talon et al. 1997; Maeder 1999). It certainly needs to be considered in the context of star formation, which is the process determining the initial angular momentum in a star (possibly in relation with problem 6; Figer et al. 1998) and its distribution function. Moreover, in addition to the effects outlined here, rotation has further implications for nucleosynthesis (Langer et al. 1999b), supernova explosions, pulsar rotation (Langer et al. 1999c; Heger et al. 1999), and even γ-ray bursters (MacFadyen & Woosley 1999)

Acknowledgments. The authors are very grateful to André Maeder, Stan Owocki, Regina Schulte-Ladbeck, Kim Venn and Stan Woosley for enlightening discussions. This work has been supported by the Deutsche Forschungsgemeinschaft through grants No. La 587/15-1 and La 587/16-1.

References

Arnett, W.D., Bahcall, J.N., Kirshner, R.P., & Woosley, S.E. 1989, ARA&A, 27, 629

Bjorkman, J.E., & Cassinelli, J.P. 1993, ApJ, 409, 429

Blaha, C., & Humphreys, R.M. 1989, AJ, 98, 1598

Braun, H., & Langer, N. 1999, in preparation for A&A

Crowther, P.A., & Bohannan, B. 1997, A&A, 317, 532

Emden, V.R. 1907, *Gaskugeln*, Leipzig

Figer, D.F., Najarro, F., Morris, M., McLean, I.S., Geballe, T.R., Ghez, A.M., & Langer, N. 1998, ApJ, 506, 384

Fitzpatrick, E.L., & Garmany, C.D. 1990, ApJ, 363, 119

Gummersbach, C.A., Zickgraf, F.-J., & Wolf, B. 1995, A&A, 302, 409

Heger, A. 1998, PhD Thesis, TU München

Heger, A., & Langer, N. 1998, A&A, 334, 210

Heger, A., Langer, N., & Woosley, S.E. 1999, ApJ, in preparation

Howarth, I.D., Siebert, K.W., Hussain, G.A.J., & Prinja, R.K. 1997, MNRAS, 284, 265

Iben, I. Jr. 1974, ARA&A, 12, 215

Kudritzki, R.P., Lennon, D.J., Haser, S.M., Puls, J., Pauldrach, A., Venn, K., & Voels, S.A. 1996, in *Science with the HST*, STScI, P. Benvenuti et al., eds.

Langer, N. 1991, A&A, 252, 669

Langer, N. 1992, A&A, 265, L17

Langer, N. 1994, in Proc. 34th Herstmonceux Conference on *Circumstellar Media in the Late Stages of Stellar Evolution*, R. Clegg et al., ed., Cambridge University Press, p. 1

Langer, N. 1998, A&A, 329, 551

Langer, N., & Maeder, A. 1995, A&A, 295, 685

Langer, N., & Heger, A. 1998, in proc. *Workshop on B[e] stars*, Paris 1997, C. Jaschek, A. M. Hubert, (eds.), Kluwer, 235

Langer, N., El Eid, M.F., & Baraffe, I. 1989, A&A, 224, L17

Langer, N., Heger, A., & García-Segura, G. 1998, RvMA, 11, 57

Langer, N., García-Segura, G., & Mac Low, M.-M. 1999a, ApJ, submitted

Langer, N., Heger, A., Woosley, S.E., & Herwig, F. 1999b, in *Nuclei in the Cosmos V*, N. Prantzos, ed., Édition Frontières, Paris, in press

Langer, N., Heger, A., & Woosley, S.E. 1999c, in proc. IAU-Symp. 193 on *Wolf-Rayet Phenomena in Massive Stars and Starburst Galaxies*, K.A. van der Hucht et al., eds., ASP Conf. Ser., San Francisco, in press

MacFadyen A., & Woosley, S.E. 1999, ApJ, submitted

Maeder A. 1999, in proc. IAU-Symp. 193 on *Wolf-Rayet Phenomena in Massive Stars and Starburst Galaxies*, K.A. van der Hucht et al., eds., ASP Conf. Ser., San Francisco, in press

Meynet, G., Maeder, A., Schaller, G., Schaerer, D., & Charbonnel, C. 1994, A&AS, 103, 97

Pasquali, A., Langer, N., Schmutz, W., Leitherer, C., Nota, A., Hubeny, I., Moffat, A.F.J., Drissen, L., & Robert, C. 1997, ApJ, 478, 340

Talon, S., Zahn, J.P., Maeder, A., & Meynet, G. 1997, A&A, 322, 209

Weaver, T.A., Zimmerman, G.B., & Woosley, S.E. 1978, ApJ, 225, 1021

Woosley, S.E., Heger, A., Weaver, T.A., & Langer, N. 1999, in *SN 1987A: Ten years after*, Eds. M.M. Phillips, N.B. Suntzeff, PASP, in press

Discussion

Nolan Walborn: In answer to a question during the talk, the two O3 stars on your 200 $M_\odot$ track are in associations. Sk−67°211 = HDE 269810 is next to a small HII region, and Mk 42 is in the 30 Dor central cluster.

Paul Crowther has found an exact spectroscopic counterpart of the LMC Ofpe/WN9 stars in the Galaxy; it has an obscure name like NS4 or some such. He has classified subgroups of this class as WN9-11 and, with Bruce Bohannan, has proposed that the galactic OIafpe stars like (and including) HD 152408 should also be classified as WN9. So there are galactic counterparts of this LMC class.

George Sonneborn: If rotation-induced mass loss can produce the rings around SN1987A, does this also place the star in the right part of the HR diagram for a single-star explosion? Does this argue for or against binary or single-star models for the 1987A progenitor?

Langer: The post-red supergiant spin-up mechanism found by Heger & Langer (1998) may certainly be relevant for shaping the inner ring; its effect on the outer rings still needs to be investigated. Since the spin-up occurs for all kinds of evolution from the Hayashi-line to hotter surface temperatures, it can not be used to discriminate between single star progenitor models and models invoking binary effects.

Philip Massey: Three comments, all concerning incompleteness:

(a) The "problem" with the main-sequence widening described by Fitzpatrick & Garmany (1990 Ap.J., 363, 119) was really dominated by the incompleteness at the hot end - not the over abundance of stars past the MS. In other words, if you expect only a tiny fraction of MS stars to be seen past the "gap" you can believe there's a problem if you greatly underestimate the number of MS stars. With more complete data, and well determined incompleteness corrections (down to 25 $M_\odot$ anyway), there doesn't seem to be any problem in the number of stars across the HRD, at least using the Geneva (Maeder's) tracks using convection overshooting (Massey et al. 1995, ApJ, 438, 188).

(b) Neither the number of blue supergiants nor the number of red supergiants are known to an order of magnitude in the Milky Way or MC's — so does the ratio really mean anything? In a recent survey for red supergiants and WR stars in NGC 6822, M31, and M33 we do see an excellent correlation in number ratio (RSG/WR) with metallicity — so maybe this is no longer a problem to solve.

(c) Finally let me note that there is one WN9/Ofpe star known in even the higher-than-solar metallicity galaxy M31; the Wolf-Rayet star OB69-WR2. Their emission is weak in terms of equivalent widths so they are hard to detect! See discussion in Massey & Johnson (1998, ApJ, 505, 793).

Langer: (a) A post-main sequence gap is predicted by all massive star models, also by the Geneva models, and it can not coincide with the observed gap between the Fitzpatrick & Garmany ledge and the red supergiant branch. Observations and theory are clearly in conflict here.

(b) The absolute numbers of blue or red supergiants are surely very uncertain. However, the relative numbers, and thus the blue/red supergiant number ratios at different metallicities, can conclusively be compared with each other (e.g., Langer & Maeder 1995, A&A, 295, 685).

Sally Oey: Your evolutionary tracks including rotation evolve at higher L. Won't this affect the evolutionary time scales and stellar winds?

Langer: Yes, it does. The life times get longer despite the increased luminosities, since the rotational mixing brings more "fuel" to the center. And the mass loss rates are considerably increased, not only due to the higher luminosities, but also due to the centrifugal force (Langer 1998, A&A, 329, 551).

New Views of the Magellanic Clouds
IAU Symposium, Vol. 190, 1999
Y.-H. Chu, N.B. Suntzeff, J.E. Hesser, and D.A. Bohlender, eds.

Massive Stars Spectroscopy in the Magellanic Clouds

K.A. Venn

Macalester College, Department of Physics & Astronomy, 1600 Grand Ave., St.Paul, MN, 55105, USA

Abstract.
Recent results from spectroscopic studies of Magellanic Cloud stars are discussed. Emphasis is placed on how to interpret CNO abundances to examine stellar evolution scenarios. Observed CNO abundances are best explained by the new rotating stellar evolution models (see Langer & Heger, these proceedings). In addition, the Rolleston et al. (these proceedings) abundances for B stars in the Magellanic Bridge are discussed.

1. Introduction

Spectroscopic observations of massive stars are used to study individual stars themselves, but also as probes of their host galaxies, and as test cases for different stellar evolution theories. Simple low (classification) resolution spectra can be used easily for important first estimates of stellar parameters. High resolution spectra and detailed model atmosphere analyses are much more labor intensive but can be used to determine more accurate stellar parameters and individual elemental abundances. To date, detailed atmospheric analyses have been performed for B-type dwarfs, giants, and supergiants, A-supergiants, F-K supergiants, and M-supergiants in the LMC and SMC. These analyses have determined abundances for elements ranging from helium and lithium, through CNO and α-elements, including Fe-group elements, and culminating in abundances for a variety of s-process and r-process elements. Although many individual elemental abundances appear to be contradictory or in poor agreement, most are in excellent agreement between the analyses, but the reader of those papers must first accept the inherent uncertainties in each analysis and must also frequently consider the *differential abundances*, e.g., abundances relative to a standard star, or the abundance ratio between two elements.

2. Spectral Types at Low-Z

The simplest way to learn more about an individual star is to study its spectral type. From this we can learn about its fundamental parameters, e.g., effective temperature, surface gravity, luminosity, and metallicity, and we can determine additional characteristics such as magnetic field strength, rotation rate, and presence of a stellar wind or circumstellar shell. The accuracy with which any one of these parameters can be determined depends critically on the comparison stars calibration scheme. Because of the lower metallicities of the Magellanic Clouds,

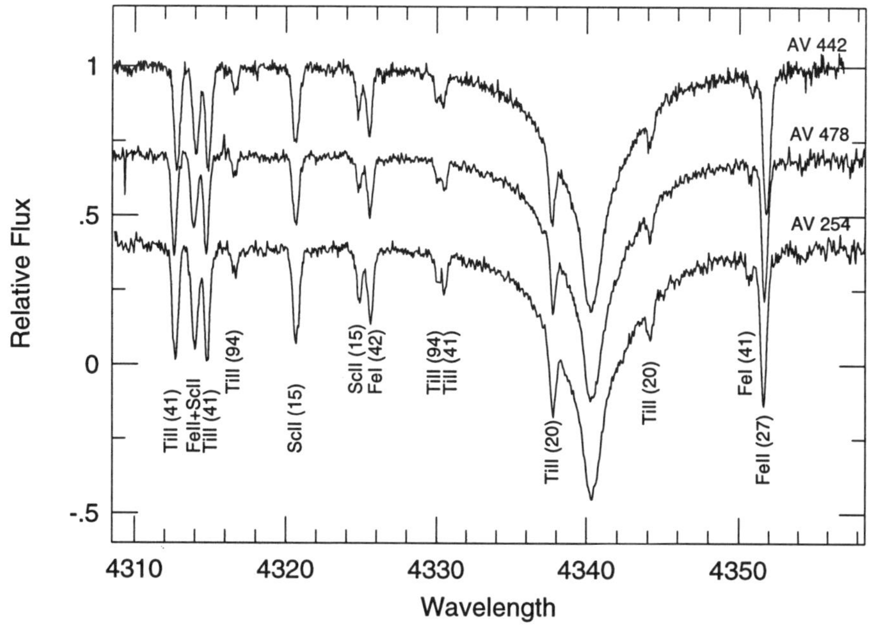

Figure 1. Spectra of three normal SMC A7 Iab supergiants near Hγ. Previously, AV 442 had been reported as a normal A3 Ia star, while the other two had been reported as "anomalous", helium-rich A3 I stars (Humphreys 1983; Humphreys et al. 1991). These claims had been partially based on a comparison of their hydrogen line profiles which are actually, as can be seen in this figure, essentially identical.

new spectral sequences based on stars in the Clouds were developed for bright B stars; in the LMC by Fitzpatrick (1991), and in the SMC by Lennon (1997). These have had two main results. Firstly, Fitzpatrick & Garmany (1990) discovered a feature, a "ledge", in the distribution of B stars in the LMC. This ledge has frequently been interpreted as a mixed population of H-core and He-core burning B-supergiants, providing evidence for blue-loop evolution scenarios for massive stars at low metallicities. Lennon also found evidence for a ledge in his refined, although sparse, SMC data set. Secondly, both Fitzpatrick and Lennon found the first evidence for BN/BC stars in the LMC and SMC, respectively. The mechanism responsible for the N and C line strength differences could be a stellar evolution effect – mixing of CNO-cycled gas into the atmosphere possibly as a previous red supergiant. This effect is thought to be metallicity sensitive, but finding these stars at three metallicities could refute that (partial mixing of gas on the main sequence is discussed below and is another, non-metallicity dependent method of achieving this phenomenon).

More recently, I have determined stellar parameters from detailed model atmosphere analyses of 10 A-type supergiants (Venn 1999) and have found that the problems in spectral typing hot stars extends to the A-types. All the A-type supergiants analyzed had been classified as too hot. In fact, some stars had been classified as anomalous — Humphreys (1983) and Humphreys et al.

(1991) had suggested that the anomalies were due to overabundances of helium in their atmospheres. These stars have identical spectra to simply cooler and less luminous stars, see Fig. 1 for an example. Misclassification of A-stars is easy to do. Line blanketing of the optical spectrum increases, such that hotter A-stars at high-Z mimic cooler A-stars at low-Z (carefully examining ionization ratios can help to avoid this problem).

3. CNO & Fe Abundances in the Clouds: Stellar and MC Evolution

Elemental abundances in the Magellanic Clouds have two main purposes: to study important metallicity effects on stellar evolution and to study the chemical history of these irregular galaxies.

With respect to stellar evolution, the main elements to be exploited are CNO. CNO are catalysts in H-burning, and any mixing between surface and interior gas results in a distinctive change in the relative ratios of C:N:O. In particular -0.15:+0.55:-0.1 are predicted by most theories due to the first dredge-up as a red supergiant star (c.f., Schaller et al. 1992). To use CNO, one must know not only the current abundance in a star, but also the initial abundance that the star had before mixing began. The latter is typically assumed to be the same as the present surface abundances in main-sequence B stars, which have undergone no mixing, and the present abundances in the H II regions. Clearly then, the B stars and H II regions are expected to have similar abundances, which is a reasonable assumption given the short lifetimes of B stars.

In the Galaxy, the B stars in the Orion nebula yield very similar results to the nebular abundance data (see Cunha & Lambert 1994). These abundances for CNO are not the same as solar, however, where the Orion abundances are approximately 0.2 dex less than in the Sun. In the Magellanic Clouds, the lower overall metallicity means that CNO are not similar to Orion or solar abundances (of course), but additionally since each element has a different main nucleosynthetic site then *C and N and O need not have the same underabundance as Fe nor as each other.*

Note: Helium and LiBeB would also be affected by mixing, but have not yet been quantitatively examined because of various difficulties with either non-LTE model atmosphere and line formation effects or because the initial abundances are not well known (e.g., the unmixed Li abundance).

3.1. Present Day CNO in the SMC

Initial CNO abundances in the SMC can be determined by examining abundances from three main-sequence B stars (Rolleston et al. 1993) and from several H II regions (Dufour 1984; Russell & Dopita 1990; Garnett et al. 1995). Adopting [Fe/H]=-0.7 for the SMC, these analysis suggest the best pristine present-day CNO abundances are [C/Fe] ~ -0.5, [N/Fe] ~ -0.8, and [O/Fe] ~ -0.2. (Recall that solar abundances are not similar to Orion abundances, where solar abundances appear $\sim$0.2 dex larger).

CNO abundances have been determined as a dataset for three types of evolved stars; B giants (Jüttner et al. 1992; Lennon et al. 1996; Korn & Wolf in these proceedings), A supergiants (Venn 1999), and F-K supergiants (Luck

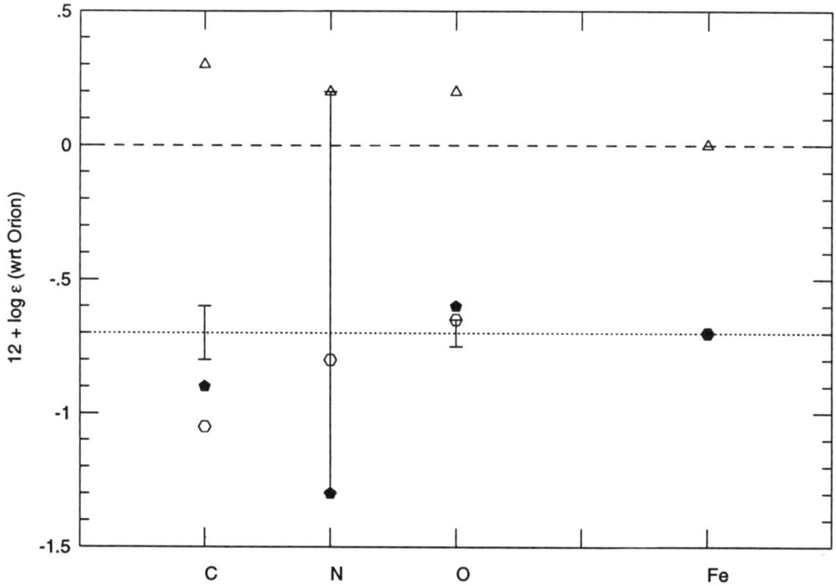

Figure 2. SMC abundances for CNO – present-day (*filled pentagons*), first-dredge-up predictions (*hollow six-sided symbols*), and the range in the abundances determined from evolved stars (see text for references). The initial abundances are dissimilar from the Fe underabundance (*dotted line*). Note also that solar abundances are dissimilar to those from Orion B star and nebular analyses which are used here at the best reference abundances (*dashed line*).

& Lambert 1992; Hill et al. 1997; Luck et al. 1998). These analyses have found [C/Fe] ~ -0.2 to -0.4, [N/Fe] ~ -1.5 to 0.0, and [O/Fe] ~ -0.2.

The first dredge-up predictions are essentially metallicity independent and are shown in Fig. 2 (relative to the present-day CNO abundances). Surprisingly, nitrogen has been so slow in its interstellar enrichment in the SMC that [N/Fe] $\sim$ 0 *after the first dredge-up and not before, as has often been assumed in the past.*

Comparing the CNO abundances in evolved stars to the present-day abundances is still confusing though. The CNO abundances do not reflect the initial abundances nor the first dredge-up abundances. Examining N specifically shows that N ranges from the initial (nebular-like) abundance through the first dredge-up enrichment and on to ~2x the predicted enrichment! This range in the N abundances is much larger than the predicted uncertainties in the analyses (even the expected non-LTE effects). [The high carbon abundance observed in the evolved stars, opposite to the predictions of the first dredge-up, may reflect uncertainties in the initial, nebular, abundance adopted, possibly due to dust grain formation].

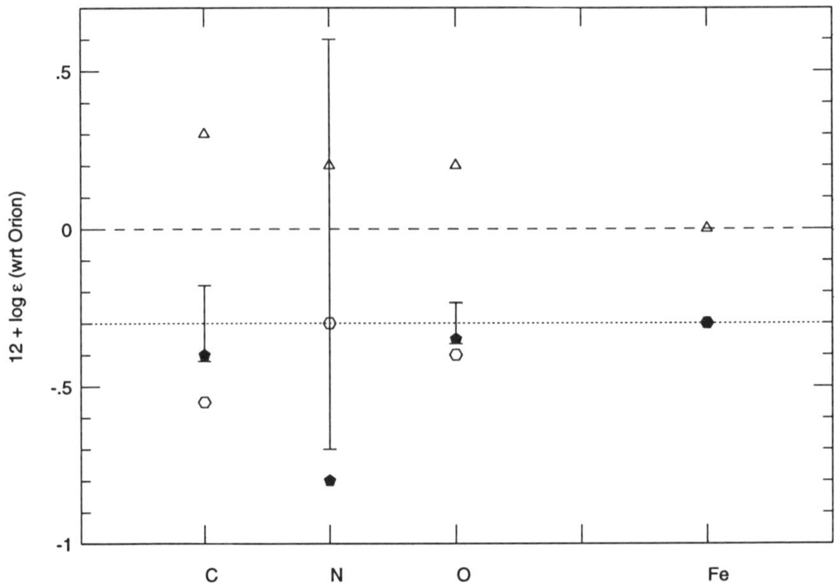

Figure 3. LMC abundances for CNO – symbols same as in Fig. 2.

3.2. Present Day CNO in the LMC

The present-day CNO abundances in the LMC can be estimated from the analyses of several H II regions (Dufour 1984; Russell & Dopita 1990; Garnett et al. 1995) and from 2 B dwarfs (Rolleston et al. 1996; Korn & Wolf in these proceedings). Again, CNO need not have the same LMC underabundance as Fe, although the underabundances of O and Fe are about the same relative to the Orion nebular and stellar abundances, i.e., [O/Fe] ~ -0.2. C and N are more underabundant than Fe, with [C/Fe] ~ -0.4 and [N/Fe] ~ -0.6. These present-day abundances are shown in Fig. 3.

The first dredge-up predictions are essentially metallicity independent and are also shown in Fig. 3. Again, as seen for the SMC, [N/Fe] ~ 0 *after the first dredge-up and not before, as has often been assumed in the past.*

CNO have been determined as a complete dataset for only a few evolved stars in the LMC; for B giants (Korn & Wolf in these proceedings) and F-G supergiants (Luck & Lambert 1992; Hill et al. 1995). In general, the CNO pattern results are similar to the SMC. These evolved stars show [C/Fe] ~ -0.2 to -0.4, [N/Fe] ~ -0.6 to $+0.7$, and [O/Fe] ~ -0.2. Thus, oxygen is similar to the present-day nebular abundance, carbon ranges from the nebular to the predicted first dredge-up abundance, and nitrogen ranges wildly from the nebular abundance to 2x the first-dredge-up prediction. Again, the range in the nitrogen abundances in these analyses are far larger than the simple model predictions.

3.3. Importance of Rotation on Stellar Evolution

The NLTE CNO abundances from evolved stars in the Magellanic Clouds prove beyond a doubt that the surfaces of these stars have undergone mixing with CNO-cycled gas from their interior. The new question raised by these results is when and how did this happen. Classical stellar evolution models predict first dredge-up mixing as a red supergiant, however the *range* in the CNO abundances do not support this alone. The range in N also shows that star-to-star variations are large – some stars showing extreme mixing and others showing little or none.

Only the newest rotating models predict significant star-to-star variations, e.g., the range in N reflects the range in rotational velocities of these massive stars and thus their success in rotational mixing on or near the main-sequence (c.f., Langer & Heger in this conference proceedings, and references therein). Whether rotational mixing happens alone or in addition to the first dredge-up remains unclear. It is probably the latter because (1) some of the enrichments of N are incredibly large (too large for just one mechanism?), and (2) most stars appear enriched to some degree in the Magellanic Clouds, which is *not* what is seen in the Galaxy (most evolved massive stars do not show strong N enrichments, c.f., Venn 1995 and references therein).

Rotation can have a huge effect on the evolution of massive stars. Theoretical tracks are used to get stellar masses & ages, and the evolution scenario is used to predict the chemical yields from stars. Also the theoretical distribution of massive stars on the HRD is used in galaxy population syntheses. To see an example of rotational effects on the evolutionary track of a 10 $M_\odot$ star, see Fig. 2 in Fliegner et al. (1996). Also, see Langer & Heger (these proceedings) for more details and effects of rotation on stellar evolution.

4. Stars in the Magellanic Bridge

The Magellanic Bridge is a ridge of H I gas between the LMC and SMC, in which evidence for on-going star formation has been found, e.g., from color-magnitude diagrams of associations of young stars (e.g., Irwin et al. 1990). Elemental abundances of normal main-sequence B stars in the Bridge have been performed (Hambly et al. 1994; Rolleston et al., in these proceedings). The abundance results suggest that C, O, Mg and Si in the Bridge B stars are less than Galactic B stars by $\sim$ 1.0 dex. This is a larger underabundance than has been found for SMC B stars of $\sim$ 0.7 dex (Rolleston et al. 1993).

How or why gas in the Magellanic Bridge could be more metal-poor than that of the SMC is a mystery, and currently not predicted from the dynamical models of the Bridge. Enrichment in the SMC to the present-day abundances is thought to have occurred $\sim$2 Gyr ago (e.g., Pagel & Tautvaišienė 1998), and yet the Bridge is thought to be gas pulled out of the SMC $\sim$0.2 Gyr ago (e.g., see Gardiner or Kunkel, these proceedings).

The abundance determinations in the Bridge B stars are very difficult because the stars analyzed so far have essentially no spectral lines for a standard detailed model atmospheres analysis (e.g., where spectral lines from different ionization stages of He or Si are used to constrain atmospheric parameters). This alone is significantly different from SMC B stars analyzed, and qualitatively suggests these stars are truly more metal-poor. Future analysis of additional stars, particularly cooler stars which should have more spectral lines, will provide better abundances of a variety of elements, and thus allow us to further examine the chemical and dynamical formation of the Bridge.

Acknowledgments. Many thanks to Danny Lennon, Robert Rolleston, Norbert Langer, and Michael Lemke for many stimulating and helpful discussions on the Magellanic Clouds and massive star abundance analyses. Much of this work has been carried out at the Universität Sternwarte München thanks to Rolf Kudritzki. Also, thanks to Clare and Evan Skillman for helping me to enjoy Vancouver Island.

References

Cunha, K.C., & Lambert, D.L. 1994, ApJ, 426, 170

Dufour R.J. 1984, in *Structure and Evolution of the Magellanic Clouds*, IAU Symp. 108, eds S. van den Bergh & K.S. de Boer (Reidel: Dordrecht), p353.

Fitzpatrick, E.L. 1991, PASP, 103, 1123

Fitzpatrick, E.L., & Garmany, C.D. 1990, ApJ, 363, 119

Fliegner, J., Langer, N., & Venn, K.A. 1996, A&A, 308, L13

Garnett, D.R., et al. 1995, ApJ, 443, 64

Hambly, N.C., Dufton, P.L., Keenan, F.P., Rolleston, W.R.J., Howarth, I.D., & Irwin, M.J. 1994, A&A, 285, 716

Hill, V., Barbuy, B., & Spite, M. 1997, A&A, 323, 461

Hill, V. 1997, A&A, 324, 435

Hill, V., Andrievsky, S., & Spite, M. 1995, A&A, 293, 347

Humphreys, R.M., Kudritzki, R.P., & Groth, H. 1991, A&A, 245, 593

Humphreys, R.M. 1983, ApJ, 265, 176

Irwin, M.J., Demers, S., & Kunkel, W.E. 1990, AJ, 99, 191

Jüttner, A., Stahl, O., Wolf, B., & Baschek, B. 1992, in *New Aspects of Magellanic Cloud Research*, eds B. Baschek, G. Klare, J. Lequeux (Springer-Verlag:Berlin), p337.

Lennon, D.J. 1997, A&A, 317, 871

Lennon, D.J., Dufton, P.L., Mazzali, P.A., Pasian, F., & Marconi, G. 1996, A&A, 314, 243

Luck, R.E., & Lambert, D.L. 1992, ApJS, 79, 303

Luck, R.E., Moffett, T.J., Barnes, T.G., & Gieren, W.P. 1998, AJ, 115, 605

Pagel, B.E.J., & Tautvaisviene, G. 1998, MNRAS, 299, 535

Rolleston, W.R.J., Dufton, P.L., Fitzsimmons, A., Howarth, I.D., & Irwin, M.J. 1993, A&A, 277, 10

Rolleston, W.R.J., Brown, P.J.F., Dufton, P.L., & Howarth, I.D. 1996, A&A, 315, 95

Russell, S.C., & Dopita, M.A. 1990, ApJS, 74, 93

Schaller, G., Schaerer, D., Meynet, G., & Maeder, A. 1992, A&AS, 96, 269

Venn, K.A. 1999, ApJ, 518, 405
Venn, K.A. 1995, ApJ, 449, 839

Discussion

Nolan Walborn: It should be noted that Norbert Langer's talk included objects above the 40-50 $M_{\odot}$ limit for a red supergiant, such as some of the Ofpe/WN9 stars, so clearly their nitrogen enhancements must be due to some other mechanism than dredge-up.

Hans Zinnecker: There is often some confusion about Z, i.e., the heavy element abundance/metallicity. Could we agree to call Z_{CNO} the "heavy element abundance" and Z_{Fe} the "metallicity"?

New Views of the Magellanic Clouds
IAU Symposium, Vol. 190, 1999
Y.-H. Chu, N.B. Suntzeff, J.E. Hesser, and D.A. Bohlender, eds.

Abundances in Magellanic Cloud Young Populous Clusters

V. Hill[1]

ESO, K. Schwarzschild Str. 2, D-85748 Garching b. München, Germany

Abstract. Detailed abundances are measured in cool supergiants in Magellanic Cloud young populous clusters (NGC 330 in the SMC, NGC 2004, NGC 2100 and NGC 1818 in the LMC) and compared to those recently observed in samples of field supergiants in each Cloud. The analysis of the field supergiants shows no evidence of any abundance dispersion among the stars, compatible with a well mixed gas in these galaxies. However, previous determinations of the metallicity of NGC 330 and NGC 1818 give values well below the respective field values. This difference is investigated using a homogeneous method for the field and cluster samples, and the abundance ratios of various elements are examined.

1. Introduction

The Magellanic Clouds' young stellar population includes a number of so-called *populous* clusters, which have no counterpart in our own galaxy. Very young (10^6 to 10^7yrs) and yet very rich, these clusters have long been targets both for cluster formation and stellar evolution studies. Among these clusters, particular attention was drawn to NGC 330 (SMC) and NGC 1818 (LMC), since early attempts to measure their metallicities indicated a low metal content (lower than that of the surrounding stars of similar ages, by factors of 5 and 4 respectively: Carney et al. 1985; Richtler et al. 1989). How could a cluster form with a metallicity *lower* than that of its surroundings? This is especially peculiar since the young population in the field of the MCs seems to be fairly well mixed: no abundance gradients or large scatter are found among the supergiants (Hill et al. 1995; Hill 1997; Luck et al. 1998), nor among the H II regions (Russell & Dopita 1990). Further spectroscopic studies confirmed that NGC 330 was indeed more metal poor than its surroundings (Spite et al. 1991). However, the reddening and hence the metallicity for this cluster has been questioned (Bessell 1991). The real metal deficiency of the MC young clusters is thus still not understood.

Our recent study of six K-type supergiants in the SMC field (Hill 1997) provides an ideal reference sample for this type of star. Taking advantage of this, we undertook a study of similar stars in four MC Young Populous Clusters. With the same models, methods and physical data being used for all stars, most of the systematic uncertainties cancel out when differential abundances (clusters

---

[1]IAG, Universidade de São Paulo, CP 3386, São Paulo 01060-970, Brazil

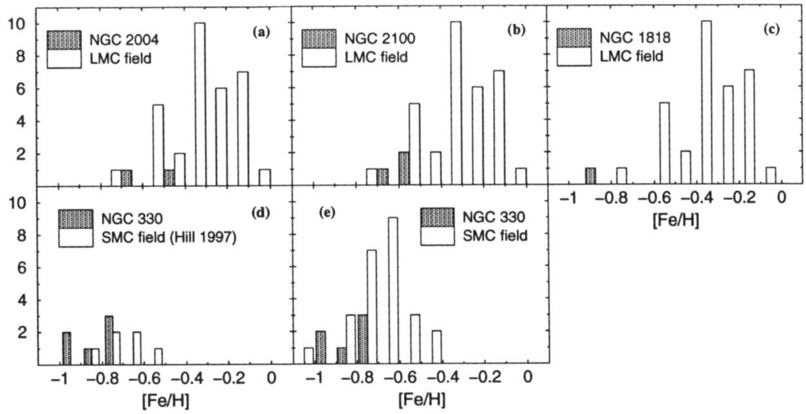

Figure 1. Histograms of the [Fe/H] distribution of cluster and field stars in the LMC (top panel) and SMC (bottom panel). (**a**) NGC 2004 (this work) and the LMC field (compilation by Luck et al. 1998.(**b**) the same for NGC 2100. (**c**) the same for NGC 1818.(**d**) NGC 330 and the SMC field (Hill 1997). (**e**) NGC 330 and the SMC field (Luck et al. 1998).

versus field) are considered. The program was conducted at the NTT (ESO) where spectra were obtained for a sample of cool supergiants in four MC young clusters: NGC 330 (~20 Myrs) in the SMC, NGC 1818 (~17 Myrs), NGC 2004 (~8 Myrs) in the LMC, and NGC 2100 (~12 Myrs) in the 30 Dor region of the LMC. The results for six stars in NGC 330 (SMC) were reported in Hill (1997b) and show that the difference, if any, between the cluster and the field stars is small and compatible with the small scatter found among the field stars. The LMC clusters are still under analysis, so the results reported here come from roughly half of the available data.

2. Cluster Versus Field Stars

The mean iron abundances found for the LMC clusters from the program stars are respectively: [Fe/H]$= -0.91$ (NGC 1818, 1 star), [Fe/H]$= -0.57$ (NGC 2004, 2 stars) and [Fe/H]$= -0.57 \pm 0.06$ (σ_{rms}) (NGC 2100, 3 stars).

The histograms in Fig.1 show that only in NGC 1818 are the field and cluster stars strikingly different. There still seems to be a systematical tendency of the four clusters (the three LMC clusters and NGC 330 in the SMC) to be more metal deficient than their surroundings, but the difference is small (<0.2 dex) and hardly significant in all cases except NGC 1818. However, since only one star in NGC 1818 has been analyzed so far, no firm conclusion can be drawn about this apparently peculiar cluster until other stars are analyzed.

Panels (a) and (b) of Fig.2 display the mean [X/Fe] ratios versus atomic number for the field (Hill 1997; Hill et al. 1995) and the LMC and SMC cluster (this work) samples. Panel (c) shows the difference between the mean for each cluster and the mean of the SMC K-type supergiants sample ([X/Fe]$_{cluster}$

- $[X/Fe]_{fieldSMC}$). This plot shows that the difference in the detailed chemical composition of the clusters and the field stars of the MCs is extremely small, seldomly exceeding the intrinsic expected uncertainty (dotted lines). The striking similarity of the abundance patterns favors a formation of young clusters from matter which evolved similarly to that of the bulk of the MCs.

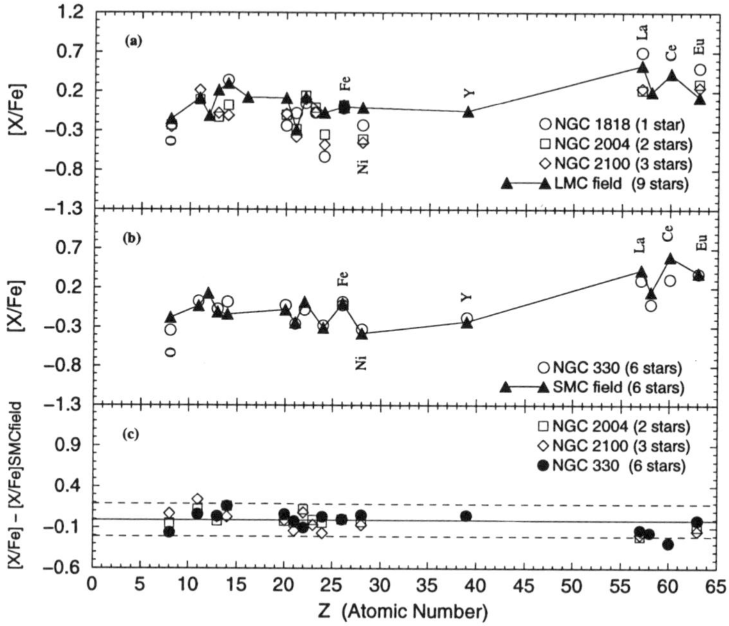

Figure 2. (a) mean abundances relative to iron of the LMC clusters and the field, plotted as a function of atomic number. (b) same as (a) for NGC330 and the surrounding SMC field. (c) Residual [X/Fe] abundances of the clusters when normalized to the SMC field K stars.

3. α Elements and Heavy Neutron Capture Elements

Oxygen and α elements are of particular interest for the study of the chemical evolution of a galaxy because they are thought to be produced in large quantities by the short-lived type II supernovae (SNIIe). This is in contrast to iron which is only marginally produced by SNIIe and more efficiently produced in the longer lived type Ia supernoave. It is now widely accepted that the α elements are not overabundant in the young matter of the Magellanic Clouds (in contrast with the Galactic dwarfs of similar metallicity). This can clearly be seen in Fig.3 (left), where both the field and cluster K supergiant [O/Fe] values are plotted for both MCs, together with Galactic data (the dashed box represents the location of the solar neighbourhood supergiants). Also striking is the *similarity of the Magellanic and Galactic supergiant oxygen to iron ratios.* This feature may be simply explained by a star formation history which occurs more slowly in the Clouds than in the Galaxy; a bursting star formation history could also produce

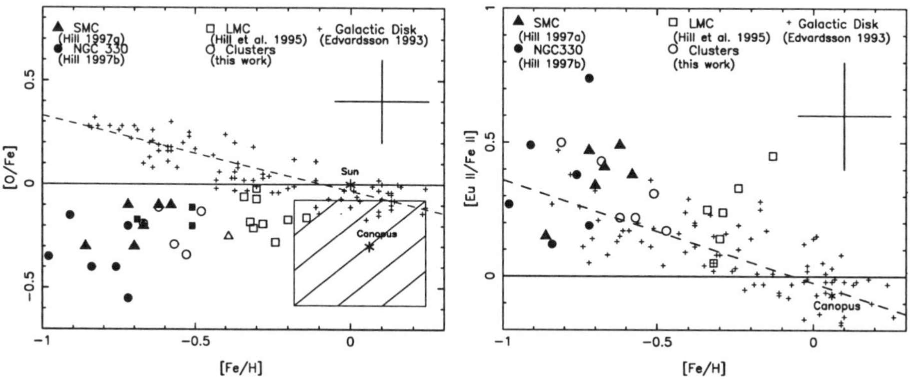

Figure 3. [O/Fe] and [Eu/Fe] ratios as a function of [Fe/H] for supergiants in both clouds (field and clusters) and in the Galaxy.

such [O/Fe] ratios in the young population. In either cases there is *no need for a IMF different in the Clouds and the Solar neighbourhood.*

On the other hand, the heavy *r*-process neutron capture elements (Europium is the best example) are also thought to be produced mainly by SNIIe. In the disk of our Galaxy, the behaviour of [Eu/Fe] indeed resembles that of the [O/Fe] (Fig.3). However, in the MCs, the heavier *s*- and *r*-process neutron capture elements are found to be significantly overabundant (by ~0.3 and ~0.4 dex respectively both in field and cluster stars). In the right panel of Fig.3 the [Eu/Fe] values for field and clusters in both MCs are plotted. The small symbols are the data for the Galactic disk (Woolf et al. 1995).

This difference of behaviour between α and r- process elements in the MCs is very difficult to understand, since both types of elements are thought to be produced in SNIIe and more work deserves to be done on this topic to unravel the true history of chemical enrichment in the Clouds.

Acknowledgments. This research was partly supported by Fapesp.

References

Bessell, M. 1991, in IAU Symp. 147, Haynes R., Milne D. (eds.), Kluwer, p.273

Carney, B.W., Janes, K.A., & Flowers, P.J. 1985, AJ, 90, 1196

Hill, V. 1997, A&A, 324, 435

Hill, V., Andrievsky, S., & Spite, M. 1995, A&A, 293, 347

Luck, E., Moffett, T., Barnes, T., & Gieren, W. 1998, ApJ, 115, 605

Richtler, T., Spite, M., & Spite, F. 1989, A&A, 225 351

Russell, S., & Dopita, M. 1990, ApJS, 74, 93

Spite F., Spite M., & Richtler, T. 1991, A&A, 252, 557

Woolf, V., Tomkin, J., & Lambert, D.L. 1995, ApJ, 453, 660

Discussion

Jay Gallagher: Can you say anything about the dispersion in abundance among the young stars in the LMC from your data?

Hill: From our 9 F supergiants in the LMC, we could see no real dispersion of abundances ($\sigma = 0.07$ dex). To my understanding, the scatter that remains in the total distribution of [Fe/H] values is due to differences in the various authors' approaches.

Dominik Bomans: You stressed the importance of differential analysis. Microturbulence is critical. How similar are the field and cluster samples in terms of microturbulence and how does it affect your results?

Hill: Both samples in SMC field and cluster display the same range of microturbulence velocity (3 - 3.5 km s^{-1}). All the more, given the range of lines of various intensities that are observed, the microturbulence velocities are derived in a rather reliable way.

Johannes Anderson: The picture of a dispersion of 0.2–0.3 dex in [Fe/H] and other ratios [X/Fe] is, of course, just what is seen in the MW disk, except for [O/Fe]. Which lines were used to determine the [O/H]?

Hill: The oxygen abundance is derived from the forbidden [O I] 6300 Å line.

New Views of the Magellanic Clouds
IAU Symposium, Vol. 190, 1999
Y.-H. Chu, N.B. Suntzeff, J.E. Hesser, and D.A. Bohlender, eds.

The New 30 Doradus

Nolan R. Walborn and Rodolfo H. Barbá

STScI, 3700 San Martin Dr., Baltimore, MD 21218, USA

Abstract. Recent optical and infrared studies of 30 Doradus have revealed a more complex and diverse star-formation history than previously recognized. Five spatially and/or temporally distinct populations have been identified, including a new generation currently being triggered by the R136 cluster in its surroundings. HST/NICMOS observations of the latter provide insights into the phenomena of triggered massive-star formation, as well as the global structure and evolution of the starburst region.

1. Stellar Populations

A groundbased, blue-violet spectral classification study of the 30 Doradus stellar content by Walborn & Blades (1997) has revealed five spatially and/or temporally distinct components:

1. The central ionizing cluster including R136 corresponds to the Carina phase of OB cluster evolution, with an age of 2-3 Myr. HST/FOS spectroscopy by Massey & Hunter (1998) has shown that two-thirds of the 60 brightest stars in and around R136 are of spectral type O3, by far the richest concentration known of these hottest and most massive young stars.

2. A younger, Orion-phase generation of age < 1 Myr, in or near the bright nebular filaments west and northeast of R136, containing heavily embedded early-O dwarfs and IR sources, has likely been triggered by the central cluster. This component is further discussed in the second section below.

3. An older population of late-O and early-B supergiants throughout the central field represents the Scorpius OB1 phase with age 4-6 Myr and has no clear structural relationship to the younger groups. These stars may simply belong to the larger young region in which 30 Doradus is embedded.

4. The previously known, older still compact cluster Hodge 301, $3'$ northwest of R136, contains A and M supergiants and corresponds to the h and χ Persei phase, of age 10 Myr.

5. A newly recognized Sco OB1-phase association surrounds the recently discovered Luminous Blue Variable R143 in the southern part of the Nebula.

Evidently, star formation has occurred in discrete events at different epochs in 30 Doradus. Global CMDs and IMFs of the region will be heterogeneous composites (see also Barbá & Walborn in these proceedings), and there are clear implications for the interpretation of more distant starbursts. This entire region would subtend 0″.5 at the Virgo Cluster.

2. HST/NICMOS Views of the New Generation

Recent optical, IR, and radio evidence for current star formation in 30 Doradus is reviewed by Rubio et al. (1998) and Walborn et al. (1999). The latter paper presents first results from an HST/NICMOS survey, which because of its 0″.1 resolution provides new insights; some further observational details are given by Barbá et al. in these proceedings. Many new IR sources, including multiple systems, clusters, and nebular structures, are found in these NICMOS images. Knots 1-3 of Walborn & Blades (1997), early O stars embedded in dense nebular knots, are all found to be compact multiple systems.

The first of the accompanying figures shows the field of Knots 1 and 2 (northeast of R136) in a WFPC2 U, V, I composite, and the second in a NICMOS J, H, K composite with the same orientation and scale; the scale bars indicate N, E and are $4'' = 1$ pc in length. Knot 1 resides at the top of a massive dust pillar oriented directly toward R136, whose summit has just been removed, exposing the newborn stellar system. Two of the brightest IR sources in 30 Dor (arrowed in the NICMOS figure), with only very faint counterparts in I, are found in an adjacent pillar with the same orientation just northeast of Knot 1. Another very bright IR source is within the head of a pillar near Knot 3 also oriented toward R136 (west of R136, not shown here), while parsec-scale jet structures have been discovered in association with Knots 2 and 3. The Knot 2 structures, arrowed in the accompanying NICMOS figure, consist of detached, nonstellar IR sources aligned on either side of the stellar system, which are interpreted as impact points of a highly collimated, possibly rotating bipolar jet on the surrounding dark clouds; there is also an H_2O maser in this field.

The outflows from young massive stars in 30 Dor are the first extragalactic examples of the phenomenon. These results establish the 30 Doradus Nebula as a prime region in which to investigate the formation and very early evolution of massive stars and multiple systems; several sequential stages and their interstellar interactions are directly observable in the optical and IR. The theme of triggered formation within the heads of extensive dust pillars oriented toward R136 is strong. In addition, these results provide further insights into the global structure and evolution of 30 Doradus, which are significant in view of its status as the best resolved extragalactic starburst. As discussed by Scowen et al. (1998), the most intense nebular emission arises from filamentary interfaces between the central cavity and surrounding molecular clouds. The IR results show that these interfaces are in fact *star-formation fronts* moving into the dark clouds. 30 Doradus is a *two-stage starburst*, in which a second generation of massive stars is being triggered around the periphery of the initial, massive central cluster. It will evolve into a giant shell H II region like N11 in the LMC (Walborn & Parker 1992) and NGC 604 in M33.

Acknowledgments. Support for this work was provided by NASA through grants numbers GO-7819.01-96A and AR-7545.01-96A from STScI, which is operated by AURA, Inc., under NASA contract NAS 5-26555.

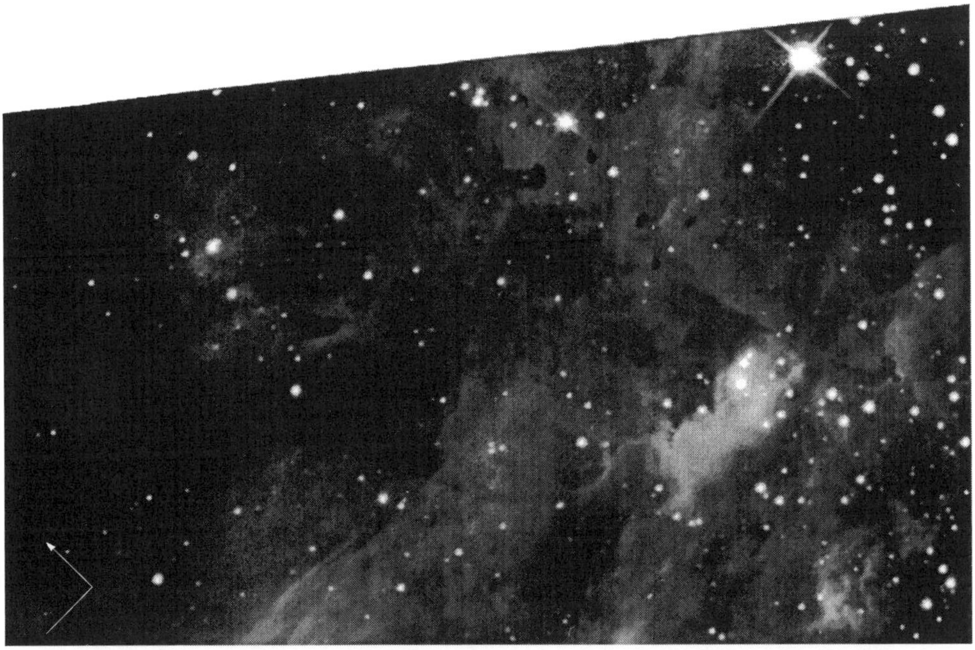

Figure 1. The field of Knots 1 and 2 northeast of R136 in a WFPC2 U, V, I composite image.

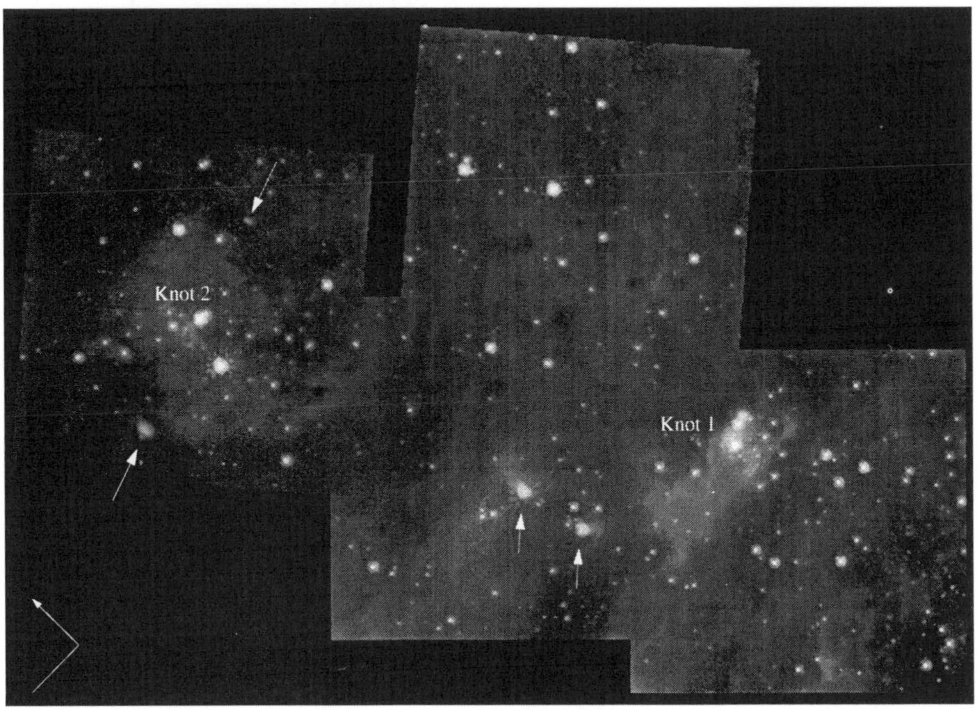

Figure 2. As in Fig 1., but in a NICMOS J, H, K composite image.

References

Rubio, M., Barbá, R.H., Walborn, N.R., Probst, R.G., García, J., & Roth, M.R. 1998, AJ, 116, 1708

Massey, P., & Hunter, D.A. 1998, ApJ, 493, 180

Scowen, P.A., et al. 1998, AJ, 116, 163

Walborn, N.R., Barbá, R.H., Brandner, W., Rubio, M., Grebel E.K., & Probst, R.G. 1999, AJ, 117, 225

Walborn, N.R., & Blades, J.C. 1997, ApJS, 112, 457

Walborn, N.R., & Parker, J.Wm. 1992, ApJ, 399, L87

New Views of the Magellanic Clouds
IAU Symposium, Vol. 190, 1999
Y.-H. Chu, N.B. Suntzeff, J.E. Hesser, and D.A. Bohlender, eds.

The Stellar Population of R136

Deidre A. Hunter

Lowell Obs., 1400 West Mars Hill Road, Flagstaff, AZ 86001 USA

Abstract. The luminous, compact star cluster R136 in the Large Magellanic Cloud is 300 times more spatially concentrated than a typical OB association and contains at least 39 O3 stars, the most massive of which is $\sim$150 $M_\odot$. Yet, the stellar IMF for 2.8–120 $M_\odot$ is normal. The massive stars are 1–2 Myrs old, and Wolf-Rayet stars are not evolved.

1. Introduction

R136 is the luminous, compact star cluster at the center of the 30 Doradus nebula in the LMC. R136 is particularly interesting for several reasons. First, it is comparable to a very young version of a globular cluster. The half-light radius of R136 is 1.7 pc (Hunter et al. 1995) while that of globular clusters is 1–8 pc (van den Bergh et al. 1991), and the mass of R136 is 6×10^4 $M_\odot$ (extrapolating the stellar initial mass function [IMF] to 0.1 $M_\odot$) compared to 4–40×10^4 $M_\odot$ for globulars (Illingworth 1976).

Secondly, R136 represents a region where many stars, especially massive stars, have formed in a very small space. Figure 1 compares the spatial concentration and richness of R136 with typical OB associations (Massey et al. 1995), and other star-forming regions (Hunter et al. 1996a,b; Dolphin & Hunter 1998). Compared to OB associations, R136 is $\sim$ 300 times more spatially concentrated and $\sim$ 10–100 times richer in stars. It has been suggested that the lower mass stars will be missing from regions where many massive stars have formed. R136 offers a place to test this idea. R136, while neither unique nor the most extreme among star clusters, is unique in being both very young and close enough to resolve into individual stars.

2. Early Stellar Population Studies of R136

R136 has intrigued astronomers for decades. It particularly gained notoriety in the early 1980's when people, puzzling over the prodigious UV output and spectra of the central object, postulated that R136a, the bright source in the center of R136, was a single star $> 10^3$ $M_\odot$ (Feitzinger et al. 1980; Cassinelli et al. 1981). The alternative explanation of a cluster of $\sim$ 30 O3 stars seemed too unreasonable by comparison (but see Melnick 1985). However, ground-based speckle observations (Weigelt & Baier 1985; Neri & Grewing 1988) and then HST observations (Weigelt et al. 1991; Campbell et al. 1992; de Marchi et al. 1993; Elson et al. 1992) showed that R136a was a collection of a dozen stars. Ground-

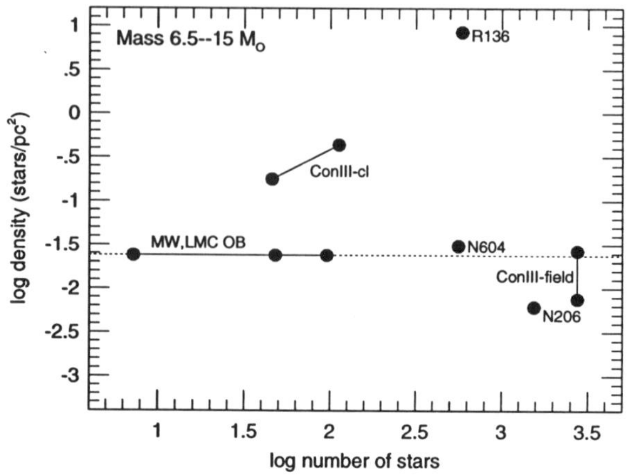

Figure 1. Richness of stars formed against their spatial concentration.

based (Walborn 1973; Moffat & Seggewiss 1983; Chu et al. 1984; Melnick 1985; Moffat et al. 1985) and HST observations (Campbell et al. 1992; Parker et al. 1995) also confirmed that R136 contained several Wolf-Rayet stars but no red supergiants. This plus stellar evolutionary models placed an age of 3–4 Myrs on the massive stars.

3. What WFPC2 Revealed

With WFPC2 on HST astronomers were finally able to measure a color-magnitude diagram for R136 (Hunter et al. 1995, 1996c), shown in Figure 2. This CMD showed a main sequence of intermediate mass stars—down to ~2.8 $M_{\odot}$, and stars down to ~1.1 $M_{\odot}$ on pre-main sequence tracks. Pre-MS stars have also been identified in ground-based near-infrared observations by Brandl et al. (1996). Isochrones show that stars on the pre-MS tracks are as young as 2 Myrs. This led to a very unexpected result: R136 did not fit the standard picture for cluster formation in which the lower mass stars form first and that, when the massive stars finally form, their strong winds disrupt the cloud and shut down star formation. On the contrary, in R136 some lower mass stars appeared to have formed 1–2 Myrs *after* the massive stars!

From the WFPC2 CMD we also learned that the intermediate mass stars were present in nearly normal proportions and that this IMF is independent of distance from the cluster's center. The lower mass limit is ≤ 2.8 $M_{\odot}$, although stars are also present to at least 1.1 $M_{\odot}$.

4. The Massive Stars

However, questions about the stellar population of R136 still remained: What is the IMF of the *massive* stars, and what is their age really? To answer these

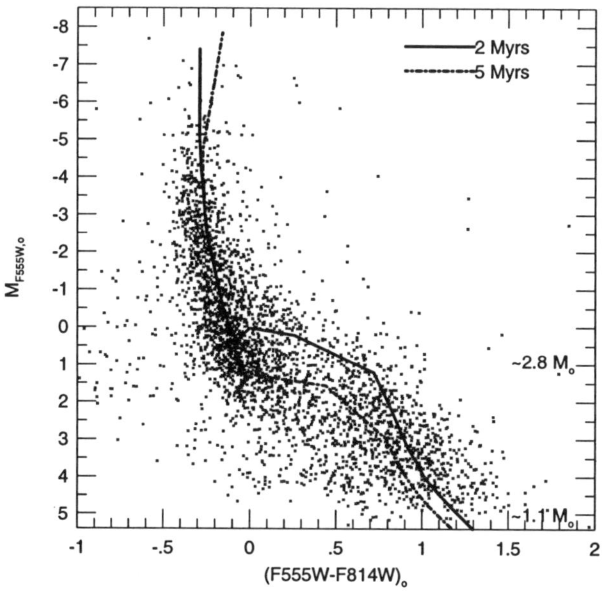

Figure 2. CMD for R136 to a radius of 5 pc from WFPC2 photometry.

questions requires spectroscopy of the individual stars, and Massey & Hunter (1998) did that. A hint of what was to come came from the results of de Koter et al. (1997) who obtained spectra of 3 stars in R136a. Two of these stars are Wolf-Rayet stars but turned out to be H-core burning.

Massey & Hunter (1998) obtained high signal-to-noise HST spectra of 65 of the bright, blue stars in R136 for the purpose of spectral classification. Including 2 of the stars observed by de Koter et al. (1997), one finds that 39 of 67 stars are O3 stars. The alternative explanation of R136 as a cluster of O3 stars was, in fact, correct! Furthermore, the highest luminosity stars turn out to be Wolf-Rayet stars that are 2.5 magnitudes too bright and are H-rich rather than H-poor. That is, as de Koter et al. found, the Wolf-Rayet stars are super Of stars rather than evolved objects. Furthermore, HR diagrams of the massive stars show that their age is 1–2 Myr, depending on which effective temperature scale for massive stars one adopts. Therefore, the massive stars are much younger than previously thought. Now R136 fits the standard scenario for star formation in a cluster: Star formation started $\geq$5 Myr ago with the lower mass stars and continued until the high mass stars formed most recently. The fact that the Wolf-Rayet stars in R136 are not evolved objects has implications as well for the "Wolf-Rayet galaxies" and their WR/O star ratio problem.

The most massive stars in R136 have masses of 140–150 $M_{\odot}$. These masses are based on extrapolating the mass–luminosity relationship from the Geneva evolutionary tracks, and so are conservative, but they are among the most massive stars known. Yet, surprisingly, the IMF slope is normal. So, the intense, rich environment of R136 did not produce an unusually shallow massive star IMF, but rather the IMF slope is just like those measured for OB associations.

Therefore, we see that R136 has a normal stellar population for $\geq 2.8\ M_\odot$ and that this IMF applies over a range of 300 in spatial concentration from OB associations to globular-like clusters. Furthermore, finding such massive stars in R136 argues that the upper stellar mass limit observed so far in star-forming regions is statistical rather than physical in nature. R136 has such massive stars simply because it is very young and very rich in stars.

5. The Future

Several questions still remain about R136 to yet intrigue astronomers. First, what are the massess of the O3 stars really since the O3 class is degenerate? STIS observations are being analyzed to address this. Second, what is the stellar population $< 2.8\ M_\odot$ in R136? NICMOS observations are aimed at this question and is the subject of the talk by Zinnecker. So, it is possible that someday we may have a complete census of the stars in this remarkable star cluster.

Acknowledgments. I wish to thank HST for grants GO-06417.02-95A and GO-07370.02-96A that funded part of the original R136 research.

References

Brandl, B., et al. 1996, ApJ, 466, 254

Campbell, B., et al. 1992, AJ, 104, 1721

Cassinelli, J.P., Mathis, J.S., & Savage, B.D. 1981, Science, 212, 1497

Chu, Y.-H., Cassinelli, J.P., & Wolfire, M.G. 1984, ApJ, 283, 560

de Koter, A., Heap, S.R., & Hubeny, I. 1997, ApJ, 477, 792

de Marchi, G., Nota, A., Leitherer, C., Ragazzoni, R., & Barbieri, C. 1993, ApJ, 419, 658

Dolphin, A.E., & Hunter, D.A. 1998, AJ, 116, 1275

Elson, R.A.W., Schade, D.J., Thomson, R.C., & Mackay, C.D. 1992, MNRAS, 258, 103

Feitzinger, J.V., Schlosser, W., Schmidt-Kaler, Th., & Winkler, Chr. 1980, A&A, 84, 50

Hunter, D.A., Baum, W.A., O'Neil, E.J., & Lynds, R. 1996a, ApJ, 456, 174

Hunter, D.A., Baum, W.A., O'Neil, E.J., & Lynds, R. 1996b, ApJ, 468, 633

Hunter, D.A., O'Neil, E.J., Lynds, R., Shaya, E., Groth, E., & Holtzman, J. 1996c, ApJ, 459, L27

Hunter, D.A., Shaya, E.J., Holtzman, J.A., Light, R.M., O'Neil, E.J., & Lynds, R. 1995, ApJ, 448, 179

Illingworth, G. 1976, ApJ, 204, 73

Massey, P., & Hunter, D.A. 1998, ApJ, 493, 180

Massey, P., Lang, C.C., Degioia-Eastwood, K., & Garmany, C.D. 1995, ApJ, 438, 188

Melnick, J. 1985, A&A, 153, 235

Moffat, A.J., & Seggewiss, W. 1983, A&A, 125, 83

Moffat, A.J., Seggewiss, W., & Shara, M.M. 1985, ApJ, 295, 109
Neri, R., & Grewing, M. 1988, A&A, 196, 338
Parker, J.Wm., Heap, S.R., & Malumuth, E.M. 1995, ApJ, 448, 705
van den Bergh, S., Morbey, C., & Pazder, J. 1991, ApJ, 375, 594
Walborn, N.R. 1973, ApJ, 182, L21
Weigelt, G., & Baier, G. 1985, A&A, 150, L18
Weigelt, G., et al. 1991, ApJ, 378, L21

Discussion

Hans Zinnecker: How many 40-80 $M_\odot$ stars (early O-stars) are present in the NGC 2070/R136 cluster and how are they distributed spatially? Are they all near the cluster centre ($R < 2$ pc)? Please draw a map with the positions of the 100 most massive stars in 30 Doradus!

Hunter: There are $\sim$ 24 stars more massive than 80 $M_\odot$. They are scattered throughout the cluster and are not disproportionally concentrated toward the center. A picture showing the location of stars with masses $>$ 100 $M_\odot$ can be found on the NOAO web page under press releases.

Daniel Wang: You mentioned that R136 is young. Do you have a new estimate of its age?

Hunter: The massive stars appear to be $1 - 2$ Myrs old, depending on which temperature scale you use for the stars. Lower mass stars appear to be on pre-MS tracks that indicate ages of 2 Myrs and older.

Sally Oey: While the expectation value of the highest-mass star per cluster increases with N (the number of stars per cluster), it must approach the upper-mass limit if it exists. Are the observations of highest-mass star and numbers consistent with your claim of the upper-mass limit being statistical, not physical, especially for R136?

Hunter: In R136 the most massive stars that we see occur where the IMF peters out in numbers. So even in R136 we are not necessarily seeing a physical upper limit to the mass of a star that can form.

New Views of the Magellanic Clouds
IAU Symposium, Vol. 190, 1999
Y.-H. Chu, N.B. Suntzeff, J.E. Hesser, and D.A. Bohlender, eds.

30 Doradus: The Low-Mass Stars

Hans Zinnecker

Astrophysikalisches Institut, An der Sternwarte 16, D-14482 Potsdam, Germany

Bernhard Brandl

Cornell University, Department of Astronomy, Ithaca, NY 14853, USA

Wolfgang Brandner

JPL/IPAC, California Institute of Technology, 4800 Oak Grove Drive, Pasadena, CA 91109, USA

Andrea Moneti

ISO Science Operations Centre — VILSPA, Villafranca, 28080 Madrid, Spain

Deidre Hunter

Lowell Observatory, 1400 W. Mars Hill Road, Flagstaff, AZ 86001, USA

Abstract. We present a first analysis of our deep NICMOS/HST F160W images of the 30 Doradus cluster, aimed at detecting the low-mass stellar population ($M \leq 2\ M_\odot$). We find that the infrared luminosity function keeps rising towards the faint end and that there is no indication of a low-mass IMF cut-off down to at least 1.5 $M_\odot$. We also find a change of slope (steepening) in the luminosity function as a function of radial distance from the cluster center. The faintest stars we have detected have H=22.5 mag, corresponding to about $M = 0.4\ M_\odot$ at an age of 2 Myr.

1. Introduction

The low-mass stellar content in the 30 Doradus cluster (NGC 2070) is of major astrophysical interest for a number of reasons:

1. 30 Dor in the LMC is a massive, metal-poor young cluster, unlike any young cluster in our Galaxy. With hundreds of OB stars present, the question arises whether low-mass star formation is suppressed or not.

2. 30 Dor appears to be like a small young globular cluster. However, only if low-mass stars are present in sufficient numbers will the cluster remain bound and later develop into a globular cluster.

3. 30 Dor may be the typical building block (unit cell) of violent star formation, now and in the high-redshift universe. Whether the IMF is low-mass deficient or not, makes a huge difference for the photometric and chemical evolution of young galaxies (e.g., Charlot et al. 1993; Zinnecker 1996; Larson 1998).

For earlier work on the IMF of this starburst cluster, we refer to Hunter et al. (1995), Brandl et al. (1996), and Massey & Hunter (1998). Their work has resolved the high-mass and intermediate-mass stellar content of the cluster ($M \geq 3\ M_\odot$), with the result that the IMF closely follows a Salpeter mass distribution (slope $\Gamma = 1.35$). The next step was to resolve the low-mass stellar population, using NICMOS on HST.

2. Observations and Data Reductions

The HST observations were carried out with the NIC2 camera (FOV $\approx 20'' \times 20''$) in the F160W filter ($1.4 - 1.8\mu$m, resolution $0.15''$) for maximum sensitivity and consisted of a 3×3 "core" mosaic centered on the cluster core, and two 3×1 "wings" extending away from the core region from two adjacent sides. In addition, the central core was also observed with NIC1 to provide improved spatial sampling of the very crowded center. Several sky observations were made in regions located $10 - 20'$ north of the cluster core. All observations were obtained in MULTIACCUM mode with an integration time of up to 896 sec (the effective integration time can be much smaller as only the shortest MULTIACCUM readouts are used; see Zinnecker & Moneti 1998 for more details and for a figure showing the layout of the observations superimposed on a NIR ground-based image). At each position four dithered images were obtained in order to remove bad pixels, cosmic rays, and the coronographic hole.

Individual frames were processed with the calnica (STSDAS) pipeline. Images were then cross-correlated in order to determine subpixel offsets, and stacked together using the DRIZZLE package. Remnant cosmic ray events, hot and bad pixels were masked out. Figure 1 shows a composite of the two wing mosaics.

Photometry was performed with DAOPHOT (Stetson 1987), and we present here preliminary results for wing2 (wing1, being much more crowded and containing many bright stars is much more difficult to analyze). The DAOPHOT analysis was performed in two runs: first the image shown in Figure 1 was analyzed, and second, the residual frame from the first run was boxcar smoothed and subtracted from the initial frame to produce a new frame free of nebulosity, and this image was analyzed. A preliminary photometric calibration was performed following the steps recommended in the STScI NICMOS photometry WWW pages. No color correction has been applied so far, and no attempt has been made to convert to a standard H-band magnitude. With this preliminary calibration, the F160W magnitude of Mk34, a bright star near the core, is 12.0, i.e. 0.20 mag larger than the H mag from Brandl et al. (1996). We consider this to be a good agreement at this stage, but that the final calibration could change by a few tenths of a magnitude.

Using this calibration we have extracted the F160W luminosity function (LF) for the wing2 field, shown in Figure 2. The main results are:

1. the faintest stars detected using our two-step DAOPHOT process are 22.5 mag (corresponding to a pre-Main Sequence mass ≈ 0.4 $M_\odot$ for a cluster age of 2 – 4 Myr and a distance modulus to the LMC of 18.5 mag).

2. the slope of the LF, i.e. log(number of stars) per unit H-magnitude, is 0.24; the slope appears to steepen slightly towards the outer edge of the field. Furthermore, there appears to be a feature in the LF at about 17.5 mag which could correspond to the turn-on to the Main Sequence.

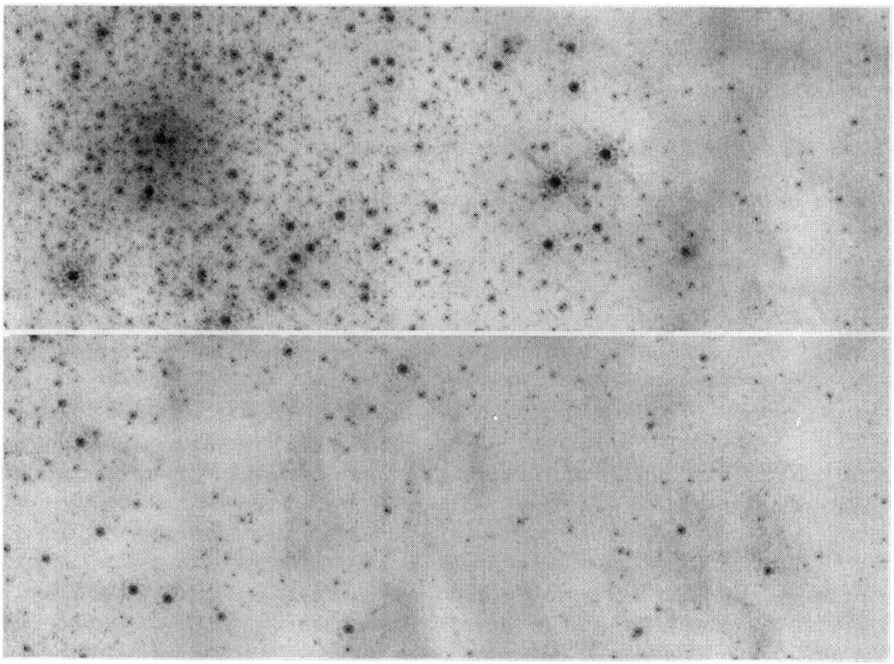

Figure 1. NICMOS/HST images (F160W filter) of the 30 Doradus cluster. Wing 1 (upper panel) contains the cluster core, wing 2 (lower panel) lies NW of the core. The horizontal width is 60″ or 15 pc.

3. Discussion

We have presented the first infrared (H-band) luminosity function of the 30 Dor cluster, albeit not in the very cluster core but at radial distances 15 – 25 pc where crowding is less severe. We have detected the first low-mass pre-MS stars below 1 $M_\odot$ (H $\approx$ 21). Although the luminosity function has so far not been corrected for incompleteness (and may therefore not exhibit the apparent turn-over, but may still be rising at the faint end), it is likely to be largely complete to H = 20. Further ignoring a minor extinction correction and a very small correction for background contamination, we can convert the slope (s) of the

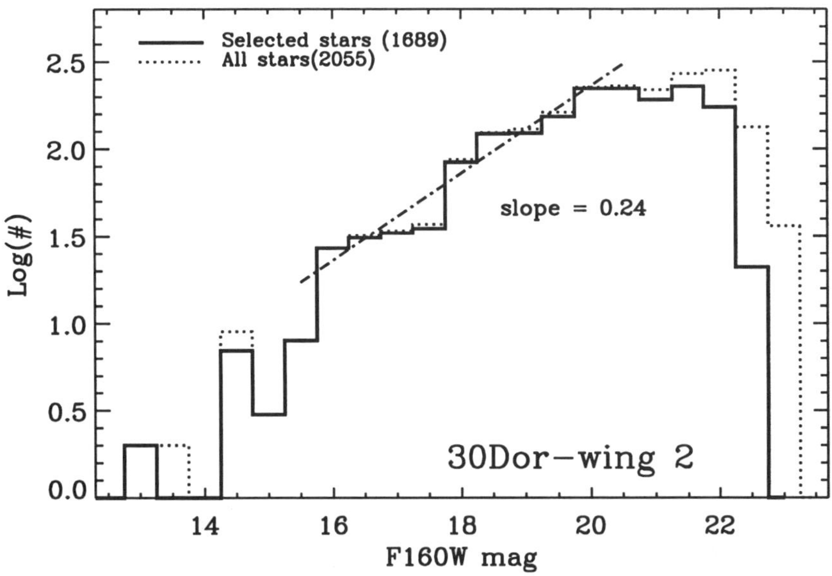

Figure 2. Infrared luminosity function (H-band, F160W) for Wing 2 in 30 Dor. Selected stars means that only stars with a magnitude error ≤ 0.2 mag were considered (full line)

luminosity function between H = 16 and H = 20 (see Figure 2) to an IMF slope (Γ), corresponding to a range of masses from 1.5 to about 10 $M_\odot$, using the time-dependent pre-MS mass-luminosity relation for the appropriate cluster age (2 Myr): $L_H \sim M^{1.4}$ (based on d'Antona & Mazzitelli 1994 tracks and dwarf colors). We find $\Gamma = 2.5 \times 1.4 \times s = 0.84$ for $s = 0.24$ (as derived from :w Figure 2). This IMF slope is very similar to that found in the young Galactic cluster NGC 3603 ($\Gamma = 0.79$) over a similar range of masses, based on adaptive optics observations (Eisenhauer et al. 1998). This is interesting, because of all young Galactic clusters NGC 3603 is the closest analog to the 30 Dor cluster in the LMC.

Acknowledgments. We thank our co-I's of the 30 Dor NICMOS project: M. McCaughrean, G.Meylan, R. Larson, M. Rosa, N. Walborn, and G. Weigelt. BB, WB, and DH acknowledge support by NASA through grant number GO7370.01-96A from STScI. HZ thanks the DFG for travel support which enabled him to participate in this enjoyable IAU Symposium in beautiful Victoria, Canada.

References

Brandl, B., Sams, B.J., Bertoldi, F., et al. 1996, ApJ, 466, 254

Charlot, S., Ferrari, F., Mathews, G.T., & Silk, J. 1993, ApJ, 419, L57

d'Antona, F., & Mazzitelli, I. 1994, ApJS, 90, 467

Eisenhauer, F., Quirrenbach, A., & Zinnecker, H. 1998, ApJ, 498, 278

Hunter, D.A., Shaya, E.J., Holtzman, J.A., et al. 1995, ApJ, 448, 179

Massey, P., & Hunter, D.A. 1998, ApJ, 493, 180

Larson, R.B. 1998, MNRAS, 301, 569

Stetson, P.B. 1987, PASP, 99, 191

Zinnecker, H., & Moneti, A. 1998, in *NICMOS and the VLT*, ESO Conf. & Workshop Proc. No. 55, p. 136 (eds. W. Freudling & R. Hook)

Zinnecker, H. 1996, in *The interplay between massive star formation, the ISM, and galaxy evolution*, eds. D. Kunth et al., p. 249

Discussion

Marc Kutner: As we are leaving the topic of star formation it is important to remember that stars are formed in molecular clouds. When we consider how some event may have caused or modified some star formation episode, we must consider the effect of the event on a parent cloud.

Zinnecker: Yes. We must also remember that molecular clouds often have a few dense cores from which star clusters or subgroups of stellar associations are forming, thus giving rise to hierarchical stellar clusterings.

Marc Kutner: How far down would the IMF have to cut off to bind the cluster (assuming a Salpeter IMF)?

Zinnecker: That depends on the mass of the original protocluster cloud and how much gas has been expelled from the cluster by the pressure of the HII regions and also massive stellar winds. The total stellar mass in the 30 Dor cluster in OBA stars down to 3 $M_\odot$ is about 2×10^4. For a Salpeter IMF the total stellar mass goes up by about a factor of 2 for a cut-off mass of 0.4 $M_\odot$. This could be sufficient to provide enough binding mass.

Nolan Walborn: To what mass does your completeness limit correspond?

Zinnecker: We think that we are about 80% complete to $M = 1.5\ M_\odot$.

Nolan Walborn: The dust, H_2, and CO seen very near to R136 is likely in front of or behind it, at distances comparable to those of the material to the west and northeast, since it is a three-dimensional structure.

Kimmo Innanen: Can you measure the velocity dispersion from the spectra of the MASSIVE stars in 30 Doradus?

Zinnecker: This is difficult because of the line broadening in massive stars. Nonetheless, I understand that the experiment has been tried by Bosch, Terlevich, Melnick, and Selman; but I don't know the results.

New Views of the Magellanic Clouds
IAU Symposium, Vol. 190, 1999
Y.-H. Chu, N.B. Suntzeff, J.E. Hesser, and D.A. Bohlender, eds.

Hot Gas and Physical Structure of 30 Dor

Q. Daniel Wang

Dept. of Physics & Astronomy, Northwestern University, 2145 Sheridan Road, Evanston, IL 60208-3112, USA

Abstract. Recent *ROSAT* and *ASCA* X-ray observations as well as *HST* images have provided new insights into the structure and evolution of the 30 Dor nebula. I review hot gas properties of the nebula and discuss related physical processes. The structure of the nebula can be understood as outflows of hot and HII gases from the parent giant molecular cloud of R136. The dynamic mixing between the two gas phases is likely a key mass loading process of the hot gas, providing a natural explanation of both temperature and density structure of the nebula.

1. Introduction

Hot gas in the 30 Dor nebula was first detected by the *Einstein* Observatory (Wang & Helfand 1991; Chu & Mac Low 1990). The nebula apparently consists of blisters of hot gas surrounded by Hα-emitting filaments. Stellar winds and supernova explosions are presumably responsible for the heating of the gas. Recently, the nebula has been observed with instruments on board the *ROSAT* and *ASCA* X-ray Observatories. X-ray images from some of these observations have been presented (Chu 1993; Wang 1995a; Wang & Gotthelf 1998), but detailed analysis of the data is still ongoing. In this paper, I review some preliminary results on hot gas properties and discuss implications on various relevant physical processes.

2. Hot Gas Properties

Fig. 1 shows a state-of-art high resolution X-ray image of the 30 Dor nebula (Wang 1999). In its core region, two point-like sources with 0.1 – 2.4 keV luminosities of $\sim 10^{36}$ ergs s^{-1} stand out and are most likely Wolf-Rayet + black hole binary systems (Wang 1995a). The presence of such systems indicates that supernovae have occurred in the recent past of 30 Dor. The X-ray contribution from the central star cluster R136 is also detected (Fig. 1). The image shows no evidence for other point-like X-ray sources which may be related to 30 Dor. The detection limit is $\sim 10^{35}$ ergs s^{-1}. There is also no correlation between the X-ray intensity and the stellar distribution. The X-ray emission is thus predominantly diffuse in origin. A three color composite image of the nebula in Hα, UV, and X-ray can be found at `http://www.astro.nwu.edu/astro/wqd/images/30d.htm`. Wang & Gotthelf (1998) have presented an overlay of X-ray contours on an Hα

image and *ASCA* SIS data in two energy bands. In addition, Chu (1993) has shown a *ROSAT* PSPC image of the nebula.

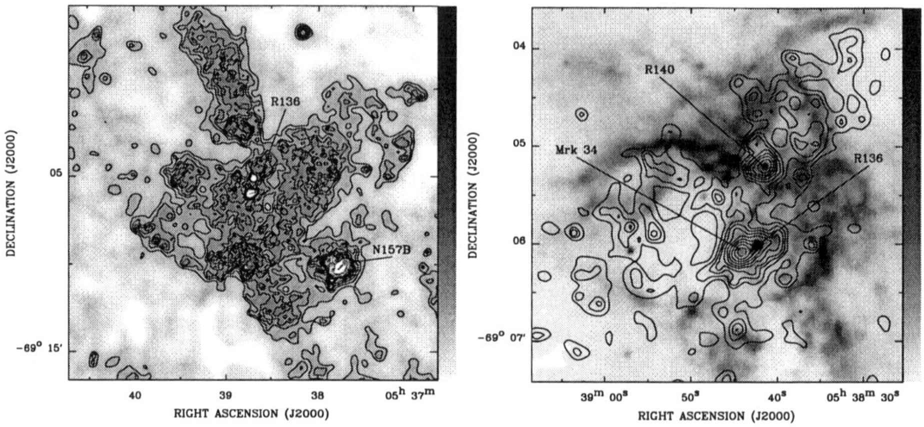

Figure 1. (left panel) X-ray image of 30 Dor. The image is a co-addition of data from *ROSAT* HRI exposures totaling 130 ks. The contours are at 0.6, 1.2, 1.9, 3.0, 4.5, 6.4, 9.1, 15, 24, 39, 63, 100, 160, 256, 407, and 647×10^{-3} counts s^{-1} $arcmin^{-2}$, above a local background of $\sim 1 \times 10^{-3}$. (right panel) The central region of 30 Dor in X-ray and Hα. The RHRI contours start at 3×10^{-3} counts s^{-1} $arcmin^{-2}$. The rest is the same as in the left panel.

These observations show that the 30 Dor nebula is an interacting complex of massive stars, dense molecular clouds, photoionized HII filaments, and diffuse hot gas. Young massive stars are concentrated in R136. But a few smaller clusters within the central region ($r \lesssim 4'$) may also contribute significantly to the energetics of the nebula. Dense molecular clouds typically appear in regions that are relatively dim in soft X-ray, apparently a result of displacement and/or X-ray absorption. The central region, in particular, shows hot gas bubbles surrounded by ionization fronts at boundaries of dense molecular clouds (Fig. 1).

An archival *ASCA* observation (Fig. 2) provides X-ray spectroscopic information about hot gas in the 30 Dor nebula (Wang 1999). The presence of individual emission lines (e.g., Ne, Mg, and Si) confirms that X-ray emission in the $0.5 - 2$ keV band is predominantly thermal in origin. The flat spectrum at higher energies, however, suggests the presence of a nonthermal component. A power law of energy slope equal to one describes this component well, although the exact spectral shape is uncertain because of the limited energy coverage and statistics of the data. Using the Raymond & Smith thermal plasma to characterize the thermal portion of the spectrum, I find that at least two temperature components are needed for a reasonable good fit. These two components characterize a range of hot gas temperature $0.2 - 1 \times 10^7$ K in the nebula. The high temperature component dominates in the energy band $\gtrsim 1.5$ keV, and arises primarily in the core region around R136 ($\lesssim 1'$; Wang & Gotthelf 1998). The

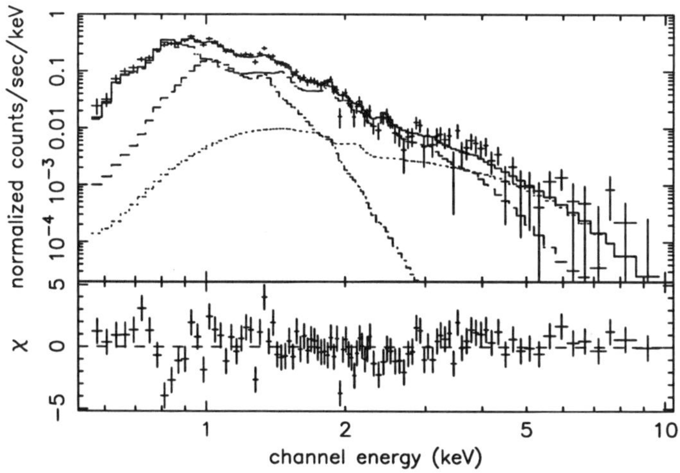

Figure 2. Average *ASCA* SIS spectrum of the X-ray emission from 30 Dor. A three-component model fit is shown as the solid histogram. The low and high temperature thermal plasma components are plotted as dot-dashed and dashed histograms, while the power law as dotted histogram.

total inferred thermal energy and mass of hot gas are $\sim 10^{52.5}$ ergs and $10^{4.6} M_\odot$, contained primarily in the low temperature component.

3. Physical Processes

The thermal energy in the hot gas is balanced by the expected mechanical energy release from stellar winds and supernovae in the nebula. The stellar winds from R136 alone have an energy release rate of $L_w \sim 3 \times 10^{39}$ ergs s^{-1} (e.g., Chu & Kennicutt 1994). Over the lifetime of R136, the integrated energy release is $\sim 10^{53}$ ergs, a large fraction of which should have been converted into the thermal energy of hot gas. There could also be tens of supernova explosions, although the exact number is difficult to estimate.

The thermal structure of the nebula is directly related to the cooling of hot gas. The radiative cooling rate of the hot gas is about $\sim 10^{39}$ ergs s^{-1}, a reasonable fraction of the estimated energy input rate. The shocked stellar wind has a temperature $\sim 10^8$ K and is not effective in X-ray emission. As hot gas moves from the core to the halo of the nebula, adiabatic cooling may be important. The average pressure p/k is about $\sim 2 \times 10^7$ K cm^{-3} in the core and is a factor of ~ 7 lower in the halo. However, the corresponding adiabatic temperature drop of a factor ~ 2 is not adequate to explain the temperature difference in the two regions. The adiabatic expansion should also rarefy the gas by a factor of ~ 3. But the measured mean hot gas densities are comparable in the two regions. Therefore, mass loading needs to be considered.

The most effective mass loading process is likely the dynamic mixing of hot gas with HII gas. While the mass efficiency of massive star formation is typically only 5 – 10% (e.g., McKee 1989), much of the parent giant molecular cloud (GMC) of the OB association NGC 2070 is being eroded via photon-evaporation. As the 30 Dor nebula has entered the so-called champagne phase (Wang 1995b), the photon-evaporated HII mass can be estimated as $\sim 0.12 M_{\odot}\ \mathrm{yr}^{-1} (S_i/10^{52}\ \mathrm{Ly\ photons\ s}^{-1})^{4/7} (t_*/10^6\ \mathrm{yr})^{2/7} (n_o/10^3\ \mathrm{cm}^{-3})^{-1/7}$, where S_i, t_*, and n_o are the ionizing flux of the OB association, its effective age with the present flux, and the mean density of the GMC (Whitworth 1979; Yorke et al. 1989). Only if $\sim$ 10% of this photon-evaporated HII gas is loaded to the shocked wind materials in the core of the nebula, can the temperature of hot gas there be explained. Over the lifetime of the nebula, the total evaporated mass is $\sim 10^5 M_{\odot}$. In order to account for the total mass of hot gas, the mass loading over the nebula must be substantial.

The dynamic mixing is evident, right within the central cavity around R136. The cavity is clearly open to the east. The radius of the cavity, as manifested by ionization fronts, is about half an arcminute, or $\sim$ 7 pc. Multi-filter *HST* WFPC2 images (Scowen et al. 1998 and this volume) clearly demonstrate the presence of evaporative flows from the ionizing fronts into the interior of the cavity. Such flows result from the strong pressure gradient (a factor of $\sim$ 10) between the ionization fronts and the interior of the cavity. Both the interaction between stellar winds and evaporative flows and the interface between the outflows of HII and hot gases are highly unstable. The dynamic mixing is bound to occur. Supernova blastwaves can further sporadically heat large amounts of HII gas. The high pressure in the core region tends to drive the heated gas out to the halo, leading to the blister-like morphology of the nebula.

Clearly, the study of 30 Dor has strong implications for the understanding of giant HII regions in general. The early mass loading, in particular, may determine hot gas properties of supergiant bubbles and galactic chimneys, which are all evolved from giant HII regions (Wang 1995b).

References

Chu, Y.-H. 1993, in *The Soft X-ray Cosmos*, AIP Conf. Proc. 313, eds: E. M. Schlegel, & R. Petre, 154

Chu, Y.-H., & Kennicutt Jr., R.C. 1994, ApJ, 425, 720

Chu, Y.-H., & Mac Low, M.-M. 1990, ApJ, 365, 510

McKee, C. F. 1989, ApJ, 345, 782

Scowen, P. A., et al. 1998, ApJ, 116, 163

Wang, Q. D. 1995a, ApJ, 453, 783

Wang, Q. D. 1995b, in *The Interplay between Massive Star Formation, the ISM and Galactic Evolution*, eds: D. Kunth, et al., 337

Wang, Q. D., & Helfand, D. J. 1991, ApJ, 370, 541

Wang, Q. D. 1999, ApJL, 510, L139

Wang, Q. D., & Gotthelf, E. V. 1998, ApJ, 494, 623

Whitworth, A. 1979, MNRAS, 186, 59

Yorke, H. W., et al. 1989, A&A, 216, 207

Discussion

George Sonneborn: What is known about the kinematics of the hot gas?

Wang: Unfortunately, the available X-ray instruments don't yet have the spectral capability to yield any information about the kinematics. In the near future, useful information may be obtained from FUV spectroscopic measurements of elements such as OVI, which are sensitive to gas at temperatures of a few times 10^5 K.

Joel Bregman: Hunter told us that the R136 cluster has an age of 1-3 Myr and that the massive stars have not evolved into SN. However, your picture requires that tens of SN have gone off. Can you reconcile these apparently contradictory results?

Wang: R136 is the main central cluster of 30 Dor. But the central region contains quite a few other star clusters, which appear to be older. The presence of the X-ray binary systems does indicate SN in the region. There are other pieces of evidence for SN, as You-Hua has argued. Energetically, my picture does not really require any SN, within the uncertainties of the present measurements.

Hans Zinnecker: I would like to make a comment. From your number of the hot, X-ray-emitting gas ($n_e = 0.2\ \mathrm{cm}^{-3}, T = 5 \times 10^6$ K), we can calculate the hot gas pressure ($\sim 10^6\ k_B$). This is 10 times higher than the internal pressure of typical dense cold proto-stellar globules ($n = 10^4\ \mathrm{cm}^{-3}, T = 10$ K). So the X-ray-emitting gas can indeed trigger secondary star formation in the 30 Dor region (cf. N. Walborn review of the generation of young stars revealed by NICMOS/HST observations).

Hans Zinnecker shows his colours during his invited review.

Part 3. Massive Stars and 30 Doradus

Section B. Poster Papers

Karen Vanlandingham and Anne Cowley comparing Tempe with Victoria during the reception.

New Views of the Magellanic Clouds
IAU Symposium, Vol. 190, 1999
Y.-H. Chu, N.B. Suntzeff, J.E. Hesser, and D.A. Bohlender, eds.

Discovery of a Stellar Association Surrounding the Massive Binary Sk−67°105 in the Large Magellanic Cloud

P. G. Ostrov[1], V. S. Niemela[1,2] and N. I. Morrell[3]

La Plata Observatory, Paseo del Bosque s/n, (1900) La Plata, Argentina

1. Introduction

Sk−67°105, a luminous O4f type stars in the Large Magellanic Cloud, is the exiting star of the H II region N 50 (DEM 193). Niemela & Morrell (1986) found this hot star to be a massive short period double lined spectroscopic binary. Because luminous OB stars are usually found in young stellar groups, we have searched for such an aggregate in the vicinity of Sk−67°105.

Here we report, as a result of our search, the discovery of a new OB association in the LMC. We also find that Sk−67°105 is the most luminous star of a small compact cluster inside this OB association.

2. Observations and Results

Our observations were acquired with the 2.15m telescope at *Complejo Astronómico El Leoncito* (CASLEO), San Juan, Argentina. Direct images of the field surrounding Sk−67°105 were obtained with a 1024 × 1024 pixel TeK CCD detector through Johnson B and V filters during two observing runs:

- 1993 October: $\sim 4' \times 4'$ field with a scale of $0.54''$ pixel^{-1} (after binning)
- 1997 November: circular field of $9'$ diameter using a focal reducer producing a scale of $0.81''$ pixel^{-1}.

The images were processed using DAOPHOT (Stetson 1987).

In Figure 1 we show the colour-magnitude diagram (CMD) based on the photometric frames obtained in 1997. This diagram reveals the presence of an OB association within the field, with Sk−67°105 as its most notorious member.

In our higher spatial resolution 1993 images Sk−67°105 appears with several fainter very close companions. Profile fitting photometry carried out with DAOPHOT reveals the existence in this compact star cluster of at least 8 stars (detected both in B and V filters) inside a circle of $10''$ radius surrounding Sk−67°105.

---

[1]Visiting astronomer, CASLEO, operated under agreement between CONICET, SeCyT, and the Universities of La Plata, Córdoba and San Juan, Argentina

[2]Member of Carrera del Investigador, CIC, Prov. Buenos Aires, Argentina

[2]Member of Carrera del Investigador, CONICET, Argentina

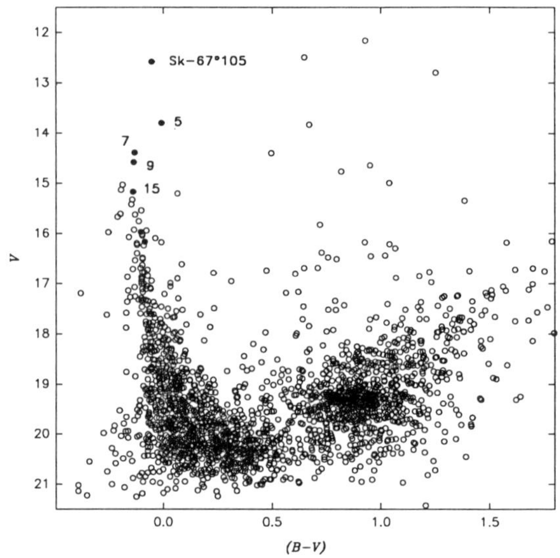

Figure 1. CMD of stars in a 9′ diameter field near Sk−67° 105. The filled symbols represent stars with known spectral types. An OB association is apparent in this CMD.

In 1998 February, we obtained spectra of some of the brightest and bluest stars in the field of the new OB Association. These data were acquired using the Boller & Chivens Cassegrain spectrograph on the 2.15 m telescope at CASLEO, with a PM 512 × 512 pixel CCD as detector. The spectra were processed and analyzed using IRAF routines. Spectral types for the observed stars were determined following Walborn & Fitzpatrick (1990). Magnitudes, colors, spectral types and radial velocities are listed in Table 1.

Table 1. Spectrophotometric data for stars in the new OB association.

ID	V	B-V	Sp.T.	R.V. (km s^{-1})
5	13.80	-0.01	B5 III	+311 ± 16
7	14.40	-0.13	B0.5 II-III	+299 ± 24
9	14.58	-0.13	B0.5 III	+296 ± 15
15	15.17	-0.14	O7 V	+330 ± 14
41	16.16	-0.08	B1 IV-V (e?)	+317 ± 21

References

Niemela, V. S., & Morrell, N. I. 1986, ApJ, 310, 715

Stetson, P. B. 1987, PASP, 99, 191

Walborn, N.R., & Fitzpatrick, E. 1990, PASP, 102, 379

New Views of the Magellanic Clouds
IAU Symposium, Vol. 190, 1999
Y.-H. Chu, N.B. Suntzeff, J.E. Hesser, and D.A. Bohlender, eds.

Ultraviolet Imaging Telescope Observations of the Magellanic Clouds

Joel Wm. Parker

Southwest Research Institute, Boulder, CO 80302, USA

Jesse K. Hill, Robert Cornett, Joan Hollis, Emily Zamkoff

Raytheon STX Corporation, NASA's GSFC, Greenbelt, MD 20771, USA

Ralph C. Bohlin

STScI, 3700 San Martin Dr., Baltimore, MD 21218, USA

Robert W. O'Connell

Astronomy Dept., Univ. of Virginia, Charlottesville, VA 22903, USA

Susan G. Neff, Andrew M. Smith, Theodore P. Stecher

Laboratory for Astronomy and Solar Physics, NASA's GSFC, Greenbelt, MD 20771, USA

Morton S. Roberts

National Radio Astronomy Observatory, Charlottesville, VA 22903, USA

Abstract.
We present an analysis of wide-field, far-ultraviolet images of the LMC and SMC obtained by the *Ultraviolet Imaging Telescope*. The photometric catalog of over 37,000 stars allows us to make large-scale, statistical studies of massive star formation in OB associations and in the field population. Our results show that: (1) the most probable slope for the initial mass function (IMF) of field stars is $\Gamma = -1.80$, slightly steeper than the Salpeter slope; and (2) there doesn't seem to be a single, unique IMF slope for stars in OB associations, with a range of values from $\Gamma = -1.0$ to -2.0. We also analyze the stellar vs. diffuse UV flux, and the population of OB star candidates in the field.

The *Ultraviolet Imaging Telescope* (*UIT*) obtained wide-field (37′ diameter), far-ultraviolet (FUV: 1300–1800 Å) images of 19 regions in the LMC and SMC. These images provide an extensive FUV mosaic of the SMC and contain numerous regions in the LMC, covering a wide range of stellar densities and current star formation activity. A total of 47 LMC/Lucke-Hodge (1970) and 37 SMC/Hodge (1985) OB associations are completely or partially included in the observed fields. Of the H II regions in the DEM catalog (Davies et al. 1976), the

UIT fields completely or partially include 102 DEM regions in the LMC and 74 DEM regions in the SMC.

In this paper we focus on the catalog of 37,333 stars in the LMC (Analyses of the *UIT* SMC data are presented by Cornett et al. [1994, 1997], and additional analyses of specific LMC regions observed by *UIT* are presented by Cheng et al. [1992], J. Hill et al. (1993, 1994), R. Hill et al. [1995], and Parker et al. [1996]), with a completeness limit of $m_{\rm UV} \approx 15$ mag and a detection limit of $m_{\rm UV} \approx 17.5$. FUV magnitudes for all the stars in these images were derived from point spread function photometry. The average uncertainty in the photometry is ~ 0.1 mag. Point-source profiles had a FWHM $\sim 3.4''$.

We divided the catalog into field stars and stars that are in DEM regions. We analyzed each of these two sets of stars independently, comparing the composite UV luminosity function of our data to UV magnitudes derived from stellar evolution and atmosphere models in order to derive the underlying stellar formation parameters. We find a most probable initial mass function (IMF) slope for the LMC field stars of $\Gamma = -1.80 \pm 0.09$, though we also find some evidence for a field star IMF slope of $\Gamma \sim -1.4$, equal to the Salpeter (1955, ApJ, 121, 161) slope. However, in the case of the stars in the DEM regions (the stars in all the regions were analyzed together as a single group), we find three IMF slopes of roughly equal likelihood: $\Gamma = -1.0$, -1.6, and -2.0.

In our analysis of the DEM regions, we find that stars typically contribute 75% of the total, integrated flux in a region, the remainder being due to a diffuse UV background and/or faint stars.

Using our data in conjunction with UV models and observations to deduce the UV magnitudes of "typical" O stars and evolved later-type stars, we estimate there are many as 1100–1500 candidate O-type stars are in the DEM regions, and 1600–2300 such stars are in the general LMC field.

This paper gives an overview of our analyses and results; the full paper with finder charts, details of the data reduction, and the complete analysis is presented by Parker et al. (1998).

References

Cheng, K.-P., et al. 1992, ApJ, 395, L29

Cornett, R. H., et al. 1994, ApJ, 430, L117

Cornett, R. H., et al. 1997, AJ, 113, 1011

Davies, R. D., Elliott, K. H., Meaburn, J. 1976, MmRAS, 81, 89 (DEM)

Hill, J. K., et al. 1993, ApJ, 413, 604

Hill, J. K., et al. 1994, ApJ, 425, 122

Hill, R. S., et al. 1995, ApJ, 446, 622

Hodge, P. W. 1985, PASP, 97, 530

Lucke, P. B., Hodge, P. W. 1970, AJ, 75, 171

Parker, J. Wm., et al. 1996, ApJ, 472, L29

Parker, J. Wm., et al. 1998 AJ, 116, 180

Salpeter, E.E. 1955, ApJ, 121, 161

New Views of the Magellanic Clouds
IAU Symposium, Vol. 190, 1999
Y.-H. Chu, N.B. Suntzeff, J.E. Hesser, and D.A. Bohlender, eds.

Comparison of Six Supersoft X-ray Binaries

A.P. Cowley[1] and P.C. Schmidtke[1]

Dept. of Physics & Astronomy, Arizona State University, Tempe, AZ 85287-1504, USA

J.B. Hutchings[1] and D. Crampton[1]

Dominion Astrophysical Observatory, HIA/NRC, Victoria, B.C., V8X 4M6, Canada

1. Introduction

The supersoft X-ray sources are a distinct class of X-ray sources identified by *ROSAT*. They are characterized by very high luminosities ($L_{bol} \sim 10^{38}$ ergs s^{-1}) and black body temperatures of kT$\sim 30 - 60$ eV. These sources are easily detected in the LMC and SMC because of the low column density of absorbing H gas. Thus, the samples found there are complete. They are much more difficult to find in the Galaxy due to soft X-ray absorption in the galactic plane.

2. Photometry, Spectroscopy, and Models

We have obtained multicolor photometry and spectra at CTIO of six supersoft X-ray binaries to intercompare their physical properties. These include three in the LMC (RX J0513.9−6951, CAL 83, and CAL 87), one in the SMC (SMC 13 = 1E0035.4−7230), and two in the Galaxy (RX J0019.8+2156 and RX J0925.7−4758).

Light curves show that their orbital periods range from $\sim 4^h$ to at least $\sim 1^d$, and they include a range of orbital inclinations (see table) from eclipsing (CAL 87) to nearly face-on (RX J0513−69, CAL 83). In addition to orbital variations, most (and perhaps all) also show long-term changes in the mean light level (e.g. CAL 83 has a range of ~1 mag (Cowley et al. 1998) and RX J0513−69 has high and low states differing by ~0.8 mag (Alcock et al. 1996)).

Spectra of all six systems show He II and O VI emission lines, and some also have phase-dependent H absorption. Equivalent widths of He II differ by $\sim 10\times$ between systems, roughly correlated with their orbital periods (and luminosities). However, the O VI lines are similar in all systems suggesting they are formed in the innermost regions of the accretion disk. H absorption lines have negative velocities indicating gas outflow with a complex geometry.

Four of the systems show high-velocity 'jets'. In CAL 83 Crampton et al. (1987) noted long-term changes in the wings of the He II and Hβ emissions and

[1] Visiting Astronomers, Cerro Tololo Inter-American Observatory. CTIO is operated by AURA, Inc. under cooperative agreement with the National Science Foundation.

suggested they might be due to outflowing material with velocities of ~1000 km s^{-1}. A 69-day period appears to fit widely scattered data from 15 years. In RX J0513−69 Crampton et al. (1996) demonstrated the presence of twin jets with outflow velocities of ~ ±4000 km s^{-1}, later confirmed by Southwell et al. (1996). Double jets with a velocity of ~ ±800 km s^{-1} are sometimes present in RX J0019+21 (Becker et al. 1998; Cowley et al. 1998). If these are bi-polar jets, their lower velocity may be due to the fact that this system is viewed at a higher inclination angle than CAL 83 or RX J0513−69. Motch (1998) has found evidence for transient jets (±5200 km s^{-1}) in RX J0925−47.

Table 1. Comparison of Supersoft Sources[a]

Name	Per. (d)	m_V (max)	Δm (orb)	K_{HeII} (km s^{-1})	i (deg)	M_V	$L_{bol}/10^{37}$ (erg s^{-1})	Notes
RX J0925	3.79?	17.1	0.2	84	30–40	~ 0	0.03–0.07	5200 km s^{-1} jets
CAL 83	1.04	16.3	0.2	35	25	−1.3	10–100	1000 km s^{-1} jets
RX J0513	0.76	16.6	0.04	11	15	−2.0	1–60	4000 km s^{-1} jets
RX J0019	0.66	12.2	0.5	67	56	+0.6	0.3–0.9	800 km s^{-1} jets
CAL 87	0.44	19.0	1.8	73	78	+0.3	6–20	eclipses
SMC 13	0.17	20.2	0.3	100	75	+1.4	0.8–2	partial eclipses

[a] Data from references in text, plus Greiner (1996) and references therein

The commonly accepted model for supersoft X-ray binaries involves a white dwarf accreting matter from a ~$2M_\odot$ companion at such a high rate that steady nuclear burning occurs (van den Heuvel et al. 1992). However, mass determinations based on emission-line velocity curves show that the mass donor has only about half the mass of the compact star, making it difficult to understand what drives the mass transfer. This suggests improved spectroscopic observations and new models are needed.

We find no evidence that the Magellanic Cloud and galactic systems differ in luminosity or other parameters as a result of differences in the ages and metallicity between these galaxies.

References

Alcock, C., et al. 1996, MNRAS, 280, L49
Becker, C.M., et al. 1998, ApJ, 506, 880
Cowley, A.P., et al. 1998, ApJ, 504, 854
Crampton, D., et al. 1987, ApJ, 321, 745
Crampton, D., et al. 1996, ApJ, 456, 320
Greiner, J. 1996, Lecture Notes in Physics, 472, 299
Motch, C. 1998, A&A, 338, L13
Southwell, K.A., et al. 1996, ApJ, 470, 1065
van den Heuvel, E.P.J., et al. 1992, A&A, 262, 97

New Views of the Magellanic Clouds
IAU Symposium, Vol. 190, 1999
Y.-H. Chu, N.B. Suntzeff, J.E. Hesser, and D.A. Bohlender, eds.

NLTE Abundances from Magellanic Cloud B Stars

Andreas J. Korn and Bernhard Wolf

Landessternwarte Königstuhl, 69 117 Heidelberg, Germany

Abstract. Nine MC young globular cluster and field stars are analysed using fully line-blanketed LTE ATLAS 9 atmospheres and the DETAIL/SURFACE non-LTE line formation code. CNO mixing seems to occur much earlier than predicted by standard theory.

1. Observations

Spectra were taken with CASPEC (R= 20 000) on the ESO 3.6m telescope at La Silla/Chile achieving a S/N of around 100. The stars show no Hα emission and are slow rotators with $v \sin i < 65$ km/s (*cf.* Reitermann et al. 1990).

2. Model Atmosphere Analysis

The method (Balmer profiles, Si II/III/IV ionisation equilibria, weak and strong O II lines) was thoroughly revised to be capable of deriving $T_{\rm eff}$, $\log g$, ξ, ε(O) and ε(Si) *simultaneously* and treating the metallicity of the underlying atmosphere in a *consistent* manner (for a full description see Gummersbach et al. 1998).

3. Results for Individual Stars

Table 1 summarizes the derived stellar parameters and abundances (in the usual notation $\varepsilon(\mathrm{X}) := \log(\mathrm{n(X)/n(H)}) + 12$). Nomenclature for cluster members is taken from Robertson (1974), for LMC field stars from Brunet et al. (1975), for SMC field stars from Azzopardi & Vigneau (1975). Brackets around ξ indicate that the microturbulence could not be co-determined simultaneously from silicon

Table 1. Derived stellar parameters

star	$T_{\rm eff}$	$\log g$	ξ	ε(He)	ε(C)	ε(N)	ε(O)	ε(Mg)	ε(Al)	ε(Si)
BRU 217 **LMC**	18100	2.66	11	10.80	%	7.60	8.42	6.90	5.80	7.12
BRU 231	17600	2.46	14	10.90	(7.80)	7.90	8.40	7.00	5.80	7.05
NGC 1818/D1	24700	4.00	0	11.15	7.65	7.59	8.46	7.35	6.05	7.10
NGC 1818/D12	16850	2.83	4	10.90	(8.10)	8.00	8.45	6.75	5.80	6.90
NGC 2004/B15	19900	3.11	(6)	10.85	7.90	%	8.35	6.78	5.90	6.74
NGC 2204/B30	23450	3.34	14	10.90	(7.60)	7.50	8.35	7.00	5.80	6.90
AV 175 **SMC**	19200	2.67	(13)	10.90	%	7.30	8.20	6.60	5.45	6.90
AV 218	23100	2.93	13	10.90	7.50	7.20	8.00	6.90	5.70	6.80
NGC 330/B30	16950	2.77	(6)	10.90	(7.30)	7.30	8.25	6.85	%	6.82
Δ(typical) = ±	1000	0.2	2	0.2 – 0.3 *dex*						

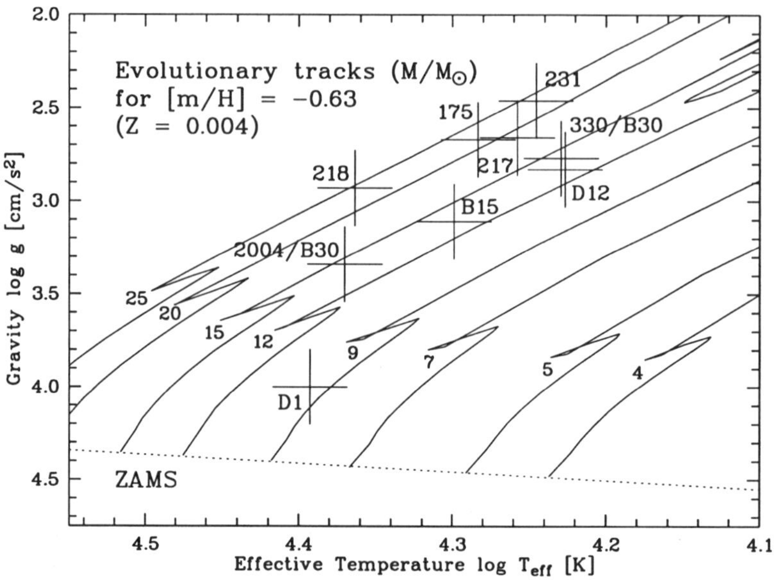

Figure 1. Kiel diagram showing the loci of the program stars with respect to evolutionary tracks according to Charbonnel et al. (1993) for a metallicity between those derived for the two Clouds.

and oxygen lines, those around ε(C) that carbon is solely based on C II 4267. Compare to MC H II data by Garnett (these proceedings)!

4. Global Results

- No significant metallicity difference between the clusters and their surrounding fields (but see Hill, these proceedings)
- No post-MS gap (stars on extended blue loops?)
- CNO mixing already on the MS (*cf.* NGC 1818/D1 in Table/Figure 1), unexplained by standard evolution, possibly rotationally induced (see contributions of Langer and Venn, these proceedings).

References

Azzopardi, M., & Vigneau, J. 1975, A&AS, 19, 271

Brunet, J.P., et al. 1975, A&AS, 21, 109

Charbonnel, et al. 1993, A&AS, 101, 415

Gummersbach, C. A., et al. 1998, A&A, 338, 881

Reitermann, A., Baschek, B., Stahl, O., & Wolf, B. 1990, A&A, 234, 109

Robertson, J.W. 1974, ApJ, 191, 67

New Views of the Magellanic Clouds
IAU Symposium, Vol. 190, 1999
Y.-H. Chu, N.B. Suntzeff, J.E. Hesser, and D.A. Bohlender, eds.

HST/WFPC2 Photometry in the 30 Doradus Nebula Beyond R136

Rodolfo H. Barbá and Nolan R. Walborn

STScI, 3700 San Martin Dr., Baltimore, MD 21218, USA

Abstract. A photometric study of the 30 Doradus Nebula beyond the core is being carried out by means of HST/WFPC2 archival images, with special attention to the numerous multiple systems and compact clusterings found there.

30 Doradus is the nearest and hence best resolved extragalactic starburst. Knowledge of its stellar content is vital to the interpretation of more distant starbursts, as well as to fundamental astrophysical problems such as the IMF, stellar mass limit, stellar evolution, and the structure of giant H II regions. Recent studies of the stellar content of 30 Doradus with HST/WFPC2 and FOS have concentrated on the central cluster core, R136 (Hunter et al. 1995 and references therein; Massey & Hunter 1998).

However, R136 and its immediate surroundings account for only a third to a half of the ionization of 30 Dor. Other very massive stars and stellar systems are distributed throughout the several-arcminute extent of the Nebula. They include objects both older and younger than R136; there is evidence that the formation of the latter has been triggered by the energetic activity of R136. So far, these important surrounding populations have been investigated only with groundbased observations (Parker 1993; Walborn & Blades 1997). In the latter spectral-classification study, five spatially and/or temporally distinct stellar components were isolated within the Nebula. But numerous multiple systems remain unresolved in these populations, particularly in the younger ones.

On the basis of HST/WFPC2 archival images (Proposals 5114, 5589, and 6122), a photometric study of the 30 Doradus Nebula surrounding R136 is being carried out, with special emphasis on the numerous multiple systems and compact clusterings found there. Of particular interest are systems in the bright nebular filaments where current massive-star formation is taking place, as revealed by both groundbased and HST/NICMOS infrared images (see Walborn & Barbá and Barbá et al. in these proceedings). Special attention is given to the objects included in the above groundbased spectral-classification study. Magnitudes and colors are derived for the newly resolved components of the multiple systems, while their ages and evolutionary status will be inferred insofar as possible. However, it is well known that the effective temperatures and masses of hot stars are degenerate when derived from photometry alone. Hence, this project is viewed as preparation for followup spatially resolved spectroscopy with HST/STIS, in order to advance our knowledge of the entire stellar content of 30 Doradus to the current state of the art, as is warranted by its unique status.

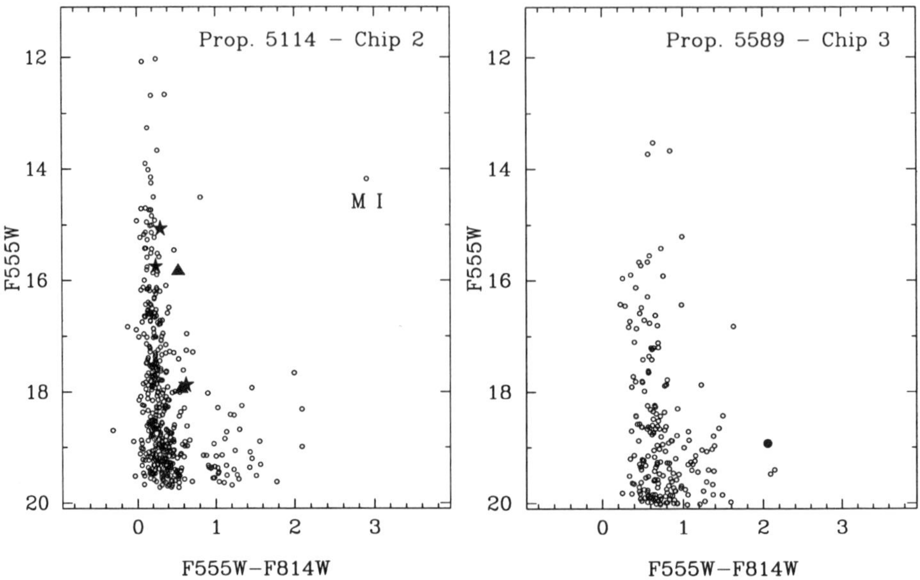

Figure 1. HST/WFPC2 color-magnitude diagrams of two areas of the 30 Doradus Nebula surrounding R136.

As an example, CMDs of two areas are shown in Fig 1. On the left is a CMD corresponding to the area E-NE of R136, where the young multiple systems Knots 1 and 2 are located. On the right is a CMD of the area S-SW. The comparison of these CMDs reveals clear differences between areas located at the same distance from R136: in the NE a main sequence is well defined, but in contrast the SW area shows a less sharp main sequence, perhaps due to higher reddening, pre-main-sequence objects, and/or fewer very high-mass objects. The M I star in the NE is likely an unrelated LMC field star in projection. Stars and triangles in the left CMD mark the Knots 1 and 2 stellar components, respectively. Note the magnitude and color similarities between the two Knots, in accordance with their spectroscopic morphologies. The black circle in the right CMD is a new IR source discovered 1′ south of R136 in HST/NICMOS images by Walborn et al. 1999.

Acknowledgments. Support for this work was provided by NASA through grants numbers GO-7819.01-96A and AR-7545.01-96A from STScI, which is operated by AURA, Inc., under NASA contract NAS 5-26555.

References

Hunter, D.A., et al. 1995, ApJ, 448, 179

Massey, P., & Hunter, D.A. 1998, ApJ, 493, 180

Parker, J.Wm. 1993, AJ, 106, 560

Walborn, N.R., et al. 1999, AJ, 117, 225

Walborn, N.R., & Blades, J.C. 1997, ApJS, 112, 457

New Views of the Magellanic Clouds
IAU Symposium, Vol. 190, 1999
Y.-H. Chu, N.B. Suntzeff, J.E. Hesser, and D.A. Bohlender, eds.

HST/NICMOS Survey in the 30 Doradus Nebular Filaments

Rodolfo H. Barbá and Nolan R. Walborn

STScI, 3700 San Martin Dr., Baltimore, MD 21218, USA

Wolfgang Brandner

IPAC, Caltech, 770 South Wilson Ave., Pasadena, CA 91125, USA

Ronald G. Probst

CTIO, NOAO, Casilla 603, La Serena, Chile

Mónica Rubio

Depto. de Astronomía, Univ. de Chile, Casilla 36-D, Santiago, Chile

Eva K. Grebel

Lick Observatory, Univ. of California, Santa Cruz, CA 95064, USA

Abstract. On the basis of ample evidence accumulated during recent years for current star formation within the 30 Doradus Nebula, which appears to have been triggered by the massive central cluster (R136) in and just beyond the curved nebular filaments surrounding it, an HST/NICMOS survey of these regions was undertaken.

In contrast with prior ideas that the 30 Doradus Nebula is essentially an evolved H II region, during recent years observational evidence of the birth of a new generation of massive stars has grown. Actually, nebular filaments encircling the core can now be recognized as very active star-forming regions, where winds and radiation from massive stars in the R136 core are interacting with surrounding molecular clouds, triggering a new stellar generation. Among that observational evidence, groundbased observations show many IR sources distributed in or beyond (with respect to R136) the nebular filaments (Rubio et al. 1998 and references therein); early O stars embedded in dense nebular knots (Walborn & Blades 1997) are spatially associated with those IR sources (Rubio et al. 1998); and WFPC2 images reveal numerous nebular microstructures, such as bright and dark dusty globules and pillars (Scowen et al. 1998; Rubio et al. 1998), symptoms of active star formation. Also, improved CO (Johansson et al. 1998) and H_2 (Probst & Rubio 1998) observations clearly demonstrate huge molecular concentrations beyond the west and northeast filaments.

On the basis of this observational evidence, HST/NICMOS observations were carried out in selected fields on the nebular filaments, under HST GO Program 7819, during February and March 1998. NIC2 broadband images in

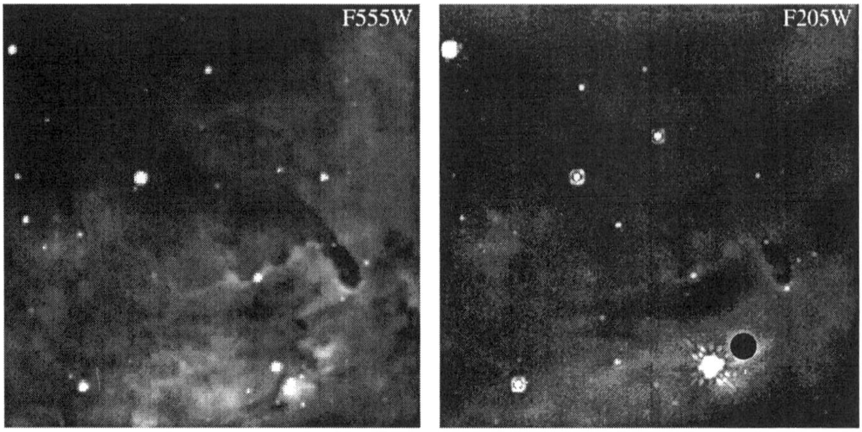

Figure 1. An example of WFPC2/F555W (left) and NIC2/F205W (right) images of filaments in the 30 Doradus Nebula.

F110W, F160W, and F205W filters were obtained in 15 fields. NIC2 narrowband images in F212N (H_2), F215N (continuum), and F216N (Brγ) were obtained in three selected fields, two in common with broadband ones. In addition, the NIC1 and NIC3 cameras were operated in parallel with the NIC2 camera. In NIC1 the F160W filter was used, and in NIC3 F187N (Paα), F190N (continuum), in addition to the same narrowband set as with NIC2. Further information about these observations is presented in these proceedings by Walborn & Barbá. A detailed morphological description of several NICMOS fields and their implications for the new view of 30 Doradus has been published (Walborn et al. 1999).

As an example, Fig. 1 shows a comparison between a WFPC2/F555W (left) and NIC2/F205W (right) image of a dark cloud shaped like a hand and "Finger of God"; it is located in a loop on the NE filament and points directly toward R136. In the NICMOS image it is possible to see several IR sources by direct comparison with the WFPC2 image. NICMOS reveals that the structure is reminiscent of the "Finger of ET", with an IR star exactly at its tip. Each image is $19\farcs2$ on a side. The dark spot in the NIC2 image corresponds to the coronographic hole.

Acknowledgments. Support for this work was provided by NASA through grants numbers GO-7819.01-96A and AR-7545.01-96A from STScI, which is operated by AURA, Inc., under NASA contract NAS 5-26555.

References

Johansson, L.E.B., et al. 1998, 331, 857

Rubio, M., et al. 1998, AJ, 116, 1708

Scowen, P.A., et al. 1998, AJ, 116, 163

Walborn, N.R., et al. 1999, AJ, 117, 225

Walborn, N.R., & Blades, J.C. 1997, ApJS, 112, 457

New Views of the Magellanic Clouds
IAU Symposium, Vol. 190, 1999
Y.-H. Chu, N.B. Suntzeff, J.E. Hesser, and D.A. Bohlender, eds.

30 Doradus as a Nearby Infrared Guide to Starbursts

M.D. Thornley[1], N.M. Förster Schreiber[1], H.W.W. Spoon, R. Genzel, D. Lutz, D. Kunze

Max-Planck-Institut für extraterr. Physik, D-85748 Garching, Germany

1. Overview

We are exploring the properties of obscured starburst galaxies, using observations of atomic fine structure lines taken with the Short Wavelength Spectrometer aboard ISO. However, it is important to ascertain how well our starburst models can recover the properties of the stellar populations in more distant starbursts. For this purpose, we use observations of a nebular "shell" in the 30 Doradus region, to show that our models reliably predict the presence of the very massive stars observed directly in the 30 Doradus region.

We use the population synthesis code STARS (Kovo, Alexander, & Sternberg 1999) to produce a composite spectral energy distribution (SED) for the starburst from a library of model stellar SEDs. Our model library contains a hybrid grid of non-LTE model atmospheres for stars with $T_{eff} > 25000$ K (Pauldrach et al. 1998), standard Kurucz atmospheres for stars with $T_{eff} < 19000$ K, and interpolated models for intermediate temperatures. The photoionization code CLOUDY (Ferland et al. 1996) is then used to predict the nebular emission created by the stellar population.

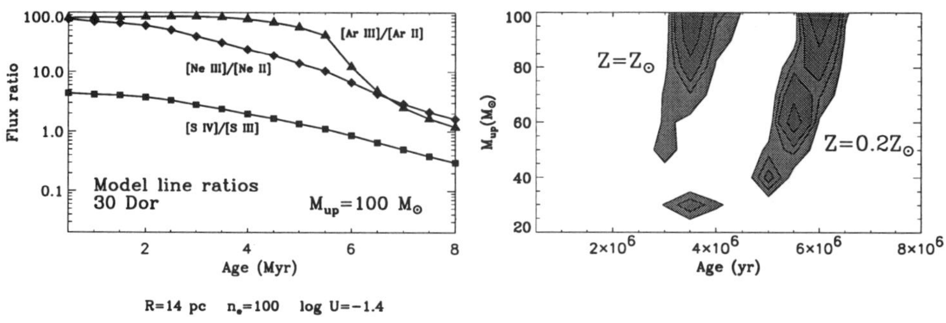

Figure 1. (left) Model line ratios ([Ar III]8.99μm/[Ar II]6.99μm, [Ne III]15.6μm/[Ne II]12.8μm, [S IV]10.5μm/[S III]18.7μm) for the conditions in 30 Doradus. (right)χ^2 fit to the three observed ratios; darker colors indicate a lower χ^2 value. Contours indicate 68.3,90,95.4, and 99% confidence levels, for two metallicities.

[1] *current addresses: MDT, NRAO (mthornle@nrao.edu); NMFS, Saclay.*

The star formation rate is assumed to decay exponentially with a 10^6 yr timescale for 30 Doradus (see Figure 1), and we have compared results for bursts with timescales of 1, 5, and 20×10^6 yr for the starburst galaxies (see Figure 2). We note that the range and variation of predicted ratios with age are dependent on the assumed star formation history.

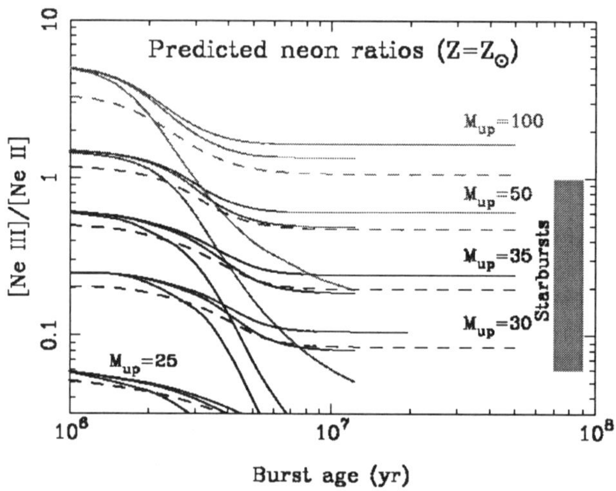

Figure 2. Predicted neon line ratios for starbursts with upper mass cutoffs between 25 and 100 $M_\odot$; for each M_{up}, ratios for timescales of 1, 5, and 20 Myr are shown. Solid lines indicate single cluster models; dashed lines show the effects of including a cluster luminosity function (e.g., McKee & Williams 1997). The range of observed values for over 25 starbursts is shown by the vertical bar at right.

According to our models, the observed nebular line ratios in 30 Doradus require stars of ~100 $M_\odot$ to be present, consistent with stellar classifications available in the literature. Using similar models on a sample of starburst galaxies suggests the presence of stars with masses greater than 25 $M_\odot$ in all sources, and the effects of aging suggest that the observed values are consistent with upper mass cutoffs in excess of 50 $M_\odot$. Thus we find no evidence for strongly truncated IMFs in starbursts.

Acknowledgments. MDT acknowledges the support of the Alexander von Humboldt-Stiftung during this work. Based on observations with ISO, an ESA project with the participation of ISAS and NASA. SWS and the ISOSDC at MPE are supported by DARA under grants 50 QI 8610 8 and 50 QI 9402 3.

References

Ferland, G.J., et al. 1996, ApJ, 461, 683

Kovo, O., Alexander, T., & Sternberg, A. 1999, in preparation

Pauldrach, A.W.A., et al. 1998, in *Properties of Hot, Luminous Stars*, (ed.) I. Howarth, ASP Conference Series, Vol. 131, 258.

New Views of the Magellanic Clouds
IAU Symposium, Vol. 190, 1999
Y.-H. Chu, N.B. Suntzeff, J.E. Hesser, and D.A. Bohlender, eds.

A High Resolution Optical Summary of the Structure and Dynamics of 30 Doradus

P. Scowen

Arizona State University, Physics & Astronomy, P.O. Box 871504, Tempe, AZ 85287-1504, USA

Y.-H. Chu, and R. Gruendl

Dept. of Astronomy, University of Illinois, Urbana, IL 61801, USA

Abstract. We present the first comparison between high resolution echelle velocity fields and a mosaic of all Hubble Space Telescope narrow-band observations of 30 Doradus.

1. Observations

Our data include 3 HST narrow-band image sets (WF/PC Hunter et al. 1995; WFPC-2 Scowen et al. 1998; WFPC-2 Chu et al. 1996) which together represent a large inner part of 30 Dor at a resolution of $0\farcs1$ pixel^{-1}. The two WFPC-2 images have R136 cluster and Hodge 301 cluster centered in the PC, respectively. CTIO 4m echelle observations were made to map the nebular kinematics in the R136 and Hodge 301 fields (marked by white rectangles in Fig. 1) The echelle observations have resolutions of 3.75 km s^{-1} and $0\farcs3$ pixel^{-1}. Iso-velocity maps are extracted and displayed in Fig. 2.

2. Discussion

Echelle iso-velocity maps reveal that gas around the R136 cluster appears blue-shifted, while gas to the SE appears redshifted. If we assume Scowen et al.'s (1998) photoevaporative flow from the walls of the HII region (normal to the local interface), the observed velocities imply that gas around the cluster is on the back side of the nebula and gas to the SE of the cluster is on the near side of the nebula. The echelle data reveal large (10-20 pc diameter) rings and arcs that canot be seen in direct image, such as the ring south of R136. The echelle data also reveal high-velocity shocked gas that is not detected in direct images.

References

Chu, Y.-H., et al. 1996, BAAS, 188, 61.04

Hunter, D.A., et al. 1995, ApJ, 448, 179

Scowen, P.A., et al. 1998, AJ, 116, 163

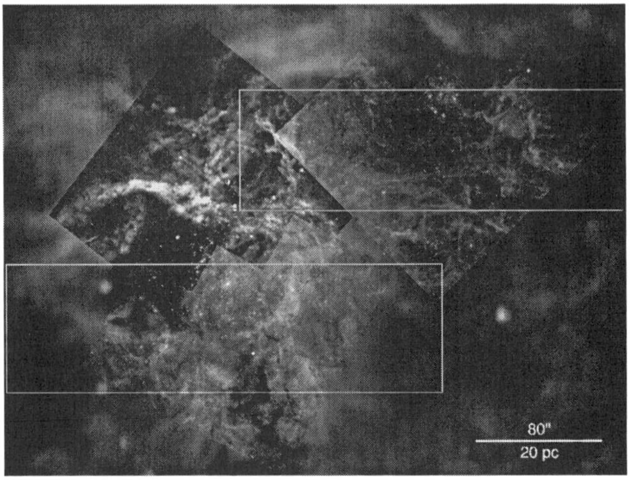

Figure 1. HST WFPC/WFPC2 images of 30 Dor. The two rectangles mark the areas, R136 and Hodge 301 fields, mapped by echelle observations.

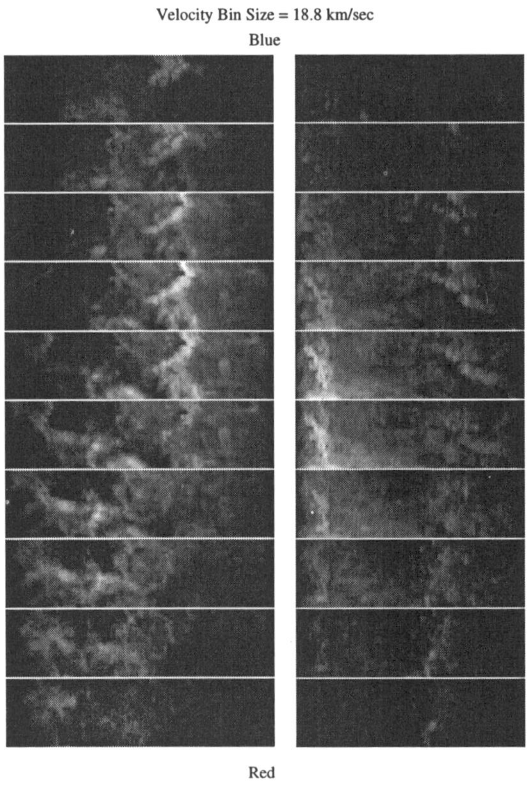

Figure 2. Iso-velocity images extracted from the echelle mapping. The left column shows the R136 field and the right Hodge 301 field.

New Views of the Magellanic Clouds
IAU Symposium, Vol. 190, 1999
Y.-H. Chu, N.B. Suntzeff, J.E. Hesser, and D.A. Bohlender, eds.

A New Investigation of the 30 Doradus Region with the Australia Telescope Compact Array

J. S. Lazendic[1]

Centre for Astronomy, University of Western Sydney, Nepean, Australia

R. F. Haynes

Australia Telescope National Facility, CSIRO, Australia

J. R. Dickel

Astronomy Department, University of Illinois, Urbana, IL 61801, USA

P. A. Jones and G. L. White

Centre for Astronomy, University of Western Sydney, Nepean, Australia

M. Costa

Physics Department, University of Tasmania, Australia

1. Introduction

30 Doradus is a giant H II region in the Large Magellanic Cloud (Bode 1801). It is the nearest extragalactic giant H II region and the location of active star formation. The complex nature of this extended region provides an excellent opportunity to investigate the impact of massive stars on the structure of the interstellar medium. Specifically the presence of supernova remnants (SNRs) is expected to play an important role.

2. A Search for Supernova Remnants within the Nebula

A number of studies of 30 Doradus have resulted in the encouraging discovery of high velocity material, which could well have originated in SNRs (e.g., Meaburn 1984; Chu & Kennicutt 1994). Since radio SNRs that "hide out" in star–forming regions would have a distinctly non–thermal radio spectrum, sensitive radio investigation at high resolution are needed to separate them from the surrounding thermal region.

Our recent image of the 30 Doradus region at a wavelength of 6 cm made with the Australia Telescope Compact Array is shown in Fig. 1. The half-power beamwidth is $1\farcs7$ and this is the most sensitive radio image ever made of the region. Our initial aim was to seek out non–thermal emission regions across the

[1]Astrophysics Department, School of Physics, University of Sydney, Sydney, Australia

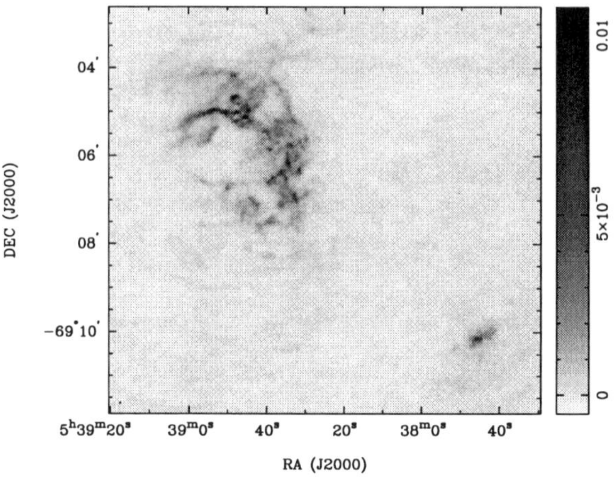

Figure 1. Radio continuum grayscale of 30 Doradus at 4.6 GHz with a HPBW of 1″.7. The peak brightness is 0.0108 Jy beam^{-1}.

field of 30 Doradus or search for small–scale, non–thermal features within the nebula itself. A comparison of the maps at different radio frequencies showed that the emission of 30 Doradus does not vary significantly with frequency, which is consistent with a thermal spectrum of this object (Lazendic 1997).

3. Hidden SNRs?

To separate more clearly possible nonthermal components from thermal emission of the HII region, we compare our radio image with an $H\alpha$ image from the U Michigan/CTIO Emission Line Survey of the LMC conducted by R.C. Smith (1997). The radio and optical images show impressive similarities but also interesting differences. A scaled difference image between the radio image and the $H\alpha$ image revealed 5 unusual small–scale features that show a possible nonthermal nature. Further high–frequency high–resolution observations are underway to investigate the properties of these features.

The supernova remnant N157B (bottom right corner in Fig. 1) was also included in the study. This Crab-type object has a radio spectral index of −0.33. Surprisingly, no polarised radio emission was detected from this SNR.

References

Bode, J. 1801, Algemeine Beschreibung und Nachweisung der Gestirne, (Berlin: Beym Verfasser)

Chu, Y.-H., & Kennicut, R. C. 1994, ApJ, 425, 720

Lazendic, J. S. 1997, BSc Honours Thesis, U. Western Sydney Nepean

Meaburn, J. 1984, MNRAS, 211, 521

Smith, R. C. 1997, `http://www.astro.lsa.umich.edu/ncels/mcsurvey.html`

Discussion

You-Hua Chu: Are the small-scale nonthermal radio source determined solely by comparison between radio and Hα images? How do you exclude the possibility of extinction effect?

Jones: Yes, the non-thermal radio features are determined by the comparison with the Hα image. We have not ruled out extinction yet. We need to compare the Hβ and Hα to determine extinction.

Ron Probst: A quick eyeball comparison of your differenced image with my H_2 image of 30 Dor appears to show that the compact 6 cm features all match compact H_2 features - except one, seen in Brγ, which is likely extinction related.

Jones: We are concerned about extinction and will compare Hβ and Hα images, and also correlate any radio features with other available data.

Rosa Williams: Is the southern feature seen near N157B real? If so, is it part of the outer shell, or is it pulsar-related?

Jones: The southern feature seen near N157B in the radio is part of the faint outer shell. There is no detected radio pulsar. The X-ray pulsar is in the middle of the radio SNR.

New Views of the Magellanic Clouds
IAU Symposium, Vol. 190, 1999
Y.-H. Chu, N.B. Suntzeff, J.E. Hesser, and D.A. Bohlender, eds.

The Low End of the Initial Mass Function for the R136 Cluster in the LMC

M. Sirianni[1]

Johns Hopkins University, Baltimore, MD 21218, USA

A. Nota[2], C. Leitherer, M. Clampin

Space Telescope Science Institute, Baltimore, MD 21218, USA

G. De Marchi

European Southern Observatory, Garching bei München, Germany

Multiple images of the R136 cluster in the Large Magellanic Cloud taken with the HST/WFPC2 and the F555W and F814W filters were combined to obtain very deep images of the cluster, with total exposure times of 1240 and 760 sec respectively. The objective of the study was to extend the luminosity function below the limit of 2.8 $M_\odot$ published by Hunter et al. (1995, 1996).

PSF fitting photometry was performed, with the conservative threshold set to discard all stars with an associated error larger than 0.2 mag in both filters. We then found 1604 stars common to both filters down to a limiting magnitude of 24.7, or ~ 1 mag deeper than the data published by Hunter et al. (1995, 1996). We generated a dereddened, color-magnitude diagram (CMD) by adopting the reddening law used by Hunter et al. (1995). Two distinct populations appear in Fig. 1 segregated in color at $(m_{F555W} - m_{F814W})_0 \simeq 0.5$.

The observed red population is very well tracked by PMS isochrones of Bernasconi (1996), and most likely consists of low mass stars (down to 0.8 $M_\odot$) still approaching the main sequence. Their ages are found to be between 1 and 10 Myr, in very good agreement with previous age determinations by Hunter et al. (1995) (4 Myr) and De Marchi et al. (1993) (3 Myr).

In order to locate the stars on the H-R diagram we converted the magnitudes and colors into the Log T_{eff} *vs* Log $L/L_\odot$ plane. We performed the completeness test for our data defining four regions surrounding the R136 cluster excluding the inner region of 4″. The four regions have different characteristics in terms of gas/dust contamination and crowding.

The Initial Mass Function (IMF), ξ, is defined as the number of star per logarithmic mass interval per unit area and its slope is given by $\Gamma = d(log\xi)/d(logM)$ where the standard IMF (Salpeter 1955) has a slope $\Gamma = -1.35$. In the case of the very young R136 cluster with coeval star formation, we can assume that the observed MF is the IMF. In order to derive the IMF of R136, we counted the number of stars falling between each pair of tracks in the HR diagram, corrected

[1] CISAS - University of Padova

[2] Affiliated to ESA,Astrophysics Division,Space Science Department of ESA.

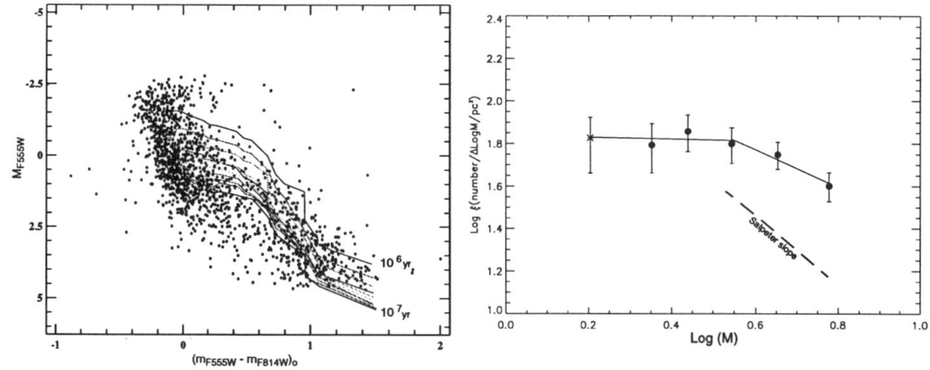

Figure 1. (left) Dereddened CMD of R136 with PMS isochrones. (right) The IMF of R136.

the number with the completeness factor and normalized the result to the width of the mass range spanned by the tracks and to the area of the observed field.

At masses of 0.8–2.75 $M_\odot$, we adopted the PMS tracks by Bernasconi (1996), while at higher masses (3–7 $M_\odot$) we used the standard evolutionary tracks of Schaller et al. (1992). In the overlap region around 3 $M_\odot$ both sets of tracks agree within the statistical errors. We used the two regions surrounding the core of R136 with the best completeness to construct the final IMF. Two different slopes are distinguishable in the IMF profile (Fig. 1): for stars in the mass range 3–7 $M_\odot$ the data points are well fitted by a slope with $\Gamma = -0.88 \pm 0.06$, while at lower masses (0.8–2.75 $M_\odot$) the IMF profile flattens out, with a derived slope $\Gamma = -0.04 \pm 0.06$. A similar effect, i.e., a deficiency of stars in the 1–2 $M_\odot$ range, is also observed by Hillenbrand (1997) in the Orion Nebula Cluster, although in that case the plateau is followed first by a steep increase between 0.5 $M_\odot$ and 0.2$M_\odot$ and then by a clear drop all the way down to the H-burning limit. Although there are several examples of a flat IMF for stars less massive than $\sim 1\ M_\odot$ (e.g., Comeron, Rieke, & Rieke (1997) in NGC 2024), only in ρ Oph have Williams et al. (1995) found a flat IMF *above* 1 $M_\odot$. The question as to whether the flattening that we observe in R136 is characteristic of this cluster or a general feature remains to be answered.

References

Bernasconi, G. 1996 A&AS, 120, 57

Comeron, F., Rieke, G. H., & Rieke, M.J. 1996, ApJ, 473, 294

De Marchi, G., et al. 1993 ApJ, 419, 658

Hillenbrand, L.A. 1997, AJ, 113.1733

Hunter, D.A., et al. 1995, ApJ, 448, 179

Hunter, D.A., et al. 1996, ApJ, 459, L27

Schaller, G., Shaerer, D., Meynet, G., & Maeder, A. 1992, A&AS 96, 269

Williams, D.M., et al. 1995, ApJ, 454, 144

(Bottom) Mary and Evangelos Kontizas listen while Jim Hesser comments on the difficulty of observing the Magellanic Clouds from Victoria with the Plaskett Telescope (top).

Part 4. Detailed Chemical Abundances

Section A. Invited Reviews

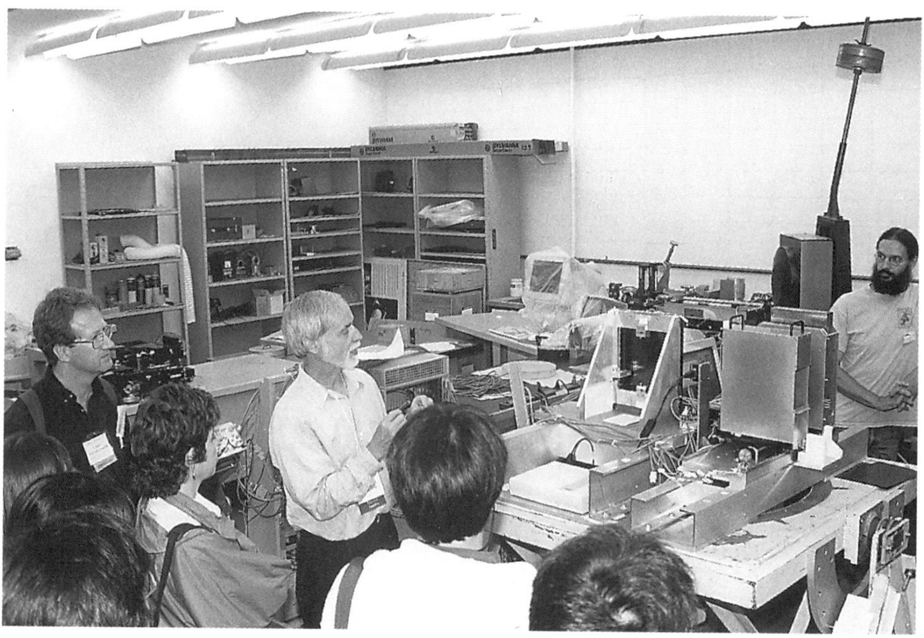

During the DAO tour Monica Rubio, Ray Haynes and You-Hua Chu examine receiver technology developed for the James Clerk Maxwell Telescope (top), while David Crampton and William Rambold demonstrate components of the Gemini Multi-Object Spectrographs.

New Views of the Magellanic Clouds
IAU Symposium, Vol. 190, 1999
Y.-H. Chu, N.B. Suntzeff, J.E. Hesser, and D.A. Bohlender, eds.

Stellar Metallicities in the Magellanic Clouds

Verne V. Smith[1]

Dept. of Physics, Univ. of Texas at El Paso, El Paso, TX 79968, USA

Abstract. The overall stellar metallicity scale as a function of age for both Magellanic Clouds reveals that they have followed a different pattern of enrichment when compared to each other as well as to the Galactic disk. In addition, an investigation into the abundances of two key elements, oxygen and europium, shows that the detailed nucleosynthesis history of these systems is unique in comparison to any Galactic population yet studied.

1. Introduction

The Large and Small Magellanic Clouds (LMC and SMC) are distinct stellar environments in comparison to the Galactic disk, halo, or the globular cluster system. In this brief review, we concentrate on two aspects of stellar abundances in the Clouds: 1) the age-metallicity trends for both the LMC and SMC, and 2) the abundance patterns of oxygen, the r-process elements and, in particular, the representative r-process element europium. These results provide insight into chemical evolution in systems quite different from the Milky Way.

2. Age-Metallicity Relations

Age-metallicity relations are fundamental properties which define the overall chemical evolution and enrichment in stellar systems. When combining data on age and metallicity in samples from the LMC and SMC, it is found that the Clouds seem to exhibit distinct histories, both in comparison to each other and to our Galaxy. Olszewski et al. (1991) and Dopita (1996) provide extensive results for LMC ages and metallicities. (Dopita's results are from planetary nebulae, while Olszewski et al. studied clusters). These studies can be supplemented by the recent results from Geisler et al. (1997) and Bica et al. (1998) for LMC clusters, who find generally similar trends. For the SMC, the most recent studies are from Da Costa & Hatzidimitriou (1998), who also analyzed clusters. Figure 1 shows a summary of ages and metallicities for the LMC and SMC, along with work from Edvardsson et al. (1993) on the Galactic disk stars. Note that for the Galactic disk we plot directly the Fe abundances, while the cluster work in both of the Clouds is based upon low-resolution spectroscopy of the Ca II

[1]McDonald Observatory, University of Texas, Austin, TX 79912

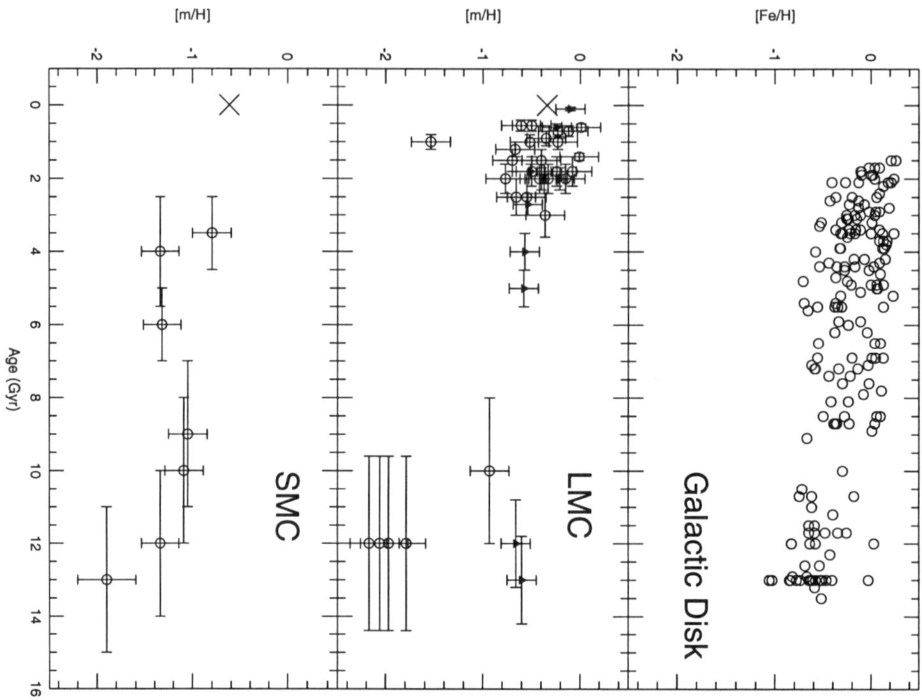

Figure 1. Age-metallicity relations for three systems. The Galactic disk data are from Edvardsson et al. (1993), while the SMC data are from clusters studied by Da Costa & Hatzidimitriou (1998). The LMC results also contain cluster work from Olszewski et al. (1991 - open circles) and Dopita (1996 - filled triangles), who studied planetary nebulae.

IR-triplet, which is calibrated to the Fe abundances of well-studied open and globular clusters (we label this [m/H]). The Dopita results are for mostly lighter alpha-elements.

The current abundances for the LMC and SMC are shown in Figure 1 represented by the large X's; these values have been culled from a number of sources in the literature, all for the youngest populations (with ages of less than 5×10^8 yr). For the LMC we have used Luck et al. (1998), Spite, Barbuy, & Spite (1993), Russell & Dopita (1992), and Richtler, Spite, & Spite (1989). For the SMC we used Luck et al. (1998), Plez, Smith, & Lambert (1993), Russell & Dopita (1992), and Spite, Spite, & Francois (1989). The average values for the current [Fe/H] values are -0.30 for the LMC and -0.62 for the SMC.

The age-metallicity relations for the three systems in Figure 1 exhibit different behaviors. The Disk abundances rise slowly but steadily from an initial [Fe/H]~ -1 towards a current [Fe/H] ~ 0, but with a fair amount of scatter at all ages (perhaps ±0.3 dex). The SMC also exhibits a slow and steady increase from [m/H]= -2 to a current value of about -0.6. The LMC, on the other hand, shows a very rapid rise in metallicity at a time of about 12 Gyr ago (from [m/H] = -2 to -1), then an almost constant value until 2 Gyr ago, when the metallicity increased from -1 to its current value of ~ -0.3. Note the apparent lack of intermediate-age clusters in the LMC, which may be related to the almost constant metallicity over this time period. Recently, Geisler et al. (1997) confirmed this gap in a more detailed survey.

3. Abundance Patterns

In the previous section we illustrated the general behavior of the buildup of metallicity in the Clouds over time. It is clear from the age-metallicity relations in Figure 1 that chemical enrichment in the Clouds has followed a different pattern than in the Galactic disk. Here we discuss the behavior of the abundances of two elements that show that the Clouds have not only undergone a different overall rate of chemical enrichment, but that the details in the abundance distributions are also different. We choose to discuss oxygen and the heavy, neutron-rich elements.

Using the references cited in Section 2, which are used to define the current metallicity, we can also define the current oxygen abundances in the Clouds: [O/Fe] for the LMC is about -0.2 and about -0.1 for the SMC. At the respective values of [Fe/H] for each Cloud, the Galactic disk values of [O/Fe] are $+0.0$ (at the LMC [Fe/H]) and $+0.2$ (at the SMC [Fe/H]). Thus, both Clouds seem to be deficient in O, relative to Fe, by about 0.2-0.3 dex in comparison to the Galactic disk. These abundances are taken from a number of studies, suggesting that this is a real effect. Oxygen is made preferentially in the most massive stars, while Fe comes from both massive, core-collapse supernovae (SN II) and the (presumably) binary SN Ia's. It is possible that in small systems, such as the Clouds, very massive stars are formed in a few starbursts (over the age of the galaxy), while lower mass stars (such as the members of the presumed binary progenitors of the SN Ia's) are formed more continuously and thus eject Fe at a more constant rate, leading to a decreased O/Fe ratio.

Another peculiar chemical signature in the LMC and SMC is found in an examination of the heavy, neutron-rich elements produced by neutron-capture nucleosynthesis. These nuclei are separated into two fairly distinct families consisting of isotopes produced by very slow neutron captures (where successive neutron captures for a given nucleus are separated by months to years - the s-process), and those isotopes produced by extremely rapid neutron captures (where dozens of successive neutron captures occur on a timescale of a few seconds - the r-process). The s-process elements are dominated by production in low- to intermediate-mass asymptotic giant branch (AGB) stars, while the r-process is believed to occur during events associated with the core collapse of a SN II. All of the various studies mentioned in Section 2 tend to find that the Clouds are somewhat overabundant in the heavy elements, relative to Fe. The overabundance pattern is characteristic of the r-process and not the s-process: this aspect of the pattern was emphasized by Plez, Smith, & Lambert (1993). In Figure 2 are summarized the abundances of the heavy elements (heavier than Sr), as tabulated by Luck et al. (1998) and Russell & Dopita (1992), for both

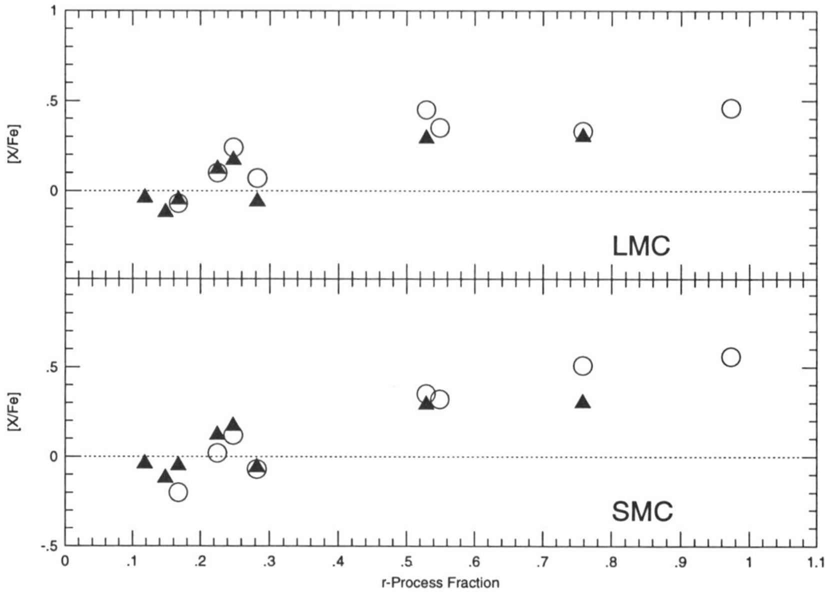

Figure 2. Heavy-element abundance ratios versus the r-process fraction for elements heavier than Sr. The open circles are from Luck et al. (1998), while the filled triangles are from Russell & Dopita (1992). The Cloud heavy-element abundances are dominated by an r-process component.

Clouds (these are the most complete and extensive studies of the heavy elements). We plot values of [X/Fe] versus the fraction of the element produced by the r-process (in the standard solar system abundance distribution), with these r-process fractions taken from Sneden et al. (1996). The trend in Figure 2 is obvious: the larger the r-process contribution, the more overabundant a particular element is in the Clouds. Heavy-element enrichment in the LMC and SMC has been dominated by the r-process when compared to the Galactic disk.

The chemical evolutionary implications of Figure 2 are not yet clear; recall from our earlier discussion of oxygen that the Clouds appear to be deficient in O/Fe. Figure 2 indicates that, contrary to the oxygen underabundance, the r-process is overabundant, yet both O and the r-process are believed to be created as a result of massive star evolution. The Magellanic Clouds are providing another piece of evidence that, to some degree, the production of oxygen and the r-process is decoupled.

The quintessential r-process element is europium, with an r-fraction of 0.97 (note the results for Eu in the top and bottom panels of Figure 2 with the largest values of r-fraction). In Figure 3 we summarize abundance results for Eu in various populations and plot [Eu/Fe] versus [Fe/H]. Europium is used as a proxy for the r-process component in these stellar populations. The open squares in Figure 3 represent the field halo stars (see references in the r-process section by Sneden in Wallerstein et al. 1997), which show a rather tight distribution of increasing [Eu/Fe] with decreasing metallicity, from [Fe/H]= +0.2 down to

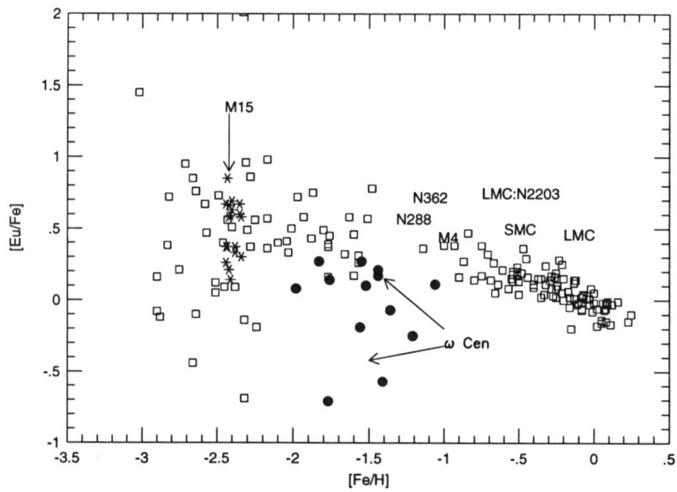

Figure 3. [Eu/Fe] as a function of [Fe/H] for various samples of stars. Open squares are Galactic field and halo stars, the six-pointed stars are from the globular cluster M15, the filled circles are from ω Cen, while the globular clusters M4 and NGC288 and NGC362 are labeled. The current LMC and SMC values are labeled, along with the LMC cluster NGC2203.

-1.0. As [Fe/H] decreases past -1.5, the scatter in [Eu/Fe] increases, becoming enormous at very low metallicities. This scatter is real and reflects differing amounts of r-process material at a given Fe abundance. Superimposed on this field star pattern, we present the results for the mono-metallicity globular cluster M15 ([Fe/H]= -2.4) from Sneden et al. (1997), which show significant scatter in [Eu/Fe]. Results for the luminous globular cluster ω Cen (Smith, Cunha, & Lambert 1995; Smith et al. 1999) are shown and stand out both for their scatter in Eu and Fe and for occupying a part of the diagram characterized by low [Eu/Fe] (relative to field stars within the [Fe/H] range of ω Cen). We also show results from Smith & Suntzeff (1999) for the globular clusters M4, NGC288 and NGC362. The current LMC value is indicated by the LMC label and we also label the current value of [Eu/Fe] for the SMC. In addition, the somewhat metal-poor LMC cluster NGC2203 is shown (Smith & Suntzeff 1999). Note how both Clouds fall above the other samples in [Eu/Fe] at a given [Fe/H]. The admittedly limited range in [Fe/H] spanned by the current LMC value and that for NGC2203 suggests that the high r-process component in the LMC has persisted over at least part of its evolution. A final conclusion from Figure 3 is not clear, but certainly points towards a qualitatively different type of chemical evolution occurring in the Clouds, especially in comparison to a system such as ω Cen.

4. Conclusions

The age-metallicity relations of the LMC and SMC indicate that chemical enrichment has followed different rates and behaviors in both Clouds: the LMC experienced a rapid enrichment more than 10 Gyr ago, followed by apparent chemical quiescence until about 2 Gyr ago, when another enrichment phase (still going on) occurred. The SMC has undergone a slower, but steadier, chemical enrichment. The current abundance of oxygen in both Clouds reveals them to be underabundant in O, relative to Fe, when compared to the Galactic disk (by about 0.2-0.3 dex). The r-process abundances, however, are strongly overabundant, relative to Fe, in the LMC and SMC. These abundances tell us that the chemical history of the Clouds is quite distinct from known Galactic populations.

Acknowledgments. We are grateful to Ed Olszewski, Nick Suntzeff, and Don Garnett for their usual helpful conversations and insights and to Katia Cunha for valuable comments on this manuscript. This research is sponsored in part by the National Science Foundation (AST96-18459).

References

Bica, E., Geisler, D., Dottori, H., Claria, J. J., Piatti, A. E., & Santos, J. F. C., Jr. 1998, AJ, 116, 723

Da Costa, G. S., & Hatzidimitriou, D. 1998, AJ, 115, 1934

Dopita, M. A. 1996, in *The Sun and Beyond*, H. C. Trung & S. Vauclair, Ed. Frontieres: Gif-sur-Yvette

Edvardsson, B., Andersen, J., Gustafsson, B., Lambert, D. L., Nissen, P. E., & Tomkin, J. 1993, A&A, 275, 101

Geisler, D., Bica, E., Dottori, H., Claria, J. J., Piatti, A. E., & Santos, J. F. C., Jr. 1997, AJ, 114, 1920

Luck, R. E., Moffett, T. J., Barnes, T. G., III, & Gieren, W. P. 1998, AJ, 115, 605

Olszewski, E. W., Schommer, R. A., Suntzeff, N. B., & Harris, H. C. 1991, AJ, 101, 515

Plez, B., Smith, V. V., & Lambert, D. L. 1993, ApJ, 418, 812

Richtler, T., Spite, M., & Spite, F. 1989, A&A, 225, 351

Russell, S. C., & Dopita, M. A. 1992, ApJ, 384, 508

Smith, V. V., Cunha, K., & Lambert, D. L. 1995, AJ, 110, 2827

Smith, V. V., & Suntzeff, N. B. 1999, in preparation

Smith, V. V., Cunha, K., Suntzeff, N. B., & Lambert, D. L. 1999, in preparation

Sneden, C., McWilliam, A., Preston, G. W., Cowan, J. J., Burris, D., & Armosky, B. J. 1996, ApJ, 467, 819

Sneden, C., Kraft, R. P., Shetrone, M. D., Smith, G. H., Langer, G. E., & Prosser, C. F. 1997, AJ, 114, 1964

Spite, F., Spite, M., & Francois, P. 1989, A&A, 210, 25

Spite, F., Barbuy, B., & Spite, M. 1993, A&A, 272, 116

Wallerstein, G., et al. 1997, Rev.Mod.Phys., 69, 995

Discussion

Sidney van den Bergh: How is one to reconcile the apparent difference between the r- and s-process yields in the LMC, SMC and Galaxy with the apparent universality of the luminosity function of star formation reported by Phil Massey this morning?

Smith: I don't have an answer to this question but just point out that we do not understand completely how mass is returned to the ISM as a function of stellar mass.

Michael Dopita: I think that we need to worry that with the lower mass of the Magellanic Clouds we may have lost a good deal of the α-process elements through galactic winds. This would be consistent with the lower yields that chemical evolution models require for these systems.

Smith: Yes, there may be selective loss of elements depending upon the velocity of the mass ejection.

Norbert Langer: I agee that one has to worry about certain elements getting selectively expelled from dwarf galaxies. However, I think this could actually explain the high s/r - ratio in ω Centauri, since the s-process elements are ejected by stars with only few km s^{-1} while the r-process stuff is released by supernovae. It may be hard to explain the low s/r-ratio or the high r/s-rate in the MCs this way.

Smith: I agree with you on these comments.

Dominik Bomans: One point not touched upon is the mixing of elements in the Clouds. Do the data give new insights in how well mixed the MCs are?

Smith: The spectroscopic results that I have reviewed suggest that the current gas in both Clouds is well-mixed. The dispersion in current abundances is $\lesssim 0.2$ dex.

New Views of the Magellanic Clouds
IAU Symposium, Vol. 190, 1999
Y.-H. Chu, N.B. Suntzeff, J.E. Hesser, and D.A. Bohlender, eds.

Element Abundances in Magellanic Cloud H II Regions from Carbon to Argon[1]

Donald R. Garnett

Astronomy Dept., University of Minnesota, Minneapolis, MN 55455 and Steward Observatory, University of Arizona, Tucson, AZ 8521

Abstract. I review measurements of heavy element abundances within H II regions in the Magellanic Clouds, highlighting in particular improved determinations of carbon abundances based on UV spectroscopy with *Hubble Space Telescope.* In general, the Magellanic Cloud H II regions show average underabundances in O, Ne, and S (relative to their Galactic counterparts) that are similar to those measured in Magellanic Cloud stars. However, comparison of stars and ionized gas shows discrepancies in C and N abundances that may be related to recently recognized mixing processes that may be operating in massive stars.

1. Introduction

Spectroscopy of photoionized nebulae provides a very useful probe of the gas phase composition of the interstellar medium (ISM), an important observational constraint for models of galaxy evolution. One motivation for studying the ionized gas phase is trivial: many H II regions are very luminous, and so the emission-line spectroscopy is relatively easy to obtain. This means that H II regions typically provide our first measurements of element abundances in galaxies other than our own, and this was the case for the Magellanic Clouds. More seriously, spectroscopy of H II regions probes the interstellar gas on a wide range of scales, from kiloparsecs down to sub-parsec scales (in the Magellanic Clouds). This allows one to analyze the homogeneity of the ISM on a variety of scales and thus to address questions of mixing and transport of stellar-enriched material in the ISM. The composition of the ISM is also a fundamental parameter in the determination of important ISM physical quantities such as the cooling function of the gas, the conversion factor from CO intensity to molecular hydrogen column density, the dust-to-gas ratio of the ISM, etc. For such issues the composition of the gas itself is more useful than the stellar abundances. Third, a comparison of the ISM composition with that of young stars can provide illumination on physical processes which may be operating in stars and gas, for example, mixing processes within stars (see Venn, this volume), or the composition of interstellar

[1]Based in part on observations with the NASA/ESA Hubble Space Telescope obtained at the Space Telescope Science Institute, which is operated by the Association of Universities for Research in Astronomy, Inc., under NASA contract NAS5-26555.

dust grains. Studies of abundances in both the ISM and in stars are thus highly complementary.

2. Abundance Data

2.1. Published Surveys

Table 1 lists the sources of H II region spectroscopy used to prepare this review. A few things can be noted about this list. First, a great deal of spectroscopic data has been accumulated for some 20 H II regions each in the LMC and SMC. At the same time, one can see that, with few exceptions, the sources are 15-25 years old, and little recent work has been done using efficient long-slit CCD spectrographs. This is a bit of a shame, since improved instrumentation often leads to new discoveries in familiar objects. In particular, the Magellanic Clouds, because of their well-constrained distances and low interstellar reddening, provide fertile grounds for comparing the observed spectrum with predictions of photoionization models and computed stellar atmospheres (e.g., see Oey & Massey 1995, Wilcots 1994a,b). Little mid-IR and far-IR spectroscopy has been done for H II regions in the Clouds as well, and we eagerly await the results of the ISO programs doing spectroscopy of H II regions in the clouds, examples of which have been shown at this conference (Thornley et al., Monetti et al., this volume).

Table 1. Bibliography of LMC and SMC H II Region Spectroscopy

Reference	Notes
Peimbert & Torres-Peimbert 1974, 1976	
Dufour 1975	
Dufour & Harlow 1977	
Pagel, Edmunds, Fosbury, & Webster 1978	
Dufour, Shields, & Talbot 1982	UV (with IUE)
Dennefeld & Stasińska 1983	
Heydari-Malayeri et al. 1985-1988	
Mathis, Chu, & Peterson 1985	30 Doradus
Rosa & Mathis 1987	30 Doradus
Russell & Dopita 1990	

Average LMC and SMC interstellar abundances for elements determined from optical spectroscopy (O, N, Ne, S, Ar) are essentially unchanged from the last review by Dufour (1990). For the rest of this review, therefore, I will concentrate on a discussion of more recent work.

2.2. A Detailed Spatially-Resolved Study of 30 Doradus

30 Doradus is certainly the best-studied H II region in the Magellanic Clouds, because it is the nearest example of a giant H II region and because of the many interesting stellar and interstellar phenomena found within the region. Some early detailed spectroscopic studies of 30 Dor hinted that the region was quite homogeneous in its composition (Mathis, Chu & Peterson 1985; Rosa & Mathis

1987). Perhaps the ultimate study can be found in the thesis of Kinkel (1993), who analyzed new and previously published spectra for 76 separate spatial positions in 30 Dor, including positions outside the bright filaments. Kinkel's work on 30 Doradus can be considered a benchmark study in the field of giant H II regions and his analysis of the 30 Dor spectra uncovered a number of noteworthy features:

1. There is an electron temperature gradient within 30 Doradus, in the sense that the derived [O III] temperature systematically *increases* as the fraction of oxygen in the form O^{+2} *decreases*. Such a variation can be understood as a combination of "radiation hardening" and less efficient cooling in the O^{+} zone, leading to higher T_e in the singly-ionized zone. The result is that the derived oxygen abundance is correlated with O^{+} fraction if a single average value for T_e is assumed. Taking into account the temperature gradient leads to uniform derived O/H across the nebula. However, the difference in O/H between the two cases is small, only 0.03 dex for 30 Doradus.

2. Derived abundances for S and Ar using standard ionization corrections for Ar^{+3} and S^{+3}, correlate with O^{+}/O, in the sense that the standard corrections appear to predict too much Ar^{+3} and S^{+3} in high-ionization regions. This may indicate problems with ionization cross sections or stellar atmosphere fluxes at high photon energies, and this remains a problem for future study.

3. An apparent enhancement of heavy element abundances at one position in 30 Doradus, reported by Mathis & Rosa (1987), was not confirmed.

The results of Kinkel's analysis show that the derived abundances across 30 Dor are remarkably uniform: $\log \mathrm{O/H} = -3.71$, with an rms dispersion of only ± 0.043 dex. Based on this uniformity, Kinkel places an upper limit of $t^2 < 10^{-3}$ (in terms of Peimbert's t^2 parameter) for electron temperature fluctuations. These results place severe limits on the effects of shocks on the observed spectra of giant H II regions, and on the pollution of the H II region by supernova ejecta.

2.3. Carbon in H II Regions from *HST* Observations

Measurements of carbon in H II regions have been much more limited than for oxygen, because the common ionization states have no strong optical emission lines; the strongest C^{+} transitions are in the UV or far-IR, while the strongest transitions of C^{+2} are in the UV. The IUE satellite permitted the first survey of carbon abundances in Magellanic Cloud H II regions (Dufour, Shields, & Talbot 1982), indicating significantly reduced C/O abundances ratios compared to the Galaxy. On the other hand, measurements of C in cool supergiants in the Clouds indicated much higher carbon abundances than the IUE results (Spite, Barbuy, & Spite 1989).

UV spectroscopy with *Hubble Space Telescope* greatly improves our ability to measure accurate C/O ratios in H II regions. The region 1600-2000 Å contains a very nice series of emission lines from C^{+2}, O^{+2}, N^{+2}, Ne^{+2}, and Si^{+2}, with similar excitation energies; similarly, the region 2100-2500 Å includes transitions from C^+, O^+, and N^+. Measurements of these UV lines together allow one to derive abundance ratios for these elements with small uncertainties from differential reddening and temperature fluctuations. My collaborators and I have been obtaining UV spectra in extragalactic H II regions, including two in the LMC and one in the SMC, to derive carbon abundances with greatly reduced uncertainties.

Our C abundances for the LMC and SMC based on *HST* data are listed in Table 2. The LMC value is the average for 30 Dor (Garnett et al. 1995) and N44C (Garnett, Galarza, & Chu 1999), while the SMC value is a single measurement for N88A (Kurt et al. 1999). Note that the SMC-N88A value is about 0.3 dex higher than the average value for SMC regions derived from *IUE* measurements (see Dufour 1990); it is not clear if this is a real difference (perhaps due to variations in grain abundances), or if it is due to systematic uncertainties. More *HST* UV spectroscopy is needed to understand this difference. Nevertheless, our measurements so far indicate with high confidence that C is significantly underabundant with respect to O in the SMC, compared to the Galaxy, while in the LMC C/O is at most slightly lower than in the solar neighborhood, taking into account a possible modest correction for C and O in dust grains.

3. Average H II Region Abundances and Comparison with Stars

Average heavy element abundances for H II regions in the Magellanic Clouds from optical/UV spectroscopy are listed in Table 2.

Table 2. Average H II Region Abundances in the Magellanic Clouds

Galaxy	log O/H	log C/H	log N/H	log Ne/H	log S/H	log Ar/H	log Si/H
LMC	−3.6	−4.1	−5.1	−4.4	−5.3	−5.8	−5.3
SMC	−4.0	−4.6	−5.5	−4.8	−5.7	−6.1	−5.7

The element underabundances in the Clouds relative to the Galaxy are shown in Figure 1 as the filled circles, along with a comparison to values for various kinds of stars in the Magellanic Clouds. For this figure I compare Magellanic Cloud objects to their solar neighborhood counterparts, i.e., LMC H II regions vs. the Orion Nebula (Esteban et al. 1998), SMC B stars vs. solar neighborhood B stars, etc. This follows the practice of Pagel (1993).

Figure 1 shows that the differential abundances for O, Ne, and S are in good agreement for both stars and H II regions (no stellar data is available for Ar, which shows the same underabundance as O-Ne-S in the H II regions). Silicon shows a curious enhancement in the H II regions relative to stars; however, Si suffers density-dependent depletion onto grains, and this differential comparison with Orion may not reflect the true abundance difference.

On the other hand, the C and N abundances show significant differences between the H II regions and the young stars. C in the H II regions seems to be in good agreement with the B stars and A supergiants in both the LMC and SMC, but disagrees significantly with the cool supergiants in the SMC and disagrees marginally with the LMC cool supergiants. Unlike Si, C shows no

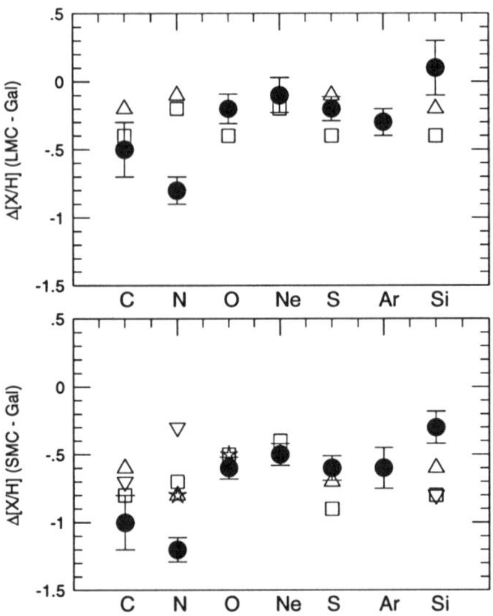

Figure 1. Comparison of H II region abundances with stellar abundances in the Magellanic Clouds. *Filled circles:* H II regions. *Open squares:* main sequence B stars (Rolleston et al. 1996; Lennon et al. 1996). *Open triangles:* cool supergiants (Hill et al. 1997; Luck et al. 1998). *Stars:* A supergiants (Venn 1999).

evidence for environment-dependent depletion (Sofia et al. 1997), so we assume that the differential abundance for the H II regions is correct. Nitrogen shows even greater discrepancies between the stars and gas, in the sense that the stars are systematically overabundant in N by up to a factor of 10 compared to the H II regions. Since nitrogen is not readily incorporated into grains (Mathis 1996), depletion effects can not be invoked to explain this discrepancy. The discrepancy with the H II regions increases as one goes from B stars to A supergiants to cool supergiants.

The differences in C and N may reflect mixing processes occurring inside massive stars, such as rotational mixing as discussed by Fliegner, Langer, & Venn (1996) and by Venn at this conference. At the same time, with a limited amount of UV data for the H II regions it is probably premature to draw firm conclusions. UV spectra for more H II regions in the Clouds are needed to rule out significant spatial variations in the carbon abundance. Deeper exposures are needed to detect the N III] 1750 Å emission multiplet, which will yield definitive nitrogen abundances. Future observations of both H II regions and stars with *HST* and large ground-based telescopes will yield tremendous insight into the properties of both stars and gas in the Clouds.

Acknowledgments. The work on carbon abundances in H II regions is supported in part by NASA-LTSARP grant NAG5-6416. Thanks go to Michael Rosa for providing me a copy of Ulrich Kinkel's fine PhD thesis. I also thank Kim

Venn for vivid discussions of stellar abundances in the Magellanic Clouds, and for pointing out a misunderstanding in my evaluation of the stellar abundances.

References

Dennefeld, M., & Stasińska, G. 1983, A&A, 118, 234

Dufour, R.J. 1975, ApJ, 195, 315

Dufour, R.J. 1990, in *Evolution in Astrophysics*, ESA SP-310, 117

Dufour, R.J., & Harlow, W.V. 1977, ApJ, 216, 706

Dufour, R.J., Shields, G.A., & Talbot, R.J. 1982, ApJ, 252, 461

Esteban, C., et al. 1998, MNRAS, 295, 401

Fliegner, J., Langer, N., & Venn, K.A. 1996, A&A, 308, L13

Garnett, D.R., et al. 1995, ApJ, 443, 64

Garnett, D.R., Galarza, V., & Chu, Y.-H. 1999, in preparation

Heydari-Malayeri, M., & Testor, G. 1985, A&A, 144, 98

Heydari-Malayeri, M., & Testor, G. 1986, A&A, 162, 180

Heydari-Malayeri, M., Niemela, V.S., & Testor, G. 1987, A&A, 184, 300

Heydari-Malayeri, M., Le Bertre, T., & Magain, P. 1988, A&A, 195, 230

Hill, V., Barbuy, B., & Spite, F. 1997, A&A, 323, 461

Kinkel, U. 1993, PhD thesis, University of Heidelberg

Kurt, C.M., et al. 1999, ApJ, 518, 246

Lennon, D.J., et al. 1996, A&A, 314, 243

Luck, R.E., Moffett, T.J., Barnes, T.G. III, & Gieren, W.P. 1998, AJ, 115, 605

Mathis, J.S. 1996, ApJ, 472, 643

Mathis, J.S., Chu, Y.-H., & Peterson, D.E. 1985, ApJ, 292, 155

Oey, M.S., & Massey, P. 1995, ApJ, 452, 210

Pagel, B.E.J. 1993, in *New Aspects of Magellanic Cloud Research*, B. Baschek, G. Klare, and J. Lequeux, Berlin: Springer-Verlag, 330

Pagel, B.E.J., et al. 1978, MNRAS, 184, 569

Peimbert, M., & Torres-Peimbert, S. 1974, ApJ, 193, 327

Peimbert, M., & Torres-Peimbert, S. 1976, ApJ, 203, 581

Rolleston, W., et al. 1996, A&A, 315, 95

Rosa, M.R., & Mathis, J.S. 1987, ApJ, 317, 163

Russell, S.C., & Dopita, M.A. 1990, ApJS, 74, 93

Sofia, U.J., Cardelli, J.A., Guerin, K.P., & Meyer, D.M. 1997, ApJ, 482, 105

Spite, M., Barbuy, B., & Spite, F. 1989, A&A, 222, 35

Venn, K.A. 1999, ApJ, 518, 405

Wilcots, E. 1994a, AJ, 107, 1338

Wilcots, E. 1994b, AJ, 108, 1674

Discussion

Giuseppe Bono: Did you estimate the helium-to-metal enrichment ratio?

Garnett: I did not look at He. The helium abundances will require new corrections for collision excitation; current estimates for electron densities are not adequate to do this accurately. Because we know that C and N abundances are under-abundant with respect to O, the $\Delta Y/\Delta Z$ relation will be slightly steeper, but not very much.

Part 4. Detailed Chemical Abundances

Section B. Poster Papers

Despina Hatzidimitriou fields questions during her SMC review.

New Views of the Magellanic Clouds
IAU Symposium, Vol. 190, 1999
Y.-H. Chu, N.B. Suntzeff, J.E. Hesser, and D.A. Bohlender, eds.

On the Star-formation History in the LMC: Observations of the Interstellar $C^{18}O/C^{17}O$ Ratio

Arto Heikkilä and Lars E.B. Johansson

Onsala Space Observatory, S-439 92 Onsala, Sweden

Hans Olofsson

Stockholm Observatory, S-133 36 Saltsjöbaden, Sweden

1. Introduction

The re-cycling of gas between stars and the interstellar medium (ISM) leads to a gradual metal-enrichment of a galaxy. Accordingly, information on the chemical evolution of a galaxy, e.g., its star-formation history (SFH), is contained in the chemical composition of the ISM. In this context, the abundance ratio of the rare oxygen isotopes, $^{18}O/^{17}O$ (usually taken as the $C^{18}O/C^{17}O$ column density ratio), appears to be a particularly promising probe of the SFH. According to present understanding of stellar nucleosynthesis, ^{17}O is mainly produced in intermediate-mass stars (say a few to ten $M_{\odot}$) while ^{18}O is synthesised in massive stars (say $>10\,M_{\odot}$) (e.g., Prantzos et al. 1996). Thus, the $^{18}O/^{17}O$ abundance ratio possibly reflects the relative number of massive stars compared to intermediate-mass stars, and thereby (qualitatively) constrains the SFH in terms of the average star-formation rate (SFR) and the initial mass-function (IMF). However, it should be remembered that the stellar nucleosynthesis of $^{17,18}O$ is not yet fully understood, leaving room for other interpretations of the $^{18}O/^{17}O$ ratio.

2. Observations

We have detected the rotational $J = 2-1$ spectral line emission of the carbon monoxide isotopomers ^{13}CO, $C^{18}O$ and $C^{17}O$ in five molecular clouds in the LMC. Four of them (N44BC, N113, N159W, N214DE) are associated with H II regions while the fifth (N132D) is associated with a supernova remnant. The $^{13}CO(J = 2-1)$ line was detected and upper limits to the $C^{18}O(J = 2-1)$ line were obtained in three additional clouds (N159S and 30Dor-10 in the SMC, and N27 in the SMC). In N27 also the $C^{17}O(J = 2-1)$ line was observed. The observations were conducted with the Swedish-ESO Submillimetre Telescope (SEST) located at La Silla, Chile.

3. Results

The abundance ratio estimates for N44BC, N113, N159S, N159W N214DE and N27 are given in Table 6 in Heikkilä et al. (1998), whereas ^{13}CO and $C^{18}O$ data in 30Dor-10 and $C^{17}O$ data in N27 can be found in Heikkilä et al. (1999). In N132D we estimate the $^{13}CO/C^{18}O$, $^{13}CO/C^{18}O$ and $C^{18}O/C^{17}O$ abundance ratios to be 25 ± 8, 44 ± 20 and 1.8 ± 0.9, respectively. The resulting averages in the LMC (five sources) then are 28 ± 4, 46 ± 8 and 1.6 ± 0.3, respectively. The $C^{18}O/C^{17}O$ in the LMC is significantly lower than in the Galactic ISM (by a factor of two) and in centres of starburst galaxies (by a factor of five).

4. Discussion

Provided that current theories of the the stellar nucleosynthesis of $^{17,18}O$ apply, the low $^{18}O/^{17}O$ ratio found in the LMC suggests that, in the past, massive stars have contributed only little to the metal enrichment of the ISM in the LMC. Such a dominance of less massive stars points towards *a low average SFR and a steep IMF in the LMC in the past.* This contrasts the present star-formation activity in some parts of the LMC, e.g., the 30 Dor area, where the SFR is clearly high and the IMF has been estimated to be close to Salpeter. On the other hand, Massey et al. (1995) have estimated a steep IMF for field stars in the Galactic halo, the LMC and the SMC. Hence, we suggest that *during a significant fraction of the past, stars in the LMC were born mainly as field stars rather than in clusters.* In fact, this agrees well with the apparent lack of medium-old (~3–14 Gyr) clusters in the LMC (e.g., Da Costa, these proceedings). Since the young clusters are more metal-rich than the old ones, the gas must have undergone metal-enrichment also during the era without cluster-formation. One may speculate that field stars possibly form due to a more gentle star-formation activity, whereas clusters originate from more violent processes, such as starbursts.

Moreover, our data indicates a correlation between the $^{18}O/^{17}O$ ratio and metallicity. This can be tested by observations of molecular clouds in the SMC. To date, we have obtained only upper limits to the $C^{17,18}O$ line emission in the SMC.

Finally, since the observed clouds are located in different parts of the LMC, the $^{18}O/^{17}O$ ratio appears to be rather constant spatially, implying a well-mixed ISM and/or a star-formation activity that took place globally, perhaps induced by a tidal interaction or by a large-scale gas-instability.

References

Heikkilä, A., Johansson, L.E.B., & Olofsson, H. 1998, A&A, 332, 493

Heikkilä, A., Johansson, L.E.B., & Olofsson H. 1999, A&A, 344, 817

Massey, P., Lang, C.C., De Gioia-Eastwood, K., & Garmany, C.D. 1995, ApJ, 438, 188

Prantzos, N., Aubert, O., & Audouze, J. 1996, A&A, 309, 760

New Views of the Magellanic Clouds
IAU Symposium, Vol. 190, 1999
Y.-H. Chu, N.B. Suntzeff, J.E. Hesser, and D.A. Bohlender, eds.

Interstellar Abundances in the Magellanic Clouds

D. E. Welty, P. C. Frisch, L. M. Hobbs, and D. G. York

University of Chicago, 5640 S. Ellis Ave., Chicago, IL 60637, USA

J. C. Blades

STScI, 3700 San Martin Dr., Baltimore, MD 20771, USA

J. T. Lauroesch

Northwestern University, Evanston, IL 60208, USA

G. Sonneborn

Goddard Space Flight Center, Code 681, Greenbelt, MD 21218, USA

Abstract. We discuss the relative gas-phase abundances found for the predominantly neutral interstellar clouds — located in the Galactic disk and halo, in the LMC or SMC, and (perhaps) in between — along the lines of sight to Sk 108 in the SMC and to SN 1987A in the LMC.

In general, relative gas-phase interstellar abundances [e.g., N(X II)/N(Zn II)] reflect both the nucleosynthetic history of the gas and any selective depletion of elements into dust grains. *HST* GHRS echelle spectra (FWHM $\sim$ 4.2 km s^{-1}) of Si II, Cr II, Mn II, Fe II, and Zn II reveal at least 25 interstellar absorption components toward Sk 108 (Welty et al. 1997). The relative abundances for the SMC components are similar to those found for clouds in the Galactic halo (Fig. 1, top). Toward SN 1987A, high-resolution (3-5 km s^{-1}) optical spectra of Ca II and Na I (Vidal-Madjar et al. 1987; Magain 1987) reveal at least 46 components, which may be associated with 10 component groups discernible in lower resolution (16-28 km s^{-1}) UV spectra of many species obtained with *IUE.* The relative abundances for the main LMC component groups toward the SN resemble those found for warm, diffuse Galactic disk clouds (Fig. 1, top); the relative abundances for the "intermediate velocity" groups are more similar to those found for the Galactic halo and SMC clouds; one group of Galactic halo clouds may be essentially dust-free (Welty et al. 1999). Since the stellar abundance ratios in the LMC and SMC are very similar those found locally in our Galaxy (for the elements we consider) (e.g., Russell & Dopita 1992), we conclude that the depletion patterns are also similar in the three galaxies — despite differences in overall metallicity, dust-to-gas ratio, and UV extinction curves. Detailed studies of additional MC lines of sight, particularly with larger $E(B-V)$ and N(H), are needed to confirm and extend that conclusion, however. Various diagnostic ratios suggest that the individual clouds in the various component groups toward SN 1987A are characterized by different physical conditions; most of the neutral

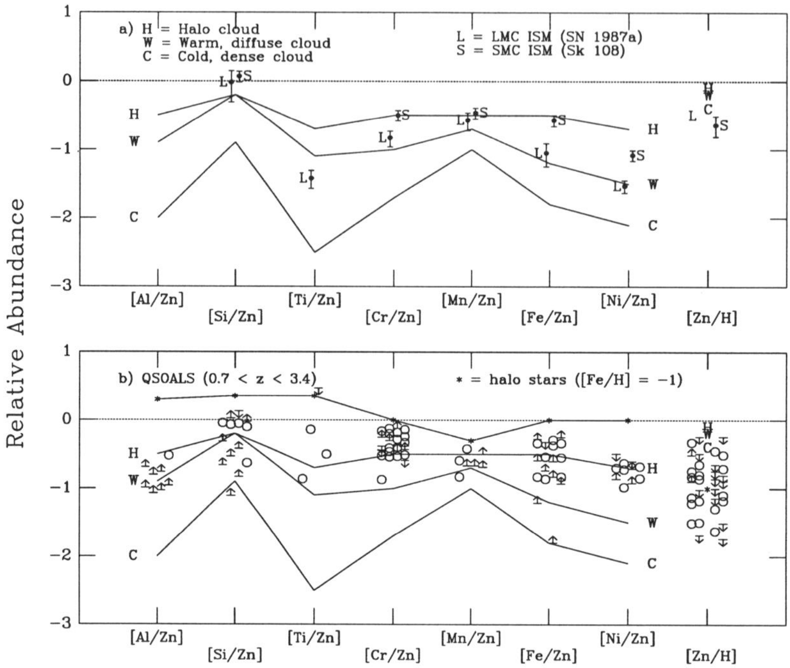

Figure 1. **(top)** Relative gas-phase abundances $\{[X/Zn] = \log[N(\mathrm{X\ II})/N(\mathrm{Zn\ II})] - (A_X - A_{Zn})_\odot\}$ for main SMC gas toward Sk 108 and main LMC gas toward SN 1987A, compared with patterns seen in Galactic cold, dense clouds (C), warm, diffuse clouds (W), and halo clouds (H). **(bottom)** Relative gas-phase abundances seen for QSO absorption-line systems are similar, on average, to those in Galactic halo and SMC clouds. Individual systems exhibit a range of patterns, reflecting differences in nucleosynthetic history and/or depletions.

LMC gas has $T \lesssim 1500$ K, however. Determination of the relationships between gas, dust, and local physical conditions in the MC will aid in understanding the absorption-line systems seen toward distant QSO's, which also are characterized by low metallicities, low dust-to-gas ratios, mixtures of neutral and ionized components, and relative gas-phase abundances similar (on average) to those in the Galactic halo and SMC clouds (Fig. 1, bottom).

References

Magain, P. 1987, Nature, 329, 606

Russell, S. C., & Dopita, M. A. 1992, ApJ, 384, 508

Vidal-Madjar, A., et al. 1987, A&A, 177, L17

Welty, D. E., Frisch, P. C., Sonneborn, G., & York, D. G. 1999, ApJ, 512, 636

Welty, D. E., Lauroesch, J. T., Blades, J. C., Hobbs, L. M., & York, D. G. 1997, ApJ, 489, 672

New Views of the Magellanic Clouds
IAU Symposium, Vol. 190, 1999
Y.-H. Chu, N.B. Suntzeff, J.E. Hesser, and D.A. Bohlender, eds.

A Laboratory of Stellar Nuleosynthesis: Isotope Ratios in the Magellanic Clouds

Yi-nan Chin

Academia Sinica Institute of Astronomy & Astrophysics, PO Box 1-87 Nankang, 11529 Taipei, Taiwan

Abstract. In radio astronomy, interstellar isotope ratios have been measured for more than two decades towards different parts of the Milky Way and central regions of some star-burst galaxies. While signals are often too weak to detect rare isotopic species in relatively distant extragalactic sources, our Galaxy only provides an environment with limited metallicity range. Obviously, this constraint can be removed by observing isotopic species in the Magellanic Clouds, located only 50 – 60 kpc away from us. We thus observed isotope ratios of hydrogen, carbon, nitrogen, oxygen, and sulfur and the results are given in Table 1.

1. Introduction

It is clear that most of the elements (except of a few which were formed soon after the Big Bang) are synthesized in the interior of stars. As the source of stellar energy, nuclear reactions lead to the formation of a great variety of elements and isotopes. Although there are many reactions and therefore many parameters involved when trying to model stellar interiors, the synthesis of isotopes can be simplified by introducing a concept discriminating between "primary" and "secondary" nuclei. While the "primary" species are synthesized and released by massive stars or by an initial stellar generation, the synthesis of "secondary" nuclei either takes place in low mass stars or in stars that are formed at later times and that possess certain "seed" nuclei. As a consequence, the primary species are released earlier into the interstellar medium than the secondary species. Even this approach may be complex, since some nuclei are synthesized in more than one reaction scheme, possible involving both "primary" and "secondary" formation. Nevertheless, we can observe isotope abundance ratios as a function of metallicity. This allows us to determine how "primary" (or "secondary") a given species is with respect to another, thus providing important constraints on stellar nucleosynthesis and "chemical" evolution.

2. Observations

The observations were carried out during several periods between 1997 and 1998 using the 15-m Swedish-ESO Submillimetre Telescope (SEST) at La Silla, Chile. The $^{12}C/^{13}C$ ratio has been derived from the line intensity ratio of HCN to $H^{13}CN$. The opacity of the HCN(1–0) line has been determined by a comparison of intensities of its hyperfine components. Similarly, the $^{14}N/^{15}N$ ratio has been obtained by additional observations of $HC^{15}N$.

The D/H ratio has been obtained from the line intensity ratio of $H^{13}CO^+$ and DCO^+, as well as $H^{13}CN$ and DCN, by applying the $^{12}C/^{13}C$ ratio derived before. Fractionation is estimated applying a detailed astrochemical model.

The oxygen $^{18}O/^{17}O$ isotope ratio has been derived from the intensities of the respective isotopic species of the CO molecule and applying the $^{12}C/^{13}C$ ratio from HCN observation. Similar to the oxygen isotope ratio, the sulfur isotope ratios have been calculated by observing CS isotopic species and applying the $^{12}C/^{13}C$ ratio.

3. Discussion

In Table 1 we present a brief summary of isotope abundance ratios obtained in the Magellanic Clouds, our Galaxy, and nearby star-burst nuclei. While the [H/D] ratio is related to cosmology, other ratios are suitable tracers of stellar nucleosynthesis. The nitrogen and oxygen ratios determined in the LMC (which represents a less evolved environment) suggest some corrections to the stellar nucleosynthesis model. For instance, the very low $^{14}N/^{15}N$ ratio indicates that ^{15}N has to be synthesized by massive stars. This is consistent with the recent massive star models which take the effects of rotation on the stellar structure and nucleosynthesis into account (Langer et al. 1999).

Table 1. Isotope abundance ratios in various astronomical environments†

Isotope	Magellanic Clouds	Galactic Center	Inner Disk	Local ISM	Solar System‡	Starburst Nuclei
$^{12}C/^{13}C$	62 ± 5	25 ± 5	40 ± 10	70 ± 10	90	≈ 50
$^{14}N/^{15}N$	114 ± 14	900 ± 200	375 ± 60	450 ± 60	272	$>> 100$
$^{16}O/^{18}O$	> 2000	250 ± 30	300 ± 30	540 ± 30	499	150 – 200
$^{18}O/^{17}O$	1.8 ± 0.4	3.5 ± 0.2	3.6 ± 0.3	3.6 ± 0.3	5.3	> 8
$^{32}S/^{34}S$	18 ± 6	—	24 ± 5	24 ± 5	22.6	—
$^{34}S/^{33}S$	3.0 ± 0.7	—	6 ± 1	6 ± 1	5.6	—

†While isotope abundance ratios of hydrogen, carbon, and oxygen are summarized by Wilson & Rood (1994), those of sulfur are measured by Chin et al. (1996).

‡The values of the Solar System represent the isotope abundance ratios in the LISM 4.5×10^9 years ago.

References

Chin, Y.-N., Henkel, C., Whiteoak, J.B., Langer, N., & Churchwell, E.B. 1996, A&A, 305, 960

Langer, N., Heger, A., & García-Segura, G. 1999, RvMA, 11, 57

Wilson, T.L., & Rood, T.R. 1994, ARA&A, 32, 191

Discussion

George Sonneborn: How did you derive the D/H value of 2×10^{-5}?

Chin: By observing the deuterated molecules DCO^+ and DCN. However, some assumptions (such as the same excitation condition) have to be made and, with a model simulation of chemical reaction, we can derive the D/H ratios. They are consistent toward different locations as well as by different molecular species. Conservatively, I would suggest this ratio to be $2 - 4 \times 10^{-5}$ in the LMC.

New Views of the Magellanic Clouds
IAU Symposium, Vol. 190, 1999
Y.-H. Chu, N.B. Suntzeff, J.E. Hesser, and D.A. Bohlender, eds.

Chemical Abundances and Physical Parameters of H II Regions in the Magellanic Clouds

R.E. Carlos Reyes[1]

Laboratório Nacional de Astrofísica, CNPq, Caixa Postal 21, CEP 37500-000, Itajubá, MG, Brazil

Seminario Permanente de Astronomía y Ciencias Espaciales, FCF, UNM San Marcos, Lima, Perú

Abstract.
I present the results of the photoionization modeling of eleven H II regions in the Magellanic Clouds (MCs). A comparison of their abundances with other systems are also presented.

1. Introduction

The chemical abundances and physical parameters of H II regions are important parameters to determine in order to understand how stars and galaxies evolve. The Magellanic Clouds offer us a unique opportunity to pursue such studies in low metalicity galaxies. In this contribution I present the results of the photoionization modeling with the code CLOUDY of five H II regions in the SMC and six in the LMC.

Optical data were collected from the literature, complemented by our own observations (Carlos Reyes et al. 1999a). UV spectra from the new IUE data bank were also used.

2. Discussion

A preliminary discussion of chemical abundances of H II regions in the MCs is presented in Carlos Reyes et al. (1999b). In Fig. 1 I show ϵ(C+N) versus [Fe/H] for K supergiants, versus [O/H] for B stars, or versus [Z_*] (geometric average of Ne/H, S/H and Ar/H) for H II regions for the data reported in Table 1. The solid line indicating [(C+N)/Fe]=[Fe/H] represents the behaviour of dwarf stars in our Galaxy. The H II region abundances in the Galaxy and the LMC are consistent with that of these young objects, whereas the H II regions in the SMC show a strong defficiency of C+N relative to K supergiants and B stars.

The physical parameters determined as the ionized mass varied in log from 1.8 to 4.5 $M_\odot$, the luminosities varied from 6.0 to 8.5 $L_\odot$, the ionization temperature varied from 4.62 to 4.72 for LMC and 4.7 for SMC (only two H II regions

[1]Visiting Astronomer and Fellowship support by CNPq (Conselho Nacional de Desenvolvimento Científico e Tecnológico), under contract 381314/97-9.

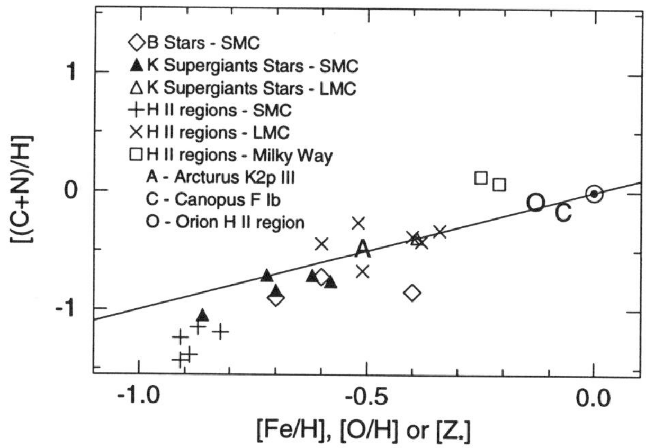

Figure 1. [(C+N)/H] versus [Fe/H] (or versus [O/H] for B stars, or $[Z_*]$ for H II regions).

in the SMC has a higher value of 4.8) and the filling factor from -4 to -1. Most of the H II regions are optically thin. Only 4 are optically thick.

Table 1. Average chemical abundances.

	ϵ(He)	ϵ(C)	ϵ(N)	ϵ(O)	ϵ(Ne)	ϵ(S)	ϵ(Ar)	C/O	N/O	$\epsilon(Z_*)$
LMC:										
H II[a]	10.93	8.03	7.37	8.33	7.73	6.82	5.97	-0.30	-0.96	6.84
K SG stars[b]	—	7.95	7.97	8.23	—	—	—	-0.28	-0.26	—
SMC:										
H II[a]	10.91	7.39	6.55	7.96	7.17	6.32	5.72	-0.57	-1.41	6.40
B Stars[b]	—	7.65	7.12	8.32	—	—	—	-0.67	-1.20	—
K SG stars[b]	—	7.57	7.52	8.03	—	—	—	-0.46	-0.51	—

[a]This work

[b]Hill et al. (1997). For H II regions of the Galaxy (Peimbert 1993; Esteban et al. 1998).

Acknowledgments. I am grateful to Dr. G. J. Ferland for kindly providing the code CLOUDY.

References

Carlos Reyes, R.E., et al. 1999a, in preparation

Carlos Reyes, R.E., et al. 1999b, in *Science with Gemini Workshop*, Florianópolis, SC, Brazil, in press

Hill, V., Barbuy, B., & Spite, M. 1997, A&A, 323, 461 and references therein

Esteban, C., et al. 1998, MNRAS, 295, 401

Peimbert, M. 1993, RevMexAA, 27, 9

Dennis Zaritsky reports on his photometric survey of the Clouds.

Part 5. Stellar Populations and Surveys

Section A. Invited Reviews

Jay Gallagher (right) tells Klaas de Boer "I'm under attack here!" (see page 579).

New Views of the Magellanic Clouds
IAU Symposium, Vol. 190, 1999
Y.-H. Chu, N.B. Suntzeff, J.E. Hesser, and D.A. Bohlender, eds.

Overview of the Field Stellar Population of the Magellanic Clouds

Bengt E. Westerlund

Astronomical Observatory, Uppsala University Box 515, S-751 20 Uppsala, Sweden

Abstract. The stellar populations in the LMC, the SMC, and the Bridge region are discussed and shown to be composed of a number of well separated generations. Asymmetries are seen in the distributions of the intermediate-age and youngest generations. The possibility is considered that all generations, apart from the oldest one, have been produced by bursts of star formation.

1. Introduction

Analytical studies of stellar populations require color-magnitude diagrams (CMD) and the luminosity functions (LF) of main-sequence (MS) and red-giant stars. Stars *below* a turn-off point at about $M_V \sim 3.5$–4.5 are useful for the determination of the stellar composition of the Clouds only if objects of different ages and metallicities can be separated. This may be difficult for low-mass stars of 10 Gyr and 1 Gyr as most isochrones for low-metal objects, $Z \leq 0.008$, appear to merge there. *Above* $M_V \sim 3.5$ isochrones differ appreciably for ages younger than ~10 Gyr and the effects of different metallicities are obvious. The structure of the red-giant clump gives important information about the composition of the older generations. The age distribution of some classes of variables and from the planetary nebulae also provide valuable clues to the underlying population.

Significant differences exist between the field populations of the LMC and the SMC. As a rule each generation in the SMC is younger than the corresponding one in the LMC. The SMC trails the LMC in evolution. The location of the various field populations in space causes no real problems in the LMC but is of major concern in the SMC. In the latter some classes of objects appear to form separate clouds in the line-of-sight. The old and intermediate-age populations define, however, a single body with, most likely, a more limited extension in the line of sight. A summary of the content of the generations in the Clouds may be found in Table 3.12 of Westerlund (1997).

2. The Youngest Populations

Most stars of the youngest generation have formed in super-associations (SAs). The supergiant shells (SGS) correspond to the SAs or sometimes form part of them. In several SAs a separation of the supergiants according to age is seen with the more evolved ones in the central regions and the OB stars in the outer parts. This may be due to a stochastic self-propagating star formation process (SSPSF) though contradictions exist. Young clusters are found inside the rim of SGS LMC-4, and the LFs for a number of small areas inside the LMC-4 indicate ages between 9 and 16 Myr without correlation with the distance to its center. One triggering event may have started star formation inside the *whole* SGS with the SSPSF acting at its rim, only (Braun et al. 1997). However, the clusters may have needed a longer time to form than single stars (Dopita 1987), and the age distribution within the SGS may have been achieved by a fast traveling SSPSF process.

Slight but important differences exist between the SAs: those in the West of the LMC have in the mean more evolved supergiants than those in the East. This causes the east-west asymmetry, seen in the FIR/6.3 cm brightness ratio – it is lower in the eastern parts – and in the diffuse X-ray emission.

2.1. The SMC Wing and the Bridge Region

The Wing extends $\sim 7°$ from the SMC core and is the beginning of a Bridge linking the SMC to the LMC. It may have been formed by tidal interaction with the LMC. HI clouds and 78 non-random groups of blue stars have been identified in it. The distances of the groups decrease from the West to the East. At the tip of the Wing the age is estimated to 10–60 Myr; further East it is 10–25 Myr. Star formation may have started in the Wing and caused a burst of formation in the Bridge (Demers & Battinelli 1998).

3. The Old and Intermediate-age Generation in the LMC

To the oldest generation, aged ≥ 10 Gyr, belong old globular clusters and a relatively rich population of RR Lyrae variables. The latter are about 1 Gyr younger and have a higher metallicity than the globular clusters. An old field population has been traced out to $\sim 7°$ from the LMC center. Photometric investigations using the *HST* have reached very faint magnitudes (see e.g., Elson, Gilmore, & Santiago 1997, hereafter EGS97; and Geha, Holtzman, & Mould 1998, hereafter GHM98). EGS97 analyze a field near the SE end of the LMC Bar, GHM98 a field near NGC1866 and two fields, N and NE of the center. The CMDs suggest similar stellar populations in these parts of the LMC.

Enhancement of the star formation rate may have occurred 6–8 Gyr ago in LMC regions East of the Bar, and about 2–3 Gyr ago in regions in the West and North (Vallenari 1997). Our reanalysis of the stellar populations in those fields and in fields observed by Battinelli & Demers (1998), EGS97, Hodge (1987), and Westerlund et al. (1995, 1998) shows that an old generation, $\sim$ 10–12 Gyr old, exists all over the LMC and that intermediate-age generations, 0.8–4 Gyr old, are found in many parts. Fields in the SW, e.g., the LW55 (Westerlund et al. 1995) and the LMC-5 (Vallenari 1997) fields deviate with steeper LFs and a lack of MS stars more luminous than $M_V \sim 1.5$–2.

Enhanced star formation at several epochs is observed in several fields. In the EGS97 field four separate generations are shown by peaks in the $(V-I)_0$ histograms for $20 \leq V \leq 22.5$ (Fig. 3 of EGS97). Those at intermediate colors represent generations aged slightly less than 1 Gyr and 2–4 Gyr.

The number of intermediate-age MS stars in a field depends upon the distance from the center of the LMC and on the star production rate in the region. Fields in (N2004) or close to presently star-forming regions (LMC-30, LMC-45, N1787 and SL196) are rich in intermediate-age stars, 0.2–0.6 Gyr (See Table 14 of Westerlund et al. 1998).

Bimodal RGCs are seen in the CMDs of many fields: one peak corresponds to red horizontal branch (RHB) stars, and one, more ill-defined, to younger, more massive and slightly more luminous stars, with ages as young as 0.3–0.6 Gyr.

The rather massive LMC Bar has contributed to establish star burst conditions by causing gas clouds to fall towards it rapidly. A burst may have occurred around 1 Gyr ago as judged from the LF of the EGS97 field and the period-luminosity (PL) relation of the long-period variables (LPV) in the central Bar. Intermediate-age stars dominate the Bar. Bar-induced star formation is seen in a number of other features: (i) A distinct group of LPVs exists with $M_{\rm MS} \sim 5$–$6\,M_\odot$ and an age around 60 Myr. (ii) The PL distribution of the Cepheids supports a burst about 60 Myr ago (Hughes & Wood 1990). (iii) There are a number of SWB II clusters, 30–70 Myr old (Dottori et al. 1996).

4. The Old and Intermediate-age Generations in the SMC

The only globular in the SMC, NGC121, is younger than those in the LMC. To the oldest generation belong the normal RR Lyrae stars, aged ≥ 10 Gyr and more metal-poor, $-1.4 \geq$ [Fe/H ≥ -2.0, than NGC121.

The lower metallicity of the SMC is a consequence of a slower evolution and shows e.g. as a lack of M giants, as odd carbon stars, and as overluminous RR Lyrae stars. The latter are 1.5 mag brighter than the normal ones (and not known in the LMC) with [Fe/H] $=-0.4$ and an age of 0.25 Gyr.

The CMD of the SMC disk (Hardy & Durand 1984, hereafter HD84) shows a young population with a MS indicating a lower age of ≤ 0.1 Gyr. The RGB matches well those of the old SMC clusters K3 and L1, aged 8–10 Gyr. A well developed RGC is seen and a relatively strong sub-giant branch; the latter indicating an intermediate-age population older than 3 Gyr. The distributions of the RGC stars in the SMC disk (HD84) and of those in the LMC disk (Hardy et al. 1984) agree well and peak at $M_V=0.3$ if the distance moduli are 18.5 and 18.9, and the absorptions $A_V=0.3$ and 0.1 for the LMC and the SMC, respectively. Also the subgiant and giant branches agree well after normalization at the peak.

About 3000 C stars have been identified in the SMC (Rebeirot et al. 1993; Morgan & Hatzidimitriou 1995; Kunkel et al. 1997). The majority is in the central 12 deg^2. The C stars in the Bridge region are few; they may be halo/disk objects. Normal C stars in the Clouds have $-6 < M_{\rm bol} < -3$ mag. In the SMC about 80 C stars with $M_{\rm bol} > -3$ mag have been identified (Rebeirot et al. 1993). No such objects have so far been found in the LMC.

The C stars belong kinematically to a spheroidal population of which also the PN may be part. Neither the C-star velocities nor those of the PN show the multiple peaks seen in samples of Population I objects. A possible velocity gradient of the C stars just East of the Bar may be interpreted as motions towards the LMC (Hardy et al. 1989).

The distribution of the RGC/HB stars in the outer parts of the SMC has been interpreted as showing that the depth varies strongly. In the far NE, ≥ 2 kpc from the center, the depth may be 15–20 kpc; in the SW 7–13 kpc. In the N and NW the corresponding values are 4–9 kpc, and between the latter and the first field the depth is 12–16 kpc. The NE feature is interpreted as lying in front of the main structure of the SMC (Hatzidimitriou & Hawkins 1989; Gardiner & Hawkins 1991; Hatzidimitriou & Cannon 1993). The apparent bimodality in the RGCs may, however, be caused by a stellar population mixture; stars from 0.3–2 Gyr to 10 Gyr may contribute. The former age group is identified by the MSs, the latter age group has been observed by Hatzidimitriou & Gardiner (1992) in fields 2–2.5 kpc from the optical center. If the RGC is composed of more than one generation, the depth of the SMC may be appreciably smaller than suggested.

References

Battinelli, P., & Demers, S. 1998, AJ, 115, 1472

Braun, J.M., Bomans, D.J., Will, J.-M., & de Boer, K.S. 1997, A&A, 328, 167

Demers, S., & Battinelli, P. 1998, AJ, 115, 154

Dopita, M.A. 1987, in IAU Symp.No.115, M. Peimbert & J. Jugaku, p.501

Dottori, H., Bica, E., Clariá, J.J., & Puerari, I. 1996, ApJ, 461, 742

Elson, R.A.W., Gilmore, G.F., & Santiago, B.X. 1997, MNRAS, 289, 157, EGS97

Gardiner, L.T., & Hawkins, M.R.S. 1991, MNRAS, 251, 174

Geha, M.C., Holtzman, J.A., Mould, J.R., et al. 1998, AJ, 115,1045, GHM98

Hardy, E., & Durand, D. 1984, ApJ, 279, 567, HD84

Hardy, E., Buonanno, R., Corsi, C.E., Janes, K.A., & Schommer, R.A. 1984, ApJ, 278, 592

Hardy, E., Suntzeff, N.B., & Azzopardi, M. 1989, ApJ, 344, 210

Hatzidimitriou, D., & Cannon, R.D. 1993, in New Aspects of Magellanic Cloud Research. R. Baschek, G. Klare, J. Lequeux: Springer-Verlag, p.17

Hatzidimitriou, D., & Gardiner, L.T. 1992, in The Stellar Population of Galaxies, eds. B. Barbuy & A. Renzini, p.157

Hatzidimitriou, D., & Hawkins, M.R.S. 1989, MNRAS, 241, 667

Hodge, P.W. 1987, PASP, 99, 730

Hughes, S.M.G., & Wood, P.R. 1990, AJ, 99, 784

Kunkel, W.E., Irwin, M.J., & Demers, S. 1997, A&AS, 122, 463

Morgan, D.H., & Hatzidimitriou, D. 1995, A&AS, 113, 539

Rebeirot, E., Azzopardi, M., & Westerlund, B.E. 1993, A&AS, 97, 603

Vallenari, A. 1997, ASP Conf.Ser. 112, 109

Westerlund, B.E. 1997, The Magellanic Clouds: Cambridge University Press

Westerlund, B.E., Linde, P., & Lyngå, G. 1995, A&A, 298, 39

Westerlund, B.E., Lundgren, K., Pettersson, B., & Koseij, E. 1998, A&A, 339, 385

Discussion

Despina Hatzidimitriou: Two comments: First, on the problem of the size of the red clump in the NE regions in the SMC: you suggested that this could be due to a young population of a few 10^8 yr. There are three pieces of evidence that I believe contradict that interpretation:

1. there do not seem to be enough young MS stars to account for the number of observed brighter clump stars;

2. if the brighter clump was indeed due to a younger population (and more metal rich according to the age-metallicity relation) then there should be some evidence of increased "size" of the RGB redwards. On the contrary, the RGB seems to be more extended blueward, just as you would expect if you just moved half the stars on the CMD upwards (distance modulus effect); and

3. there is magnitude (distance)-radial velocity correlation which can be more naturally understood in terms of a structural rather than an age effect.

The second comment refers to your reference to the age of the field population in the SMC varying with location. What I think is actually happening is that the mixture of older and younger populations is a function of radius. It is not a variation on the time when a major period of SF occurred.

Westerlund:

1. The younger part of the RGC should only be about 1/10 of the older. Maybe there is enough of a MS for this.

2. I have seen theoretical CMDs for 400 Myr and 4 Gyr with $Z = 0.008$ which represent the situation rather well without any obvious change in colour. (See also the poster by Harris & Zaritsky)

3. I cannot comment on this at present as I do not know it in detail.

New Views of the Magellanic Clouds
IAU Symposium, Vol. 190, 1999
Y.-H. Chu, N.B. Suntzeff, J.E. Hesser, and D.A. Bohlender, eds.

The LMC Intermediate and Old Populations

Edward W. Olszewski

Steward Observatory, University of Arizona, 933 N. Cherry, Tucson, AZ 85721, USA

Abstract. This review concentrates on the oldest populations in the LMC. Specifically, the LMC has a large extent and mass, it has star clusters coeval with those in the Milky Way halo, and more work needs to be done before we are convinced that there are similarly-aged field stars in the LMC. Further, all studied fields in the Clouds are different from Milky Way halo fields. Observations with the new generation of large telescopes can clear up the uncertainties noted in this review.

1. Preamble

Before we can discuss the halo population of the LMC we have to agree on what we mean by that term. It is not semantics, because the connotations of the term "halo" rest upon how we think the Milky Way evolved. Those connotations may not apply to other galaxies. Does the LMC have any clusters or field stars that are demonstrably the same age and metallicity as a typical metal-poor "old halo" globular cluster in the Milky Way? Since LMC field stars have now been traced to large distances from the LMC bar, does "halo" mean "far from the center," or "in an extended z-direction distribution" and/or "supported by its own velocity dispersion?" Even with such limited definitions it is difficult to place the oldest stars in the LMC comfortably into a simple definition of halo. This discussion has particular relevance in 1998 because the interpretation of the microlensing results (there was heated discussion at the Third Stromlo Symposium in August 1998) depends on the structures of the Magellanic Clouds themselves. We need to know if the LMC has no halo or a spherical halo, or if the LMC has significant outer structure or warps. Guides to the earlier literature can be found in Olszewski et al. (1996) or in Westerlund (1997).

2. Mass and Extent of LMC

What do we know about the mass and extent of the LMC? While we tend to think of the LMC as only a few kpc in size (the "SN 1987A" view), a very large size, perhaps $14° \times 16°$ ($1°$ is about 0.9 kpc), of the LMC has been known for a long time (summarized in de Vaucouleurs & Freeman 1973). In more recent times, Bothun & Thompson (1988) used a "parking lot" CCD camera to measure $D_{25}=9.1°$, a disk scale length of $1.67°$, and a diameter of at least $10°$. Irwin (1991) estimated a size of $23° \times 17°$ from star counts, and showed that the "halos" of the two Clouds essentially overlap. Nikolaev & Weinberg (1998) have estimated the LMC tidal radius (~11 kpc) based on USNO-A starcounts and on prototype 2MASS star counts. They then derive an LMC mass of $>1\times10^{10}$ $M_{\odot}$, given that the Milky Way's mass is $>10^{12}$ $M_{\odot}$.

Suntzeff et al. (1992) showed that the Reticulum (11.4° from LMC bar, or 10 kpc) and NGC 1841 (14.9°, 13 kpc) clusters had velocities consistent with LMC membership. Schommer et al. (1992) showed a sparsely populated rotation curve out to the distance of Reticulum. If one assumes that the rotation curve is flat, or if one uses the velocity dispersion of the oldest clusters, one derives a "massive" LMC of 1-2×10^{10} $M_{\odot}$. Such a galaxy has a luminous halo or dark halo or both. Kunkel et al. (1997) have measured the velocities of hundreds of newly discovered C stars out to beyond 10 kpc and have estimated an LMC mass of 0.6-1.5×10^{10} $M_{\odot}$. The velocity dispersion, ~30 km s^{-1}, of the oldest star clusters does imply a distribution with a z scale height of of a few kpc. It is known, for example in Walker (1991), that some clusters are at different distances from the fields on which they are projected, or are at different distances from the best-fit planar model. It was also noted by Schommer et al. (1992) that the velocities of some of the oldest clusters projected on the bar of the LMC cannot fit the planar model. The velocity dispersion of the Kunkel et al. C stars, 15 km s^{-1}, however, is not what is expected for a spherical LMC halo. These stars are presumably drawn from an intermediate-age population.

Other tracers of the z extent of the LMC halo may be old long-period variables with dispersion 33 km s^{-1} (Hughes et al. 1991, but see discussion in Olszewski et al. 1996), or the old, rather uncertain result (Freeman, unpublished) of an RR Lyrae velocity dispersion of ~50 km s^{-1}. One can also use the dispersion of the magnitudes of RR Lyrae stars in the older photographic surveys (Graham 1973, see also Walker 1991), or the dispersion in magnitudes of the clump giants. A change in stellar populations may make an important change in the measured magnitude dispersions, although this effect is not well quantified. The more modern all-LMC surveys, if well calibrated, could place useful limits on the depth of the LMC. We await those results eagerly since the older work is based on too few stars to place meaningful limits on the depth of the LMC.

Finding stars distant from the LMC center does not guarantee that we are seeing a halo population. Freeman et al. (1983) first noted that all populations, including that of the oldest clusters, were in a rotating system. Schommer et al. (1992) measured high-quality velocities for these and other objects, and showed that all objects, regardless of age, have the same disk rotation curve. This result is very different from that of the Milky Way or of M33, for example (Schommer et al. 1991). Is this set of strange kinematics the result of late formation of "halo objects," or is there another explanation (Weinberg 1999)?

3. Ages of the Oldest Clusters

3.1. Direct Age Determinations

Olsen et al. (1998, 1999) and Johnson et al. (1999) have used WFPC2 images to derive the ages of eight of the oldest clusters in the LMC. They age-dated the clusters by comparing the CMDs to those of galactic globulars of the same metallicity. No differences were found, leading both groups to deduce that the oldest LMC clusters are the same age as the oldest Milky Way clusters to ±1 Gyr. Although there have been papers claiming, based on color-magnitude diagrams, that some of these LMC clusters were indeed the same age (or younger, for that matter) as Milky Way clusters, the papers by Olsen et al. and Johnson et al. are the first that have the photometric accuracy to make that claim at the ± 1-2 Gyr level. These two papers show that 8 old LMC clusters observed to date with HST, both in the LMC bar and in the outer regions of the LMC, are indistinguishable in age from the older Milky Way clusters.

Many authors, myself included, had expected that these clusters would be 2-3 Gyr younger than Milky Way globulars (see Da Costa 1993 and Olszewski et al. 1996 as examples). Da Costa (1999) argues that some of these LMC clusters have horizontal branch morphologies consistent with their derived ages, while the radially more distant LMC clusters seem to show the same sort of second-parameter effects found in outer halo clusters in the Milky Way. He also argues that some of these 8 clusters measured with HST have very similar HB morphologies but abundances that vary by 0.7 dex, and therefore wonders if metallicity is even the first parameter in the LMC.

3.2. Is the Set of Oldest Clusters Complete?

There are several reasons to believe that the list of 13 old clusters found in Table 6 of Suntzeff et al. (1992) is complete. First, estimates of LMC cluster destruction timescales (see discussion in Olszewski 1993) show that, unlike the case for the Milky Way, even modest LMC clusters can survive a Hubble time. Second, Mateo (1993) has shown that fading models predict that the brighter of the current blue, young clusters will still be luminous enough to be in cluster catalogs at an age of 15 Gyr. Third, there is a demonstrable cluster age gap (see Olszewski et al. 1996), even if the edges of that gap are uncertain at the ~1 Gyr level (Sarajedini 1998). Fourth, Olszewski and collaborators (Olszewski 1993) showed that, for random clusters without age determinations, there are fewer clusters than expected at the metallicity of clusters that would be in the age gap. And fifth, Da Costa (1991) and Geisler et al. (1997) found no new age-gap clusters in their searches. The Geisler et al. study targeted clusters thought to be in the age gap from integrated photometry and from Olszewski and collaborators' metal abundances. We can therefore conclude that the Olsen et al. and Johnson et al. results are representative of the LMC old cluster system as a whole. The cluster age gap remains important for comparing LMC star formation histories as derived from clusters and from field stars.

4. Ages of the RR Lyraes

RR Lyrae stars are old, but are they ancient? If we see an RR Lyrae star in the LMC, can we conclude that it is "15 Gyr old", and can we conclude that it is the same age as an RR Lyrae star in one of the ancient clusters? Could we conclude that there is a group of RR Lyraes in the LMC field that is demonstrably younger than the cluster RR Lyraes, thus showing that deriving the LMC star formation history from clusters is incorrect?

NGC 121 in the SMC is demonstrably not of "globular cluster age" (Stryker et al. 1985) yet it contains RR Lyraes. There are metal-rich, thick-disk-kinematics RR Lyraes in the Milky Way (Taam et al. 1976) that might be thought to be younger than halo RR Lyraes. There are RR Lyraes in the "young halo" Milky Way globulars (see Da Costa & Armandroff 1995 for a good list of "young" halo clusters, and see the "Catalog of Variable Stars in Globular Clusters" for the list of variables). Alcock et al. (1996) claim a dearth of high amplitude RR Lyraes in the LMC field, and that what RR Lyraes there are must have formed from a population with a red horizontal branch of fairly low abundance. We must therefore conclude that the presence of an RR Lyrae indicates that the system is "old," as opposed to "ancient." Without additional information, such as the line of sight velocity dispersion of RR Lyraes as a function of abundance, or the relative fraction of blue horizontal branch stars in the LMC field, or some light-curve age diagnostic, we cannot tell if the RR Lyraes are part of the very oldest population in the LMC.

5. Ages of the Oldest Stars

Are there any field stars that are demonstrably ancient? One way to attack this problem is to study the color distribution of the clump giants and horizontal branch stars. From the color-magnitude diagrams of LMC and SMC clusters, we can see how the red clump changes into the red horizontal branch as a cluster ages (see Udalski 1998). Gardiner and Hatzidimitriou (1992) used an older version of this relation to derive the fraction of ancient horizontal branch stars in the SMC, concluding that 7% of the SMC was ancient (this percentage drops substantially if the modern calibration is used, see discussion in Olszewski et al. 1996). Smecker-Hane et al. (1994) showed the existence of a red clump and a HB in the Carina dSph, differing [mostly] in magnitude, and coming from populations of differing ages. With good-enough photometry and simple-enough populations, one should be able to quantify the age distribution of the old LMC as seen by the red clump and red horizontal branch.

Finding the LMC blue horizontal branch is more problematic. The BHB is easily hidden by the main sequence of a younger population. There are no LMC fields with a possible BHB that do not also have a younger population. One must therefore rely on the part of the BHB that is distinct from the main sequence, thus requiring better photometry than is available with the ongoing LMC surveys.

We therefore conclude that we have yet to see the population II horizontal branch in the LMC, even if it exists. Searching for old subgiants in the LMC field, with or without the use of HST (Walker 1999 is one example), may be the best way to prove definitively whether the oldest LMC field star population is "old" or "ancient."

6. Cautionary Tale: The LMC as a not-so-simple System

Since the dawn of the CCD age the vast majority of the deep photometry of the Clouds has been in the LMC. Probably the major reason is that the LMC was considered simple, while the SMC was considered complicated and crowded.

However, it now looks as though neither galaxy is well behaved, at least in terms of a halo population, although the SMC might be the simpler galaxy after all. NGC 2257 in the LMC is ~8 degress N of the bar. The surface brightness in the field surrounding this cluster is ~27 mag arcsec^{-2}. We (see Olszewski 1993) have measured metal abundances of 36 stars in this field. Twenty-one are more metal rich than [Fe/H]=−1.0. Of the fifteen more metal poor than [Fe/H]=−1.0, only seven are more metal poor than [Fe/H]=−1.5. The startling aspect of this field is the high abundance of most of the stars. This must be among the simplest fields in the LMC, yet it is very different from what one would have predicted for a pure halo field. We conclude that the distribution of metallicity in "HST" fields needs to be derived from ground-based spectroscopy of giants in those fields (actually, in larger fields surrounding the tiny HST fields) before one can derive the star formation history. The metallicity distribution of the SMC field NGC 121 (Suntzeff et al. 1986, 1999) seems to me to be equally non-halo-like. It has proven difficult in practice to find fields in the Clouds that are anything like the halo of the Milky Way. It will be fascinating to see which connotations of the word "halo" must be abandoned to find the oldest, least-confined-to-a-plane populations in each Cloud.

7. Concluding remarks

The review of Olszewski et al. (1996) concluded with a substantial list of questions that need to be answered before we have a good understanding of the older populations in the Clouds. Many of these problems can now or soon be addressed with the new surveys (MACHO, EROS, Great Circle Camera, OGLE), the new telescopes (Gemini, VLT, Magellan, etc.), and the new multi-object spectrographs (Hydra south, 2dF, etc.), and with HST.

Acknowledgments. EO was partially supported by the NSF. Since this symposium honored Sidney van den Bergh, and the DAO by association, let me thank him and them for the great environment in which to be a postdoc, and for being examples of good scientists and people.

References

Alcock, C., et al. 1996, AJ, 111, 1146

Bothun, G.D., & Thompson, I.B. 1988, AJ, 96, 877

Da Costa, G.S. 1991, in "The Magellanic Clouds," IAU Symp. 148, edited by R. Haynes & D. Milne, p. 183

Da Costa, G.S. 1993, in "The Globular Cluster-Galaxy Connection," edited by G.H. Smith & J.P. Brodie, ASP Conf Series 48, p.363

Da Costa, G.S., & Armandroff, T.E. 1995, AJ, 109, 2533

Da Costa, G.S. 1999, this volume

De Vaucouleurs, G., & Freeman, K.C. 1973, Vistas Astron., 14, 163

Freeman, K.C., Illingworth, G., & Oemler, A. 1983, ApJ, 272, 488

Gardiner, L.T., & Hatzidimitriou, D. 1992, MNRAS, 257, 195

Geisler, D., Bica, E., Dottori, H., Claria, J.J., Piatti, A.E., & Santor, J.F.C., Jr. 1997, AJ, 114, 1920

Graham, J.A. 1973, in "Variable Stars in Globular Clusters and Related Systems," edited by J.D. Fernie, p.120

Hughes, S.M.G., Wood, P.R., & Reid, N. 1991, AJ, 101, 1304

Irwin, M.J. 1991, in "The Magellanic Clouds," IAU Symp. 148, edited by R. Haynes & D. Milne, p. 453

Johnson, J.A., Bolte, M., Bond, H.E., Hesser, J.E., de Oliveira, C.M., Richer, H.B., Stetson, P.B., & VandenBerg, D.A. 1999, this volume

Kunkel. W.E., Demers, S., & Irwin, M.J. 1997, ApJ, 488, L129

Mateo, M. 1993, in "The Globular Cluster-Galaxy Connection," edited by G.H. Smith & J.P. Brodie, ASP Conf Series 48, p. 387

Nikolaev, S., Weinberg, M.D. 1998, BAAS, 192, #55.03

Olsen, K.A.G., Hodge, P.W., Mateo, M., Olszewski, E.W., Schommer, R.A., Suntzeff, N.B., & Walker, A.R. 1998, MNRAS, 300, 665

Olsen, K.A.G., Hodge, P.W., Mateo, M., Olszewski, E.W., Schommer, R.A., Suntzeff, N.B., & Walker, A.R. 1999, this volume

Olszewski, E.W. 1993, in "The Globular Cluster-Galaxy Connection," edited by G.H. Smith & J.P. Brodie, ASP Conf Series 48, p.351

Olszewski, E.W., Suntzeff, N.B., & Mateo, M. 1996, ARA&A, 34, 511

Sarajedini, A. 1998, AJ, 116, 738

Schommer, R.A., Christian, C.A., Caldwell, N., Bothun, G.D., & Huchra, J. 1991, AJ, 101, 873

Schommer, R.A., Olszewski, E.W., Suntzeff, N.B., & Harris, H.C. 1992, AJ, 103, 447

Smecker-Hame, T.A., Stetson, P.B., Hesser, J.E., & Lehnert, M.D. 1994, AJ, 108, 507

Stryker, L.L., Da Costa, G.S., & Mould, J.R. 1985, ApJ, 298, 544

Suntzeff, N.B., Friel, E., Klemola, A.A., Kraft, R.P., & Graham, J.A. 1986, AJ, 91 275

Suntzeff, N.B., Schommer, R.A., Olszewski, E.W., & Walker, A.R. 1992, AJ, 104, 1743

Suntzeff, N.B., Walker, A.R., Smith, V.V., Kraft, R.P., & Klemola, A. 1999, this volume

Taam, R.E., Kraft, R.P., & Suntaeff, N.B. 1976, ApJ, 207, 201

Udalski, A. 1998, AcA, 48, 383

Walker, A.R. 1991, in "The Magellanic Clouds," IAU Symp. 148, edited by R. Haynes & D. Milne, p.307

Walker, A.R. 1999, this volume

Weinberg, M.D. 1999, in proceedings of Third Stromlo Symposium

Westerlund, B.E. 1997, "The Magellanic Clouds," Cambridge Astrophysics Series #29

Discussion

Jim Hesser: Our poster makes some points relevant to the robustness of conclusions about the relative ages of the oldest clusters in the Magellanic Clouds and the Galaxy. We report differential ages for four clusters in the outer halo of the Milky Way at roughly twice the distance of the Clouds. For NGC2419, an [Fe/H] $= -2.2$ cluster, we see no evidence for an age difference relative to the nearby cluster M92. As Harris et al. (1997) pointed out, this suggests that a burst of star and/or star cluster formation occurred over a protogalactic volume of some 200 Kpc diameter. For three intermediate metallicity outer halo clusters with stubby red horizontal branches, so-called second parameter clusters, we find evidence that they are somewhat younger than nearby clusters of comparable metallicity in the sense of Lee, Demarque and Zinn. However these conclusions do depend fairly sensitively on knowing what the relative abundance ratios ([Fe/H], [α/Fe], etc.) are in the distant and nearby clusters where we have information of very different quality: high-dispersion spectroscopy for many stars in the nearer objects; and, for more distant clusters, low-dispersion spectroscopy – often just of the α-element Ca – for a few of their most luminous giants, integrated-light photometry or CMD comparisons. To know how the oldest Cloud clusters really compare to Galactic globulars in age we are going to have to get more reliable spectroscopic information for them.

New Views of the Magellanic Clouds
IAU Symposium, Vol. 190, 1999
Y.-H. Chu, N.B. Suntzeff, J.E. Hesser, and D.A. Bohlender, eds.

Intermediate-age and Old Populations in the Small Magellanic Cloud

D. Hatzidimitriou

Physics Department, University of Crete, P.O. Box 2208, Heraklion 71003, Crete, Greece

Abstract. The star formation history and kinematics of intermediate-age and old populations in the Small Magellanic Cloud are the subject of this brief review. New results confirm the occurrence of a period of enhanced SF about 8-10 Gyr ago in the SMC. Younger populations are confirmed to be more centrally concentrated than older ones. The SMC, at least in the areas studied (which *do not* include the Bar) seems to have had a different history of star formation from the LMC. Although the kinematics of the SMC have been extensively studied, it is still unclear if there are distinct Bar, disk and halo components, or if all populations share the same kinematics.

1. Introduction

As a dwarf irregular galaxy distinguished by its membership of a closely interacting system of galaxies, the Small Magellanic Cloud (SMC) is able to provide us with insights into the evolutionary development of dwarf irregular galaxies as well as the role of external dynamical interactions in stimulating star formation (SF) and modifying dynamical evolution.

Intermediate-age and old populations constitute most of the mass locked in stars in the SMC, and therefore understanding their history of formation is of the utmost importance. Ideally, one would like to address the following still unresolved questions:

- What is the age of the first (identifiable) generation of stars in the SMC? Is it as old as the oldest population in the Galaxy?
- Has SF been more or less continuous, or did it proceed in several major bursts? If the latter is the case what is the duration and strength of these bursts? In the Large Magellanic Cloud (LMC), such a global burst of SF occurred some 2-4 Gyr ago (e.g., Elson et al. 1997). Did a similar event take place in the SMC?
- Does the SF history depend on location?
- Is there any correspondence between major events in the SF history of the SMC and close passages with either the LMC or the Galaxy?
- Is there a kinematic halo in the SMC?

- Is the so-called Bar of the SMC a distinct dynamical entity, and if yes when was it formed?
- How has the dynamical state of the SMC evolved with time?

In the following, some recent developments will be briefly presented (for an older comprehensive review see Olszewski et al. 1996).

2. Star Formation History of Intermediate-age and Old Populations

Some general characteristics of the star formation (SF) history can be readily obtained (Hatzidimitriou 1997) by comparing the mean SF rate (total mass in stars divided by 15 Gyr) $< SFR >= 0.09 M_{\odot}\ yr^{-1}$, with the current SF rate (derived from Hα emission measurements) $SFR_o = 0.05 M_{\odot}\ yr^{-1}$ and the "initial" SF rate (estimated from the RR-Lyrae variable star content) $SFR_i = 0.03 - 0.08 M_{\odot}$ yr^{-1}. All these values are very similar within a factor of 2 (as is the case in other dwarf irregulars), and they are all a factor of 5 lower than the corresponding values for the LMC. It is noteworthy that the $< SFR >$ is almost a factor of 2 higher than the current SFR_o, suggesting that there may have been some period(s) of increased SF activity at intermediate ages. There is also little doubt that at earlier times active SF extended to larger distances from the Galactic center than it does at the present time. This is clearly demonstrated in Figure 1, where the median age of the field populations (derived from the clump-age indicator of Hatzidimitriou 1991) is plotted against projected radial distance from the optical center of the SMC. The same effect is apparent in the distribution maps of clump/horizontal branch stars and young main sequence stars produced by Gardiner & Hatzidimitriou (1992, hereafter GH92), of carbon stars (Morgan & Hatzidimitriou 1995) and of RR-Lyrae variables (Hatzidimitriou 1994). This radial aging -also seen in other dwarf irregulars- is probably reflecting a gradual decrease of the gas density in the outer regions below a critical value necessary for SF. The issue will become more complicated if the SMC is proved to be a system consisting of different components (disk, halo, and Bar) superimposed, rather than a system where all populations share similar kinematics (see next section).

Table 1. Field Color-Magnitude Diagrams in the SMC.

Source	V_{lim}	**Area** (sq.deg)	**references**
UKST	$\simeq$20.5	130	Gardiner & Hatzidimitriou 1992
OGLE	$\simeq$21.2	2.4	Udalski et al. 1998
CTIO	$\simeq$23.5	0.06	Hatzidimitriou et al. 1999
CTIO	$\simeq$23.5	0.015	Suntzeff et al. 1999
HST	$\simeq$27.0	0.001	Holtzman et al. 1999

Derivation of the detailed SF history is not an easy task. It requires well-sampled and sufficiently deep color-magnitude diagrams (CMDs), as well as adequate knowledge of the age-metallicity relation (cf. Da Costa 1999) and of the

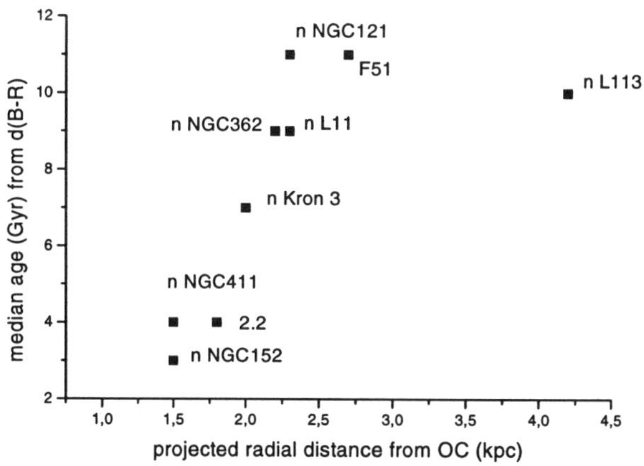

Figure 1. Median age of field populations as a function of projected radial distance from the optical center. Most points are for fields near star clusters (see references in GH92). The point marked as 2.2 corresponds to the CTIO 4-m observations of an inner disk field (Hatzidimitriou et al. 1999+) shown in Fig.2, and the point marked as F51 refers to the area F51:2.4 studied in detail in GH92.

metallicity distribution, because the CMDs are degenerate in age and metallicity. Range in reddening and line-of-sight distance must also be taken into account. Table 1 summarizes recent color-magnitude-diagram (CMD) studies of field regions in the SMC. A compilation of studies prior to 1992, which include mostly regions close to star clusters, can be found for example in GH92.

All recent studies in Table 1 are still in progress. Preliminary analysis of the Hatzidimitriou et al. (1999) data (see Figure 2 for an example) indicates that at least the disk fields studied lack the strong intermediate-age component (2-4 Gyr) found to be prominent in the LMC. Applying the Bertelli et al. (1992) method, we derived the R_1, R_2 and R_3 ratios for the four fields studied. These ratios depend in different ways on the relative strengths of populations of different ages, and they are compared with SF history models by Bertelli et al. It is remarkable how different these ratios are (Table 2) in the SMC (from Hatzidimitriou et al. 1999) from the LMC (from Bertelli et al. 1992).

Table 2. Comparison of the R-ratios in the SMC and the LMC field

	R_1	R_2	R_3
SMC	1.6-1.8	1.2	8.2-30.6
LMC	2.6-5.0	1.9-3.3	3.7-8.5

They also seem to suggest that a significant period of SF (of uncertain duration) occurred more than 8 Gyr ago in the SMC fields studied. There is a very marked difference in the history of SF of the inner and outer disk fields, as

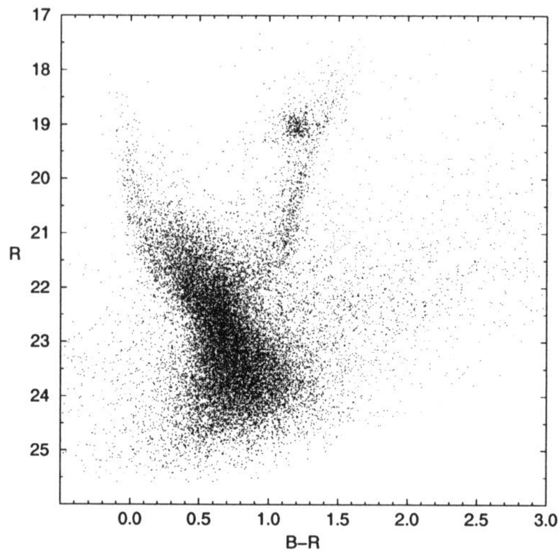

Figure 2. Color-magnitude diagram of a 200 square-arcmin field in the inner disk of the SMC, taken with the CTIO 4m-telescope (Hatzidimitriou, Smecker-Hane & Croke, 1999, in preparation).

exemplified by a comparison of the main sequence luminosity functions in the two regions (Figure 3). The outer field is older in the mean, as expected from earlier studies. Suntzeff et al. (1999) have produced a similar deep CMD near NGC 121, also using the CTIO 4-m telescope. Their main conclusion is that at least in this area active SF occurred between $\simeq$ 15 Gyr ago (the age of old Galactic globular clusters) and 9 Gyr ago, in agreement with the Hatzidimitriou et al. result.

3. Kinematics

In recent years there has been a significant qualitative and quantitative improvement of kinematical studies of young and intermediate age populations in the SMC. A compilation of these studies is presented in Table 3.

The intermediate age and old populations (upper panel of Table 3) do not seem to participate in any rotational motion (Hardy et al. 1989; Hatzidimitriou et al. 1997).

There is clear evidence of streaming motions in certain locations: along the Wing (Hardy et al. 1989), and along the line-of-sight in the NE outer regions (Hatzidimitriou et al. 1993). These streaming motions are believed to result from tidal interactions with the Large Magellanic Cloud and the Galaxy. It is important to note at this point that intermediate-age stars have been recently discovered in the inter-cloud "Bridge" (e.g., Kunkel et al. 1997), thus confirming the tidal origin of the feature.

Except for these distinct cases, the radial velocity dispersion values of Table 3 indicate that within the observational uncertainties all populations, including young "Pop I", have similar kinematics, regardless of age and location. However,

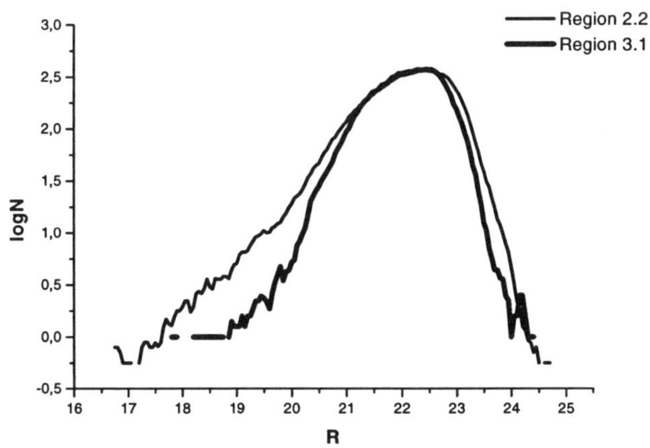

Figure 3. Main Sequence Luminosity Functions for two regions studied with the CTIO 4m-telescope. The two regions are located in the inner (2.2) and outer (3.1) disk (halo?) west of the Bar.

careful analysis of the existing data and comparison with N-body simulation predictions (Gardiner & Hatzidimitriou 1999+) seem to favor halo kinematics for the older stars. Clarification of this very important issue has to await new extended observations. Lastly, it must be emphasized that there is no kinematical information about the older stellar component, traced by RR-Lyrae variables. Time consuming as it may be, identification and subsequent kinematical study of RR-Lyraes, with good spatial coverage, will be instrumental in deciphering the dynamics of the SMC.

Table 3. Recent kinematical studies of field populations in the SMC.

Type	Region	n	<RV>	σV	references
clump stars	NE outer	30	151(6)	33(4)	Hatzidimitriou et al. 1993
				18(5)[a]	
red giants	NW outer	36	135	26	Suntzeff et al. 1999
planetary nebulae	central	44	146(4)	25(3)	Dopita et al. 1985
carbon stars	central	131	148(2)	27(2)	Hardy et al. 1989
carbon stars	outer	70	149(3)	25(2)	Hatzidimitriou et al. 1997
		62[b]	145(3)	21(2)	
star clusters		7	138(6)	16(4)	Da Costa & Hatzidimitriou 1998
PopI	central				Maurice et al. 1989
cepheids	main	61	149(3)	22(3)	Mathewson et al. 1988
HI shells	main	501	155(1)	25(1)	Staveley-Smith et al. 1997

[a] Same as previous row, but after having removed the radial velocity versus line-of-sight distance correlation.

[b] Same as previous row, but without including stars belonging to the Wing.

Acknowledgments. The author is grateful to B.F. Croke and T. Smecker-Hane who were instrumental in analysing the CTIO-4m data in time for the meeting. This work was partially supported by a grant from the American Astronomical Society.

References

Bertelli, G., Mateo, M., Chiosi, C., & Bressan, A. 1992, ApJS, 388, 400

Da Costa, G.S. 1999, in this volume

Da Costa, G.S., & Hatzidimitriou, D. 1998, AJ, 115, 1934

Dopita, M.A., Ford, H.C., Lawrence, C.J., et al. 1985, ApJ, 296, 390

Elson, R.A.W., Gilmore, G.F., & Santiago, B.X. 1997, MNRAS, 289, 157

Gardiner, L.T., & Hatzidimitriou, D. 1992, MNRAS, 257, 195

Hardy, E., Suntzeff, N.B., & Azzopardi, M. 1989, ApJ, 344, 210

Hatzidimitriou, D. 1991, MNRAS, 251, 545

Hatzidimitriou, D. 1994, in Astronomy from Wide-Field Imaging, eds. H.T. MacGillivray et al., Dordrecht:Kluwer, 489

Hatzidimitriou, D. 1997, in Star Formation Near and Far, eds. S. Holt & L.G. Mundy, AIP: New York, 561

Hatzidimitriou, D., Cannon, R.D., & Hawkins, M.R.S. 1993, MNRAS, 241, 667

Hatzidimitriou, D., Croke, B.F., Morgan, D.H., & Cannon, R.D. 1997, A&AS, 122, 507

Hatzidimitriou, D., Smecker-Hane, T., & Croke, B.F. 1999, in preparation

Holtzman J.A., Mould J.R., & Gallagher, J.S. 1999, in this volume

Kunkel, W.E., Irwin, M.J., & Demers, S. 1997, ApJ, 488, 129

Maurice, E., Martin, N., & Bouchet, P. 1989, A&AS, 78, 445

Mathewson, D.S., Ford, V.L., & Visvanathan, N. 1988, ApJ, 333, 617

Morgan, D.H., & Hatzidimitriou, D. 1995, A&AS, 113, 539

Olszewski E.W., Suntzeff N.B., & Mateo M. 1996, ARA&A, 34, 511

Staveley-Smith L., Sault R.J., Hatzidimitriou D., Kesteven, M.J., & McConnell D. 1997, MNRAS, 289, 225

Suntzeff, N.B., Walker, A.R., Smith, V.V., Kraft, R.P., Klemola, A., & Stetson, P.B. 1999, in this volume

Udalski, A., Szymanski, M., Kubiak, M., Pietrzynski, G., Wozniak, P., & Zebrun K. 1998, AcA, 48, 147

Discussion

Lance Gardiner: What significance do you attach to the similar velocity dispersions seen in the H I, Cepheids, carbon stars and HB/clump? I believe that they are largely coincidental - the H I is affected by the kinematic effects of star formation activity, whereas we are probably seeing a kinematic halo in the outer parts.

Hatzidimitriou: That is one possibility. Or, the velocity dispersion in the solar regions could be increased because of streaming motion in some locations. Or, it could be that all populations share the same kinematics.

Carme Gallart: I would like to comment on your reference to the synthetic CMD technique to recover SFH's, and to the fact that there are a large number of parameters to be taken into account. I want to remark that it is difficult and a lot of work but it can be done. I would like to refer to the work I am presenting on Leo I, where I explore a considerable volume of the parameter space, including IMF, Z(t) and binary function, and I do find a "unique" and statistically significant solution. This kind of work can be done for the MC if good enough data is used. Deep CTIO 4m data, for example, is perfect for this purpose.

Hatzidimitriou: I agree. I just wanted to emphasize that whatever solutions one derives from the CMD's, they must be consistent with the age-metallicity relation, element ratios (e.g. [O/Fe]) etc.

Hans Zinnecker: The finding that the LMC experienced a significant increase in its star formation rate 2-4 Gyr ago while the SMC did not provokes the following question: was there an LMC/SMC close encounter then or where exactly were the two Magellanic Clouds with respect to each other at this critical time? Do we know?

Hatzidimitriou: I am not all that certain that there is no enhancement of the SF rate in the central regions of the SMC ("Bar") at 2-4 Gyr ago. We just do not know. Different models predict somewhat different epochs for past close encounters. So, we do not really know.

New Views of the Magellanic Clouds
IAU Symposium, Vol. 190, 1999
Y.-H. Chu, N.B. Suntzeff, J.E. Hesser, and D.A. Bohlender, eds.

Star Formation History of the Disk of the Large Magellanic Cloud

J. S. Gallagher and Andrew A. Cole

Dept. Astron., U. Wisconsin, Madison, WI 53706, USA

J. Holtzman

Dept. Astron., New Mexico State U., Las Cruces, NM 88003, USA

T. Smecker-Hane

Dept. Phys. & Astron., U. California, Irvine, CA 92697, USA

Abstract. New optical observations of field stars in the Large Magellanic Cloud from the ground and with the *Hubble Space Telescope* provide the basis for investigations of the history of field star formation in our nearest galactic neighbor. Results from WFPC2 photometry of field stars indicate that the stellar initial mass function for stars around 1 $M_{\odot}$ is similar to that in the solar neighborhood. Long-term variations occurred in the average SFR with enhanced activity in the past 2-4 Gyr, a fallow period about 2-10 Gyr in the past, and an initial phase of astration ~12 Gyr ago. The recent increase in field star formation is less than that for the birth of compact star clusters; the cluster formation rate does not give a complete evolutionary picture.

1. Introduction

As the nearest galaxy the LMC provides an excellent opportunity to quantitatively test our ideas about the evolution of complex astrophysical systems; including stellar evolution, characteristic of star cluster populations, distribution of metals, importance of bars as dynamical sub-systems, and effects from proximity to the giant Milky Way spiral galaxy (Westerlund 1997). This brief review focuses on the history of the stellar populations in the field of the LMC disk. Since stars may be captured in the LMC bar, this component could have a different history which is not yet properly explored. Understanding the history of the main disk of the LMC is fundamental to our knowledge of the LMC's evolution and to the interpretation of properties of blue galaxies seen in the distant universe at substantial lookback times (Ellis 1997). If we cannot decode the history of star formation in the nearby LMC, then we must be concerned about the credibility of our analyses of distant galaxies (Tolstoy 1998).

A variety of measurements can be used to measure star formation rates (SFR) and histories (SFH) in galaxies. Integrated $SFRs$ are determined, for example, from Hα, far infrared, or ultraviolet luminosities (Kennicutt 1998), while the SFH depend on measures of ratios between younger and older stars, as measured by broad band colors or Hα emission equivalent widths. The LMC has the advantage that individual stars can be observed; thus we can apply the techniques of "Galactic astronomy" to another system. By observing the distributions of stars as functions of age, mass, and metallicity, detailed information about the SFH of the LMC over cosmic times scales is obtained (Hodge 1989). In this paper we consider some aspects of this problem from the perspective of modern ground-based and WFPC2 digital photometry of stars in the LMC.

2. Measuring Resolved Stars

Images of any object suffer from a variety of defects and as a result we cannot perfectly measure characteristics of the target. Crowding in the dense star fields of even a nearby galaxy like the LMC results in overlaps between stars; faint objects are lost and the accuracies of colors and magnitudes are reduced. In the common case of confusion between a star on the red giant branch (RGB) and one on the main sequence (MS), the resulting image would land in the gap between the MS and RGB on an observed color-magnitude diagram (CMD). The extreme example of this effect occurs in physical binary stars, where the components cannot be resolved. An equal mass the binary has the color of a single star but is 0.75 mag too bright. With increasing mass ratio the binary becomes redder in color and closer in brightness to the primary. Fortunately these effects can be handled statistically. However, close binary evolution also produces stars with unusual properties that are not included in single star evolutionary models (e.g., blue stragglers in star clusters).

Additional uncertainties can arise in even excellent data from a variety of factors. Background galaxies are often compact and confused with stars, especially at very faint magnitudes. Photometric transformations to standard systems have limits to their precision for observed stars and for the models. Even when these are very accurate, they may suffer from non-unique conversions between observables and stellar parameters, as in age-metallicity degeneracies on the RGB (Da Costa 1998). Observed colors and magnitudes are also influenced by dust within the LMC and in the Galactic foreground.

Given these complexities in interpreting LMC CMDs, is this still a worthwhile enterprise? Yes, because this is one of the best approaches that we have; other methods based on integrated properties simply hide many of the difficulties (e.g., Hodge 1989; Tolstoy & Saha 1996; Grebel 1997; Mateo 1998).

Observational progress can still be made in improving the quality of CMDs. Data from WFPC2 have minimal crowding problems in most of the LMC; the WFPC2's 0.1 arcsec pixel is 0.02 pc and it can reach to V$\sim$27 (or M_V $\approx$8; corresponding to 0.6 $M_{\odot}$) at the LMC. Multicolor measurements, particularly in the blue and infrared, allow effects of extinction to be traced (Harris et al. 1997; Romaniello 1998), as do the high angular resolution HI maps of Kim (1998). Intermediate band filter observations can play an important role in giving a first look at metal abundances in large populations of stars (e.g., Ardeberg et al. 1997; Cole et al. 1999). Figure 1 shows an example of a modern LMC disk field CMD derived from observations with WFPC2.

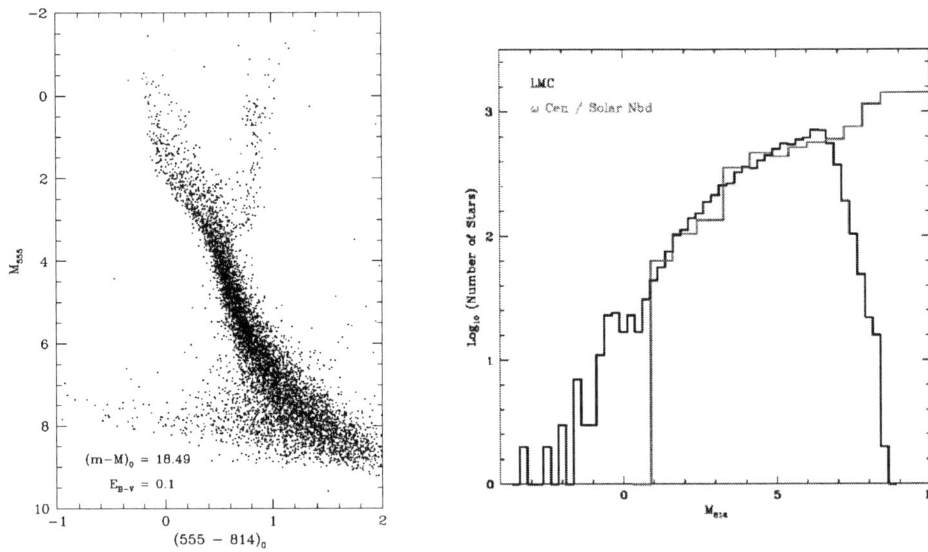

Figure 1. *Left:* mean disk LMC CMD in the WFPC2 VI-bands from the three fields studied in Geha et al. (1998). Note the MS band, lower RGB, red clump at $M_{F555W} \approx 0.5$, and subgiants between the RGB and MS for older stars. *Right:* The LMC LF compared with that of Galactic star clusters (Elson et al. 1995; von Hippel et al. 1996).

3. Interpretation of LMC CMDs

3.1. Stellar Luminosity Functions

The numerical dominance of stars on the main sequence (MS) is obvious in Figure 1. Stellar luminosity functions (LF) can therefore be measured with unusual ease from such observations; there are no serious effects from variations in sample volumes with magnitude, as in Galactic studies (Mighell & Butcher 1992). Even though we observe only small patches of the LMC disk with each WFPC2 pointing, orbital phase mixing randomizes the locations of disk stars in a few LMC rotation periods or ~1 Gyr (Gallagher et al. 1996).

Results for MS luminosity functions in disk fields are given by Holtzman et al. (1997; H97) and Geha et al (1998; G98). These papers show that *the MS LF in the LMC is similar to that in the solar neighborhood, or to those of metal-rich globular clusters below the MS turnoff (Figure 1).* G98 also demonstrate that the LMC LF is relatively constant with position in the LMC disk.

Interpreting the MS LF depends on both the SFH and IMF (e.g., Scalo 1986). Thus it is always possible to trade between the form of the IMF and the SFH to some degree. G98, however, showed that for a Salpeter IMF ($n(m)dm \propto m^{-2.35}dm$) over the observed range of stellar masses, models of the LMC SFH with highly enhanced recent SFRs (e.g., the models suggested by Bertelli et al. (1992) or derived from the birthrates of star clusters) have too many intermediate age stars. Good evidence exists that the recent SFH for the LMC field stellar population differs from that for compact star clusters, which probably survive as bound objects for long time periods; *the enhancement in field SFR in the past 2-4 Gyr is less extreme than that implied by the cluster formation rate (see also van den Bergh 1998).*

G98 conclude their study with an acceptable SFH model which has constant SFR for ~2-10 Gyr before the present, a modest enhancement in the SFR during galaxy formation ~10-12 Gyr ago, and a factor of 3 increase in SFR during the past 2 Gyr. Further tests of this LMC SFH model await the larger sample of field stars observed with WFPC2 by Smecker-Hane et al. (1999), as well as better information on chemical element abundances in stars with intermediate and old ages.

3.2. CMD Model Fitting

Additional information beyond that contained in the MS LF can be derived by fitting the three dimensional distribution of stellar densities on a CMD (the Hess diagram). The observed LMC field star CMD is a sum of a large number of population components which can be characterized by the SFR as a function of time and metallicity (Hodge 1989), where we begin with the probably realistic assumption that the IMF has remained approximately constant over time. An example of the complications introduced by such composite populations is provided by the structure of the RGB (e.g., Da Costa 1998; Gallagher et al. 1998); blue parts of an RGB can be made either by old, very metal-poor stars or younger, more massive stars. In a complex stellar population fits must be made to both the MS and RGB to distinguish between these options.

Numerical techniques for fitting CMDs are still being developed. A key point concerns the weighting of model fits. In the WFPC2 VI-band studies of LMC disk fields, the distributions of stars just below the least luminous MS turnoff, with $M_V \approx 4$–5, tend to drive SFH model fits to the entire CMD[1]. Thus we are in effect fitting only a modest range in stellar mass rather then the entire CMD. Techniques such as those of Tolstoy & Saha (1996) allow a more global approach to CMD fitting, as do the numerical approaches of G98, Han (1998), and Hernandez et al. (1998).

All of these methods depend on the fidelity of the stellar evolution models, and on the correct choice of distance and extinction to the field being studied. The former point is currently a major issue. Evolution tracks from stellar models from the various groups disagree in detail, a natural result of differing choices, e.g., for convective core overshoot or mixing. In addition the current models often exclude "second-order" effects which exist in real stellar populations, such as stellar rotation; these influence, e.g., the MS band (Gallagher et al. 1996).

Of course when we attempt to fit stars in different evolutionary phases additional problems emerge. For example, a variety of studies suggest that colors of the RGB in intermediate age and metallicity stellar models are too red in all but the most recent stellar evolution models (Gallagher et al. 1996; Elson et al. 1997; Girardi et al. 1999). Thus to get a fit to the observed RGB, the SFH models must include old, metal-poor populations that are not necessarily

---

[1]This occurs because fitting weights scale roughly as the color error divided by the square root of the number of stars, and this peaks just below the last MS turnoff in WFPC2 data; Fig. 1.

present, as indicated by the location and stellar densities at low luminosity MS turn offs. But with the increasing sophistication of stellar evolution and atmosphere models and large amounts of new stellar data from observations of LMC star clusters with the *HST* and soon with the 8-m VLT and Gemini Telescopes, the reliability of the LMC CMD fitting process will rapidly improve.

4. Discussion

In addition to the deep studies emphasized in this paper, we will also benefit from shallower, wide angle surveys in the optical and infrared (Zaritsky et al. 1997); these are yielding nearly complete information on more luminous stars and patterns of interstellar dust obscuration and reddening within the LMC.

Even so our understanding of the SFH of the LMC disk remains a work in progress. However, several key points already stand out:

1. The stellar LF and mid-mass range (0.7–2 $M_\odot$) IMF for the disk of the LMC are similar to their Galactic equivalents. The average stellar IMF in galactic disks seems to be roughly constant over a wide range of conditions.

2. Building from Butcher's (1977) initial explorations, we are beginning to obtain reasonable empirical models for the SFH of the LMC. While these are consistent with those derived from the chemical evolution of the LMC (Dopita 1999; Pagel & Tautvaišienė 1998), they do *not* agree with standard models for galactic evolution. The LMC appears to have experienced neither smooth evolution with constant SFRs nor extreme, short duration starbursts. Three major epochs of star formation are found during which *average* SFRs were roughly constant, albeit at different levels. The physical origins of this behavior are not known, although interactions with the Milky Way, SMC, and their surrounding interstellar material remain prime suspects in this unsolved case.

3. The recent SFH of the stellar field in the LMC disk shows less extreme time variations than the birth rate of compact star clusters (G98; van den Bergh 1998).

We now have exciting opportunities to extend LMC SFH studies to a wider range of issues. We still need to measure how the ratio of intermediate-to-old age stars varies with radius in the LMC disk and to see if the bar has a distinct evolutionary history. Preliminary investigations suggest that the LMC, like several other galaxies with dwarf-like structures, has flatter radial gradients in its older stellar components. Thus it would have been a diffuse object in its youth (van den Bergh 1991), unlike the compact blue faint galaxies. This leaves the possible formation of the bar as the last likely repository for a stellar population that could have formed in an intense starburst. We will no doubt have interesting new results on the SFH for the next LMC Symposium.

Acknowledgments. We thank many of our colleagues for discussions of the Magellanic Clouds and related topics as well as sharing their results: R. Dohm-Palmer, R. Elson, E. Grebel, P. Hodge, D. Hunter, P. Linde, P. Massey, J. Mould, E. Skillman, E. Tolstoy, B. Wakker, and especially Sidney van den

Bergh. JSG, AAC, and JAH were supported in part by NASA under contract NAS 7-1260 to JPL for the WFPC2 Investigation Definition Team.

References

Ardeberg, A., Gufstafsson, B., Linde, P., & Nissen, P. E. 1997, A&A, 322, L13

Bertelli, G. Mateo, M., Chiosi, C., & Bressan, A. 1992, ApJ, 388, 400

Butcher, H. 1977, ApJ, 216, 372

Cole, A. A., Smecker-Hane, T., & Gallagher, J. S. 1999, in this volume

Da Costa, G. S. 1998, in Stellar Astrophysics for the Local Group, eds. A. Aparicio, A. Herrero, F. Sánchez (Cambridge, Cambrdige University Press), p. 351

Dopita, M. 1999, in this volume

Ellis, R. C. 1997, ARA&A, 35, 389

Elson, R. A., Gilmore, G. F., Santiago, B. X., & Casertano, S. 1995, AJ, 110, 682

Elson, R., Gilmore, G. F., & Santiago, B. X. 1997, MNRAS, 295, 860

Gallagher, J. S., et al. 1996, ApJ, 466, 732

Gallagher, J. S., et al. 1998, AJ, 115, 1869.

Geha, M.C., et al. 1998, AJ, 115, 1045

Girardi, L., Bressan, A., Bertelli, G., & Chiosi, C. 1999, in preparation

Grebel, E. 1997, RvMA, 10, 29

Han, M. 1998, private communication

Harris, J., Zaritsky, D., & Thompson, I. 1997, AJ, 114, 1933

Hernandez, X., Valls-Gabaud, D., & Gilmore, G. 1998, MNRAS, 304, 705

Hodge, P. 1989, ARA&A, 27, 139

Holtzman, J., et al. 1997, AJ, 113, 656

Kennicutt, R. C. 1998, ARA&A, 36, 189

Kim, S. 1998, Ph.D. Thesis, The Australian National University

Mateo, M. 1998, ARAA, 36, 435

Mighell, K. J. & Butcher, H. R. 1992, A&A, 255, 26

Pagel, B.E.J., & Tautvaišienė 1998, MNRAS, 299, 535

Romaniello, M. 1998, Ph.D. thesis, University of Pisa

Scalo, J. 1986, FCPh, 11, 1

Smecker-Hane, T., et al. 1999, in this volume.

Tolstoy, E. 1998, in Dwarf Galaxies and Cosmology, eds. T. X. Thuan et al., p.213

Tolstoy, E., & Saha, A. 1996, ApJ, 462, 672

van den Bergh, S. 1991, ApJ, 369, 1

van den Bergh, S. 1998, ApJ, 507, L39

von Hippel, T., et al. 1996, AJ, 112, 192

Westerlund, B. 1997, The Magellanic Clouds, (Cambridge, Cambridge U. Press)

Zaristky, D., Harris, J., & Thompson, I. 1997, AJ 114, 1002

Discussion

Carme Gallart: I would like to point out that the comparison of isochrones with CMD's of complex (not single age) stellar systems can be quite misleading because a) the Z in the isochrones are discrete and not necessarily adequate for the particular set of data (small changes in Z lead to large changes in posiiton in the CMD) and b) the isochrones don't give information on the lifetime of the stars along it.

Gallagher: I think we agree - isochrones provide a means for a first look at a data set, but are not a proper tool for fitting galactic Hess diagrams.

New Views of the Magellanic Clouds
IAU Symposium, Vol. 190, 1999
Y.-H. Chu, N.B. Suntzeff, J.E. Hesser, and D.A. Bohlender, eds.

Results from the EROS Microlensing Survey

J.P. Beaulieu

Kapteyn Institute, Groningen, The Netherlands

H.J.G.L.M. Lamers, W.J. de Wit

SRON and Astronomical Institute, Utrecht, The Netherlands

Abstract.

We present a status report of the original EROS and the on-going EROS-2 microlensing surveys, which were created to search for dark matter in the Galactic halo via microlensing effects on LMC/SMC stars. Microlensing surveys provide long-term systematic observations of millions of stars in both Clouds which yield a unique database of stellar photometry and variability. We review the results obtained on pulsating stars and detail the similarities and differences of the Magellanic Cloud Herbig Ae/Be stars with their Galactic counterparts.

1. Observations Made By the EROS Microlensing Survey

EROS (Expérience de Recherche d'Objets Sombres) is a French collaboration between astronomers and particle physicists to search for baryonic dark matter in the Galactic halo by the microlensing (Paczyǹski 1986) of stars in the Magellanic Clouds (MCs). A compact object in the Galactic halo passing close enough to the line of sight to a background star in the Magellanic Clouds creates an increase in the apparent brightness of the star. This phenomenon occurs during a chance alignment of the observer, the deflector, and the background star. If one assumes a standard co-rotating halo, the time scale τ_0 of a microlensing event is given by the relation $\tau_0 = 70\sqrt{M/M_{\odot}}$ days where M is the mass of the deflector. To be sensitive to a wide range of mass for compact objects $(10^{-7} - 10^{-1}M_{\odot})$ two complementary approaches have been developed.

The first approach involved photographic monitoring of 6.4 million stars over a 5 × 5 degree field using the ESO Schmidt telescope from 1990 to 1994. When possible, the plates were taken once a night in B_J and R_C. Isolated stars can be found down to magnitude 20 with typical photometric uncertainties of 0.3 mag.

The second approach was sensitive to low mass compact objects. Between 1991 and 1994 about 15000 CCD images were taken with two broad bandpass filters B_E and R_E of an area of 1 x 0.4 degrees centered on the bar of the LMC. An SMC field of 1 x 0.4 degrees was observed between 1993-1995 and about 6000 images were taken. Light curves of 270,000 stars exist with up to 48 points per night!

EROS has been upgraded recently and entered the EROS-2 phase. EROS-2 uses the dedicated Marly 1m telescope at ESO La Silla. The prime focus is equipped with a focal reducer and a dichroic beam splitter with a mosaic of eight CCD 2048 × 2048 in each channel. The total field is 0.7 × 1.4 deg. The data collection started in July 1996 for a baseline observing period of 4 years. EROS-2 observes about 20 million stars a night, and monitors the LMC, SMC, the Galactic Bulge and some fields in the spiral arms (with approximately a daily sampling rate).

2. Microlensing Results

The initial success of microlensing surveys was to show that the experiment was feasible. Indeed, the first microlensing events towards the LMC were announced almost simultaneously in October 1993 by the two competing teams EROS (Aubourg et al. 1993) and MACHO (Alcock et al. 1993, reported by Cook in this symposium). Now, about 18 events have been observed toward the LMC and the SMC. Despite the fact that some of the microlensing candidates turned out to be variable stars (EROS1 is a Be star, Beaulieu et al. 1995; EROS2 is an A0-2 eclipsing binary, Ansari et al. 1995; EROS-97-SMC-1,also known as MACHO-97-SMC-1, is an eclipsing binary, Palanque et al. 1998), the vast majority are most likely due to genuine microlensing. Moreover, we still do not know an intrinsic physical mechanism to mimic the observed photometric variation, and the microlensing interpretation is still favored.

EROS and MACHO measured consistent optical depths for gravitational lensing toward the LMC of $\tau_{LMC} = 0.82^{+1.1}_{-0.5}\ 10^{-7}$ (Renault et al. 1997a) and $\tau_{LMC} = 2.9^{+1.4}_{-0.9}\ 10^{-7}$ (Alcock et al. 1997), respectively. A subsequent combined EROS MACHO study gave $\tau_{LMC} = 2.1^{+1.3}_{-0.8}\ 10^{-7}$ (Alcock et al. 1998). The expected optical depth to gravitational lensing towards the LMC due to known stellar populations is expected to be smaller than $0.5\ 10^{-7}$. The generally favored hypothesis is that all the observed microlensing are due to lenses in the halo. We should underline that in most cases, we do not have direct access to the different parameters of the system. The parameters are degenerate and therefore we do not know where the lens is (in halo or in the Clouds). Based on likelihood analysis it is concluded that half of the dark halo of our Galaxy is composed of MACHOs with a mass of $\sim 0.5 M_{\odot}$. If they are old white dwarfs it would lead to various serious problems involving the cooling rate of white dwarves, shape of the IMF for the first generation of stars, star formation rate, etc. If not, what are they? Do they belong to the halo?

On theoretical ground, Zhao (1998) had suggested the elegant idea of a tidal streamer front of the LMC as the cause of the microlensing signal. Zaritsky & Lin (1997) claimed that a vertical extension of the red clump observed in LMC CMDs (VRC hereafter) was indeed the signature of such a streamer located at 35 kpc from the Galaxy. Beaulieu & Sackett (1998) confirmed the detection of the VRC, but showed that it could result from stellar evolution within the LMC, i.e. younger clump stars. The observed colors and densities in the CMDs are reproduced with a very simple model. Moreover, a radial velocity survey of VRC stars (Ibata, Lewis, & Beaulieu 1998) showed that they have the same systemic velocity as the LMC field stars, and the same velocity dispersion. The chance of a tidal streamer having these kinematic properties are indeed low. The VRC is not the signature of a tidal streamer. However, it would be interesting to do an extensive search of tidal streamers very close to the LMC.

Recent studies towards the SMC by EROS (Palanque et al. 1998; Afonso et al. 1998) or by PLANET (Albrow et al. 1998) suggest that the events for which the degeneracy between the parameters have been partially broken are due to lensing in the Clouds themselves, not to halo lenses. The LMC and SMC self-lensing may have been systematically underestimated.

3. Variable Star Results

The microlensing surveys generate a very large homogeneous database of stars observed over a long period of time in the two Magellanic Clouds. They provide gold mines for stellar studies (see also results from the MACHO collaboration Welch et al., Alves et al. this volume). We will briefly review the results on pulsating variables, novae, and eclipsing binaries. Finally, we will present a status report of our study on pre-main sequence stars in the Magellanic Clouds.

3.1. Pulsating Variables, Novae, Eclipsing Binaries

We have systematically searched the EROS database for variable stars using the AoV method (Schwarzenberg-Czerny 1989) and have built catalogues of variable stars in the LMC and in the SMC. A differential study of about 550 Cepheids from LMC and SMC has been conducted to derive constraints on stellar pulsation theory (position of resonance centers and beat Cepheids) and to explore the subtle metallicity effect on the PL relation (Beaulieu et al. 1995, 1997a, 1997b; Sasselov et al. 1997). A larger sample of Cepheids from LMC and SMC has been observed as a pilot project of the EROS-2 survey (Bauer et al. 1998; Marquette, Graff, Ripepi et al. this volume).

The first results on ~2000 LMC RR Lyraes from EROS Schmidt plates has been presented by Hill & Beaulieu (1997). The period histograms of RR Lyrae from the bar and from the field are similar. From the field they confirm the histograms found in the literature. The metallicity derived for these stars suggests that the old population in the field might have formed a little earlier than the oldest cluster in the LMC.

Grison et al. (1995) give a catalogue of 79 eclipsing binaries in the bar of the LMC ranging from close to well-detached type. These binaries can be used for accurate distance determination of the LMC or as a test of stellar pulsation models.

De Laverny et al. (1998) report the observation of a slow nova in the SMC. Due to the sampling rate of the observations (2 points/ 20 min), low amplitude short time scale variation has been observed at maximum. This nova is a DQ Her type, similar to Galactic novae of the same class. A large helium enhancement in the shell is found, and C and O enrichments are suspected.

Beaulieu et al. (1997) and Kollàth et al. (1997) discuss the properties and compute linear and hydrodynamic models of a particular variable star observed in the bar of the LMC. It has 2 modes of pulsation coupled by a 3:2 resonance, shows a strong $H\alpha$ emission, and has a spectral type of late O to early B. Its nature remains a puzzle. Is it a pre-main sequence star or post-AGB?

3.2. Pre-main Sequence Stars in the Magellanic Clouds

The process of star formation of low mass stars and their subsequent evolution to the main sequence in external galaxies is essentially unknown. All the studies of star formation in external galaxies so far have been restricted to massive star clusters such as the 30 Dor region in the LMC. This is because massive young stars are very luminous and the HII regions are signposts of recent massive star formation. The population of low mass stars can be studied in such cluster by deep photometry. In this way, the low mass population of the 30 Dor and around SN1987A has been studied (see Panagia, Walborn, Zinnecker and references in this volume). However, such studies are always biased to conditions in massive star clusters which are not characteristic for the formation and the IMF of the majority of the less massive stars.

Using data from the EROS survey in the bar of the LMC, we discovered a population of Herbig Ae/Be stars (HAeBe stars hereafter) on the basis of their characteristic irregular photometric variations and their colour. Subsequent optical spectra confirmed the spectral types B, and the presence of strong $H\alpha$ emission. The first seven have been published by Beaulieu et al. (1996), Lamers et al. (1998), and the new 15 HAeBe stars by de Wit et al. (1999). The LMC HAeBe stars are located in a large star-forming region in the Bar with strong IR excess at IRAS wavelengths. Two stars with very similar characteristics were found in the SMC in a pilot search using EROS-2 data. These stars are the analogues of the Galactic pre-main sequence HAeBe stars.

The fields near the HAeBe stars are ideally suited for the study of the low mass star formation, because they contain many T Tauri stars. In fact, a study of these fields at ESO in BVRI and Hα down to 19.5^m reveals colour-magnitude diagrams that contain hundreds of emission line objects (HAeBe and T Tauri stars) down to the detection limit. Only a few dozen emission lines objects are observed in the control fields outside the star forming region. The magnitudes of the optical study limit the determination of the IMF to $M_v \simeq +0.5^m$ or $M_* \simeq 3M_\odot$.

The location of the Galactic PMS in the HRD shows an upper envelope called "the birth-line". It marks the location of the stars at the end of the fast accretion rate when they become optically visible through the dust. The birth-line has been predicted for different accretion rates by Palla & Stahler (1993) and Bernasconi & Maeder (1996). There are no Galactic HAeBe stars with $M_* > 10M_\odot$, with the exception of a few stars which have uncertain distances. Five of the six HAeBe stars in the LMC studied in detail by Lamers et al. (1998) have luminosities above the observed and predicted Galactic birth-line by factors 3 to 10 (see Figure 1)! Does the birth-line depend on metallicity? If so, we should find a systematic trend from SMC, LMC to Galaxy. What could be the reason? Is the accretion rate dependent on metallicity? Faster accretion would result in a higher birth-line. Or does dust play a more crucial role than predicted in determining the luminosity of the star when its optical radiation emerges? (A lower metallicity may result in a smaller dust/gas ratio).

Galactic HAeBe stars show irregular photometric variations with amplitudes of typically a few tenths of a magnitude and occasionally a deep minimum up to 2 mag. Normally, Galactic HAeBe stars are "redder-when-fainter". This is explained by irregularity of the surrounding dust clouds orbiting the star,

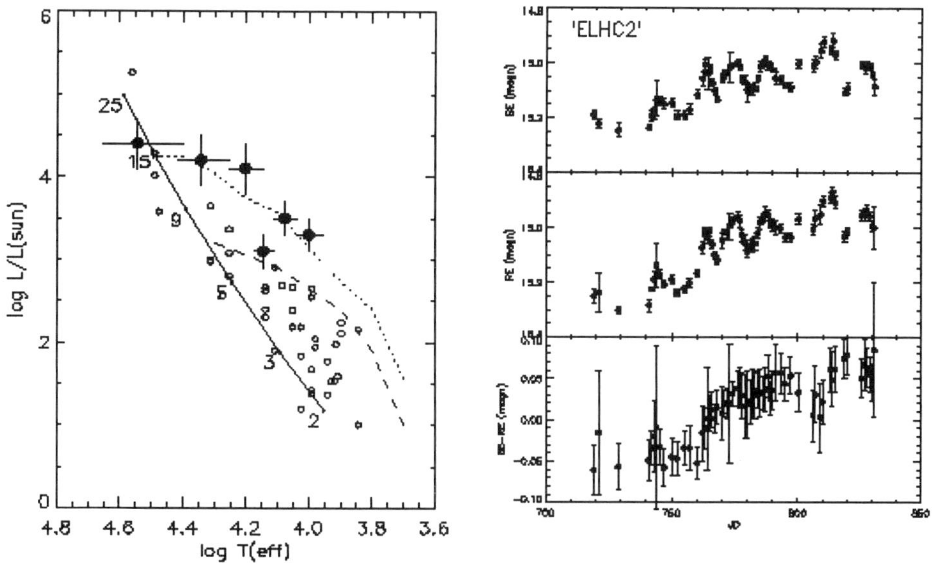

Figure 1. **Left figure**: The location of the first six LMC HAeBe stars in the HRD (points) with the Galactic HAeBe stars (circles). The LMC HAeBe stars are well above the galactic upper limit ("birth-line"). Dashed line: predicted birth-line for an accretion rate $10^{-5} M_\odot \ yr^{-1}$ and dotted line for $10^{-4} M_\odot \ yr^{-1}$ from Palla & Stahler (1993). **Right figure**: The light curve and the colour curve of the LMC HAeBe star ELHC2. The time is in days since Jan 1 1990. Notice the irregular variations and the curious fact the star is "bluer-when-fainter". (From Lamers et al. 1998).

and confirmed by simultaneous photometric and polarimetric observations (e.g., Grinin 1994). The LMC and SMC HAeBe stars show a peculiar irregular colour magnitude variation: they are "bluer-when-fainter" (see star ELHC2 in figure 1). This reverse colour effect might be due to the contribution of a reflection nebula. The dust clouds that produce the light curves are small compared to the reflection nebula and may diminish the light of the star, while most of the (blue) scattered light from the reflection nebula is unattenuated. The net result is a decrease in brightness at all wavelengths, but slightly less in B than in V. This effect rarely occurs in observations of Galactic HAeBe stars, (only during very deep minima) because the photometric aperture is much smaller than the projected reflection nebulae.

We suggest that the bluing effects of LMC/SMC HAeBes is because we are observing the stars at 50 kpc, and therefore not resolving their associated reflecting nebulae. For instance, the reflecting nebulae around V380 Ori about 1 arcmin in diameter would be only 0.6 arcsec projected in the LMC. So the scattered blue light from the nebulae will contribute significantly to the EROS photometry. We have modeled this effect by scaling the photometric variations of the HAeBe star V380 Ori and its nebulae to a distance of the LMC. The "redder-when-fainter" behavior of V380 becomes a bluer when fainter behavior.

Acknowledgments. This work is based on observation held at ESO la Silla. JPB is grateful to IAU for financial support in attending IAU190, and to members of the EROS and PLANET collaborations.

References

Afonso, C., et al. (EROS coll.) 1998, A&A, 337, L17

Albrow, M., et al. (PLANET coll.) 1998, ApJ, 509, 687

Alcock, C., et al. (MACHO coll.) 1993, Nature, 365, 621

Alcock, C., et al. (MACHO coll.) 1997, ApJ, 486, 697

Alcock, C., et al. (MACHO coll.) 1998, ApJ, 499, 9

Ansari, R., et al. (EROS coll.) 1995, A&A, 299, L21

Aubourg, E., et al. (EROS coll.) 1993, Nature, 365, 623

Beaulieu, J.P., et al. (EROS coll.) 1995, A&A, 299, 168

Beaulieu, J.P., et al. (EROS coll.) 1995, A&A, 303, 137

Beaulieu, J.P., & Sasselov, D.D. 1997, in Variable stars and the astrophysical return of microlensing surveys, eds Ferlet R. & Maillard J.P., editions frontieres.

Beaulieu, J.P., et al. (EROS coll.) 1996, Science, 272, 995

Beaulieu, J.P., et al. (EROS coll.) 1997a, A&A, 318, L47

Beaulieu, J.P., et al. (EROS coll.) 1997b, A&A, 321, L5

Beaulieu, J.P., & Sackett, P.D. 1998, AJ, 116, 209

Bernasconi, P., & Maeder, A. 1996, A&A, 307, 829

de Laverny, P., et al. (EROS coll.) 1998, A&A, 335, L93

Grinin, V.P. 1994, in Nature and evolutionary status of Herbig Ae/Be stars ASP Conf Ser. Vol. 62, ed. P.S. Thé.

Grison, P., et al. (EROS coll.) 1995, A&A, 109, 445

Hill, V. & Beaulieu, J.P. 1997, in Variable stars and the astrophysical return of microlensing surveys, eds Ferlet R. & Maillard J.P., editions frontieres.

Ibata, R.A., Lewis, G.F., & Beaulieu, J.P. 1998, ApJ, 509, L29

Lamers, H.J.G.L.M., Beaulieu, J.P., & de Wit, W.J. 1998, A&A, 341, 827L

Palanque-Delabrouille, N., et al. (EROS-2 coll.) 1998, A&A, 332, 1

Palla, F. & Stahler, S.W. 1993, ApJ, 418, 414

Renault, C., et al. (EROS coll.) 1998a, A&A, 324, L69

Renault, C., et al. (EROS coll.) 1998b, A&A, 329, 522

Sasselov, et al. (EROS coll.) 1997, A&A, 324, 471

Welch, D.L., et al. 1999 (MACHO coll.), this volume

Zhao, H.S. 1998, MNRAS, 294, 139

Zaritsky, D., & Lin, D. 1997, AJ, 114, 254

Discussion

Despina Hatzidimitriou: What is the status of analysing the EROS data for RR-Lyrae's in the SMC?

Beaulieu: We have light curves for 5.3 million stars in the SMC. We haven't done the absolute calibration yet. We haven't searched for RR Lyraes. In the LMC, we have catalogued 10,500 RR Lyraes among the 13 million stars.

Hans Zinnicker: Congratulations on your discovery of pre-main sequence stars in the Magellanic Clouds! Let me also refer you to the poster by Brandner et al. who use the technique of comparing deep CCD Hα images with deep CCD R images to detect Hα emitting pre-Main sequence stars in the Magellanic Clouds, e.g. in NGC 346.

New Views of the Magellanic Clouds
IAU Symposium, Vol. 190, 1999
Y.-H. Chu, N.B. Suntzeff, J.E. Hesser, and D.A. Bohlender, eds.

The Magellanic Clouds Photometric Survey

Dennis Zaritsky, Eva K. Grebel, Jason Harris

UCO/Lick Observatory, UCSC, Santa Cruz, CA, 95064, USA

Ian Thompson

Carnegie Observatories, 813 Santa Barbara St., Pasadena, CA, 91101, USA

Abstract. We present a brief description of the Magellanic Clouds Photometric Survey. In particular, we discuss examples of how the new generation of densely populated CMDs provide new information than that available previously.

1. Introduction

The Magellanic Clouds Photometric Survey is a digital UBVI survey of the central $8° \times 8°$ of the Large Cloud and $4° \times 4°$ of the Small Cloud. The data will help address a variety of issues and provide, among other products, a differential reddening map, cluster census (which will include estimates of richness, morphology, and age), and a map of the field star formation history.

2. Observations and Data Reduction

The data are acquired using the 40-inch Swope telescope at Las Campanas and the Great Circle Camera (Zaritsky, Shectman, & Bredthauer 1996). The pixel scale is $0''.7$ mm^{-1} and the typical seeing is around $1''.5$. The observing mode is drift scanning and we produce scans that are $\sim 2° \times 24'$. The effective exposure time as any one object drifts across the field of view is about 4 minutes. Each scan overlaps adjacent scans, which provides comparison data with which to verify astrometry and photometry. The survey will require 40 clear nights for completion and roughly half is done.

The scans are reduced using an automated pipeline constructed around DAOPHOT (Stetson 1987). The basic description of the reduction procedure is presented by Zaritsky, Harris, & Thompson (1997). The stellar catalog is 50% complete at V = 21.5 (although this limit is highly sensitive to crowding). Astrometric uncertainties are generally $\leq 0''.5$, sufficiently good for multiobject spectroscopy.

The catalog contains RA, Dec, U, σ_U, B, σ_B, V, σ_V, I, and σ_I. A star is only included in the catalog if both B and V magnitudes are measured. U and I magnitudes are not required, and fewer stars (generally around 50 to 75% of the total) have U or I magnitudes. The photometry is placed on the Landolt system

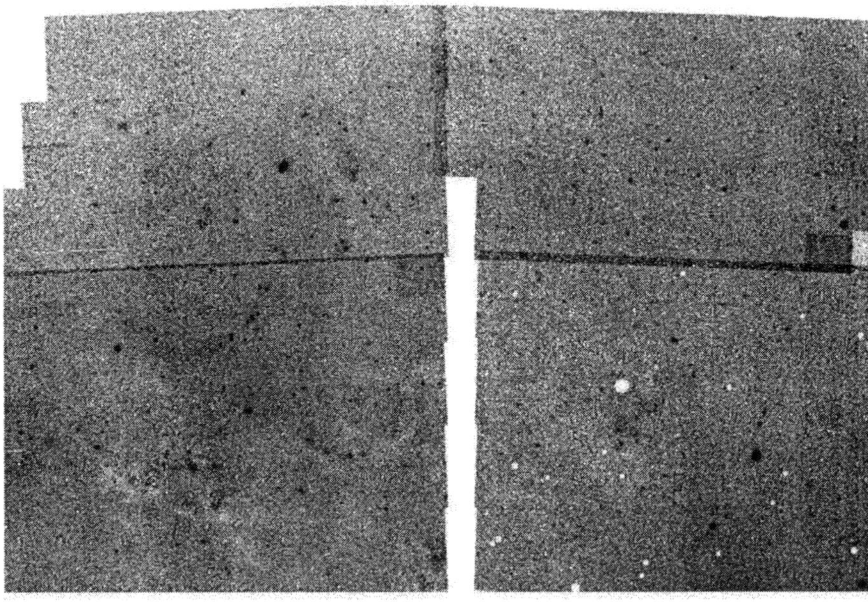

Figure 1. The stellar density within the area of the LMC for which we have reduced data. The $4^\circ \times 4^\circ$ region lies directly north of the bar (N up, E left). Darker pixels correspond to denser regions. White circles are areas in which saturated stars have been removed.

(Landolt 1992) and astrometry is derived from a comparison to the Magellanic Catalogue of Stars (Tucholke et al. 1996).

3. Discussion

So far, we have reduced data from an area of $\sim 4^\circ \times 4^\circ$ in the LMC that is located above the bar. Figure 1 shows the stellar density map of that region based on 4 million stars with measured B and V magnitudes (and $V < 21$). The most visible structure is the Constellation III/LMC IV region (in the middle of the left side of the Figure). Some artifacts of the survey are also evident (there are (1) some horizontal and vertical stripes of higher density illustrating regions where stars from overlapping scans have not yet been matched, (2) some gaps between scans where the data from the first year was judged to be of unusable quality, and (3) a few small patches, such as in the far right of the area, where a problem with the reduction pipeline caused an overdensity or underdensity of identified stars). All of these artifacts are easily removed, but illustrate some of the quality-control issued raised in such a survey.

The Hess diagram for the stars in Figure 1 is presented in Figure 2. The major features (upper main sequence, red clump, and red giant branch) are visible in the left panel, while more subtle features (red supergiants, foreground disk stars and blue supergiants) can be seen in the right panel. Color-magnitude

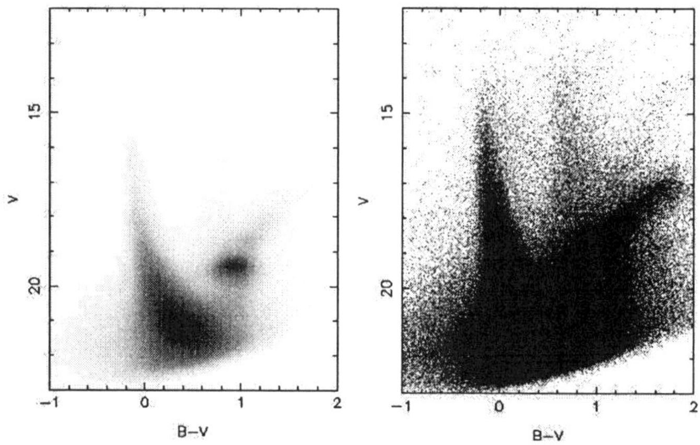

Figure 2. The B-V Hess diagram (low contrast on the left and high contrast on the right) based on 4 million stars in the region shown in Figure 1.

diagrams constructed from such a large number of stars illuminate poorly populated phases of stellar evolution and therefore not only provide information on the nature of stellar populations in the Clouds, but also empirical constraints on our theoretical understanding of stellar evolution. For example, a weak clump is seen on the giant branch above the red clump (termed the RGBB, Zaritsky & Lin 1997, but identified with a evolutionary phase on the AGB, Gallart 1998).

A second feature noted by Zaritsky & Lin was the vertical extension of the red clump, termed the VRC. The existence of this feature has been confirmed (Beaulieu & Sackett 1998), but the interpretation is controversial. Aside from the interesting implications that this feature may have for models of stellar evolution and the star formation history of the LMC, it may also provide information on the line of sight distribution of red clump stars near the LMC, which would in turn affect estimates of the microlensing optical depth to the LMC. Unfortunately, the answer to the nature of the VRC is unlikely to be straightforward.

The two competing interpretations, (1) a foreground distribution of red clump stars (Zaritsky & Lin 1997) and (2) red clump stars of age $\sim$ 0.5 Gyr (Beaulieu & Sackett 1998) make differing predictions. The latter predicts that the VRC stars should have characteristics similar to upper main sequence stars (e.g., low velocity dispersion and strong correlation with sites of recent star formation). Using the data presented in Figure 1, we can test whether the VRC stars are distributed like the $\sim$ 0.5 Gyr old main sequence stars by calculating the ratio of the two populations (Figure 3). The behavior of the ratio with local density of main sequence stars suggests that some fraction of the VRC stars traces the main sequence stars (since the ratio does not asymptotically approach zero), but that the remainder do not trace the main sequence stars (since the number of VRC stars is independent of main sequence stellar density as the

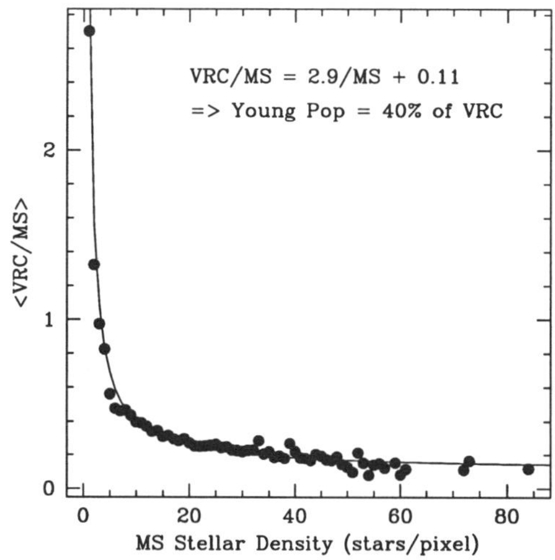

Figure 3. The ratio of VRC stellar density to main sequence stars with $19 < V < 19.5$. We fit a two-component model where the two components are (1) a uniformly distributed VRC and (2) a directly correlated VRC component.

density approaches zero). A simple model suggests that 60% of the VRC stars may be unassociated with recent star formation. Supporting evidence for this is given by the large velocity dispersion (~ 20 km s^{-1}) measured for the VRC stars (Zaritsky et al. 1998), which is consistent with a stellar population of several Gyr age (Meatheringham et al. 1988). Therefore, neither hypothesis on the nature of the VRC appears to account for the entire population. Nevertheless, the large databases produced by surveys like the Magellanic Cloud Photometric Survey are invaluable in discovering such features and will be critical in determining their nature.

Acknowledgments. We gratefully acknowledge financial support from a variety of sources (NSF grant AST-9619576, NASA LTSA grant NAG-5-3501, the Dudley Observatory, the Packard Foundation, and the Sloan Foundation).

References

Beaulieu, J.-P., & Sackett, P.D. 1998, AJ, 116, 209

Gallart, C. 1998, ApJ, 495, 43

Landolt, A.U. 1992, AJ, 104, 340

Meatheringham, S.J., et al. 1988, ApJ, 327, 651

Stetson, P.B. 1987, PASP, 99, 191

Tucholke, H.-J., de Boer, K.S., & Seitter, W.C. 1996, A&AS, 119, 91

Zaritsky, D., Harris, J., & Thompson, I. 1997, AJ, 114, 1002

Zaritsky, D., & Lin, D.N.C. 1997, AJ, 114, 2545

Zaritsky, D., Shectman, S.A., & Bredthauer, G. 1996, PASP, 108, 104

New Views of the Magellanic Clouds
IAU Symposium, Vol. 190, 1999
Y.-H. Chu, N.B. Suntzeff, J.E. Hesser, and D.A. Bohlender, eds.

Carbon Stars in the Magellanic Clouds

M. Azzopardi

IGRAP, Observatoire de Marseille, 2 Place Le Verrier, F-13248 Marseille Cedex 4, France

Abstract. Thanks to their pronounced and well recognizable spectral features, carbon stars can be identified easily in Galactic halo and galaxies. In this paper we will review the extensive searches for carbon stars in the Magellanic Clouds with emphasis on the recent work devoted to the carbon stars of low luminosity and the more classical carbon stars populating the outer regions of these systems.

1. Introduction

The red giant stars are of special interest for the study of the morphology, stellar evolution and kinematics of the galaxies. In addition, the ratio of carbon (C) stars to the late-type M stars, and possibly the faint-end of the C star luminosity function, give information on the metallicity of the parent system. The envelopes and the atmospheres of Asymptotic Giant Branch (AGB) stars are enriched with molecules formed out of either oxygen or carbon, which make their spectroscopic and/or photometric detection possible in external galaxies as far away as the M81/NGC 2403 group. The excess carbon with regard to oxygen (C/O>1) in C stars allows their identification mainly through their strong CN and Swan C_2 bands. These readily recognized features, combined with the proximity of the Magellanic Clouds, have contributed significantly to the success of the Magellanic Cloud C star surveys for the past four decades. Note that the large extent in the sky of these galaxies and the general relative intrinsic brightness of the classical C stars make the wide field photographic plate surveys still competitive, even in the era of CCD detectors, so long as CCD mosaic cameras are not yet in current use.

2. Early Carbon Star Surveys

The discovery of nine C star candidates in the Large Magellanic Cloud (LMC) was first reported by Westerlund (1960). It was the result of a photometric (R, I) and near-infrared objective-prism surveys ($\approx$ 2100 Å/mm dispersion at the A-band) of four central regions in the LMC, carried out by the author in the late 1950s using the 50/65/173 cm Schmidt telescope at Uppsala Southern Station, Mt Stromlo Observatory. An extension of this survey (I >13) to the major part of the LMC (62.5 sq. deg.) led to the identification of 302 C star candidates (Westerlund 1964; Westerlund et al. 1978). Meanwhile, the CTIO blue-green Curtis-Schmidt telescope objective-prism search (1360 Å/mm dispersion at Hγ) by Sanduleak & Philip (1977) of part of the area previously surveyed by Westerlund led to the detection of 474 LMC C star candidates of which 399 were newly discovered objects. This result shows clearly, that although the near-infrared low dispersion objective-prism technique favours detection of the late N-type C stars displaying strong blends of CN-bands near $\lambda\lambda$7910, 8100 and 8320 Å, the strength of the Swan C_2 bands at $\lambda\lambda$4737, 5165 and 5635 Å more than compensates the loss of flux in the blue-green spectral domain of C stars, hence does not introduce this bias while allowing detection of the hotter ones. Subsequent detailed spectroscopic and photometric studies by Crabtree et al. (1976) and Richer et al. (1979) of selected candidates drawn from those lists were to confirm the carbon star nature and provide information on the physical properties of the first LMC C stars.

Deeper (I >17) near-infrared grism searches (2350 Å/mm and 1700 Å/mm dispersions at the A-band) for C and late M-type stars in selected fields of the Magellanic Clouds were carried out by Blanco & McCarthy using a photographic camera equipped with a Schott RG630 filter at the prime focus of the CTIO 4.0-m reflector: 1045 C star candidates in the LMC and 860 ones in the Small Magellanic Cloud (SMC) were detected in $\approx$6.2 and $\approx$6.8 sq. deg. total fields, respectively (Blanco et al. 1980; Blanco & McCarthy 1983, 1990). Carbon star-count isopleths based on the surface densities determined for these objects in 52 LMC sample areas and 37 SMC ones led Blanco & McCarthy (1983) to estimate the total numbers of cool C stars as 11000 in the LMC and 2900 in the SMC. Also, study of the distributions of M giants led these authors to investigate the ratios of the frequencies of C and M-type stars in the Magellanic Clouds. They found that the C/M ratio, i.e. (number of C stars)/(number of giants of type M6 or later) is 2.2±0.1 in the LMC and appears to be constant throughout the galaxy; for the SMC, this ratio is 19.2±0.8 in the central regions and decreases continuously outwards to reach 4.7±0.4 in its periphery. If C/M is related to metallicity alone – not also a function of the red giant age – this lets us ascertain that no significant metallicity gradient exists in the LMC, while for the SMC an increase of metallicity toward the external regions is indicated at the time, 3-5 Gyr ago, when the red giants were formed.

An extensive photographic slitless spectroscopy survey for field C stars in the SMC was achieved by us (see Azzopardi & Westerlund 1984) at the prime focus of the ESO 3.6-m telescope. We used the wide field corrector equipped with a blue-green grism (2200 Å/mm dispersion) in combination with a Schott GG435 filter and IIIa-J plates in order to restrict the instrumental spectral range to a useful domain (4300–5300 Å) and keep the number of spectrum overlaps as low as possible. The main body of the SMC – except its western regions – was covered by 13 partially overlapping fields, 0.78 deg^2 each. Adopting the C star detection technique innovated by Sanduleak & Philip (1977) (search for Swan C_2 band features) we found 1707 C star candidates (Rebeirot et al. 1993). Our C star detection is basically complementary to the near-infrared one (search for CN bands) by Blanco and associates (see McCarthy 1987). The overall surface distribution of the SMC C stars shows that they seem to belong to a spheroidal-like system, in fairly good agreement with the smoothed isophotes of the SMC in red light (see de Vaucouleurs & Freeman 1972).

3. Low-Luminosity Carbon Stars in the Magellanic Clouds

The relation between the carbon abundance and a colour-equivalent given by the spectrophotometry of almost all C stars provided by the catalogue of Rebeirot et al. (1993) allowed us to detect natural groups and especially, to identify a sample ($\approx 5\%$ of the set) of faint and relatively blue objects. Subsequent ESO medium resolution spectroscopy and JHK photometry of 50 of them led Westerlund et al. (1992, 1995) to discover SMC C stars at the highest temperature and lowest luminosity ever found in an external galaxy. Up to now the faint-end of the SMC C star luminosity function inferred from 161 stars is $M_{\rm bol} \approx -1.4$ mag, assuming $(m-M)_0 = 18.8$ as the absorption-free distance modulus for the parent system (see Azzopardi 1994).

Using the same selection criteria, we are now looking for faint C stars in the LMC from the spectrophotometry of about 1200 objects identified in eight selected fields which have been obtained – as a backup programme – during our observations of the SMC. Work is in progress, but so far we have not found any C star fainter than $M_{\rm bol} = -3.3$ mag. This preliminary result as well as our recent discovery of low-luminosity C stars in the Fornax dwarf spheroidal galaxy – resembling those we had previously identified in the SMC, another sligthly metal-poor galaxy – corroborates the theoretical predictions leading to the expectation that the mean luminosity of C stars should be lower in the more metal-deficient system. According to Frantsman (1997, and these proceedings) the low-luminosity C stars may be in the early AGB evolutionary stage and should be "formed as a result of the mass transfer in close binary systems while the primary is a C star on the thermally-pulsing AGB stage" (see also in these proceedings the paper by Marigo & Girardi and references quoted therein).

4. Carbon Stars in the Magellanic Cloud Haloes

Investigations of the dynamical interactions between the components of the Galaxy-LMC-SMC system for a better understanding of the morphology of the SMC and the LMC as well as the attempts to determine the total mass of these galaxies has incited searches for suitable kinematic tracers in the outer regions of the Magellanic Clouds. Jointly, the discovery of CH stars in the outer halo of the LMC by Hartwick & Cowley (1988), strengthened by their intrinsic brightness (see Feast & Whitelock 1992), has clearly emphasized the usefulness of the carbon-type stars for this kind of study. In consequence extensive surveys for C stars in the outer haloes of the Magellanic Clouds, including the inter-Cloud region, and subsequent spectroscopic observations for kinematic purposes have come out recently.

The UK Schmidt Telescope (UKST) objective-prism (800 Å/mm dispersion near 4300 Å) survey of C stars in the outer parts of the SMC by Morgan & Hatzidimitriou (1995), covering a total area of $\approx$220 sq. deg. on the sky, resulted in the detection of 1185 additional candidates. Coupling this survey with that of the inner regions of the SMC by Rebeirot et al. (1993), Morgan & Hatzidimitriou estimated that the total number of C stars in the SMC is $\simeq$ 3060. Radial velocities of a sample of 71 C stars selected in the outer parts of the SMC allowed Hatzidimitriou et al. (1997) to determine the mean heliocentric velocity of these stars to be 149.3 ± 3.0 km/s with a velocity dispersion of 25.2 ± 2.1 km/s which lead to a SMC mass of $\simeq 1.2\ 10^9 M_\odot$. Similar searches by Demers et al. (1993) and Kunkel et al. (1997a) in the inter-Cloud region and in the Magellanic Cloud haloes from the photometry of UKST B_J and R plates (19 fields) resulted in the identification of over one thousand C stars. Radial velocities of 759 C stars on the periphery of the LMC led Kunkel et al. (1997b) to determine the rotation curve from 3 to 12 kpc and the mass contained within a 5 kpc radius, $6.2 \pm 0.9\ 10^9 M_\odot$ (see also the papers by Demers and Kunkel in these proceedings).

References

Azzopardi, M. 1994, in The Local Group: Comparative and Global properties, A. Layden, R.C. Smith & J. Storm, ESO Conference and Workshop Proceedings No. 51, 129

Azzopardi, M., & Westerlund, B.E. 1984, The Messenger, 36, 12

Blanco, V.M., & McCarthy, M.F. 1983, AJ, 88, 1442

Blanco, V.M., & McCarthy, M.F. 1990, AJ, 100, 674

Blanco, V.M., McCarthy, M.F., & Blanco, B.M. 1980, ApJ, 252, 133

Crabtree, D.R., Richer, H.B., & Westerlund, B.E. 1976, ApJ, 203, L81

Demers, S., Irwin, M.J., & Kunkel, W.E. 1993, MNRAS, 260, 103

de Vaucouleurs, G., & Freeman, K.C. 1972, Vistas in Astronomy, 14 163

Feast, M.W., & Whitelock, P.A. 1992, MNRAS, 259, 6

Frantzman, J. 1997, A&A, 319, 511

Hartwick, F.D.A., & Cowley, A.P. 1988, ApJ, 334, 135

Hatzidimitriou, D., Croke, B.F., Morgan, D.H., & Cannon, R.D 1997, A&AS, 122, 507

Kunkel, W.E., Irwin, M.J., & Demers, S. 1997a, A&AS, 122, 463

Kunkel, W.E., Demers, S., Irwin, M.J., & Albert, L. 1997b, ApJ, 488, L129

McCarthy, M.F. 1987, in Stellar Evolution and Dynamics of the Outer Halo of the Galaxy, M. Azzopardi & F. Matteucci, ESO Conference and Workshop Proceedings No. 27, 203

Morgan D.H., & Hatzidimitriou, D. 1995, A&AS, 113, 539

Rebeirot, E., Azzopardi, M., & Westerlund, B.E. 1993, A&AS, 97 603

Richer, H.B., Olander, N., & Westerlund, B.E. 1979, ApJ, 230, 724

Sanduleak, N., & Philip, A.G.D. 1977, Publ. Warner Swasey Obs. 2, 105

Westerlund, B.E. 1960, Uppsala Astron. Obs. Ann. 4, No. 7

Westerlund, B.E. 1964, in IAU/URSI Symp. No. 20, F.J. Kerr & A.W. Roggers, Australian Academy of Science, Canberra, Australia, 239

Westerlund, B.E., Olander, N., Richer, H.B., & Crabtree, D.R. 1978, A&AS, 31, 61

Westerlund, B.E., Azzopardi, M., Breysacher, J., & Rebeirot, E. 1992, A&A, 260, L4

Westerlund, B.E., Azzopardi, M., Breysacher, J., & Rebeirot, E. 1995, A&A, 303, 107

New Views of the Magellanic Clouds
IAU Symposium, Vol. 190, 1999
Y.-H. Chu, N.B. Suntzeff, J.E. Hesser, and D.A. Bohlender, eds.

AGB Stars in the Large Magellanic Cloud

C. Loup

Institut d'Astrophysique de Paris, CNRS, 98 bis Bd. Arago, 75014 Paris, France

Abstract.
This paper reviews the luminosity distribution of AGB stars in the Large Magellanic Cloud. I briefly summarize the history and results of the previous surveys, and then present a first analysis of the DENIS survey in the IJK$_S$ bands, and ISOCAM observations at 4.5 and 12 μm.

1. Brief History of Previous Surveys

Searches for AGB stars and red supergiants in the Large Magellanic Cloud started in the 1960's, and were based on optical/I-band photographic and objective prism surveys. One of the most important surveys was the deep survey by Blanco et al. (hereafter BMB) in 1980. Compared to the previous work which generally surveyed the whole LMC, they observed only 3 small regions in the LMC and 2 in the SMC, each of 23′ in diameter. However, the I limiting magnitude was much deeper at 17 compared to 14 in the early work. As a result, BMB found 50 times more AGB stars than previously known. Their survey was later extended to 52 regions in the LMC (Blanco & McCarthy 1990, hereafter BM). A key result was the derivation of the true luminosity function of LMC carbon stars (see e.g., Richer 1981, and Costa & Frogel 1996). This result was not consistent with the theory of the time which predicted a much higher luminosity for the onset of the formation of C stars. The theory has since been refined by taking into account physical phenomena such as high mass-loss rates and convective "overshooting" (see e.g., Bertelli et al. 1985).

These types of surveys, including those devoted to the search for Long-Period Variables (LPVs, see e.g., Hughes 1989), do not find AGB stars with high mass-loss rates which can be shrouded by dust in the optical. These surveys could only detect AGB stars with little or no mass-loss. An early explanation for the lack of high luminosity C stars as predicted by theories was that AGB stars all lose mass at a high rate when they evolve to high luminosity, and were just missed by BMB. The JHK survey of Frogel & Richer (1983), however, failed to find such very luminous and obscured sources. In 1986, IRAS was used to search for "dust obscured" AGB stars and red supergiants. To date, about 50 obscured AGB stars have been discovered with the IRAS satellite (see e.g., Reid et al. 1990; Wood et al. 1992; Zijlstra et al. 1996) and it is clear that some very luminous obscured AGB stars exist. Their numbers are small and do not substantially change the C-star luminosity function found by BMB.

Despite the impressive work performed by several groups between the 60's and the beginning of the 90's, our knowledge of the LMC AGB population is still very incomplete. Optical carbon stars and luminous (mainly luminosity type I and II) optically discovered M stars are well sampled over the LMC. But the bulk of M stars on the E-AGB and TP-AGB phases may not have been found, except in the BMB's Bar West field (Frogel & Blanco 1990). Searches for LPV's seem to be reasonably complete for optical Miras, but were not sensitive enough to extract semi-regular and irregular variables. For TP-AGB stars with high mass-loss rates, only the brightest and most luminous ones could have been detected by IRAS (Loup et al. 1997), missing obscured AGB stars with luminosities lower than about $10^4\,L_{\odot}$. Finally, AGB stars with intermediate mass-loss rates, well known and numerous in our Galaxy, would have been completely missed both by optical/I-band and IRAS searches.

2. First Results From the DENIS and ISO Surveys

New surveys are underway which address the deficiencies in the previous surveys. The EROS2 and MACHO experiments can detect almost every optical AGB variable in the LMC whatever the variability type due to very good time sampling and a reasonable spatial coverage of the data. In the near-infrared, the two sky surveys DENIS (the southern hemisphere survey in IJK_S) and 2MASS (the whole sky in JHK_S) have sufficiently faint detection limits to find the shrouded AGB stars in the LMC.

Two studies based on DENIS data have been performed, one in the BMB's Bar West and Optical Center fields (Loup 1998), and one in a 20 sq-deg field. These results were presented in these proceedings by Cioni et al. (1999) Here we summarize the main results. The DENIS data contain all but the most obscured TP-AGB stars (with J–$K_S > 4$) while the E-AGB and RGB stars are found to $M_{bol} \sim -2.7$. C- and M-type stars with low mass-loss rates follow two distinct sequences in the DENIS color-color diagram : (I–J) is a good indicator of the M subtype and varies from 0.9 to $\lesssim 4$ (influence of TiO and VO bands), whereas optical C stars are traced by (J–K_S) from 0.8 to 2.2 (C_2, CN, and CO bands). The luminosity functions for these two samples contain 1200 and 10,000 AGB stars respetively. The RGB tip is located around $M_{bol} \simeq -3.45$. There is a clear bump in the luminosity distribution above the RGB-tip: this bump is formed by C stars and by the M stars with spectral types later than M5, i.e., by the TP-AGB stars. Statistically, we find that optical C stars, M5 to M10 stars, and AGB stars with intermediate and high mass-loss rates, have similar luminosity distributions.

The DENIS survey has been complemented by a mini-survey performed with the camera of the ISO satellite at 4.5 and 12 μm in 0.8 sq.degrees of the LMC bar region in order to detect sources with high mass-loss rates. To date 0.14 sq-deg have been fully analyzed. As expected, IRAS missed about 50% of the obscured AGB stars. The bolometric luminosities for these stars have been derived using both DENIS and ISO data. It confirms the previous result from DENIS data, i.e. that TP-AGB stars with high mass-loss rates are not significantly more luminous than optical C stars. This can be seen in Figure 1 where sources detected by ISO are drawn with squares, and sources only seen

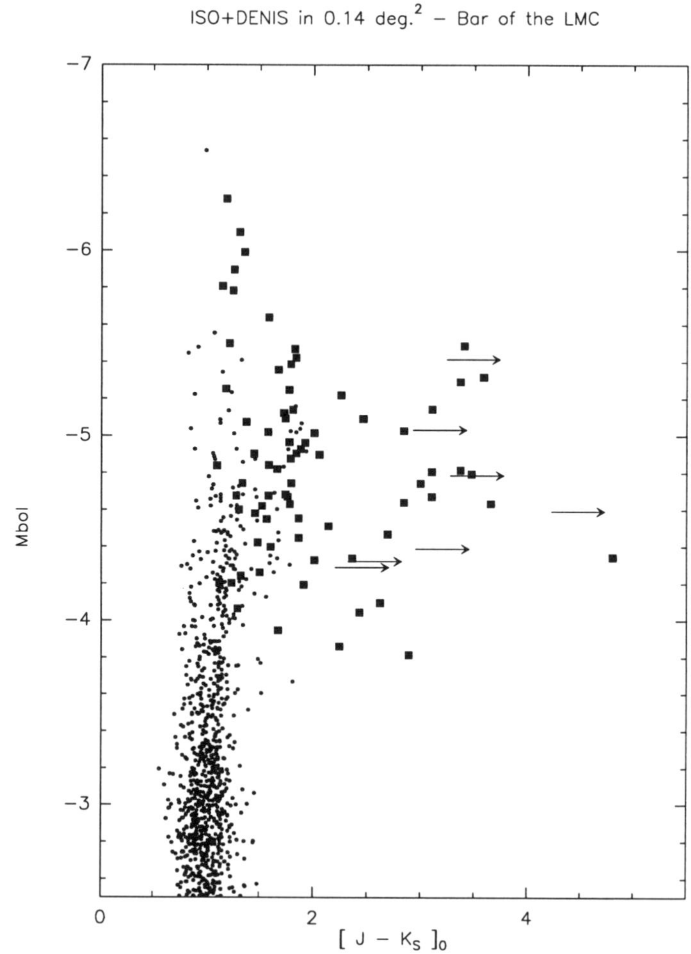

Figure 1. Luminosity distribution of AGB stars from ISO and DENIS data. See text at the end of section 2.

with DENIS by dots; the reddest sources (K_S−12 from 4 to 7.5) are not detected in J (arrows).

Acknowledgments. The work presented on the ISO observations is a collaboration with J. Blommaert & N. Trams (ESA), M-R. Cioni & H. Habing (Leiden Observatory), M. Groenewegen (MPI), E. Josselin & A. Omont (IAP), F. Kerschbaum (Wien observatory), and J. van Loon & L. Waters (University of Amsterdam). This paper made use of observations obtained at the European Southern Observatory as part of the DENIS project. DENIS has been financially supported by the Science Plan of the Commission of the European Communities; by the Ministère de l'Education Nationale, the Centre National de la Recherche Scientifique, and the Institut National des Sciences de l'Univers in France; by the Expertise Centrum for Astronomical Image Processing in the Netherlands;

by the Consiglio Nazionale delle Ricerche in Italy; by the Fonds zur Forderung der Wissenschaftlichen Forschung and the Bundesministerium für Wissenschaft und Forchung in Austria; by the state of Baden–Württemberg in Germany; and by the Foundation for the Development of Scientific Research of the State of São Paulo in Brazil.

References

Bertelli, G., Bressan, A.G., & Chiosi, C. 1985, A&A, 150, 33

Blanco, V.M., & McCarthy, M.F. 1990, AJ, 100, 674 (BM)

Blanco, V.M., McCarthy, M.F., & Blanco, B.M. 1980, ApJ, 242, 938 (BMB)

Cioni, M-R., Habing, H.J., Loup, C., & Epchtein, N. 1999, this volume

Costa, E., & Frogel, J.A. 1996, AJ, 112, 2607

Epchtein, N., et al. 1997, *The Messenger*, 87, 27

Frogel, J.A., & Richer, H.B. 1983, ApJ, 275, 84

Frogel, J.A., & Blanco, V.M. 1990, ApJ, 365, 168.

Groenewegen, M.A.T., Whitelock, P.A., Smith, C.H., & Kerschbaum, F. 1998, MNRAS, 293, 18

Hughes, S.M.G. 1989, AJ, 97, 1634

Loup, C., Zijlstra, A.A., Waters, L.B.F.M., & Groenewegen, M.A.T. 1997, A&AS, 125, 419

Loup, C., Duc, P-A. Fouqué, P., Bertin, E., & Epchtein, N. 1998, in *The Impact of Near-Infrared Sky Surveys on Galactic and Extragalactic Astronomy* p.115, editor N. Epchtein, Kluwer Academic Publishers

Reid, I.N., Tinney, C., & Mould, J. 1990, ApJ, 348, 98

Reid, I.N. 1991, ApJ, 382, 143

Richer, H.B. 1981, ApJ, 243, 744

Wood, P.R., Whiteoack, J.B., Hughes, S,M,G,, Bessell, M.S., Gardner, F., & Hyland, A.R. 1992, ApJ, 397, 552

Zijlstra, A.A., Loup, C., Waters, L.B.F.M., Whitelock, P.A., & Guglielmo, F. 1996, MNRAS, 279, 32

Discussion

Michael Feast: Does the $I-J/J-K$ diagram distinguish between M giants and M-type Miras?

Loup: Very likely yes, because the $(I-J/J-K)$ diagram is a very good tracer of the M spectral index, and the Mira variables are in general of later spectral type than non-variable M giants. We did not check yet precisely, but this work will be done.

New Views of the Magellanic Clouds
IAU Symposium, Vol. 190, 1999
Y.-H. Chu, N.B. Suntzeff, J.E. Hesser, and D.A. Bohlender, eds.

Planetary Nebulae in the Magellanic Clouds

Michael A. Dopita

Mt. Stromlo & Siding Spring Observatory, Institute of Advanced Studies, The Australian National University, Australia

Abstract. The proximity, accurately known distance and low line-of-sight reddening give the ideal circumstances to pursue studies of individual stellar populations in the Magellanic Clouds. Here we show how our understanding of the evolution and chemical composition of the planetary nebulae in the Magellanic Clouds has been impacted by imaging and UV spectroscopic studies using the Hubble Space Telescope. Images provide sizes, internal morphological structure, absolute fluxes, and dynamical ages, while spectra allow us to place the central stars accurately on the H-R Diagram, and we can also examine the details of the evolution, of mass- and age- dependent chemical dredge-up processes, and infer the star-formation history of the Magellanic Clouds.

1. Introduction

The final evolution of stars from the tip of the Asymptotic Giant Branch (AGB) through the planetary nebula (PNe) stage represents a fascinating stage of stellar evolution. For Galactic objects, quantitative studies are beset by uncertainties in the PN distance scale. However, this problem is eliminated in the Magellanic Clouds, which provide a population of PN at known distance and having low field reddening. From the ground, this population has been studied in detail. Data on spectra, line fluxes, densities, expansion velocities and kinematics have all been accumulated in the last few years (see the review by Barlow 1989, also Dopita et al. 1985a,b; Meatheringham et al. 1988a,b; Meatheringham & Dopita 1991a,b). In addition, accurate positions have been obtained (Leisy et al. 1997), and the IUE satellite has provided valuable data on important dredge-up species such as C (Aller 1987; Leisy & Dennefeld 1996).

A key observational parameter is the size and internal morphology of the nebula. However, the nebulae are typically too small in angular extent and from the ground only fragmentary data has been obtained (Jacoby 1980; Wood et al. 1987; Jacoby et al. 1990; Jacoby & Kaler 1993; Wood, et al. 1986). The imaging capability of the Hubble Space Telescope (HST) is well suited not only to measure the size but also to resolve details of the internal structure of the Magellanic Cloud PNe (Blades et al. 1992; Stanghellini et al. 1999). Furthermore, the poweful spectroscopic capability of the HST is ideal for the study of the UV spectra of these PNe. Here we briefly describe the main results of our integrated imaging and spectroscopy program (Dopita et al.; Vassiliadis et al.; Papers I - VI; *see references*).

2. Evolutionary Theory

PNe are derived from low- and intermediate- mass stars. The exact initial mass limit (above which the star will explode as a supernova) is still somewhat uncertain, but is most likely 6 - 8 $M_\odot$. However, PNe more typically represent an intermediate-age stellar population, derived from stars having initial masses in the range 1 - 3 $M_\odot$. Stars evolve to the PNe stage at the end of their nuclear-burning lifetime, when the unburnt envelope mass becomes so low that it reveals the hot core of the star that remains. The theory of this phase hs made rapid progress recently (Schönberner 1983, Weidemann 1987; Alongi et al. 1993; Vassiliadis & Wood 1993; Bressan et al. 1993). Because of their low mass, the lifetime of the precursor star can be a considerable fraction of the age of a galaxy. From the Vassiliadis & Wood (1993) models:

$$\tau \sim 11.0\,(m/M_\odot)^{-3.1} + 0.46\,(m/M_\odot)^{-4.6}\,\mathrm{Gyr} \qquad (1)$$

The shell of gas, ionised in the PNe phase, was thrown off from the central star during the asymptotic giant branch (AGB) phase of evolution, assisted by the large radiation pressure. It is also enhanced by the shell flashes which result from burning helium under electron-degenerate conditions. As a result, the mass-loss rate is a very steep function of mass. This serves to channel a wide range of initial stellar masses (1 - 8$M_\odot$) into a rather narrow range of final (White Dwarf) stellar mass (0.55 - 1.4$M_\odot$). The relationship between initial mass, m, and final core mass, m_{core}, can be expressed as (Vassiliadis & Wood 1993; Marigo, Bressan, & Chiosi 1996):

$$m_{core}/M_\odot = 0.524 + 0.0438\,(m/M_\odot) + 0.0095\,(m/M_\odot)^2 \qquad (2)$$

Since the central star does not proceed beyond helium burning, and since much of the unburnt envelope is lost, the PNe provides a time-capsule of ionised material which has been chemically unchanged since the star formed, at least as far as the heavy elements are concerned. In principle, PNe may be used to probe the history of chemical enrichment in galaxies, although so far this has done very extensively (but see Dopita et al. 1997). For the lighter elements, and the heaviest elements, the material ejected as a PNe shell is not entirely pristine. A variety of mass- and metallicity- dependent convective phases dredge up partially hydrogen-burnt material from the boundary of the core during both the giant and asymptotic giant phases of evolution (Iben & Renzini 1983; Renzini & Voli 1981). These are:

- The first dredge-up in the red giant phase, produced by the penetration of the convective envelope into regions which are partially CNO-burnt. The dredged-up material is mixed throughout the envelope, with an enhancement of ^{13}C and ^{14}N, and a decrease in ^{12}C.

- The second dredge-up in the early AGB stars of $m > 3 - 5M_\odot$, occurring when the hydrogen-burning shell extinguishes. This time envelope enhancements of ^{4}He, ^{14}N and ^{13}C are produced.

- The third dredge-up in the thermally-pulsing AGB phase. After each He-burning shell flash, the convective envelope dips down, dredging up material rich in ^{4}He, ^{12}C and the $s-$process elements.

- Finally, the so-called *hot-bottom burning* occurs in the more massive AGB stars ($m > 3M_{\odot}$) when convection in the stellar envelope cycles matter through the hydrogen-burning shell during the inter-pulse phase. Significant ^{14}N, and possibly ^{4}He production, may result.

3. Photoionisation Modeling of Magellanic Cloud PNe

For each PN, the UV spectrophotometry obtained with HST (Vassiliadis et al. 1996, 1998) was combined with ground-based data (Meatheringham & Dopita 1991a,b) to generate a self consistent set of de-reddened line intensities extending from Ly-α to beyond 7300Å in the red. In some cases, the S/N ratio of the UV data is even sufficient to study mass-loss in the central star (Bianchi et al. 1997)! Extremely valuable constraints to the modeling are given by the HST imaging data (Papers III and VI) which consist of images in the [O III] λ 5007Å line, chosen because it is generally the brightest emission line in the optical. These images not only provide the size, but also give detailed structural information which can be used as input to detailed photoionisation modeling analysis.

These data are interpreted (Papers I, II and V) using the generalised modeling code MAPPINGS II (Sutherland & Dopita 1992). The goal of this self-consistent nebular modeling is to match the modeled nebular size, density, and structure with observations, measure the degree of optical thinness of the nebula, derive the luminosity of the central star matching both the absolute luminosity of the nebula and the observed stellar continuum flux in the UV, estimate the stellar temperature and the ionisation parameter which matches both the degree of ionisation and excitation, and finally to derive the nebular abundances which matches the global line spectrum. Ideally, a fully self consistent photoionisation model should have a continuous, fully three dimensional density distribution. However, the observational constraints are still insufficient to define such a structure. In the actual modeling, a two zone model was generally adopted.

4. Hydrogen - or Helium - Burners?

If we simply compare the radius of the PNe with its position on the H-R diagram, we find (as expected) that the smallest PNe tend to lie on the earlier parts of the evolutionary tracks, while larger objects are generally located where older PNe are expected to be found. However, at any point on the H-R diagram, a wide spread in sizes is found, which suggests that a single evolutionary model is inadequate to explain the observations. Likewise, if we use the size and expansion velocity to compute a simple dynamical age, we find a wide range of values at any point on the H-R diagram, which also points to heterogeneous evolution. Furthermore, the dynamical age computed in this way is always smaller, and generally much shorter than evolution theory would indicate.

From a theoretical viewpoint, it is clear that the evolution of the central star and of its nebular shell depends critically upon whether it is He- or H- burning. He- burning stars are those which leave the AGB near the peak of the shell flash, while H- burners turn off the AGB between thermal pulses. In terms of relative numbers of these stars, we would expect to see only about 10-20% of He-burners. However, the observability of the PNe is strongly affected by the evolution. The H-burning PNe are initially more luminous than their He- burning counterparts. However, the time available for evolution of the PN shell is generally much shorter for the H-burners, and they fade rapidly; before they have input much energy to accelerate the nebular shell into expansion. Nebulae around H- burners therefore rapidly evolve from small and bright to small and faint. On the other hand, the evolution of He- burners across the H-R diagram is slow and lazy, as hydrogen is re-ignited before the final fading of the central star. These stars put a lot more energy into the nebular shell while it is fairly bright, and so greater numbers of PNe will be observed in the bright phase, and these will be phyically much larger.

It is thus possible (Paper IV) to separate the Magellanic Cloud PN into H- burner and He- burner classes. At least half, and possibly that the majority of the luminous PNn in the LMC are associated with stars which left the AGB as He- burners - a viewpoint that would not have been generally accepted in the past (e.g. Schönberner 1983; Blöker & Schönberner 1991).

5. The Age-Metallicity Relation

From the position of the central stars on the theoretical $log(L/L_{\odot}) : log(T_{eff})$ Hertzsprung-Russell (H-R) Diagram, we determine the current core mass, the original mass, and the age of the star. These can then be used in conjunction with the observed mean abundance relative to solar of the α-process elements (O, Ne, Ar and S), [α/H], to derive the metallicity : age relationship for the LMC.

This reveals that in the LMC there was a long period of quiescence between ~15 Gyr and ~4 Gyr ago, during which the metallicity remained close to SMC values. There is no evidence in this sample of any "halo" abundance objects. About 1-3 Gyr ago, there appears to have been a strong burst of star formation which more than doubled the metallicity. In recent times, star formation and the rate of chemical evolution slowed again. This is in remarkable agreement with the data on both field stars and clusters presented in this conference. For example, the latest cluster data (Girardi et al. 1995), shows a long period of quiescence, followed by an large rate of cluster formation 1-2 Gyr ago. An "age gap" is also seen in the colour magnitude diagrams of individual clusters (Da Costa 1991). Recent HST C-M data on a field in the outer disk of the LMC also show an age gap, with a strong burst in star formation starting ~ 2.3 Gyr ago (Gallagher et al. 1996).

6. Heavy Element Dredge-up

For the LMC our detailed abundance analyses (Dopita 1997; Paper V), shows a striking and systematic trend in the abundances of He, C and N with α-process abundance and / or stellar mass or age. A strong enhancement in C is found for old stars low mass. This indicates that the third dredge-up of C is important for such stars, as is well known from studies of carbon stars. An important result is that the sum of the C + N abundances shows little systematic trend, in agreement with the results of Kaler & Jacoby (1990, 1991). This indicates that the third dredge up is significant at all masses.

In younger stars of higher mass and abundance, "hot-bottom" burning appears to be operating efficiently to produce the very high N abundances, large N/C ratios and high He abundances observed, with most of the dredged-up C having being converted to N. These show the typical bipolar or butterfly morphology of the Peimbert's Type I; always suspected of being derived from relatively massive precursor stars. There is no sign of the expected dredge up of ^{22}Ne .

7. Conclusions

Imaging and spectroscopic studies of the PN in the Magellanic Clouds has permitted us to resolve details of the internal nebular structure, to obtain high signal to noise UV spectra of both the nebula and the central star; sufficient to examine the P-Cygni profiles of the mass-losing central stars in some cases. In conjuction with ground-based spectrophotometry, absolute flux, expansion velocity and density information, this data set permits fully self consistent diameters, ages, masses, and abundances to be derived for the nebulae, and the central stars can be accurately placed on the H-R Diagram. Thus, we have been able to examine the details of the evolution, and of the mass- and age- dependent dredge-up processes in a way which is just not possible for Galactic PN.

Acknowledgments. This paper summarises the efforts of all the observing team, Emanuel Vassiliadis, Peter Wood & Steve Meatheringham at Mt Stromlo, Ralph Bohlin at the Space Telecope Science Institute, Holland Ford at the Johns Hopkins University, Pat Harrington at the University of Maryland, and Ted Stecher and Steve Maran at Goddard Space Flight Centre. This paper is based upon observations with the NASA/ESA Hubble Space Telescope, obtained at the Space Telescope Science Institute, which is operated by the Association of Universities for Research in Astronomy, Inc., under NASA contract NAS5-26555. The author wishes to acknowledge travel and publication support under a Major Grant from the International Science and Technology Division of the (Australian) Department of Industry, Science and Technology.

References

Aller, L.H., et al. 1987, ApJ, 320, 159

Alongi, M., et al. 1993, A&AS, 97, 851

Barlow, M. 1989 in Planetary Nebulae, ed. S. Torres-Peimbert, Kluwer : Dordrecht, p319

Bianchi, L., Vassiliadis, E., & Dopita, M.A. 1997, ApJ, 480, 290

Blades, J.C., et al. 1992, ApJ, 398, L41

Blöker, T., & Schönberner, D. 1990, A&A, 240, L11
Bressan, A., Fagotto, F., Bertelli, G., & Chiosi, C. 1993, A&AS, 100, 647
Da Costa, G.S. 1991, in The Magellanic Clouds , eds. R. Haynes & D. Milne, Kluwer: Dordrecht, p183
Dopita, M.A., et al. 1985a, ApJ, 296, 390
Dopita, M.A., Ford, H. C., & Webster, B.L. 1985b, ApJ, 297, 593
Dopita, M.A., et al. 1993, ApJ, 418, 804 (Paper I)
Dopita, M.A., Jacoby, G.H., & Vassiliadis, E. 1992, ApJ, 389, 27
Dopita, M.A., et al. 1988, ApJ, 327, 651
Dopita, M.A., et al. 1994, ApJ, 426, 150 (Paper II).
Dopita, M.A., et al. 1996, ApJ, 460, 320 (Paper IV)
Dopita, M.A., et al. 1997, ApJ, 474, 188 (Paper V).
Gallagher, J.S., et al. 1996, ApJ, 466, 732
Girardi, L., Chiosi, C., Bertelli, G., & Bressan, A. 1995, A&A, 298, 87
Iben, I. Jr., & Renzini, A. 1983, A&A, 21, 271
Jacoby, G.H. 1980, ApJS, 42, 1
Jacoby, G.H., Ciardullo, R., & Walker, A.R. 1990, ApJ, 365, 471
Jacoby, G.H., & Kaler, J.B. 1993, ApJ, 417, 209
Kaler, J.B., & Jacoby, G.H. 1990, ApJ, 362, 491
Kaler, J.B., & Jacoby, G.H. 1991, ApJ, 382, 134
Leisy, P., & Dennefeld, M. 1996, A&AS, 116, 95
Leisy, P., Dennefeld, M., Alard, C., & Guibert, J. 1997, A&AS, 121, 407
Marigo, P., Bressan, A., & Chiosi, C. 1996, A&A, 313, 545
Meatheringham, S.J., & Dopita, M.A. 1991a, ApJS, 75, 407
Meatheringham, S.J., & Dopita, M.A. 1991b, ApJS, 76, 1085
Meatheringham, S.J., et al. 1988a, ApJ, 327, 651
Meatheringham, S.J., Dopita, M.A., & Morgan, D.H. 1988b, ApJ, 329, 166
Renzini A., & Voli, M. 1981, A&A, 94, 175
Schönberner, D. 1983, ApJ, 272, 708
Stanghellini et al. 1999, ApJ, 510, 687
Sutherland, R.S., & Dopita, M.A. 1993, ApJS, 88, 253
Vassiliadis, E., et al. 1996, ApJS, 105, 375 (Paper III)
Vassiliadis, E., et al. 1997, ApJS, 114, 237 (Paper VI)
Vassiliadis, E., et al. 1998, ApJ, 503, 253 (Paper VII)
Vassiliadis, E., & Wood, P.R. 1993, ApJ, 413, 641
Wiedemann, V. 1987, A&A, 188, 74
Wood, P.R., Bessell, M.S., & Dopita, M.A. 1986, ApJ, 311, 632
Wood, P.R., et al. 1987, ApJ, 320, 178

Discussion

Norbert Langer: Do you find a systematic difference in the PN morphology when you compare the hydrogen burners with the helium burners?

Dopita: No, and indeed we would not expect to, since they are both drawn from the same stellar population.

Dominik Bomans: In the He-burner diagram one stood out with large size and location above the 2.5 $M_\odot$ tracks. Is it just chance to pick it or is there a special reason for this object?

Dopita: No, this is simply the faintest, oldest, and most evolved object in the sample.

Hans Zinnecker: You said PN theory is by and large in good shape, but there are problems in detail. What, then, are your recommendations for future theoretical work?

Dopita: Since we now have a reasonable age-metallicity relation for the LMC it would be very helpful to have stellar evolutionary tracks for these ages and initial metallicities. In addition, the observations can now be used to constrain the efficiencies of the various dredge-up episodes, which will tell us about the physics of the late phases of evolution. For example, there is no observational evidence for the dredge-up of ^{22}Ne predicted by current models such as those of Marigo.

Part 5. Stellar Populations and Surveys

Section B. Poster Papers

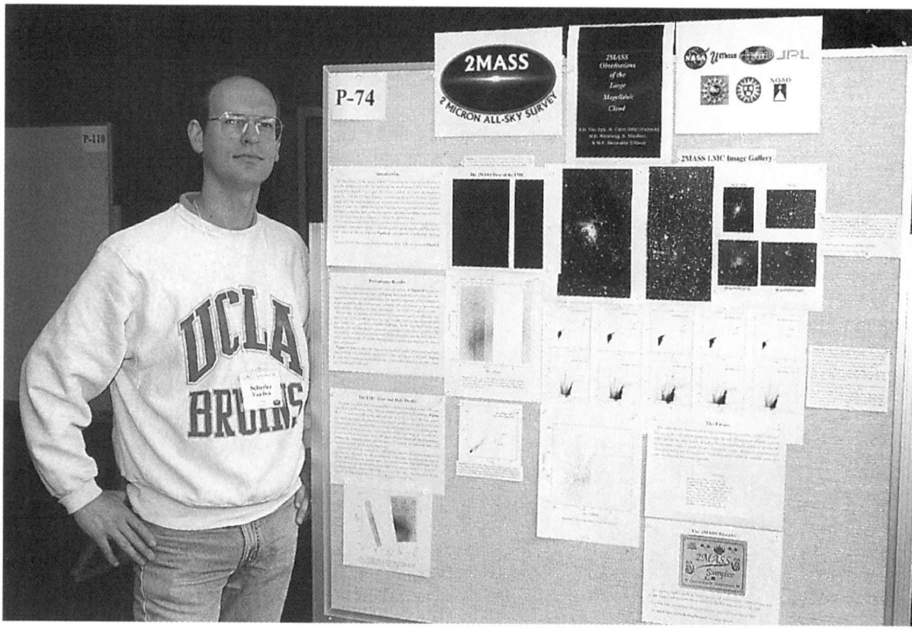

(Top) Schuyler Van Dyk with his 2MASS poster, and (bottom) Steve Majewski answering questions on his poster.

New Views of the Magellanic Clouds
IAU Symposium, Vol. 190, 1999
Y.-H. Chu, N.B. Suntzeff, J.E. Hesser, and D.A. Bohlender, eds.

A Photometric Survey of the LMC Field near NGC 2257

A. R. Walker, R. A. Schommer, N.B. Suntzeff

Cerro Tololo Inter-American Observatory, NOAO, Casilla 603, La Serena, Chile

P.W. Hodge

University of Washington, Dept. Astr., Box 351580, Seattle, WA 98195, USA

M. Mateo

University of Michigan, Dept. Astr., 821 Dennison Bldg., Ann Arbor, MI 48109-1090, USA

E. W. Olszewski

University of Arizona, Steward Obs., Tucson, AZ 85721, USA

Abstract.
Star formation in the outermost regions of the LMC, several kpc from the center, has been dormant for some Gyr. With deep HST and wide-field ground-based imaging surveys, sufficient stars can be measured to accurately delineate the critical turn-off and subgiant branch regions of the CMD, and thus provide a picture of the first 10 Gyr of star formation in the LMC.

The LMC field in the vicinity of NGC 2257 has been subject to several studies over the last 15 years. There exists photometry, spectroscopy of RGB stars and proper motions; results which can be compared with detailed studies of more inner fields (e.g., Elson, Gilmore, & Santiago 1997; Holtzman et al. 1997).

Our group has HST WFPC2 Cycle 6 images of four fields near NGC 2257, and CTIO 4-m prime focus imaging datasets covering 0.25 deg^2 with extension to 5 deg^2 scheduled. For each HST pointing total exposure times were 2000s in F555W (V) and 2400s in F814W (I). The fields are sparse and there are typically 50-80 stars on each WF CCD frame, contamination by non-stellar objects is thought to be negligible brighter than $V \sim 25, I \sim 24$. The resultant CMD is plotted in Figure 1. Approximately 850 stars are easily identifiable as being LMC members, and show a main sequence turn-off at $V \sim 22$ with a well-populated subgiant branch, plus a younger MS extending to approximately $V = 21.2$. The red giant clump is presumably associated with the brighter turn-off. After adopting $E(B - V) = 0.06$ and a distance modulus for the LMC of 18.55 mag (Walker 1999), a fit was made with Bertelli et al. (1994) isochrones. The old turn-off and SGB is well-fit with a 12 Gyr isochrone and metallicity [Fe/H]$= -1.6$.

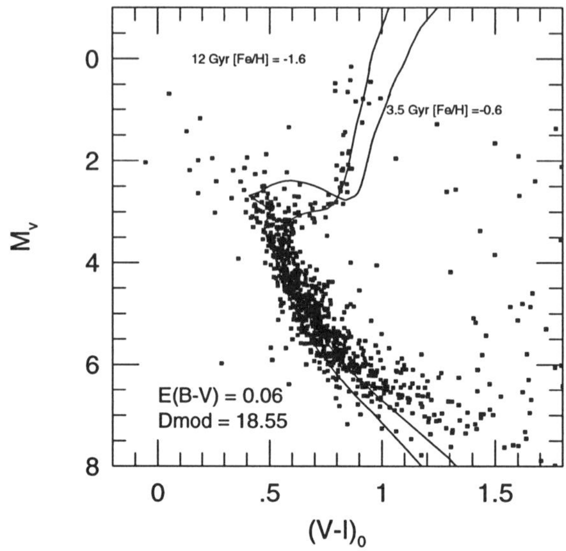

Figure 1. NGC 2257 Field - 4 WF (HST) pointings

An even more metal-poor isochrone would be needed to force the isochrone to lie more central on the MS, indirect evidence that this is not likely is that the LMC field RR Lyraes do not appear to be more metal-poor in the mean than [Fe/H]~ -1.6. A 12 Gyr population is old enough to generate RR Lyraes and there may not be any need for a minority even more ancient ($\sim$ 15 Gyr) population.

The stars making up the bulk of the MS are clearly younger and more metal rich, an isochrone with age 3.5 Gyr and [Fe/H]$= -0.6$ is a possible fit. However the match at the turn- off is clearly not very good, and possibly a distribution of ages and metallicities would be more apt (although we and others have found difficulties fitting these isochones in general, in the sense that the theoretical RGB's are too red cf the MS). Presumably this population corresponds to the [Fe/H]$= -0.5$ (in the mean) red giants studied spectroscopically by Olszewski (1993).

References

Bertelli, G., Bressan, A., Chiosi, C., Fagotto, F., & Nasi, E. 1994, A&AS, 106, 276

Elson, R.A.W., Gilmore, G.F., & Santiago, B.X. 1997, MNRAS, 289, 157

Holtzman, J., et al. 1997, AJ, 113, 656

Olszewski, E. 1993, in *The Globular Cluster-Galaxy Connection*, eds. Smith, G.H., & Brodie, J.P., ASP Conf. Ser. No. 48, publ. ASP., p. 351

Walker, A.R. 1999, in *Post Hipparcos Cosmic Candles*, ed. Caputo, F., & Heck, A., publ. Kluwer, p.125

New Views of the Magellanic Clouds
IAU Symposium, Vol. 190, 1999
Y.-H. Chu, N.B. Suntzeff, J.E. Hesser, and D.A. Bohlender, eds.

A Photometric Survey of Field Stars in the Large Magellanic Cloud: Probing its Star-Formation History[1]

T. A. Smecker-Hane[2]

Dept. of Physics & Astronomy, Univ. of California, Irvine, CA 92697, USA

J. S. Gallagher, III, Andrew Cole

Astronomy Dept., Univ. of Wisconsin, Madison, WI 53706, USA

P. B. Stetson

National Res. Council of Canada, Herzberg Inst. of Astrophysics, Dominion Astrophysical Obs., Victoria, BC V8X 4M6, Canada

E. Tolstoy

Space Telescope-European Coordinating Facility, Karl-Schwarzschild-Str. 2, D-85748 Garching bei Muenchen, Germany

1. Introduction

The Large Magellanic Cloud (LMC) is unique among galaxies in the Local Group in that it is the most massive non-spiral, is relatively gas-rich, and is actively forming stars. Determining its star-formation rate (SFR) as a function of time will be a cornerstone in our understanding of galaxy evolution. The best method of deriving a galaxy's past SFR is to compare the densities of stars in a color-magnitude diagram (CMD), a Hess diagram, with model Hess diagrams. The LMC has a complex stellar population with ages ranging from 0 to $\sim$ 14 Gyr and metallicities from $-2 \lesssim$ [Fe/H] $\lesssim -0.4$, and deriving its SFR and simultaneously constraining model input parameters (distance, age-metallicity relation, reddening, and stellar models) requires well-populated CMDs that span the magnitude range $15 \leq V \leq 24$. Although existing CMDs of field stars in the LMC show tantalizing evidence for a significant burst of star formation that occurred $\sim$ 3 Gyr ago (for examples, see Westerlund et al. 1995; Vallenari et al. 1996; Elson, et al. 1997; Gallagher et al. 1999, and references therein), estimates of the enhancement in the SFR vary from factors of 3 to 50. This uncertainty is caused by the relatively large photometric errors that plague crowded ground-based images, and the small number statistics that plague CMDs created from single Wide Field Planetary Camera 2 (WFPC2) images.

---

[1] Based on observations with the Cerro Tololo Inter-American Observatory and the NASA/ESA Hubble Space Telescope. CTIO is operated by AURA, Inc. under cooperative agreement with the National Science Foundation. The Space Telescope Science Institute is operated by AURA, Inc. under NASA contract 5-26555.

[2] Partial funding provided by NSF grant AST-9619460 to TSH.

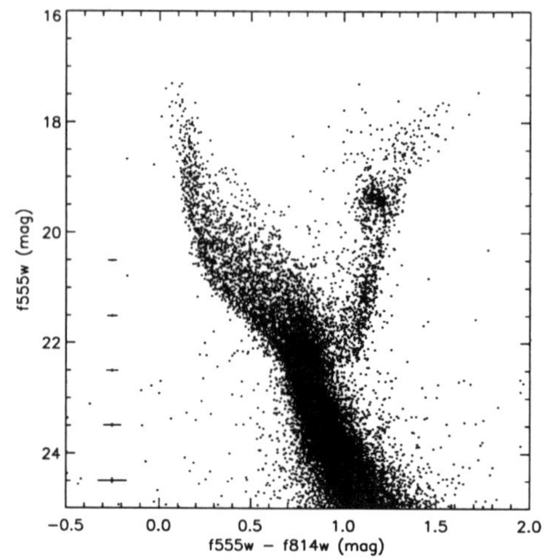

Figure 1. A CMD containing $\sim$ 28,000 stars from one WFPC2 pointing in the LMC bar. (Zero-points have been added to the instrumental magnitudes to approximate V and V-I magnitudes.) Notice the structure in the main-sequence turnoff region at $f555w = 21.5$ and 22 mag, signs of discontinuities in the SFR and/or age-metallicity relation.

2. Our Photometric Survey of LMC Areas

We are combining multicolor imaging of six areas of the LMC obtained with the Cerro-Tololo Inter-American Observatory 1.5-meter telescope with deeper, higher-spatial resolution, WFPC2 imaging in two of these fields. One area lies at the center of the bar (see Figure 1) and one lies in the disk approximately 2° from the center. Crucial ground-based observations provide data on stars with V $\lesssim$ 18.5 that are saturated in our WFPC2 images, an accurate calibration of WFPC2 photometry, and help quantify problems such as WFPC2's charge-transfer effect. Our goal is to create CMDs for numerous areas that each will contain $\sim 10^4$ stars in the critical magnitude range $20 \leq V \leq 23$. By modeling Hess diagrams with such a large sample, we can measure a 20% change in the star-formation rate averaged over 1 Gyr intervals at approximately the 2-sigma level. By comparing numerous fields in the bar and disk, we can begin to disentangle the histories of the LMC's three distinct components: the bar, the disk, and the thick disk/halo.

References

Elson, R.A.W., Gilmore, G.F., & Santiago, B.X. 1997, MNRAS, 289, 157

Gallagher, J.S., III, et al. 1999, this volume

Vallenari, A., et al. 1996, A&A, 309, 367

Westerlund, B.E., Linde, P., & Lyngå, G. 1995, A&A, 298, 39

New Views of the Magellanic Clouds
IAU Symposium, Vol. 190, 1999
Y.-H. Chu, N.B. Suntzeff, J.E. Hesser, and D.A. Bohlender, eds.

Isochrone Probability Functions for Old Stellar Systems

Peter A. Bergbusch

Department of Physics, University of Regina,
Regina, Saskatchewan, Canada, S4S 0A2

Abstract. The isochrone probability function (IPF) is derived from the slope of the mass–distance relation on an isochrone in $\mathcal{M}_\odot$/mag, where the distance along the isochrone is computed with respect to some arbitrary, well-defined point. IPFs contain the information needed to calculate both luminosity functions and color functions, and they provide a straightforward way of generating synthetic stellar populations.

1. Mathematical Formalism

The mathematical formalism for the derivation of accurate IPFs parallels that for the derivation of luminosity functions presented in Bergbusch & VandenBerg (1992) (hereafter BV92), except that the distance along an isochrone is used as the observable coordinate instead of luminosity.

In the $\log L$–$\log T_{\rm eff}$ plane, the distance along an isochrone is defined by

$$d\mathcal{D} = \left[c_1 \left(d \log L\right)^2 + c_2 \left(d \log T_{\rm eff}\right)^2\right]^{1/2}, \quad (1)$$

where c_1 and c_2 are chosen to stretch the subgiant branch (SGB) and to obtain a convenient range of distances. The slope of the mass–distance relation is then obtained from

$$\frac{d\mathcal{D}}{d\mathcal{M}} = \left[c_1 \left(\frac{\partial \log L}{\partial \mathcal{M}}\right)^2 + c_2 \left(\frac{\partial \log T_{\rm eff}}{\partial \mathcal{M}}\right)^2\right]^{1/2}. \quad (2)$$

Consequently, equation (6) from BV92 is recast as

$$\phi(\mathcal{D}) = N(\mathcal{M}) \left(\left.\frac{d\mathcal{D}}{d\mathcal{M}}\right|_t\right)^{-1}, \quad (3)$$

where $\phi(\mathcal{D})$ is the isochrone probability function, and the evaluation of the partial derivatives in equation (2) is performed as described in BV92.

In the observer's plane, distance is calculated with respect to a point on the SGB 0.05 mag redward of the main sequence turnoff (regardless of color index). A convenient definition of distance is

$$dD = \left[dM_i^2 + 16\, dC_{ik}^2\right]^{1/2}, \quad (4)$$

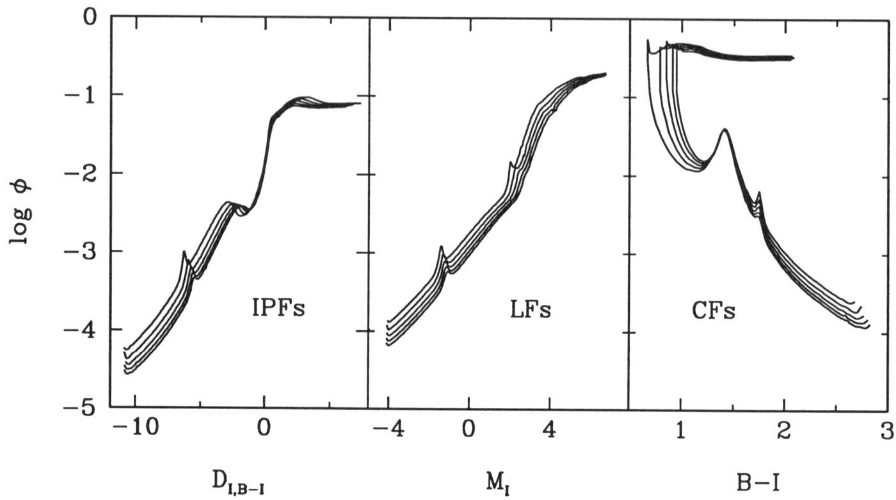

Figure 1. IPFs, LFs, and CFs for the ages 8–16(2) Gyr and [Fe/H] = −2.14, plotted for the I passband and the $B-I$ color index.

where M_i is the absolute magnitude in the $i^{\rm th}$ passband and C_{ik} is the color index for passbands i and k. A $\mathcal{D}$–D calibration is used to translate bin limits from the observer's plane to the theoretical plane, and integrations across bins are performed on the theoretical plane.

2. Results

The Akima spline (Akima 1970) has been utilized throughout the interpolation relations to obtain the isochrone variables (mass, luminosity, effective temperature) and their derivatives. The leftmost panel of Fig. 1 shows the differential IPFs on the observer's plane for [Fe/H] = −2.14, binned at 0.05 mag intervals, transformed to the observer's plane. The luminosity functions (LFs) plotted in the middle panel (in 0.05 mag bins) and the color functions (CFs) plotted in the rightmost panel (in 0.02 mag bins) were both derived directly from the IPF. The small wiggles which appear throughout the IPFs, LFs, and CFs are mostly due to inadequacies in the bolometric corrections and in the color–$T_{\rm eff}$ relations. As pointed out by Bergbusch & VandenBerg (1997), the CF of the subgiant region can be used to provide a distance-independent constraint on cluster ages.

References

Akima, H. 1970, JACM, 17, 589

Bergbusch, P., & VandenBerg, D. A. 1992, ApJS, 81, 163

Bergbusch, P., & VandenBerg, D. A. 1997, AJ, 114, 2604

New Views of the Magellanic Clouds
IAU Symposium, Vol. 190, 1999
Y.-H. Chu, N.B. Suntzeff, J.E. Hesser, and D.A. Bohlender, eds.

The Star Formation History of the Magellanic Clouds: A Preliminary Report

Jason Harris, Dennis Zaritsky and Eva K. Grebel

UCO/Lick Observatory and Department of Astronomy and Astrophysics, University of California, Santa Cruz, CA 95064, USA

Ian Thompson

Observatories of the Carnegie Institute of Washington, 813 Santa Barbara Street, Pasadena, CA 91101, USA

Abstract. We present a method to determine the star formation history (SFH) of a mixed stellar population, based on an iterative maximum Likelihood comparison of stellar photometry to model color-magnitude diagrams. We demonstrate the algorithm on a subregion of the Large Magellanic Cloud, observed as part of our ongoing Magellanic Clouds Photometric Survey. We will eventually perform this analysis on hundreds of regions in both Clouds, resulting in a homogeneous SFH map of these galaxies.

1. Introduction

The evolution of a galaxy is defined by its star formation history (SFH): the star formation rate as a function of time, position and metallicity throughout the galaxy. Understanding the detailed SFH of resolved local group galaxies can potentially lead to a greater understanding of a variety of astrophysical problems, including the evolution of distant, unresolved galaxies, the propagation of star formation, and the role of galactic gravitational interactions.

2. Data and Method

We have *UBVI* photometry for approximately 57,000 stars in a randomly selected $24' \times 24'$ region of the Large Magellanic Cloud, from our ongoing Magellanic Clouds Photometric Survey (for details, see Zaritsky et al. 1997). The survey will eventually provide *UBVI* photometry of tens of millions of stars in each of the Clouds.

We determine the SFH by matching the observed photometric distribution of stars as closely as possible using stellar evolutionary models. The Padua isochrones (Bertelli et al. 1994) are used because they provide wide coverage in age and metallicity. Every point along each isochrone is assigned a relative occupation probability (OP) based on a Salpeter IMF. The OP of each isochrone point is then distributed along the reddening line according to the observed red-

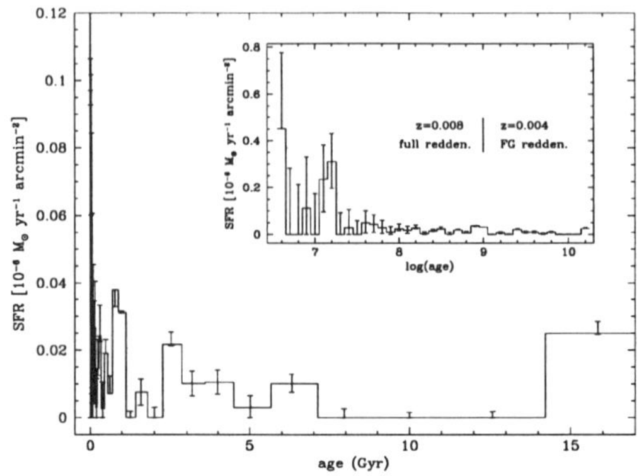

Figure 1. The derived SFH of the LMC region.

dening statistics (Harris et al. 1997) and error-blurred according to results from artificial star tests. These procsses result in a library of model Hess diagrams. One can then simply combine Hess diagrams linearly, assigning each an independent amplitude, to form a composite Hess diagram that can be compared to the data. The set of isochrone amplitudes describes the SFH of the model. To reduce the number of degrees of freedom in the model, we have imposed a chemical enrichment law, so that there is only one possible metallicity at each age. Eventually, we will allow metallicity choices at every epoch.

We use an "amoeba" algorithm to find the set of isochrone amplitudes that produces the best match to the observed Hess diagram. It starts at a randomly selected point in parameter space, and computes the Likelihood of the resulting model. It then takes a step in each parameter dimension, and by recomputing the Likelihood at these new locations, determines the local Likelihood gradient. A step in the direction of the gradient is taken, and the procedure iterates until a maximum Likelihood is found.

3. Results

The derived SFH of this region is shown in Figure 1, a plot of the isochrone amplitudes of the most likely model. It indicates an ancient burst of star formation, followed by 7 Gyr of quiescence, followed by an epoch of increasing, more or less continuous star formation activity that is continuing today. The error bars indicate the 95% confidence limits on each model parameter, but we believe that these underestimate the true uncertainty associated with each amplitude.

References

Bertelli, G., et al. 1994, A&AS, 106, 275

Harris, J., Zaritsky, D., & Thompson, I. 1997, AJ, 114, 1933

Zaritsky, D., Harris, J., & Thompson, I. 1997, AJ, 114, 1002

New Views of the Magellanic Clouds
IAU Symposium, Vol. 190, 1999
Y.-H. Chu, N.B. Suntzeff, J.E. Hesser, and D.A. Bohlender, eds.

Peculiarities of the Stellar Population of Bars in the LMC and Other Barred Galaxies

A.S. Gusev

Sternberg Astronomical Institute, Universitetsky Pr., 13, 119899, Moscow B-234, Russia

Abstract.
The age distributions in the bar and disk of the Large Magellanic Cloud are studied using data for 262 clusters with known ages, of which 89 are located in the bar. The spatial distributions for clusters in different age groups are investigated. Epochs of active star formation are determined for the bar and disk of the LMC. The ages of the stellar populations in the bar differ from that for the galaxy as a whole: in the bar, there is a deficit of intermediate-age clusters ($5 \times 10^8 - 6 \times 10^9$ yr) and an excess of young clusters with ages $3 \times 10^7 - 5 \times 10^8$ yr. This indicates that the star formation history in the bar differs from that in regions outside the bar at the same distances from the center of the galaxy. Data on the composition of the old stellar populations in the bar and the disk of the LMC and the spatial distribution of intermediate-age clusters in the galaxy suggest that the LMC bar formed $(2 - 6) \times 10^9$ years ago. Deviation of colors of bars from the normal color sequence of galaxies on several two-color diagrams is found using multicolor surface photometry for eight barred galaxies. Using results of evolutionary synthesis modeling, I can model the abnormal colors of bars by an intermediate-age star ($\sim 10^9$ yr) deficiency. I find that the deficiency of an intermediate-age stellar population is a general property of bars.

There are many studies of multi-color photometry of both early-type and late-type barred galaxies in which deviations of the color indices of bars from the normal color sequence for galaxies are noted in the two-color diagrams. These deviations increase toward the ends of the bars. This can be explained by a deficit of intermediate-age stars (or an excess of old and young stars) in the bars of these galaxies (Esipov et al. 1993). The LMC is a unique example of a barred galaxy in which we can estimate differences between the stellar populations in the bar and disk by directly comparing the properties of clusters with known ages in the bar and in the galaxy as a whole.

The age distribution of clusters of our sample in the LMC bar differs from one in the LMC disk. Figure 1a shows distributions of the number of clusters per unit age interval (dN/dt) in the bar and disk of the LMC for our sample and the data of Bica et al. (1992, 1996). This distribution characterizes and compares the star formation history in the bar and the galaxy as a whole. As one can see in Figure 1a, there is a deficit of old and young clusters and an excess of clusters with ages $3 \times 10^7 - 5 \times 10^8$ yr and of very young clusters with ages less than 2×10^6 yr in the LMC bar. Evidently the star formation history

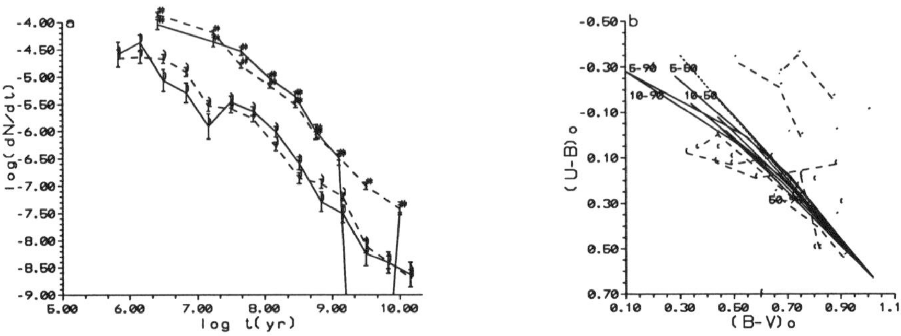

Figure 1. (a) Distribution of the number of clusters per unit age interval in the bar (solid curve) and the disk (dashed curve) of the LMC as a function of age, based on our data (asterisks) and the data of Bica et al. 1992, 1996 (squares). Poisson errors are indicated. All data are normalized. The curves for the data of Bica et al. are shifted upward by $\log(dN/dt) = 1.0$. (b) Model sequences (solid lines) for systems with no intermediate-age stars (a range of ages is marked by numbers in Gyr) for the $(U-B)_0-(B-V)_0$ two-color diagram. The short dashed lines are the normal color sequence of galaxies. The observed colors for galaxy bars (asterisks) are indicated. The dashed curves show how color indices change along semi-major axis of bars.

in the bar differs from that for the disk. This is true for regions in and outside the bar at the same distances from the LMC center.

I studied the color indices in eight early- and late-type barred galaxies along the bar major axis, and also the general color indices for 67 barred galaxies. The bar color indices corrected for Galactic and inner absorption are shown in the two-color diagram $(U-B)_0-(B-V)_0$ in Figure 1b. In this figure, the color indices of bars clearly separate from the normal color sequence (NCS) of galaxies.

Using a modified PEGASE program, I studied the color indices of the stellar system with no intermediate-age star formation ($5 \times 10^8 - 9 \times 10^9$ yr) using evolutionary synthesis modeling. Stellar systems with different SFR(t) were examined and the results are presented in Figure 1b. This figure shows that a deficiency of intermediate-ages stellar population (with ages $(1-9) \times 10^9$ years) is observed in bars. The same result is seen in other two-color diagrams.

These results show that a deficiency of intermediate-age stellar populations is a general property of bars.

Acknowledgments. I would like to thank A.V. Zasov, Yu.N. Efremov, and V.V. Kravtsov for advice and comments. I am most grateful to the organizers of the Symposium for financial support.

References

Bica, E., Claria, J.J., & Dottori, H. 1992, AJ, 1992, 103, 1859

Bica, E., et al. 1996, A&AS, 102, 57

Esipov, V.F., Zasov, A.V., & Popravko, G.V. 1993, AZh, 70, 1

New Views of the Magellanic Clouds
IAU Symposium, Vol. 190, 1999
Y.-H. Chu, N.B. Suntzeff, J.E. Hesser, and D.A. Bohlender, eds.

The Star Formation History and IMF in the LMC and SMC from Deep HST Imaging

J. Holtzman

New Mexico State University, Box 30001, Dept. 4500, Las Cruces, NM 88003, USA

J. R. Mould

MSSO/SSO, Private Bag, Weston Creek P.O., 2611 Canberra, ACT, Australia

J. S. Gallagher III

University of Wisconsin, Dept. of Astronomy, 475 N. Charter St., Madison, WI 53706, USA

Abstract. We present deep photometry to $V \sim 27.5$ obtained with the HST in several fields in the LMC and the SMC. We derive luminosity functions for the faintest stars which are *consistent with* an initial mass function similar to that of the solar neighborhood, although moderate variations are not excluded. We discuss implications of these observations for the star formation history in these regions of the LMC and SMC.

1. Observations and Results

Four fields have been observed in the LMC with the WFPC2 for several thousand seconds in F555W and F814W. Three of the fields are several degrees from the LMC center; one is located in the bar. In the SMC, two separate fields have been observed which are just adjacent to the bar.

Color-magnitude diagrams in all of the fields are roughly similar, showing broad upper main sequences which imply roughly constant star formation over the past few Gyr. There is no evidence for discrete bursts of star formation within the past few Gyr. The giant branches are bluer than expected from the Padova stellar models for metallicities expected for the LMC; Geneva isochrones predict *even redder* giant branches. The main sequence around the oldest turnoff is relatively blue as well. One would need field stars which have lower metallicity than clusters of a comparable age to populate this region if the stellar models and photometry are correct.

The luminosity functions for the fields are similar, after correction for differing levels of completeness which arise from different exposure times, crowding, and image quality.

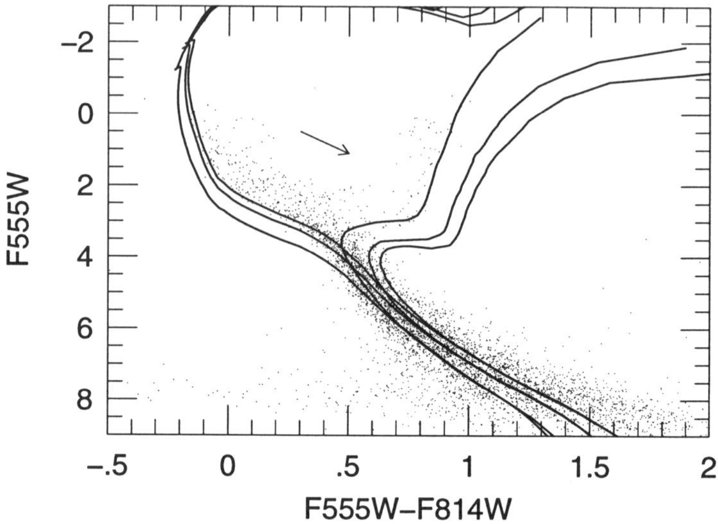

Figure 1. The color-magnitude diagram of one outer LMC field. Isochrones are for 1 and 10 Gyr with Z=0.0004, 0.004, and 0.008.

2. Discussion

Various techniques have been discussed to derive star formation histories from the distribution of stars in a color-magnitude diagram. However, in the presence of systematic errors in either the stellar models or the observations, it is difficult (and dangerous) to infer the star formation history using this method. This is especially true if areas in the CMD are weighted by the number of stars; errors in densely populated regions will drive the derived star formation history parameters. In addition, one must insure that results derived from the color-magnitude diagram are consistent with known information about the stellar metallicity distribution and the age-metallicity relation.

As an example, if the full LMC color-magnitude diagram is fit, the blue mid-main sequence requires the existence of a significant quite metal-poor component of the population, which is unlikely to be real.

Fitting the luminosity function alone bypasses some of the potential problems, because systematic errors in the magnitudes (not colors) are likely to be relatively small. Clearly, however, information is lost when only fitting the luminosity function! We compensate for this somewhat by constraining the models to use the age-metallicity relations of Pagel & Tautvaisiene (1998, MNRAS 299, 535) for the Clouds, which match observations rather well. We find that, if a solar neighborhood IMF is assumed, a moderate component of "older" stars is required, with $\geq$ 40% of LMC stars having ages older than 3-4 Gyr. This would imply a different star formation history for field stars than for LMC clusters.

We have also compared the predicted luminosity function from the star formation histories derived by Pagel & Tautvaisiene with our observations. These predicted LFs match the observed LMC LF reasonably well, but appear to have too many young stars for the SMC.

Discussion

Johannes Andersen: Results will certainly depend on your choice of models; for example, the published colours of the Padova models are redder than others for the same models, which would lead you to fit more metal poor models.

How do you deal with reddening in the CMD fit?

Holtzman: I have been adopting single values of the reddening for all of the fields.

Carme Gallart: What would you think would be the effect of including binaries in the models on the amount of star formation that you find in the epoch between about 3 and 10 Gyr?

Holtzman: If you assume that binaries are drawn independently from the same IMF, it makes little difference for the relatively massive stars ($> 0.8\ M_\odot$) seen in the LMC. If binaries preferentially have mass ratios near unity, it can make a significant difference.

Knut Olsen: Are your solutions of star formation histories derived from the luminosity function? If so, there seem to be problems with degenerate effects of all of the parameters that are allowed to vary.

Holtzman: I have fit both the Hess diagram and the luminosity function. Fits to the former appear to be affected by systematic differences between the models and observations, which led me to just fit the luminosity functions. Clearly, however, you are correct that this throws away a lot of information and may lead to degenerate solutions.

New Views of the Magellanic Clouds
IAU Symposium, Vol. 190, 1999
Y.-H. Chu, N.B. Suntzeff, J.E. Hesser, and D.A. Bohlender, eds.

HST FOC and Broad Band Colors for Young and Intermediate Simple Stellar Populations

E. Brocato

Oss. Astr. Collurania, Via M. Maggini, I-64100 Teramo, Italy

V. Castellani

Dip. di Fisica, Univ. Pisa, Piazza Torricelli 2, I-56126 Pisa, Italy

G. Raimondo

Oss. Astr. Collurania, Via M. Maggini, I-64100 Teramo, Italy

M. Romaniello

STScI, 3700 San Martin Drive, Baltimore, MD 21218, USA

Abstract. We present theoretical HST and broad band colors from young and intermediate stellar populations with age ranging from t = 8 Myr to t = 5 Gyr and three different metallicities Z =0.02, 0.006 and 0.001. We suggest the combined use of HST UV colors and IR colors (V-K in particular) to disentangle the metallicity-age effect in integrated colors of young stellar populations (t $\leq$ 1 Gyr).

1. Introduction

The challenge of interpreting integrated spectra and/or colors of distant objects (galaxies and stellar clusters) in terms of stellar ages and chemical compositions was gathered approaching the problem of population synthesis both from a theoretical and an observational point of view.

The homogeneity of adopted library of stellar evolutionary tracks is the basic assumption which determines the predicted integrated properties of a Simple Stellar Population (SSP). To cover the main evolutionary phases expected, we will rely on the extensive set of stellar models (Teramo-Pisa-Frascati cooperation) presented for high (Brocato & Castellani 1993), intermediate (Cassisi, Castellani & Straniero 1994) and low mass (Straniero & Chieffi 1991; Castellani, Chieffi & Straniero 1992) stars.

2. Discussion

By using this homogeneous set of stellar evolutionary tracks as computed under canonical (no overshooting) assumptions, we computed theoretical HST and

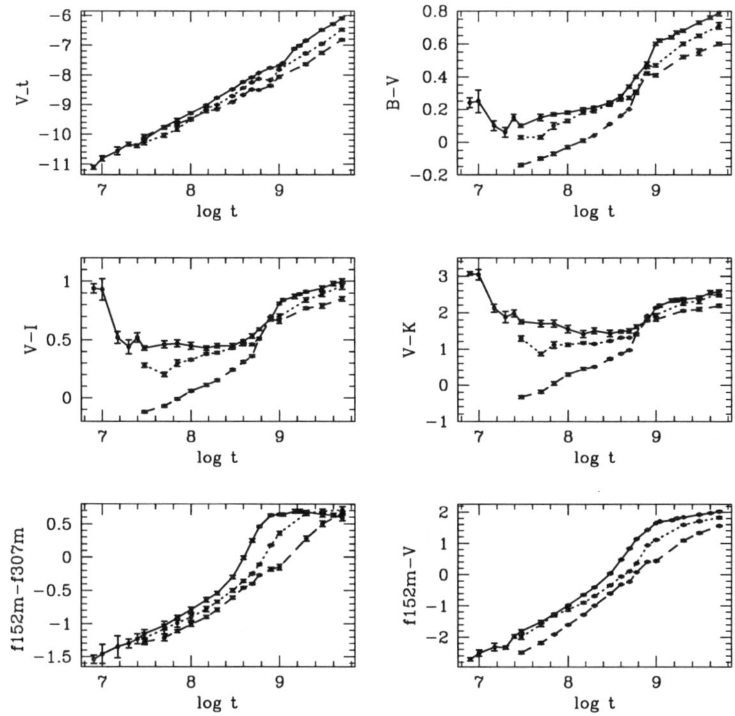

Figure 1. Time evolution of integrated V magnitude and colors for three different assumption on metallicities: Z = 0.02 (solid lines), Z = 0.006 (dotted lines) and Z = 0.001 (dashed lines).

broad band colors from population synthesis models, covering the range of cluster ages from t = 8 Myr to t = 5 Gyr for three different metallicities Z = 0.02, 0.006, and 0.001 (Fig. 1).

We find that for F152M−F307M < 0.5 and F170M−F278M < 0.5 (which means ages lower than 1 Gyr) the HST UV colors can still be used to infer reliable indication on the age of distant clusters. On these basis, we present a calibration of the HST UV two-colors (F152M−F307M, F170M−F278M) in terms of cluster ages for the three above quoted metallicities. Theoretical predictions for the time behavior of broad band colors ($UBVK$) appear in good agreement with data for LMC stellar clusters with known age. We suggest that the combined use of HST UV colors and near-IR colors ($V - K$ in particular) should disentangle the metallicity-age effect in integrated colors of young stellar populations (t $\leq$ 1 Gyr). For this reason the NICMOS camera on board HST, particularly its F222M filter, is expected to be a powerful tool in interpreting the integrated colors of distant objects in term of age and metallicity.

MC stellar clusters represent a very fundamental test to check the reliability of population synthesis models. Unfortunately, age and metallicity determinations of single clusters available up to now are based, with few exceptions, on indirect calibrations. Thus, an extensive observational work with the goal of deriving accurate and precise CMDs for a large sample of MC stellar cluster would be of invaluable help to check the reliability of population synthesis results.

References

Brocato E., & Castellani V. 1993, ApJ, 410, 99

Cassisi S., Castellani V., & Straniero O. 1994, A&A, 282, 753

Castellani V.,Chieffi A., & Straniero O. 1992, ApJS, 78, 517

Straniero O., & Chieffi A. 1991, ApJS, 76, 525

New Views of the Magellanic Clouds
IAU Symposium, Vol. 190, 1999
Y.-H. Chu, N.B. Suntzeff, J.E. Hesser, and D.A. Bohlender, eds.

The LMC Centre Unfolding

A. Ardeberg, P. Linde

Lund Observatory, Box 43, SE-221 00 Lund, Sweden

B. Gustafsson

Uppsala Astronomical Observatory, Box 515, S-751 20 Uppsala, Sweden

Abstract. We study star formation history and chemical evolution in the LMC Bar centre. With the HST PC and WFC plus uvby photometry, we find 30% of the stars younger than 200 My and evidence for reduced star formation 3 to 0.2 Gy ago. The overall [Fe/H] is -0.5. Many stars older than 4 Gy contradict a very young Bar.

1. Introduction

In the Bar of the Large Magellanic Cloud (LMC), stellar density and distance have prevented safe studies of star formation, stellar abundance and evolution. Image crowding prohibited precision turn-off photometry in colour magnitude diagrams (CMD), with spectroscopic abundance studies feasible for the brightest stars only. These data shed little light on star formation and chemical evolution.

2. Star formation and metallicity

The Hubble Space Telescope (HST) and its planetary and wide field camera (WFPC2) dramatically improve our studies of individual faint LMC Bar stars. This is due to the high image quality and combined PC and WFC image scales and fields of view. With the HST PC and uvby photometry, we derived a detailed CMD and metal abundances for the centre of the LMC Bar. We studied individual stars with V(y)<24 (Ardeberg et al. 1997, 1999).

The PC study showed star formation over up to 9 Gy. It gave proof that close to 30% of the stars in the Bar centre are less than 500 My. There was evidence of a halt in star formation starting at least 2 Gy ago and ending with the recent burst of star formation. More exact determinations were impossible due to weak statistics of brighter stars, a consequence of the small PC field.

Since our first studies, based on PC data only, we added data from one WFC field. It covers 6400 arcsec^2, as compared to the PC field of 1200 arcsec^2. By combining PC and WFC data, we can profit from best spatial resolution, highest photometric accuracy and favourable number statistics.

For the luminosity function, we developed a new method, including stars individually unidentified but statistically measurable. The LF agrees well with the solar vicinity one, even in details. See Fig. 1, left. It is still rising at V=26. In both respects, it resembles the LF of LMC structures outside the Bar.

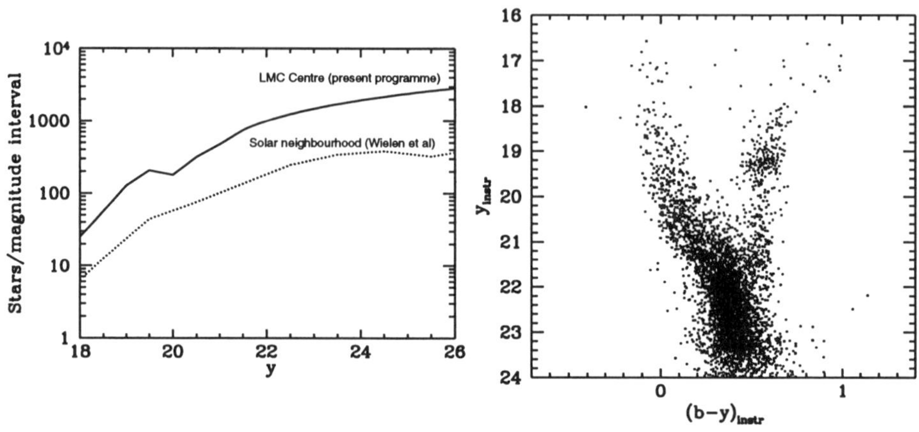

Figure 1. Left: Luminosity function compared to solar neighbourhood. Right: CMD from WF3 field. 6000 stars are included.

Fig. 1 (right) shows a tentative CMD. Of somewhat lower quality compared to PC data, for the brighter stars the WFC has important advantages. Comprising 6000 stars brighter than V(y)=24, the WFC CMD presents far better statistics than the PC CMD for V(y)<22. The improvement is even larger for stars around and above the upper turn-off region. Noteworthy is the tip of the main sequence as is the clump and the CMD morphology. The combined PC and WFC data improve population definition and star formation history, especially for stars younger than 3-4 Gy. With our new data, using the techniques described in Ardeberg et al. (1997), we improve our estimate of the upper age limit for the youngest population from 500 to 200 My. We confirm that 30% of the stars belong to this population. We improve the lower age limit of reduced star formation from 2 to 3 Gy. The improved CMD makes our morphology based estimate of metallicity safer while remaining at [Fe/H] = -0.5, close to the average result from our m_1 versus (b-y) data obtained from PC observations. The latter data point to an evolutionary trend in [Fe/H] over the latest 6 Gy, an evolution so far unknown to any detail.

3. Concluding remarks

Our results are highly relevant for the lively discussion of the evolution of the LMC Bar. We find little support for the current tendency (van den Bergh, in this volume) to classify the Bar as a young part of the LMC. This is contradicted by both our age determinations of stellar populations and by the trend in [Fe/H] noted for the same stars. Further reduction and analysis will improve our CMD statistics and [Fe/H] data, giving us a better handle on evolution.

References

Ardeberg, A., Gustafsson, B., Linde, P., & Nissen, P.-E., 1997, A&A, 322, L13

Ardeberg, A., et al. 1999, in IAU Symp. No. 184, in press

New Views of the Magellanic Clouds
IAU Symposium, Vol. 190, 1999
Y.-H. Chu, N.B. Suntzeff, J.E. Hesser, and D.A. Bohlender, eds.

The Centre of the LMC Bar – Analysis of HST Data

P. Linde, A. Ardeberg

Lund Observatory, Box 43, SE-221 00 Lund, Sweden

B. Gustafsson

Uppsala Astronomical Observatory, Box 515, S-751 20 Uppsala, Sweden

Abstract. We used the HST to obtain high resolution images of the LMC Bar centre. Through careful data analysis, taking into account cleaning, dithering and crowded field decomposition, we show that accurate uvby photometry can be achieved for F dwarfs, reaching stars around the lower turn-off region, as faint as V=22.

1. Introduction

To resolve individual main sequence stars in the core of the LMC Bar, we have used a total of 14 hours of exposure with the Hubble Space Telescope (HST) WFPC2 uvby filters. This allowed high precision photometry of F-G stars with $19.5<V<22.5$. Additionally, stars down to V=26 were included for luminosity function studies. Ardeberg et al. (1997) compares ground and space data, emphasising the necessity of high resolution observations for detailed LMC studies. To preserve the inherent high quality of the HST data, dedicated image analysis is necessary.

2. Cosmic Ray Image Deteroriation and Image Dithering

The space environment of the HST strongly affects long exposures, due to numerous impacts of charged particles on the CCD detectors. Efficient detection and elimination of these events are fundamental for good photometry. Fig. 1 (upper left) shows an example of detected impacts (no stars). With a modified median-filtering algorithm on several separate images of the same field, successful removal of the cosmic rays is possible.

Due to the under-sampling in WFPC2 images, accurate photometry is difficult in crowded fields, especially with the WF cameras. Through application of image dithering, it is possible to decrease this problem. We designed a dithering pattern such that each of 4 consecutive exposures were offset by a quarter of a pixel, simultaneously in x and y directions for both the PC and the WFC. Subsequent image restoration gives a significantly smaller apparent pixel size.

In Fig. 1, the lower left panel shows a sub field (from a single exposure) before filtering, the lower right panel after cleaning, dithering and summation of 16 exposures. The upper right panel shows the final full field.

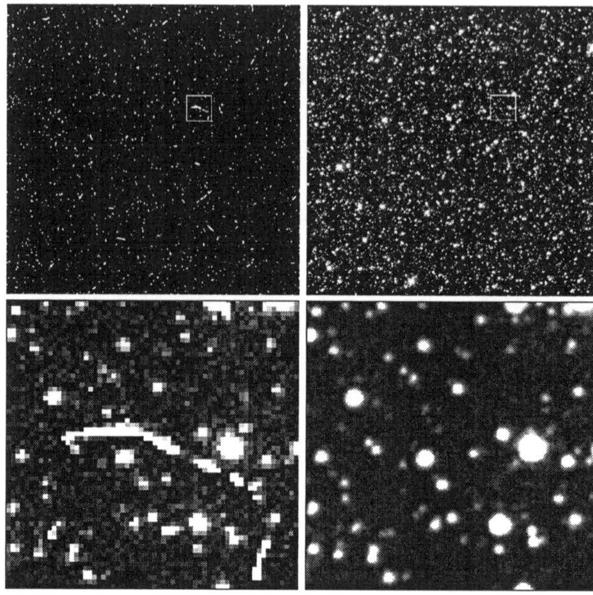

Figure 1. Upper left: Detected cosmic ray events in a single 800 sec WF3 exposure. Upper right: Complete WF3 b field. Lower left: Detail from single WF3 b exposure. Lower right: Same field after filtering, dithering and adding 16 exposures.

3. Photometric Performance

PC and WFC photometric accuracy for our field is shown in Fig. 2. To evaluate photometric improvement from dithering, 16 PC images from the same field have been analysed, both as a set of 16 individual images and as four sets of dithered images. The accuracy is significantly improved (see Linde and Snel, 1997).

References

Ardeberg, A., Gustafsson, B., Linde, P., & Nissen, P.-E., 1997, A&A, 322, L13

Linde, P., & Snel, R. 1997, ADASS VI, A.S.P. Conference Series, Vol. 125, eds. Gareth Hunt and H. E. Payne, p. 431

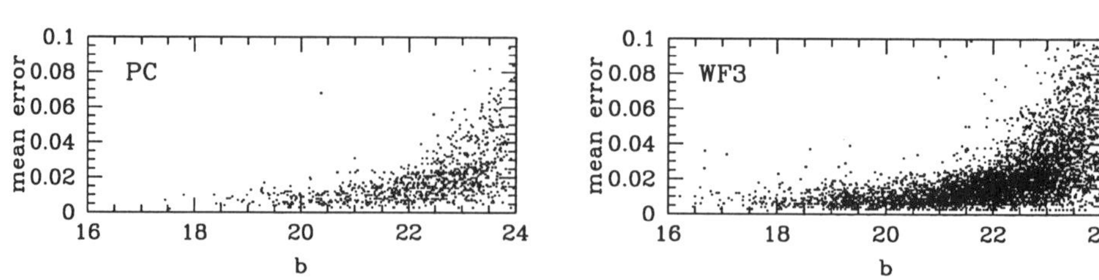

Figure 2. Mean error as function of b for the PC and the WF3.

New Views of the Magellanic Clouds
IAU Symposium, Vol. 190, 1999
Y.-H. Chu, N.B. Suntzeff, J.E. Hesser, and D.A. Bohlender, eds.

New R Coronae Borealis Stars in the LMC, Discovered in the MACHO Photometry Database

Geoffrey C. Clayton

Department of Physics and Astronomy, Louisiana State University, Baton Rouge, LA 70803, USA

D. Kilkenny

South African Astronomical Observatory, P.O. Box 9, Observatory 7935, South Africa

D.L. Welch

Department of Physics and Astronomy, McMaster University, Hamilton, ON L8S 4M1, Canada

and The MACHO Collaboration

Abstract. An extensive search has been made in the MACHO Photometry Database for new R Coronae Borealis (RCB) stars in the LMC. Only three such stars were known previously. Eleven new LMC RCB stars have been spectroscopically confirmed so far.

1. Introduction

We report the discovery of several new R Coronae Borealis (RCB) stars in the Large Magellanic Cloud (LMC) using the MACHO project photometry database (Alcock et al. 1996, 1999). RCB stars have a wide range of temperature but they can be divided simply into three subgroups, cool (5000 K), warm (7000 K) and hot (20,000 K). The identification of at least eleven new stars has now been confirmed spectroscopically. Eight are cool RCB stars characterized by very strong Swan bands of C_2, and violet bands of CN, and weak or absent Balmer lines, G-band and $^{12}C^{13}C$ bands. Two are warm RCB stars having weaker Swan bands. One star is an example of a hot RCB star of which only 3 were previously known to exist in the Galaxy and none in the LMC. All of the stars have shown deep declines in brightness typical of RCB stars (Clayton 1996). The discovery of these new stars increases the number of known RCB stars in the LMC by almost a factor of five to 14 and demonstrates the utility of the MACHO photometric database for the discovery of new RCB stars.

2. Absolute Luminosity of RCB Stars

Two interesting trends are present in the characteristics of the new stars. First, they tend to be significantly fainter at maximum light than the three previously known LMC RCB stars. They seem to have absolute luminosities, M_V, about a magnitude fainter than the other three stars. Second, the stars discovered thus far are mostly members of the cool subgroup of the RCB stars. Most RCB stars in the Galaxy fall in the warm subgroup. The new observations presented here suggest that there is a wider range of absolute luminosity than given by the canonical, $M_V = -4$ to -5 mag. Therefore, the absolute luminosities of RCB stars are now uncertain since they are based solely on the LMC stars. The HIPPARCOS data for the Galactic RCB stars provided only lower limits on distances, so the LMC sample is the only source of distances and absolute luminosities (Cottrell & Lawson 1998).

3. RCB Stars as Final Helium Shell Flash Stars

The evolutionary history of RCB stars remains mysterious. There are two major evolutionary models for the origin of RCB stars, the Double Degenerate and the Final Helium Shell Flash. In the Final Flash model, there is a close relationship between RCB stars and planetary nebulae. Only about 10% of RCB stars are known to have shells but no systematic survey has been done. A link between RCB stars and planetary nebulae if it can be established will be an important step forward in understanding the evolution of post-AGB stars.

Only three stars have been observed while going through a final helium shell flash since the advent of modern instrumentation, FG Sagittae (in 1894), V605 Aql (in 1919) and Sakurai's Object (in 1996). The only spectrum taken of V605 Aql while it was in its cool giant phase in 1921 is identical to a typical cool RCB star and very similar to the present spectrum of Sakurai's Object (Clayton & De Marco 1997). Both show strong Swan bands of C_2 and violet CN bands as well as weak hydrogen features and little or no evidence of ^{13}C. V605 Aql today provides a unique opportunity to investigate the evolution of a Final Flash star over a period of 75 years. The star is very faint but shows a WC spectrum. It has been suggested that it has already returned to its pre-outburst brightness in disagreement with model timescales which predict several thousand years. However, it is more likely that the star is still 10^3 - 10^4 L$\odot$ but obscured by dust. V605 Aql lies at the centre of the planetary nebula, A58, of which the central knot is hydrogen deficient.

References

Alcock, C., et al. 1996, Ap.J, 470, 583

Alcock, C., et al. 1999, in preparation

Clayton, G.C. 1996 , PASP, 108, 225

Clayton, G.C., & De Marco, O. 1997, AJ, 114, 2679

Cottrell, P.L., & lawson, W.A. 1998, PASA, 15, 179

New Views of the Magellanic Clouds
IAU Symposium, Vol. 190, 1999
Y.-H. Chu, N.B. Suntzeff, J.E. Hesser, and D.A. Bohlender, eds.

2MASS Observations of the Large Magellanic Cloud

S.D. Van Dyk, R. Cutri

IPAC/Caltech, Pasadena, CA 91125, USA

M.D. Weinberg, S. Nikolaev, M.F. Skrutskie

Dept. of Physics & Astronomy, Univ. of Massachusetts, Amherst, MA 01003, USA

Abstract. The Two Micron All Sky Survey (2MASS) has observed a portion of the Large Magellanic Cloud (LMC) at J (1.25 μm), H (1.65 μm), and K_s (2.17 μm), as part of its routine nightly operations with an automated 1.3-m telescope at CTIO, Chile. The camera observes the sky in the three channels simultaneously, using 256 × 256 HgCdTe detector arrays. The survey samples the sky in 6° × 8$'$.3 scans. The 2MASS Production Processing System provides final atlas images and source extractions with precise photometric calibration and astrometric positions. The survey's 10σ sensitivity is 15.8 mag at J, 15.1 at H, and 14.3 at K_s. 2MASS will ultimately detect $\sim 10^7$ point sources in the LMC and will support analyses of the ages, luminosity and mass functions, and metallicities of the red stellar populations and a census of AGB and carbon stars, as well as extinction maps, across the LMC. This work presents an initial analysis of ~20 square degrees.

1. Introduction

The LMC was observed by 2MASS as part of routine nightly southern operations in 1998 March and April. These observations covered mostly the eastern half of the galaxy and required ~4 hours total observing time. Sample 2MASS Atlas Images of selected regions of the LMC can be found in the Image Gallery at `http://www.ipac.caltech.edu/2mass/`.

2. Preliminary Results

Production processing resulted in $\sim 10^6$ source extractions. In Figure 1 we show the color-color and color-magnitude diagrams, respectively. The observed dispersion is due to differences in ages, metallicity, and, predominantly, crowding. On each diagram we have overlaid isochrones from Bertelli et al. (1994) for the appropriate metallicities, assuming $m - M = 18.5$ (Panagia et al. 1991).

The diagrams are dominated by intermediate-age K and M giants, with ages ~2 Gyr (for $Z = 0.008$) or ~5–6 Gyr (for $Z = 0.004$). In the inner galaxy young main sequence stars, and blue and red supergiants are apparent, with ages to

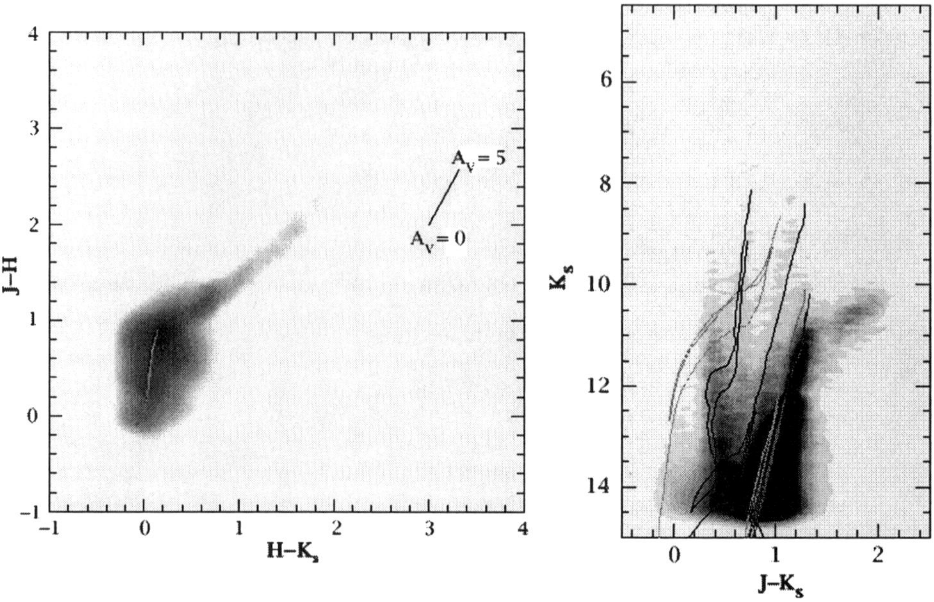

Figure 1. Color-color and color-magnitude (Hess) diagrams for the LMC, based on 2MASS source extractions. Isochrones have been overlaid, representing various LMC populations and foreground dwarfs and giants.

$\gtrsim$100 Myr. There are also bright AGB stars, carbon stars, and dust-enshrouded AGB stars; conservative estimates are ~2500 C stars and 550 obscured AGB stars. An ancient population, with $Z = 0.001$, is not obvious from our diagrams.

The stellar halo of the LMC can be modeled to derive a lower limit on the LMC mass. We have selected twelve $0\overset{\circ}{.}5 \times 0\overset{\circ}{.}5$ subfields $2^\circ - 5^\circ$ from the LMC center. We have applied the Maximum Likelihood method to parameterized models of the spatial density and luminosity function, assuming both a Gaussian and a power-law ($\gamma = 2$) density profile. Both density profile models give a tidal radius of 10.8 kpc and $M_{\rm LMC} \gtrsim 1.0 \times 10^{10}\ M_\odot$, consistent with the star counts of the LMC giants. Additionally, we find an optical depth due to microlensing of $\tau \sim 1.7 \times 10^{-7}$. The fraction of the halo mass in the form of MACHOs is ~0.25.

References

Bertelli, G., et al. 1994, A&A, 106, 275

Panagia, N., et al. 1991, ApJ, 380, L23

Discussion

Hans Zinnecker: It is reassuring that ZMASS and DENIS find the same results! After all, science is not only about discovery, but also about independent confirmation.

Cecile Loup: DENIS has observed the whole central part of the LMC. In total about 50 square degrees. Data will get to the public in about one year.

New Views of the Magellanic Clouds
IAU Symposium, Vol. 190, 1999
Y.-H. Chu, N.B. Suntzeff, J.E. Hesser, and D.A. Bohlender, eds.

Pre-main-sequence Stars in the SMC and LMC

Wolfgang Brandner

CalTech - JPL/IPAC, Pasadena, CA 91125, USA

Eva K. Grebel

Lick Observatory, Univ. of California, Santa Cruz, CA 95064, USA

Hans Zinnecker

Astrophysikalisches Institut, Potsdam, Germany

Bernhard Brandl

Department of Astronomy, Cornell University, Ithaca, NY 14853, USA

Abstract.
We present first results of a survey for pre-main-sequence stars in the Magellanic Clouds. Our search concentrated on NGC 346, the most prominent OB association in the Small Magellanic Cloud, and on the 30 Dor starburst cluster in the Large Magellanic Cloud. The identification of the young low- to intermediate-mass stellar population in the SMC and LMC allows us to study whether or not these populations formed simultaneously with high-mass stars, and to what an extent lower metallicity affects the low-mass IMF. We can also evaluate the duration of star formation in a starburst region.

1. Introduction and Observations

We followed two strategies to search for pre-main-sequence stars in the Magellanic Clouds. In NGC 346, we selected a $2' \times 2'$ field slightly off the brightest stars of the association and in a region with a relatively small amount of diffuse nebulosity. This field was observed with the ESO/NTT in the optical under subarcsec seeing conditions. The total exposure time of our NGC 346 Deep Field amounts to 300 minutes in Hα, and 16 minutes in the V and R band. Pre-main-sequence star candidates were identified based on their Hα emission, and their position in a V vs. V-R color-magnitude diagram.

The crowded environment in the starburst cluster in 30 Dor requires higher spatial resolution, and thus a different observing strategy. We used HST/NICMOS (see contribution by Zinnecker et al.) to obtain deep H band images of the central $1' \times 1'$ of the cluster and of two $20'' \times 1'$ wings extending outwards (effective exposure time 60 min at each position). In addition, 4 comparison fields outside the starburst cluster, but still within the 30 Dor region were observed.

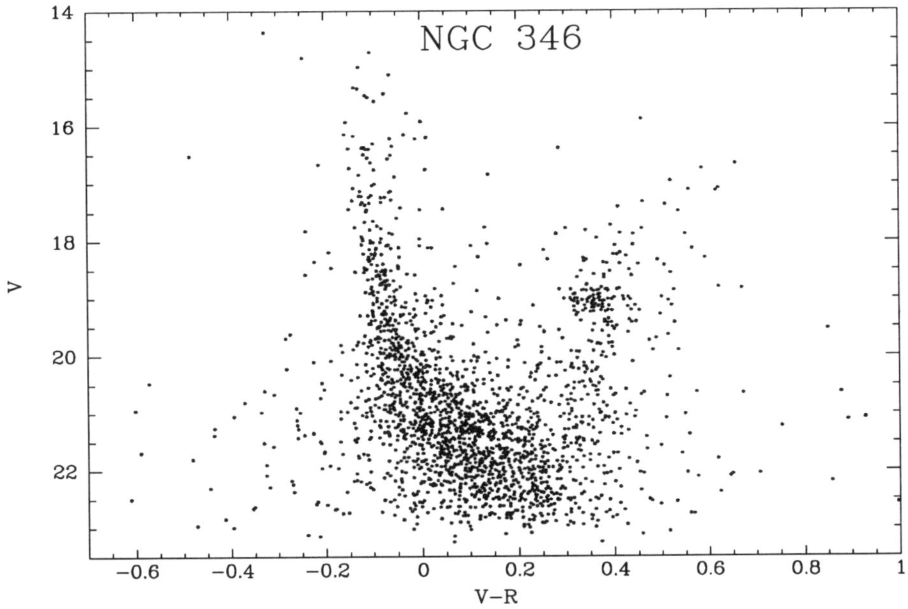

Figure 1. The V vs. V–R color-magnitude diagram shows the young population of NGC 346 (blue main-sequence) and the underlying intermediate age field population (red clump).

2. Results and Discussion

In NGC 346, we reached a limiting magnitude of $22^{\text{m}}_{\cdot}5$ in broad and narrow-band filters (see Figure 1), which enables us to detect main-sequence stars down to a spectral type of F5 (1.4 $M_\odot$) and – for an age of 2 Myr – also slightly more massive pre-main-sequence stars (at this particular age, pre-main-sequence stars with a mass between 1 $M_\odot$ and 2 $M_\odot$ are actually less luminous than their main-sequence counterparts with the same mass). Objects with Hα emission were identified by comparing the observations in the R-band and the Hα filter. Even on spatial scales of $0\farcs5$ (best seeing value in any single 30min exposure in Hα), variations in the diffuse background proved to be the limiting factor for identifying pre-main-sequence stars based on their Hα emission. Nevertheless, we detected about 150 objects with excess emission in Hα among 2500 stars. The Hα emitting stars are good candidates for pre-main-sequence Herbig AeBe stars. We intend follow-up observations using the Focal Reducer/low dispersion Spectrograph (FORS) at the VLT.

In 30 Dor, observed luminosity functions on and off the cluster were compared with theoretical luminosity functions. We were able to get an independent age estimate, to identify low-mass pre-main-sequence stars, and to compute the slope of the IMF. A detailed description of the 30 Dor results can be found in the contribution by Zinnecker et al. (this volume).

Acknowledgments. WB and BB acknowledge support under NASA/STScI grant GO-7370.01-96A.

New Views of the Magellanic Clouds
IAU Symposium, Vol. 190, 1999
Y.-H. Chu, N.B. Suntzeff, J.E. Hesser, and D.A. Bohlender, eds.

Strömgren Photometry of Field Red Giants in the LMC

Andrew A. Cole, John S. Gallagher

Astronomy Dept., University of Wisconsin-Madison, 5534 Sterling Hall, 475 N. Charter St., Madison, WI, 53706, USA

Tammy A. Smecker-Hane

Dept. of Physics and Astronomy, 4129 Reines Hall, University of California, Irvine, CA 92697, USA

Abstract. We present here preliminary results of our program to obtain rough abundance estimates for $> 10^4$ red giant branch and red clump stars in four fields of the LMC. The derived abundance distributions will be a strong constraint on models of the LMC's star-formation history based on field star color-magnitude diagrams.

1. Introduction

An essential ingredient to color-magnitude diagram (CMD) studies of field star populations is the availability of independent metallicity information, owing to the degeneracy of CMD morphology in the age-metallicity plane (e.g., Holtzman et al. 1999). Intermediate-band photometry is a practical way to obtain approximate abundances for the tens of thousands of stars which are conatined in a single CCD image of the LMC field. With abundance information for the red giant and red clump stars in our broad-band CMDs (see Smecker-Hane et al. 1999), it will be possible to derive in a self-consistent way the star-formation history and age-metallicity relation for the field star populations of the Large Magellanic Cloud.

2. Data: Observations, Reductions, and Photometry

Our exposure catalog for this program is shown in Table 1. We observed four fields centered at the locations of fields studied with WFPC2 by Smecker-Hane et al. (1999) and Holtzman et al. (1997). Photometric standards from the list of Richtler (1990) were observed extensively each night. All images were reduced within IRAF and photometry obtained using the IRAF/DAOPHOT routines. We corrected our photometry for an extinction $E_{B-V} = 0.09$ mag (Burstein & Heiles 1982) using the extinction curve of Cardelli et al. (1989).

Table 1. Summary of CTIO 1.5-m Observations, 14'.8 × 14'.8 FOV.

Field	R.A.	Dec.	Date	Exposure time (sec)		
	(B1950.0)			v	b	y
Disk 1	05:11:22	-71:17:27	28 Nov 97	4 × 1300	2 × 1300	2 × 1200
Disk 2	05:14:33	-65:21:00	29 Nov 97	3 × 1300	2 × 1300	2 × 1200
Disk 3	05:58:36	-68:21:26	29 Nov 97	3 × 1300	2 × 1300	2 × 1200
Bar	05:24:47	-69:49:01	1 Dec 97	2 × 1300	1300 + 1000	1200 + 900

3. Color-Metallicity Plots

According to the calibration of Grebel & Richtler (1992), isoabundance contours for luminosity class I-III stars in the range $0.4 \leq (b-y) \leq 1.1$ are linear in the Strömgren $(m_1, b-y)$ plane. Using this calibration, we have determined a mean metallicity [Fe/H] $= -0.8 \pm 0.3$ for a subsample of 1,007 RGB stars in our Disk-1 field. 90% of the red giants in this field lie within the range $-1.6 \leq$ [Fe/H] ≤ -0.2.

4. Discussion

Although the RGB stars behave as expected, the red clump apparently obeys a different color-metallicity relation than that of Grebel & Richtler (1992). The reasons for this are still being explored. These stars are important to our analysis because 80% of the stars available to us lie in the region of the clump.

Our preliminary results for the RGB agree well with the prediction of Pagel & Tautvaišienė (1998). However the significance of this is difficult to assess because we are forced to exclude the red clump from our analysis at the present time and our sample is thus incomplete for low metallicities.

Analysis of three further fields is in progress, and we are attempting to model the behavior of the red clump in the $(m_1, b-y)$ plane using stellar evolutionary tracks and model atmospheres. Refinement of the color-metallicity calibration is also underway.

References

Burstein, D., & Heiles, C. 1982, AJ, 87, 1165

Cardelli, J.A., Clayton, G.C., & Mathis, J.S. 1989, ApJ, 345, 245

Grebel, E.K., & Richtler, T. 1992, A&A, 253, 359

Holtzman, J.A., et al. 1997, AJ, 113, 656

Holtzman, J.A., Mould, J.R., Gallagher, J.S., & WFPC2 IDT 1999, this volume

Pagel, B.E.J., & Tautvaišienė, G. 1998, MNRAS, 299, 535

Richtler, T. 1990, A&AS, 86, 103

Smecker-Hane, T.A., Gallagher, J.S., Cole, A.A., Tolstoy, E., & Stetson, P.B. 1999, this volume

New Views of the Magellanic Clouds
IAU Symposium, Vol. 190, 1999
Y.-H. Chu, N.B. Suntzeff, J.E. Hesser, and D.A. Bohlender, eds.

Radiation-Pressure Ejection of Planetary Nebulae in Asymptotic-Giant-Branch Stars

A.V. Sweigart

NASA Goddard Space Flight Center, Code 681, Greenbelt, MD 20771, USA

Abstract. We have investigated the possibility that radiation pressure might trigger planetary nebula (PN) ejection during helium-shell flashes in asymptotic-giant-branch (AGB) stars. We find that the outward flux at the base of the hydrogen envelope during a flash will reach the Eddington limit when the envelope mass $M_{\rm env}$ falls below a critical value that depends on the core mass $M_{\rm H}$ and composition. These results may help to explain the helium-burning PN nuclei found in the Magellanic Clouds.

1. Description of Radiation-Pressure Instability

We have computed extensive AGB evolutionary sequences for heavy-element abundances Z = 0.01716 (solar) and 0.002 in order to study a radiation-pressure instability for ejecting a PN during a helium-shell flash. The existence of such an instability was previously confirmed by Wood & Faulkner (1986) but only for more massive AGB stars with large core masses ($M_{\rm H} > 0.86\,M_\odot$). In contrast, we find this instability at core masses as small as $M_{\rm H} \approx 0.6\,M_\odot$ due, in part, to our use of the new OPAL opacities (Rogers & Iglesias 1992).

Results for a typical sequence are presented in the left panels of Figure 1. The strong flashes in this sequence cause the hydrogen shell to expand outward to very low temperatures and densities. This, in turn, lowers the gas pressure and hence the minimum value $\beta_{\rm min}$ of the ratio β of gas to total pressure near the base of the hydrogen envelope. Most importantly, we note from Figure 1 that $\beta_{\rm min}$ decreases monotonically from flash to flash and eventually goes to 0, implying that the entire envelope is then supported by radiation pressure.

During the final flashes in Figure 1 the temperature within the hydrogen shell falls to log T = 5.3 at which there is a well-known peak in the OPAL opacities. When the shell reaches this peak, its opacity suddenly increases, and, as a result, its Eddington luminosity abruptly drops. This occurs at a time when the outward flux within the shell is considerably enhanced by the flash. As the shell continues to cool, its opacity increases further, thereby forcing an even greater expansion and driving β down to even smaller values. The star thus encounters an "opacity catastrophe" whose likely outcome is envelope ejection.

The present sequences show that there is a critical $M_{\rm env}$ at which radiation pressure will support the hydrogen envelope during a flash (see right panel of Figure 1). This critical $M_{\rm env}$ increases linearly with $M_{\rm H}$ and decreases with decreasing metallicity, since the peak in the OPAL opacities is then smaller.

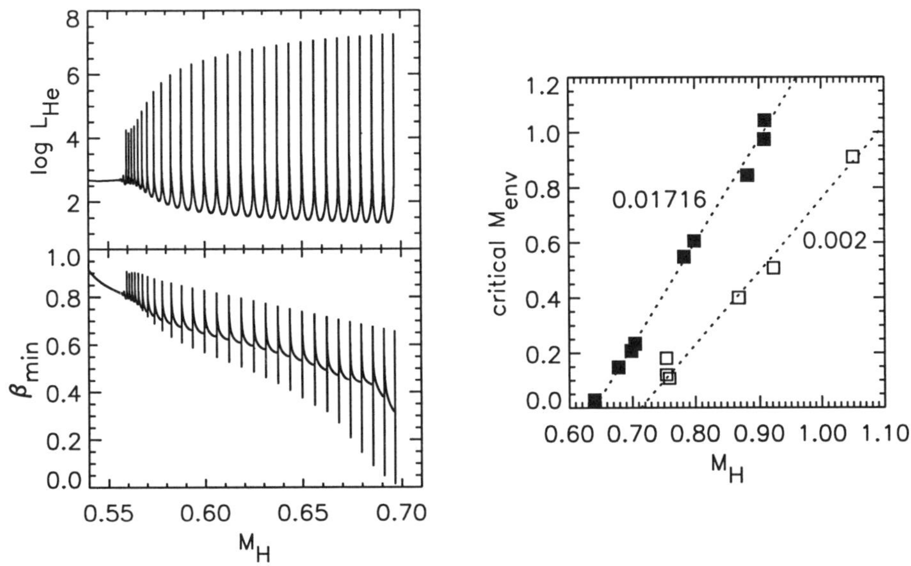

Figure 1. Helium-burning luminosity L_{He} in solar units (upper left panel) and minimum value β_{min} of the ratio β of gas to total pressure near the base of the hydrogen envelope (lower left panel) as a function of the core mass M_H during the helium-shell flashes of a solar metallicity AGB star. The right panel shows the dependence of the critical envelope mass M_{env} at which β goes to 0 during a flash on M_H for two heavy-element abundances: Z = 0.01716 (solid squares) and 0.002 (open squares). The dashed lines are linear fits to the model data.

The critical values of M_{env} in Figure 1 represent the minimum envelope mass at which an AGB star can undergo a flash without β going to 0. These results depend, however, on the extent of 3rd dredge-up in the model calculations.

The minimum in β during a flash occurs at the time when 3rd dredge-up is most likely. We suggest that such dredge-up might trigger PN ejection by increasing the opacity and thereby lowering the Eddington luminosity at the base of the envelope, especially at low metallicites where dredge-up is favored by current theoretical models. This might explain why the majority of halo PN's and most non-Type I PN's in the Magellanic Clouds are carbon rich. It might also explain why the PN K648 in M15 has a high carbon abundance despite the lack of AGB carbon stars in this globular cluster. We suggest therefore that the high carbon abundance of K648 was the cause of its PN ejection.

References

Rogers, F. J., & Iglesias, C. A. 1992, ApJS, 79, 507

Wood, P. R., & Faulkner, D. J. 1986, ApJ, 307, 659

New Views of the Magellanic Clouds
IAU Symposium, Vol. 190, 1999
Y.-H. Chu, N.B. Suntzeff, J.E. Hesser, and D.A. Bohlender, eds.

The Carbon Star Luminosity Functions in the Magellanic Clouds

Paola Marigo

Max-Planck-Institut für Astrophysik, Garching bei München, Germany & Department of Astronomy, University of Padova, Italy

Léo Girardi

Max-Planck-Institut für Astrophysik, Garching bei München, Germany

Abstract. We address the question of reproducing the observed luminosity functions of carbon stars (CSLF) in both the LMC and SMC, characterized by quite different features (i.e., faint end and peak location), using a new formalism for the third dredge-up in TP-AGB models.

The formation of single carbon stars is usually ascribed to the third dredge-up at thermal pulses in AGB stars. Complete evolutionary models are still affected by a large degree of uncertainty in describing this process, and in general fail to describe the formation of faint low-mass carbon stars. In this context, synthetic TP-AGB models can provide useful indications.

Marigo et al. (1996, 1998) and Marigo (1998) developed an updated semi-analytical code to describe the evolution of TP-AGB stars. In the latter models, the dredge-up is described by means of two parameters: (1) the efficiency parameter $\lambda = \frac{\Delta M_{\rm dred}}{\Delta M_{\rm c}}$, defined as the fraction of the core mass increment over an inter-flash period which is dredged-up to the surface at the subsequent thermal pulse; (2) the temperature parameter $T_{\rm b}^{\rm dred}$, expressing the minimum value of temperature at the base of the convective envelope for the occurrence of dredge-up at the stage of the post-flash luminosity maximum.

If λ is of common use in synthetic TP-AGB models (e.g. Groenewegen & de Jong 1993; Marigo et al. 1996), the introduction of $T_{\rm b}^{\rm dred}$ represents an improvement, as it allows us to infer the dependence of the minimum core mass for dredge-up to occur in M and Z. Moreover, no ad hoc assumption is then needed to determine when the dredge-up stops in the final stages of AGB evolution.

The models couple analytical relations derived from detailed full AGB models (Wagenhuber & Groenewegen 1998) and integrations through the complete envelope relying on the most recent input physics. They account for the occurrence of hot-bottom burning in the most massive TP-AGB stars ($M > 3.5M_{\odot}$), and the possible break-down of the $M_{\rm c} - L$ relation (see Marigo 1998). The adopted mass-loss prescription is that by Vassiliadis & Wood (1993).

The formation of carbon stars is investigated with the aid of these models. The dredge-up parameters λ and $T_{\rm b}^{\rm dred}$ are determined by demanding that the theoretical luminosity functions of carbon stars, calculated for initial metallicity $Z = 0.008$ and $Z = 0.004$, suitably fit the distributions observed in the LMC and SMC, respectively.

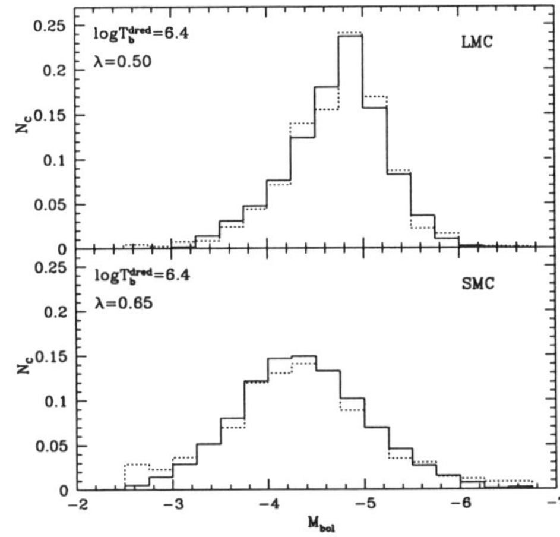

Figure 1. Observed LFs of carbon stars (dotted lines) in both Clouds, and the theoretical best fits (solid lines). The data are from Costa & Frogel (1996) for the LMC, and from Rebeirot et al. (1993; see also Groenewegen 1997) for the SMC.

Our best fitting model for the LMC is obtained with $\log T_b^{\rm dred} = 6.4$ and $\lambda = 0.50$, and is shown in Fig. 1. It turns out that the parameter $T_b^{\rm dred}$ essentially determines the faint end of the CSLF, that is the lowest luminosity level at which C-stars are expected to form. The efficiency λ essentially controls the peak location of the CSLF. We checked the sensitivity of the results to the assumed history of star formation (SFR). The faint wing and peak location are not affected by possible variations in the SFR occurring at ages older than $\sim 6\times10^8$ yr. On the contrary, it follows that the high luminosity tail could reflect the details of the recent (from 0.6 to 0.1 Gyr ago) history of star formation.

Interestingly, an excellent fit to the CSLF of the SMC is obtained from the $Z = 0.004$ models, with similar parameters, $\log T_b^{\rm dred} = 6.4$ and $\lambda = 0.65$, and a constant SFR (Fig. 1). This result is in agreement with other theoretical indications that the third dredge-up is more efficient at lower metallicities.

References

Costa, E., & Frogel, J.A. 1996, AJ, 112, 2607

Groenewegen, M.A.T., & de Jong, T. 1993, A&A, 267, 410

Groenewegen, M.A.T. 1997, presented in: ISO's view on Stellar Evolution, Noordwijkerhout, The Netherlands (astro-ph/9708011)

Marigo, P., Bressan, A., & Chiosi, C. 1996, A&A, 313, 545

Marigo, P., Bressan, A., & Chiosi, C. 1998, A&A, 331, 564

Marigo, P. 1998, A&A, 340, 463

Rebeirot, E., Azzopardi, M., & Westerlund, B.E. 1993, A&AS, 97, 603

Vassiliadis, E., & Wood, P.R. 1993, ApJ, 413, 641

Wagenhuber, J., & Groenewegen, M. 1998, A&A, 340, 183

New Views of the Magellanic Clouds
IAU Symposium, Vol. 190, 1999
Y.-H. Chu, N.B. Suntzeff, J.E. Hesser, and D.A. Bohlender, eds.

Red Giant Stars in Magellanic Cloud Clusters: Constraining Population Synthesis Models

Léo Girardi

Max-Planck-Institut für Astrophysik, Garching bei München, Germany

Abstract. We have developed a population synthesis code, aimed mainly at the interpretation of CMDs. Magellanic Cloud clusters have been largely used for the test/calibration of the underlying stellar tracks.

We computed a large set of stellar evolutionary tracks (Girardi et al. 1999), replacing the Bertelli et al.'s (1994) database for $M \leq 7 M_{\odot}$. The mass resolution has been improved, and the input physics updated. The complete mass range goes from 0.15 to $120 M_{\odot}$. The new sets are available for $Z = 0.001, 0.004, 0.008, 0.019, 0.03$, and more are in progress. From these tracks, theoretical isochrones are constructed for any intermediate value of age and Z.

Some critical checks have been performed using Magellanic Cloud clusters:

The **luminosity of clump stars** presents a minimum at ages of ~ 1 Gyr, just prior to the onset of degenerate helium cores, which is 0.4 mag fainter than the clump of old stars. Our models reproduce well this minimum, first observed by Corsi et al. (1994) using the clump against turn-off magnitude for LMC clusters (see Fig. 10 in Girardi & Bertelli 1998). The models reproduce also the behaviour of the clump I-magnitude in clusters of ages from 2 to 10 Gyr, as measured by Udalski (1998). As expected (Fig. 1 in Girardi et al. 1998), in this age range the clump magnitude is almost constant.

The **AGB-tip magnitude** is mainly determined by the mass-loss rates adopted during this evolutionary phase. We have tested four different mass-loss prescriptions in our models (Girardi & Bertelli 1998). The maximum AGB luminosity observed in LMC clusters of different ages, together with the initial–final mass relation in the solar vicinity, allowed us to rule out some of the formulations. Models with Vassiliadis & Wood's (1993) mass-loss provide the best results, describing well also the location of carbon stars in the $M_{\rm bol}$–age plane (Marigo et al. 1996), and the C-star luminosity functions in both Clouds (Marigo & Girardi, this meeting). The latter models, including the formation of C-stars and the 'over-luminosity' caused by hot-bottom burning, are now being implemented in our codes.

The age distribution of LMC clusters reflects the occurrence of episodic cluster formation in this galaxy (Girardi et al. 1995). Thus, the information given by AGB stars in clusters is not complete over the entire age range of their progenitors (Marigo et al. 1996). Data on **integrated $V-K$ colours** of LMC clusters are more complete in age, and indicate that hot-bottom burning is probably operating for $M > 3 M_{\odot}$ (Girardi & Bertelli 1998); this is the simplest way to explain the lack of AGB stars (and hence blue colours) in clusters slightly older than 10^8 yr.

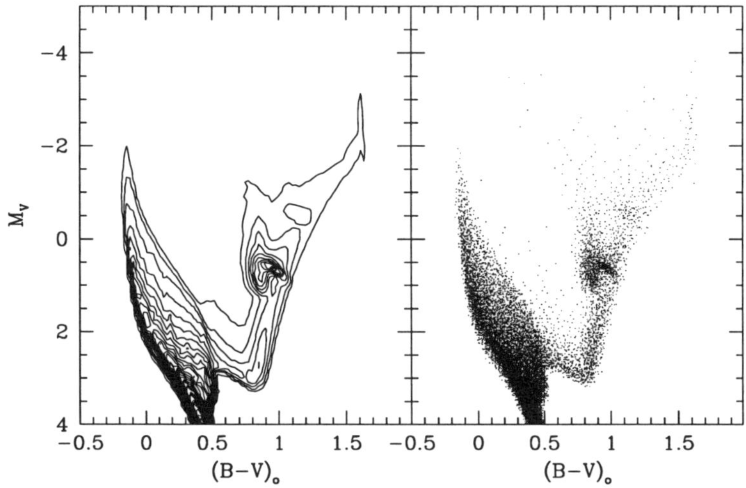

Figure 1. Theoretical distribution of stars in the M_V vs. $B-V$ diagram, for a simple model meant to represent the LMC field. **Left:** Hess diagram, with contour levels at (density)$^{1/2}$ intervals; **Right:** synthetic CMD with 2.5×10^4 stars.

Synthetic CMDs are being constructed for different histories of star formation and chemical enrichment. They reproduce well some fine details of observed CMDs (e.g., the width of the main sequence, the vertical extension of the red clump, the clump of early-AGB stars, the RGB bump), and contain TP-AGB stars in realistic proportions. Fig. 1 shows an example.

An interesting new feature in these models is **the faint extension of the red clump**. It has a solid theoretical basis, is present on the *Hipparcos* CMD of nearby stars (Girardi et al. 1998), and may correspond to the faint clumps recently observed in some fields of the LMC (Geisler et al., this meeting). A detailed description of this feature is provided in a forthcoming paper.

Acknowledgments. This work is funded by the Alexander von Humboldt-Stiftung. I thank the IAU for a travel grant to attend this meeting.

References

Bertelli, G., Bressan, A., Chiosi, C., et al. 1994, A&AS, 106, 275

Corsi, C.E., Buonanno, R., Fusi Pecci, F., et al. 1994, MNRAS, 271, 385

Girardi, L., Chiosi, C., Bertelli, G., & Bressan, A. 1995, A&A, 298, 87

Girardi, L., & Bertelli, G. 1998, MNRAS, 300, 533

Girardi, L., Groenewegen, M.A.T., Weiss, A., et al. 1998, MNRAS, 301, 149

Girardi, L., Bressan, A., Bertelli, G., & Chiosi, C. 1999, in preparation

Marigo, P., Girardi, L., & Chiosi, C. 1996, A&A, 316, L1

Udalski, A. 1998, AcA, 48, 383

Vassiliadis, E., & Wood, P.R. 1993, ApJ, 413, 641

Discussion

Carme Gallart: Is the lower part of the RGB of your new tracks bluer than in the previous tracks?

Girardi: Yes, the present evolutionary tracks have a slightly bluer RGB than the previous Padova ones (Bertelli et al. 1994). This is due to the use of new tables of low-temperature opacities.

John Gallagher: Do your models reproduce the upper RGB in the Corsi et al. sample of MC star clusters?

Girardi: I don't remember exactly how the RGB compares with the observed ones. (I intend to present such a comparison in the next months.)

Jason Harris: Do you have any models that do not produce the faint extension of the Red Clump (since this feature is not observed in many fields)?

Girardi: No. All He-burning models slightly more massive than 2 $M_{\odot}$ burn helium at fainter luminosities. This behaviour can also be found in several other sets of evolutionary tracks in the literature. Notice that the faint clump should not be present in fields which did not form stars about 1 Gyr ago.

New Views of the Magellanic Clouds
IAU Symposium, Vol. 190, 1999
Y.-H. Chu, N.B. Suntzeff, J.E. Hesser, and D.A. Bohlender, eds.

ISO Observations of Compact HII Regions in the Large Magellanic Cloud[1]

A. Moneti[2], R.J. Laureijs

ISO Data Centre, VILSPA, 28080 Spain

J.M. van der Hulst

Kapteyn Astronomical Institute, 9700-AV Groningen, The Netherlands

F. Israel, P.P. van der Werf

Sterrewacht, 2300-RA Leiden, The Netherlands

Abstract.
With the detection of strong PAH features and H_2 emission in selected knots of the N159, N11A, and 30 Dor regions in the LMC, we present the first results of a study that is part of a coordinated Guaranteed Time ISO programme to investigate star formation in the Magellanic Clouds. The PAH features have different ratios than the ones in Galactic reflection nebulae.

1. Introduction

The targets (Table 1) were selected on the basis of their compactness and their associations with CO (e.g., Johansson et al. 1998) and with H_2 (Israel & Koornneef 1991; van der Werf, unpublished data). The 30 Dor reference position was chosen to represent typical PDRs associated with fairly evolved H II regions. All data were processed with the instruments' respective Interactive Analysis packages, and using the latest calibration files available (May 1998 versions); see Table 2 for the instrumental configurations.

2. Results

All of the PHT-S spectra (Fig. 1a) show the PAH features and the fine structure lines of [Ar III] 9.0 μm and [S IV] 10.52 μm. The intensities are in excellent agreement with the PHT-P 10 μm photometry, once the results are corrected

---

[1]Based on observations with ISO, and ESA project funded by ESA member states (especially the PI countries: France, Germany, the Netherlands, and the United Kingdom) and with the participation of ISAS and NASA.

[2]On contract from SERCo F.M.

Table 1 Target coordinates

Source	RA (J2000)	Dec (J2000)
N159-5	05 40 04.2	−69 44 42.9
N159-S	05 40 00.8	−69 47 18.7
N11A	04 57 16.3	−66 23 22.8
Dor_ref	05 38 49.9	−69 04 48.3

Table 2 Instument setup

Instr.	Range	Apt (″)
PHT-S	6–12 μm	24 x 24
PHT-P	3.5–25 μm	52 ø
SWS	17 μm	14 x 27
SWS	28 μm	20 x 27

for the different aperture sizes, indicating that the emission arises over a regions smaller than the PHT-S 24″ square aperture.

The PAH features and the continua have very similar shapes, indicating comparably hard radiation fields in the four targets. The intensities are indicative of the beam-filling factors, which are the highest in the 30 Dor reference position. In Fig. 1b the 30 Dor spectrum is compared to that of the Galactic reflection nebula NGC7023, without any renormalization. The 6.2, 7.7, and 8.5 μm features are much stronger in the Galaxy relative to the 11.3 μm feature. The surface brightnesses of the 11.3 μm feature is nearly the same in the two sources, indicating similar beam-filling factors.

H_2 was detected in N11A, N159-5 and in the 30 Dor reference position. A detailed analysis of these results, including a comparison with the CO line strength from Johansson et al. (1998), is in progress.

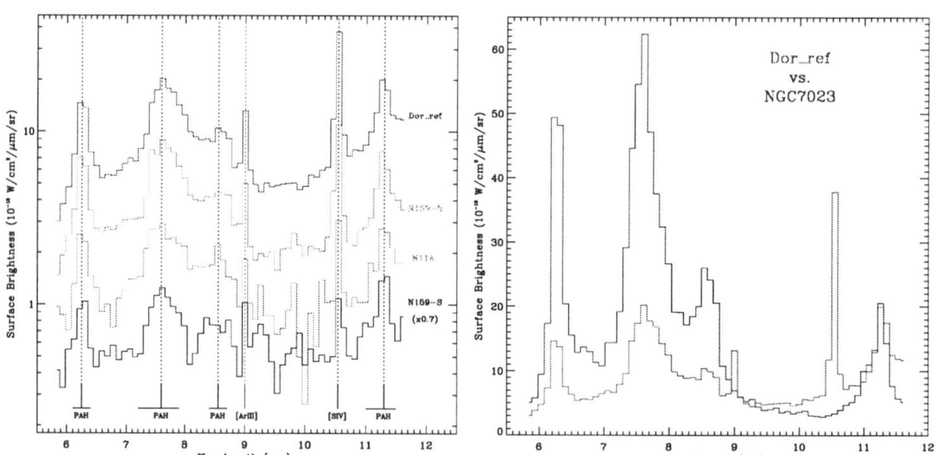

Figure 1. PHT-S spectra (left) and a comparison of the 30 Dor spectrum with that of the Galactic reflection nebula NGC 7023 (right).

References

Israel, F.P., & Koornneef, J. 1991, A&A, 250, 475

Johansson, L.E.B., et al. 1998, A&A, 331, 857

New Views of the Magellanic Clouds
IAU Symposium, Vol. 190, 1999
Y.-H. Chu, N.B. Suntzeff, J.E. Hesser, and D.A. Bohlender, eds.

The Initial/Final Mass Relation for Stars with Different Initial Chemical Compositions

Ju. Frantsman

Astronomical Institute, Latvian University, Raina Blvd. 19, Riga, LV-1586, LATVIA

Abstract.
The initial/final mass relation (M_i–M_f) for small and intermediate mass stars depends strongly on the initial chemical composition (Z). Two factors play an important role in this effect: the dependence on Z of mass loss rates and the carbon-oxygen core initial mass. For heavy element abundances $Z = 0.001$ and $Z = 0.02$ the greatest differences of M_f and M_i are found between $1.5 < M_i/M_\odot < 4$ and may be as great as $0.1\ M_\odot$.

The M_i–M_f relation for stars is important because:

1. The amount of matter returned to the interstellar medium by mass loss depends on this relation. This matter is enriched by the products of the stellar thermonuclear reactions. An accurate estimate of this relationship is crucial for the modeling of the chemical evolution of galaxies.
2. The determination of the mass distribution of white dwarfs (WD) depends on this relation. According to some models, such phenomena as SN Ia and millisecond pulsars in globular clusters depend on this mass distribution.

To estimate an accurate M_i–M_f relation, we must carefully choose accurate analytical expressions which will allow us to calculate the initial mass of the C–O core, the stellar luminosity, and the mass loss rate as a function of chemical composition.

In our calculations the mass loss was represented by Reimers (1975) law. The observations, however, show that apart from the conventional stellar wind, some other mechanism (apart from the planetary nebulae ejection) also operates during the AGB phase, substantially increasing the mass loss. In our calculations, an abrupt tenfold jump in the mass loss rate for the AGB stars reaching a certain luminosity is suggested (α=1 for $M_{\rm bol} > -5\overset{\rm m}{.}5$ and α=10 for $M_{\rm bol} < -5\overset{\rm m}{.}5$).

Table 1 contains the numerical results of the M_i–M_f calculations. A notable feature is a significant dependence of these relations on the metallicity in the region between $M_i = 1.5M_\odot$ and $M_i = 4M_\odot$. Between $1M_\odot$ and $2.5M_\odot$ our results are similar to the results of Weidemann (1987). For larger (M_i, M_f) values, we find that Weidemann's results are significantly smaller than ours.

The strong dependence of M_i–M_f relation on the initial chemical composition must significantly influence the mass distributions of the WDs. Such distributions were calculated by the population simulation method using a Salpeter initial mass function and constant stellar birth rate. These distributions strongly depend on the initial chemical composition. In case of small metallicities the distribution is shifted to the greater masses, significantly raising the number of WDs with masses greater than 0.7 $M_\odot$. This fact may be very important in considering the sub-Chandrasekhar WD explosions as the mechanism of SN Ia phenomenon.

Table 1. The initial/final mass relation for different initial chemical compositions and mass loss rates

M_i	M_f^a	M_f^b	M_f^c
1	0.57	0.56	0.56
2	0.65	0.62	0.71
3	0.69	0.65	0.82
4	0.81	0.79	0.87
5	0.94	0.91	0.95
6	1.01	0.98	1.03
7	1.10	1.05	1.11

[a] Z=0.02, α=1 if $M_{bol} > -5\overset{m}{.}2$ and α=10 if $M_{bol} < -5\overset{m}{.}2$
[b] Z=0.02, α=1 if $M_{bol} > -5\overset{m}{.}0$ and α=10 if $M_{bol} < -5\overset{m}{.}0$
[c] Z=0.001, α=1 if $M_{bol} > -5\overset{m}{.}2$ and α=10 if $M_{bol} < -5\overset{m}{.}2$

References

Blocker, T. 1995, A&A, 299, 755
Dopita, M.A., Vassiliadis, E., Wood, P. R., et al. 1997, ApJ, 474, 188
Forrestini, M., & Charbonnel, C. 1997, A&AS, 123, 241
Girardi, L., Chiosi, C., el al. 1995, A&A, 298, 87
Groenewegen, M. A. T., & de Jong, T. 1993, A&A, 267, 410
Koester, D., & Reimers, D. 1996, A&A, 313, 810
Lattanzio, J. C., & Malaney, R. A. 1989, ApJ, 347, 989
Marigo, P., Bressan, A., & Chiosi, C. 1996, A&A, 313, 545
Reid, N., Tinney, C., & Mould, J. 1990, ApJ, 348, 98
Reimers, D. 1975, Mem.Soc.Roy.Sci., Liege, 8, 369
Renzini, A., & Voli, M. 1981, A&A, 94, 175
Vassiliadis, E., & Wood, P.R. 1993, ApJ, 413, 641
Weidemann, V. 1987, A&A, 188, 74
Westerlund, B. E., Azzopardi, M., et al. 1991, A&AS, 91, 425

New Views of the Magellanic Clouds
IAU Symposium, Vol. 190, 1999
Y.-H. Chu, N.B. Suntzeff, J.E. Hesser, and D.A. Bohlender, eds.

The Formation of Carbon Stars in the Magellanic Clouds from Mass Transfer in Close Binaries

Ju. Frantsman

Astronomical Institute, Latvian University, Raina Blvd. 19, Riga, LV-1586, LATVIA

Abstract. The presence of the carbon stars in the MC with luminosities higher or lower than predicted for thermally–pulsing (TP) AGB stars, can be explained by processes that happen during the early AGB (E–AGB) stage. I examine this assumption by means of a population simulation technique. I find that there must be TP–AGB C and S stars in the MC that formed as a result of mass transfer in binary systems. Their presence may influence the age determinations of MC clusters.

The classical N–type carbon stars are associated with the TP–AGB stage. There are two groups of carbon stars whose luminosity and effective temperatures indicate that these stars are not on this stage.

1. Both theory and observations show that the minimum luminosity of carbon stars in the TP–AGB stage is about $M_{bol} = -3^{m}.0$. But according to *JHK* photometry (Westerlund et al. 1992, 1995) and a low–dispersion spectroscopic survey (Rebeirot et al. 1993) of carbon stars, the SMC N–type stars fainter than $M_{bol} = -3^{m}.0$ are rather common. These stars reach luminosities as low as $M_{bol} = -1^{m}.0$

2. Suntzeff et al. (1993) have reviewed the results of photometry of LMC carbon stars which earlier were discovered as CH–stars (Hartwick & Cowley 1988; Cowley & Hartwick 1991). These carbon stars are significantly bluer than ordinary carbon stars in the LMC. The mean bolometric magnitude is $M_{bol} = -5^{m}.3$ while the brightest members reach $M_{bol} = -6^{m}.2$.

We assume that these stars are in an E–AGB evolutionary phase and are the result of evolution of close binary systems. The carbon-enriched medium was transferred from a carbon TP–AGB star to its secondary companion which is at an earlier stage of evolution. The secondary becomes a carbon star, and remains so during its subsequent evolution up the AGB, including the E–AGB and TP–AGB stages.

The populations of AGB stars formed as a result of both single star evolution and mass transfer in close binary systems have been theoretically modeled by the method of synthetic evolution. The details are described by Frantsman (1998). The simulations of the carbon star population (luminosity functions) are presented in Figure 3 of Frantsman (1997). The most significant result is a substantial extension of the carbon star luminosity towards the lower luminosities as compared to the TP–AGB carbon stars. There are two reasons why the faint carbon stars in the MC have been discovered only recently: i) they are apparently fainter in comparison with TP–AGB stars and ii) they have higher effective temperatures. It is likely that E–AGB carbon stars cannot be formed in the Galaxy because of high initial heavy element abundance. Another result is that a few high luminosity E–AGB stars ($M_{bol} = -5^m$) must exist. It is interesting to note that E–AGB stars can reach high luminosities only if the heavy element abundance is low, which is typical for the MC but not for the Galaxy. Only massive E–AGB stars reach high luminosity, comparable to the TP–AGB star luminosity.

The N–type carbon star population in the MC is probably not homogeneous but consists of objects which belong to two different stages of evolution. The maximum luminosities of these groups are comparable, but effective temperatures differ significantly. The borders of regions occupied by stars on the E–AGB and TP–AGB phases in the HR diagram also are presented by Frantsman (1997) in his Figure 5.

During the evolution along AGB through the thermal pulsing stage, the stars remain at constant luminosity, and therefore the differences in the luminosities of carbon stars in the MC must be rather small. But there are some clusters where the most luminous carbon stars are much more luminous than the rest. Such high luminosity stars (in comparison with other TP–AGB stars) are found mainly in rich clusters where many TP–AGB stars are observed. This must affect the age determinations of a cluster if the luminosity of the TP–AGB is used.

References

Cowley, A.P., & Hartwick, F.D.A. 1991, ApJ, 373, 80

Frantsman, Ju. 1997, A&A, 319, 511

Frantsman, Ju. 1998, this volume "The initial–final mass relation for stars with different initial chemical composition"

Hartwick, F.D.A., & Cowley, A.P. 1988, ApJ, 334, 135

Rebeirot, E., Azzopardi, M., & Westerlund, B.E. 1993, A&AS,97, 603

Suntzeff, N.B., et al. 1993, PASP, 105,350

Westerlund, B.E., Azzopardi, M., Breysacher, J., & Rebeirot, E. 1992, A&AS, 91, 425

Westerlund, B.E., Azzopardi, M., Breysacher, J., & Rebeirot, E. 1995, A&A, 303, 107

New Views of the Magellanic Clouds
IAU Symposium, Vol. 190, 1999
Y.-H. Chu, N.B. Suntzeff, J.E. Hesser, and D.A. Bohlender, eds.

Expected Differences between AGB Stars in the LMC and the SMC Due to Differences in Chemical Composition

Ju. Frantsman

Astronomical Institute, Latvian University, Raina Blvd. 19, Riga, LV-1586, LATVIA

Abstract.
Certain aspects of the AGB population, such as the relative number of M and N stars, the mass loss rates, and the initial masses of carbon–oxygen cores, depend on the initial heavy element abundance Z. I have calculated synthetic populations of AGB stars for different initial Z values taking into consideration the evolution of single and close binary stars. I present the results of population syntheses of AGB stars in clusters as a function of different initial chemical compositions. The relation for the tip luminosity of AGB stars versus cluster age as a function of Z is presented and is used to determine the ages for a number of clusters in the LMC and the SMC, including clusters with no previous age determinations.

Population simulations show that for low heavy element abundance ($Z = 0.001$) few M stars are formed with respect to the number of carbon stars. However, the total number of all AGB stars in clusters is not affected by the initial chemical composition.

As a result of the evolution of close binary components after the mass exchange, an increase in the range of limiting values of the thermal pulsing AGB star luminosities is expected. The difference between the maximum luminosity on the AGB of single star and the luminosity of a star after a mass exchange event in a close binary system may be as great as 1 magnitude for very young clusters. The specific value is a function of the chemical composition. For old clusters the difference is small.

The AGB star models predict that the maximum luminosity reached along the AGB is a function of the stellar initial mass. Therefore, knowing the bolometric magnitude of the brightest star in a cluster, one can derive the initial mass of star now on the AGB, and thus the age of the cluster. This method was used by Frantsman (1988). I have redetermined the cluster ages using the new results of AGB modeling taking into account the differences of initial chemical composition. Table 1 shows my results. The first column denotes the cluster, the second - heavy element abundance Z, the third – the luminosity of the most luminous star in the cluster (Westerlund et al. 1991), the fourth column - cluster ages t determined by calibration curves, and the last column – the ages T, determined by classical methods.

Table 1. The ages of the MC clusters. The age t was obtained using the maximum luminosity of the cluster AGB stars. The age T was obtained by classical methods.

Cluster	Z	M_{bol}	$t(10^9)yr$	$T(10^9)yr$
NGC 1783	0.008	-5.15	0.28	0.2
				0.9
NGC 1831	0.02	-5.23	0.22	0.55
				0.186
NGC 1846	0.008	-5.14	0.28	
NGC 1852	0.008	-5.20	0.27	
NGC 1866	0.008	-6.04	0.10	0.09
NGC 1916	0.008	-4.97	0.3–1.2	
NGC 1978	0.008	-5.18	0.25	2.1
				0.45
NGC 1987	0.008	-5.28	0.25	0.48
NGC 2107	0.008	-5.73	0.18	0.22
NGC 2108	0.008	-5.46	0.22	0.52
NGC 2121		-5.36	0.24	0.39
			0.47	0.7
NGC 2134		-5.72	0.18	0.087
			0.19	0.19
NGC 2154	0.008	-5.27	0.26	
NGC 2190	0.015	-5.05	0.25	0.81
				1.0
NGC 2209	0.008	-5.36	0.24	0.91
	0.002		0.45	0.7
NGC 2213	0.008	-5.30	0.25	0.9-1.6
NGC 2231	0.008–0.01	-4.91	0.32–1.0	0.55
				1.2
NGC 152	0.002	-5.02	2.29	0.5
				1.3
NGC 339	0.002	-4.83	3.24	3.0
NGC 411	0.002	-4.91	2.82	1.5
				1.8
NGC 419	0.002	-5.19	0.79	0.67
L11	0.002	-4.79	3.47	3.5
L44	0.002	-4.93	2.69	
L47	0.002	-5.28	0.40	
L113	0.002	-4.80	3.47	4.0
				5.0
K3	0.002	-4.51	5.75	4.5
				8.0

References

Frantsman, Ju.L. 1988, Ap&SS, 145, 251

Westerlund, B.E., et al. 1991, A&AS, 91, 425

New Views of the Magellanic Clouds
IAU Symposium, Vol. 190, 1999
Y.-H. Chu, N.B. Suntzeff, J.E. Hesser, and D.A. Bohlender, eds.

AGB Stars in the Large Magellanic Cloud as Seen with DeNIS

M.R. Cioni, H.J. Habing

Sterrewacht Leiden, Postbus 9513, 2300 RA Leiden, The Netherlands

C. Loup

IAP, 98 bis Boulevard Arago, F-75014 Paris, France

N. Epchtein and the DeNIS consortium

Obs. de Coté d'Azur, Dep. Fresnel BP 4229, 06304 Nice Cedex 04, France

Abstract.
We present infrared photometry of LMC stars taken from a region of 2.5° in right ascension from the DeNIS (Deep Near Infrared Southern Sky Survey) survey.

1. Observations

The DeNIS experiment (Epchtein et al. 1997) has obtained simultanous photometry in three photometric bands: I (0.8 μm), J (1.25 μm) and K_s (2.15 μm), during an observing campaign from the end of 1996 to the beginning of 1997.

2. Discussion

We have statistically subtracted the Galactic component towards the Magellanic Clouds. Figure 1 shows the corrected colour-magnitude diagram of the LMC sources that characterize the Asymptotic Giant Branch evolutionary stage (O-rich and C-rich AGB stars - Loup et al. 1998 and the Loup contribution to this conference).

References

Epchtein, N., et al. 1997, The Messenger, 87, 27

Loup, C., et al. 1998, in The Impact of Near-Infrared Sky Surveys on Galactic and Extragalactic Astronomy, N. Epchtein, Kluwer Acadmic Publishers, 115

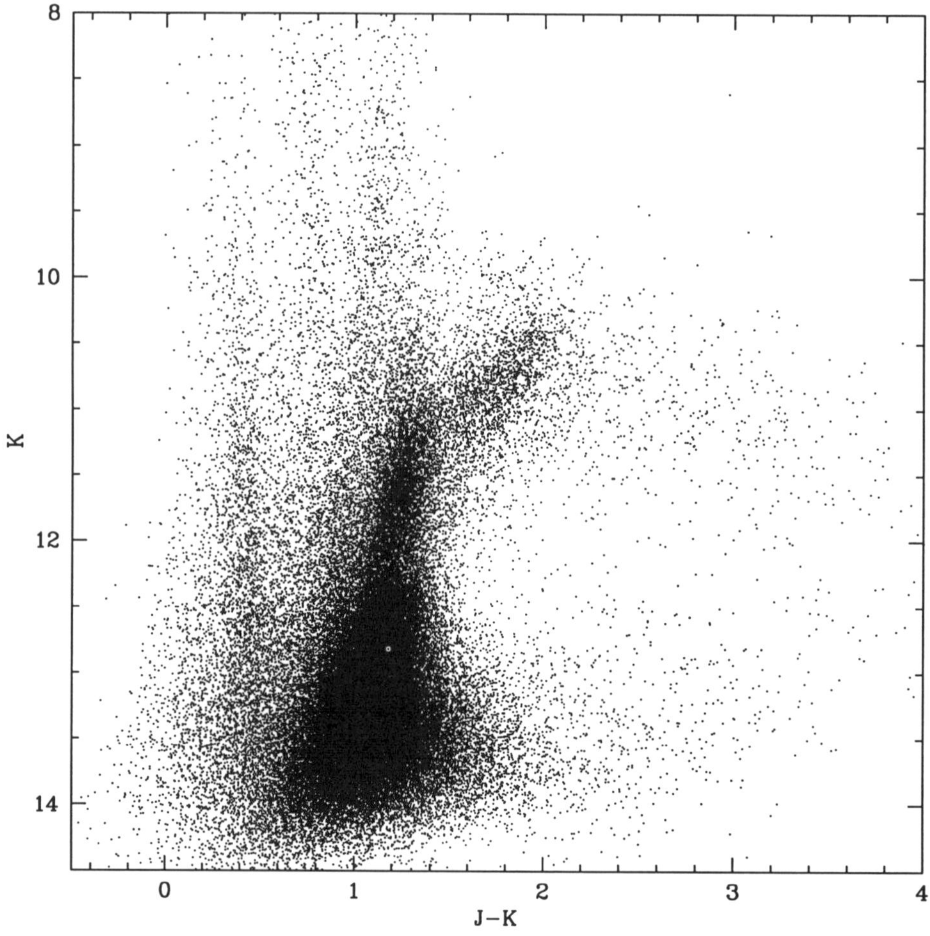

Figure 1. Colour-magnitude diagram of $\sim 10^5$ sources in the Large Magellanic Cloud.

New Views of the Magellanic Clouds
IAU Symposium, Vol. 190, 1999
Y.-H. Chu, N.B. Suntzeff, J.E. Hesser, and D.A. Bohlender, eds.

Obscured Asymptotic Giant Branch Stars in the Magellanic Clouds

Jacco Th. van Loon

Astronomical Institute, University of Amsterdam, Kruislaan 403, NL-1098 SJ Amsterdam, The Netherlands

Abstract.
I present results from an ongoing study of the evolution and mass loss of AGB stars with optically thick circumstellar envelopes in the Magellanic Clouds.

1. Introduction

Asymptotic Giant Branch (AGB) stars are highly evolved stars with intermediate progenitor masses and can be studied in great detail in the Magellanic Clouds, avoiding disadvantages in the Milky Way such as unknown distances or severe interstellar extinction. Because of the differences in metallicity between these galaxies, several aspects of stellar evolution can be studied for their metallicity dependence. Alternatively, the presence and properties of an AGB population in a galaxy reflect on the metallicity and star formation history in the past 10^{8-10} yr.

The most important phenomena that (can) occur to a star when it ascends the AGB are Long Period Variability (LPV), the transformation into a carbon star, and mass loss at rates up to 10^{-5} $M_\odot$ yr^{-1} or more. The stars with the highest mass-loss rates are obscured from the optical by their dust shells, but can still be bright at infrared wavelengths. We have searched for AGB counterparts of IRAS point sources in the direction of the Magellanic Clouds. Carbon stars had not been found at luminosities higher than $M_{\rm bol} \sim -6$ mag in optical surveys (e.g., Iben 1981; Hughes & Wood 1990; Costa & Frogel 1996), but they might be found among obscured AGB stars.

Obscured AGB stars have very red infrared colors, typically $(H-K)\gtrsim 1$ mag. The near-infrared color depends on the mass-loss rate, but also on the dust-to-gas ratio and the luminosity, and on the expansion velocity $v_{\rm exp}$ of the shell. We can measure $v_{\rm exp}$ from circumstellar maser emission profiles. These velocities are also an excellent means for determining the kinematics and mass distributions in the Magellanic Clouds.

2. Results

We successfully searched for near-infrared AGB stars as counterparts of IRAS point sources, mainly in the LMC (Loup et al. 1997; Zijlstra et al. 1996; van Loon et al. 1997, 1998a). Together with already known obscured AGB stars and the recent search in the SMC by Groenewegen & Blommaert (1998), the samples of obscured AGB stars consist now of $\sim$ 50 stars in the LMC and $\sim$ 25 in the SMC. In the LMC there are $\sim$ 3 times as many IRAS sources left unexplored. IRAS picked up sources with $S_{25\mu}\gtrsim0.1$ Jy. With ISO we were able to study bolometrically fainter or less obscured stars with $S_{25\mu}\gtrsim0.01$ Jy.

Using color characteristics and infrared spectroscopy from the ground (ESO, SAAO, CTIO) and space (IRAS, ISO), we established the chemical composition for many of the obscured AGB stars in the LMC. We found the most luminous carbon star in the Magellanic Clouds, IRAS04496−6958 with $M_{\rm bol} = -6.8$ mag, and another candidate carbon star with $M_{\rm bol} \sim -7.0$ mag. Thus carbon stars do exist up to the AGB tip luminosity. Due to mass loss the stellar mantle is reduced, thereby switching off "Hot Bottom Burning" that otherwise prevents massive stars from becoming carbon stars.

The optical depth of the circumstellar dust shell increases with mass-loss rate, but also with decreasing luminosity of the underlying star. Very obscured AGB stars exist in both the LMC and the SMC, despite their low metallicities. Assuming a dust-to-gas ratio proportional to the metallicity, the near-infrared colors of the LMC and SMC stars suggest very similar mass-loss rates. However, the outflow velocities in the shell are expected to decrease with decreasing metallicity, causing an increase of the optical depth. Hence SMC stars may in fact have lower mass-loss rates than LMC stars.

Expansion velocities based on measurements of OH profiles alone are not very reliable, as is shown by our recent SiO and H_2O maser detections (van Loon et al. 1996, 1998b) of known OH masers in the LMC (Wood et al. 1992). We argue that the expansion velocities may have been generally underestimated, and more maser searches need to be performed in the Magellanic Clouds before conclusions can be drawn on the metallicity dependence of the expansion velocity and the mass-loss rate.

Acknowledgments. I wish to thank the organizers for a pleasant and interesting conference and for a travel grant. E para o anjinho Joana: muito obrigado, por tudo.

References

Costa, E., & Frogel, J.A. 1996, AJ, 112, 2607

Groenewegen, M.A.T., & Blommaert, J.A.D.L. 1998, A&A, 332, 25

Hughes, S.M.G., & Wood, P.R. 1990, AJ, 99, 784

Iben, I. 1981, in Physical Processes in Red Giants, I. Iben Jr. & A. Renzini, Dordrecht: Reidel, 3

Loup, C., et al. 1997, A&AS, 125, 419

van Loon, J.Th., et al. 1996, A&A, 306, L29

van Loon, J.Th., et al. 1997, A&A, 325, 585

van Loon, J.Th., et al. 1998a, A&A, 329, 169

van Loon, J.Th., et al. 1998b, A&A, 337, 141

Wood, P.R., et al. 1992, ApJ, 397, 552

Zijlstra, A.A., et al. 1996, MNRAS, 279, 32

New Views of the Magellanic Clouds
IAU Symposium, Vol. 190, 1999
Y.-H. Chu, N.B. Suntzeff, J.E. Hesser, and D.A. Bohlender, eds.

Kinematics of F to M Supergiants in the 30 Dor and Shapley II Regions of the Large Magellanic Cloud

E. Maurice, N. Martin

Observatoire de Marseille 2, Place Le Verrier F-13248 Marseille Cedex 04, France

G. Testor, M.C. Lortet

Observatoire de Meudon, DAEC, F-92195 Meudon Cedex, France

Abstract.
Accurate radial velocities have been obtained with the CORAVEL photoelectric scanner at ESO, La Silla, for 36 F-M supergiants in the 30 Dor and Shapley II regions of the LMC (5^h20^m < RA(2000) < 5^h50^m, $-70°09'$ < Dec (2000) < $-68°30'$). On the basis of these data, we revisit the content and spatial extent of the four kinematical groups of F-M supergiants discovered in this area by Prévot et al. 1989. Each is the oldest part of a larger star forming region, recognized by young HII regions or CO clouds, and/or a surrounding Hα + [NII] superbubble. We discuss various gas tracers including the interstellar sodium and calcium lines.

In order to study the structure in depth and age of this interesting region of the LMC, we used different sets of data: accurate radial velocities of F-M supergiants, the HI survey at Parkes telescope by Rohlfs et al. (1984) and Luks et al. (1992), and additional HI data of better sensivity at 9 positions by Meaburn et al. (1987). These data along with the high resolution NaI and CaII ISM observations in front of 35 OB stars allowed us to correct the description of the HI gas distribution in the quoted survey (especially towards the S-E of the field and the HI poor center of Shapley II). In addition, OB stars with known velocity and data on CO clouds (Cohen et al. 1988; Johansson et al. 1998) and the associated dust allowed a first hint on the depth structure in the region under study. We obtained revised parameters of the four kinematical groups of F-M supergiants and we attempted to relate each of them to a larger star formation region.

- Shapley II (276.3 ±3.5 (s.e.) km s^{-1}, N = 41) inside the superbubble LMC3 (Meaburn 1980) and 30 Dor 1 (288.7 ±3.9 km s^{-1}, N = 11) are related to one or two distinct extended HI features of similar velocities in which the star formation nearly exhausted the gas.
- 30 Dor 2 (266.4 ±4.4 km s^{-1}, N = 26) extends from -69° 30′ to LH 90 and the core of the 30 Dor nebula and probably to the North and East to Sk−68°135, Sk−69°270, Sk−69°247. L1 designates this complex. It is obscured in all its S-E part by a large amount of dust linked to CO clouds.

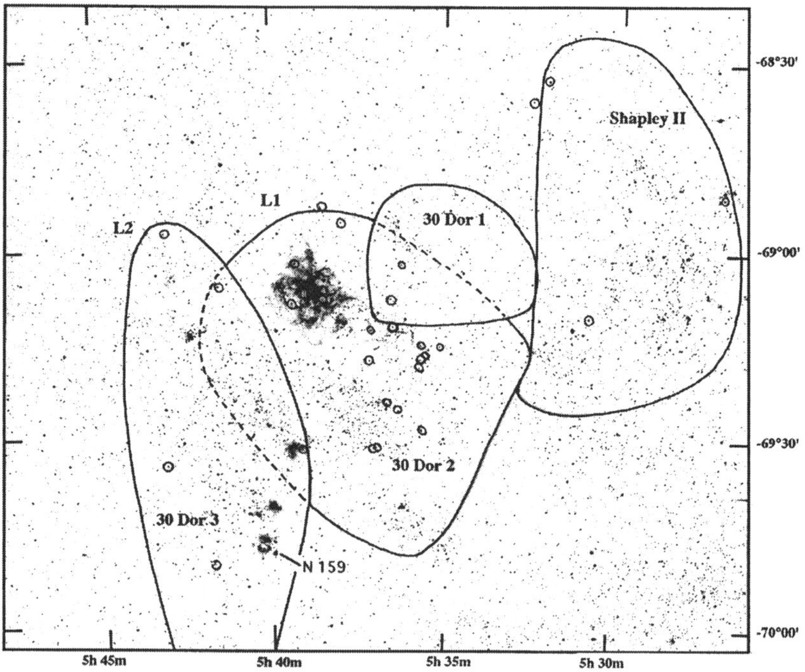

Figure 1. Schematic view of the structure of the 30 Dor and Shapley II regions. The four kinematical groups of F-M supergiants, Shapley II, 30 Dor 1, 30 Dor 2 and 30 Dor 3 are indicated inside the (very approximate) boundaries of the associated star forming regions. Open circles indicate the OB stars we used.

• 30 Dor 3 (249.7 ±5.1 km s^{-1}, N = 14) is the oldest part of a star forming region orientated North-South, here labelled L2, rich in young HII regions and CO clouds around N159 (Johansson et al. 1998).

We show that L1 lies behind L2 and behind the HI cloud related to 30 Dor1. More details on this work will be given in a forthcoming paper.

References

Cohen, R.S., et al. 1988, ApJ, 331, L95

Johansson, L.E.B., et al. 1998, A&A, 331, 857

Luks, T., & Rohlfs, K., 1992, A&A, 263, 41

Meaburn, J. 1980, MNRAS, 192, 365

Meaburn, J., Marston, A.P., McGee, R.X., & Newton, L.M. 1987, MNRAS, 225, 591.

Prévot, L., Rousseau, J., & Martin, N. 1989, A&A, 225, 303

Rohlfs, K., Kreitschmann, J., Siegmar, B.C., & Feitzinger, J.V. 1984, A&A, 137, 343.

New Views of the Magellanic Clouds
IAU Symposium, Vol. 190, 1999
Y.-H. Chu, N.B. Suntzeff, J.E. Hesser, and D.A. Bohlender, eds.

Analyses of the LMC Novae

Karen M. Vanlandingham, Greg J. Schwarz, Sumner Starrfield

Department of Physics & Astronomy, Arizona State University, Tempe, AZ 85287-1504, USA

Peter H. Hauschildt

Department of Physics and Astronomy, University of Georgia, Athens, GA 30602-2451, USA

Steven N. Shore

Department of Physics and Astronomy, Indiana University South Bend, 1700 Mishawaka Ave, South Bend, IN 46634-7111, USA

George Sonneborn

Laboratory for Astronomy and Solar Physics, Code 681, NASA/GSFC, Greenbelt, MD 20771, USA

Abstract.

In the past 10 years, 6 classical novae have been observed in the Large Magellanic Cloud (LMC). We have begun a study of these objects using ultraviolet spectra obtained by IUE and optical spectra from nova surveys. We are using the results of this study to further our understanding of novae and stellar evolution.

Our study includes analysis of both the early, optically thick spectra using model atmospheres (Hauschildt et al. 1992), and the later nebular spectra using optimization of photoionization codes (Ferland 1996; James & Roos 1993). By analysing all the LMC novae in a consistent manner, we can compare their individual results and use their combined properties to calibrate Galactic novae. In addition, our studies can be used to determine the elemental abundances of the nova ejecta, the amount of mass ejected, and the contribution of novae to the ISM abundances. To date we have analysed Nova LMC 1988#1 (Schwarz et al. 1998) and Nova LMC 1990#1 (Vanlandingham et al. 1999), and have obtained preliminary results for Nova LMC 1991. The results of this work are presented in this poster.

1. Introduction

The metal content of the LMC is known to be sub-solar and varies as a function of location within the cloud. A detailed abundance analysis of the ejecta of the LMC novae provides important information concerning the effect of initial metal abundances on energetics of the nova outburst. Since the distance to the LMC is well known, many important parameters of the outburst, such as the luminosity, can be absolutely determined. In addition, both galactic and extragalactic novae have been proposed as potential standard candles. Recent work by Della Valle & Livio (1995) has improved on the standard relations by including novae from the LMC and M31. Unfortunately, the dependence of the nova outburst on metallicity has not been well-studied. Recent theoretical work by Starrfield et al. (1998) indicates that the luminosity of the outburst increases with decreasing metal abundances. If there is a dependence of luminosity on metallicity, it will have to be determined before novae can be used as standard candles.

2. Results

Our nebular analysis results for LMC 1990#1 show that all elements which are represented in our emission line spectrum are enhanced relative to solar material. We find a high N/O ratio for LMC 1990#1, which is indicative of an outburst on a high mass white dwarf (Starrfield et al. 1992). Our estimate of the ejected mass is $M_{ej} \simeq 4 \times 10^{-4} M_{\odot}$.

We have determined the high Z metal abundances (*e.g.*, iron) of LMC 1988 #1 and LMC 1991 from model atmosphere analyses of optical and UV spectra obtained near visual maximum. We find an iron abundance of 1/3 solar in LMC 1988#1, typical of the LMC, but $Fe = 0.1Fe_{\odot}$ in LMC 1991. This is consistent with the locations of these novae within the LMC. The visual lightcurves derived from the best fit synthetic spectra are in excellent agreement with the observed lightcurves when both novae were optically thick. In addition, the bolometric lightcurves show that LMC 1988#1 and LMC 1991 were significantly brighter than the Eddington luminosity for a 1.4 $M_{\odot}$ white dwarf.

References

Della Valle, M., & Livio, M. 1995, ApJ, 452, 704

Ferland, G.J. 1996, Department of Physics & Astronomy Internal Report (University of Kentucky)

Hauschildt, P.H., Wehrese, R., Starrfield, S., & Shaviv, G. 1992, ApJ, 393, 307

James, F., & Roos M. 1993, MINUIT, CERN Program Library D506

Livio, M. 1992, ApJ, 393, 516

Schwarz, G.J., Hauschildt, P.H., Starrfield, S., & Sonneborn, G. 1998, MNRAS, 300, 931

Starrfield, S., Shore, S.N., Sparks, W.M., Sonneborn, G., Truran, J.W., & Politano, M. 1992, ApJ, 391, L71

Starrfield, S., Truran, J.W., Wiescher, M.C., & Sparks, W.M. 1998, MNRAS, 296, 502

Vanlandingham, K.M., Starrfield, S., Shore, S.N., & Sonneborn, G. 1999, MNRAS, submitted

New Views of the Magellanic Clouds
IAU Symposium, Vol. 190, 1999
Y.-H. Chu, N.B. Suntzeff, J.E. Hesser, and D.A. Bohlender, eds.

Properties of a Proper-Motion Selected Sample of Giants in the Small Magellanic Cloud Near NGC 121

Nicholas B. Suntzeff, Alistair R. Walker

Cerro Tololo Inter-American Observatory, La Serena, Chile

Verne V. Smith

University of Texas at El Paso, Texas, USA

Robert P. Kraft, Arnold Klemola

UCO, UC Santa Cruz, California, USA

Peter B. Stetson

Dominion Astrophysical Observatory, Victoria, BC, Canada

We observed a sample of zero proper-motion stars ($\mu < 0\farcs50$ cent^{-1}) from a field previously studied by Suntzeff et al. (1986). This field is 2°.5 NW (2.6 kpc) of the center of the SMC. We obtained spectra for $\sim$ 40 stars in the region of the Ca II infrared triplet using the CTIO Argus fiber-fed spectrograph. We also obtained Argus echelle spectra of a single order at 6300Å with $R = 18000$ in one run. The low-dispersion spectra were reduced to metallicities based on the Ca II equivalent widths using the Da Costa & Armandroff (1995) technique and the metallicity scale from Zinn & West (1984). The typical abundance error is 0.12 dex. For half the sample, we have echelle velocities which are accurate to 1.5 km s^{-1}. For the rest of the sample, the low-dispersion data yield single-observation velocities accurate to about 5 km s^{-1} based on repeat observations.

The average properties of the sample are:

$$\overline{[\mathrm{Fe/H}]} = -1.33,\ \sigma = 0.42,\ \mathrm{N} = 35$$
$$\overline{\mathrm{v}} = 135.4\ \mathrm{km\ s^{-1}},\ \sigma = 25.7\ \mathrm{km\ s^{-1}},\ \mathrm{N} = 36$$

We compare these results to the data of Da Costa & Hatzidimitriou (1998).

In Fig. 1 we plot a VI cmd from CTIO 4m data. The $(V - I)$ zero point is provisional. We also plot the M5 ridge line from Sandquist et al. (1996) and Johnson & Bolte (1998). The M5 ridge line has been moved to the distance of the SMC using the relative V magnitudes of the RR Lyraes in M5 and the field of the SMC adjusted for reddening differences. No metallicity correction is needed. The M5 ridge line forms the lower envelope of the subgiant branch (SGB). The width of the SGB in V seems to indicate that star formation proceeded for about $\sim$ 6 Gyrs after the corresponding age of M5, and then essentially stopped. If the most metal-rich stars are the youngest (which we have not shown), then the field metallicity stopped at about [Fe/H]=–0.5. From the lack of a well defined HB and given the low mean metallicity, there is no evidence for a populous old component.

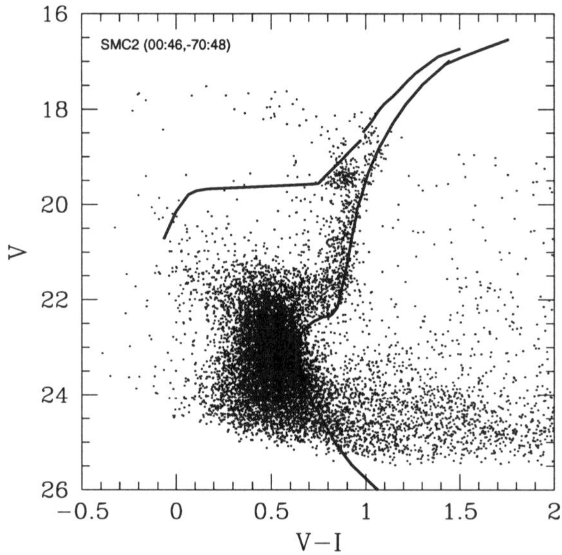

Figure 1. Color magnitude diagram of the SMC field and the ridge line for M5 moved to the distance and reddening of the SMC.

The summary of our results:

- The mean metallicity of the SMC field near NGC 121 is [Fe/H] = −1.3 with a real dispersion of 0.4 dex. The most metal rich stars are at −0.5 dex and the most metal-poor stars at −2.1 dex in a sample of 35 stars.
- The velocity dispersion of the sample is 25 km s^{-1}. The dispersion of various kinematic samples in the SMC is independent of age.
- The RGB colors are consistent with Galactic globular clusters of similar metallicity. The field population, however, does not have an extended horizontal branch, and the main sequence turnoff extends 1 magnitude brighter than the turnoff in M5. Evidently *active* star formation in this region started roughly at the age of M5 and extended for about 6 Gyrs, and then stopped.

References

Da Costa, G.S., & Armandroff, T.E. 1995, AJ, 109, 2533

Da Costa, G.S., & Hatzidimitriou, D. 1998, AJ, 115, 1934

Johnson, J.A., & Bolte, M. 1998, AJ, 115, 693

Sandquist, E.L., Bolte, M., Stetson, P.B., & Hesser, J.E. 1996, ApJ, 470, 910

Suntzeff, N.B., Friel, E., Klemola, A., et al. 1986, AJ, 91, 275

Zinn, R., & West, M.J. 1984, ApJS, 55, 45

Part 6. Stellar Clusters

Section A. Invited Reviews

(Top) Sidney van den Bergh and Gary Da Costa and (bottom) Sally Oey and Don Garnett exchanging reactions to Symposium presentations.

New Views of the Magellanic Clouds
IAU Symposium, Vol. 190, 1999
Y.-H. Chu, N.B. Suntzeff, J.E. Hesser, and D.A. Bohlender, eds.

Star Clusters in the Magellanic Clouds

G. S. Da Costa
Mt. Stromlo & Siding Spring Observatories, Private Bag, Weston Creek Post Office, ACT 2611, Australia

Abstract. Recent results for the old and intermediate-age star clusters of the Magellanic Clouds are reviewed. Highlights include new evidence that the LMC old clusters are as old the Galaxy's halo globular clusters and the persistence of the LMC *cluster* "Age Gap" despite *field star* evidence for significant star formation during the cluster age gap epoch. For the SMC new data confirm the lack of significant change in cluster abundances with age prior to ~4 Gyr ago.

1. The Old LMC Clusters

The current sample of LMC clusters considered analogues of the Galactic halo globular clusters consists of 13 objects (Suntzeff et al. 1992). Like their Galactic halo counterparts, for existing abundance estimates the LMC clusters show no radial abundance gradient (e.g., Da Costa 1993). Yet kinematically the LMC cluster system is quite different from that of the Galactic halo. It shows "disk-like" motions with $V_{circ} \approx 60$ kms^{-1} and $\sigma \approx 25$ kms^{-1} (Schommer et al. 1992). This kinematic difference might suggest that the LMC old clusters are somewhat younger than the Galactic halo clusters. Such a suggestion, however, is difficult to substantiate from the ground because, while ground-based estimates of LMC old cluster main sequence turnoff luminosities do indicate similar ages to Galactic halo clusters, they lack the ± 1 Gyr precision required to be certain of any age differences. Horizontal branch (HB) morphology – abundance diagrams, on the other hand, do suggest that some LMC old clusters are younger than the inner Galactic halo clusters (e.g., Da Costa 1993). However, the interpretation of such diagrams in terms of age differences is a process fraught with uncertainty.

The question of the age of the LMC old clusters relative to Galactic halo clusters, however, can be settled with WFPC2 data. Two groups report their results in these proceedings. Olsen et al. (1998 and these proceedings, hereafter O98) derive age estimates from main sequence photometry for five "inner" clusters while Johnson et al. (these proceedings) studied three "outer" objects. Both groups find the same result: *the LMC clusters are as old as their Galactic halo counterparts and there is no evidence for any age range in excess of ~1 Gyr.* Thus the proto-Galaxy and the proto-LMC began forming stars at similar epochs.

The new results have two interesting consequences. First, O98 have increased significantly the number of LMC old clusters that can be plotted in the HB morphology – abundance diagram. If we adopt the "age is the second parameter" interpretation of this diagram, then the location of the clusters NGC 1898, 1754, 2019, 2005 and Hodge 11 (see O98, Fig. 17a) suggests that these clusters should have the same age, and moreover, that this age should be comparable to that of the inner Galactic halo clusters. This prediction is in excellent accord with the new WFPC2 results. The same interpretation applied to the clusters Reticulum, NGC 1466, 2257 and 1841 would require these clusters to be ~3 Gyr younger than the first group. However, from the Johnson et al. results we know that this younger age interpretation is not correct, at least as regards NGC 1466 and NGC 2257. This suggests strongly that the "age is the second parameter" interpretation of HB morphology – abundance diagrams is not always valid. This is consistent with recent WFPC2 results for the Galactic halo, in which outer halo clusters with red HBs have been found to be somewhat younger than inner halo clusters of similar abundance, but the observed age difference is somewhat less than that expected from the HB morphology difference (see Hesser et al., these proceedings).

The second consequence follows from the abundances derived by O98. They apply the "simultaneous metallicity and reddening" technique which, given their well defined cluster giant branches, should result in abundances that are more reliable than existing estimates. These latter estimates come mostly from CaII spectroscopy but are often based on only a single star per cluster (Olszewski et al. 1991). For NGC 1754, 1835 and 1898 the agreement between the two techniques is satisfactory (and this is also the case for the three clusters studied by Johnson et al.), but for NGC 2005 and 2019 the new abundances are ~0.6 dex higher than the earlier estimates. If we now use the O98 abundances in a plot of abundance versus projected radial distance for the LMC clusters, as is illustrated in Fig. 1 (cf. the upper panel of Fig. 3 of Da Costa 1993), it is evident that there is now support for the existence of a *radial abundance gradient in the LMC old cluster system.* Given the implications of such a result, it is important to have spectroscopic confirmation of the O98 abundances, and an improved abundance estimate for NGC 1916, the innermost metal-poor system. O98 could not derive an abundance for this cluster because of differential reddening; the abundance in Fig. 1 is based on a CaII measurement for a single cluster star only.

2. The LMC Cluster Age Gap – An Update

Geha et al. (1998, see also many contributions in these proceedings) have used deep WFPC2 images to analyze the LMC field star formation history. They find that their data are most consistent with a star formation rate that is approximately constant for most of the LMC's history but which increases by a factor of ~3 approximately 2 Gyr ago. This history produces a stellar population with approximately equal numbers of stars older and younger than ~4 Gyr. This result contrasts strongly with the observed *cluster* age distribution in which only a single cluster with an age between ~3 Gyr and the age of the oldest clusters is known, despite the existence of numerous 1 – 3 Gyr old clusters. Consequently, unless there is a mechanism to disrupt LMC clusters older than ~3 Gyr, which seems unlikely, the star formation history implied by the (relatively luminous) LMC cluster population is *not* that followed by the field population. Given the

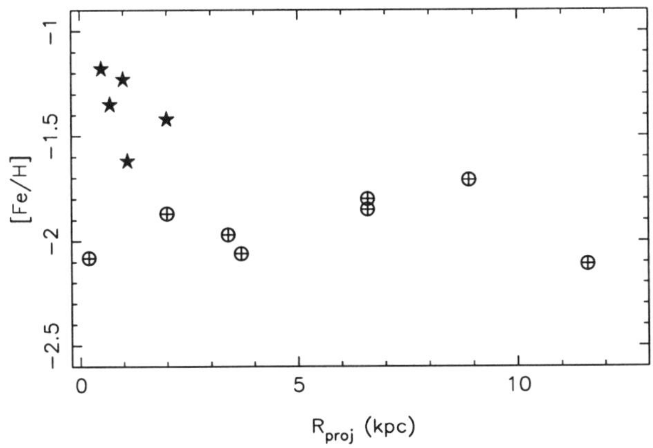

Figure 1. Abundance versus projected radial distance for the LMC old cluster population. Star symbols represent new abundances for five inner clusters from Olsen et al. (1998), while the circled plus sign symbols are abundances from Suntzeff et al. (1992 and references therein).

implications of this result, it is not surprising that there have been additional searches for LMC star clusters that fall in the "Age Gap".

The first of these is that of Geisler et al. (1997, see also these proceedings) in which candidate clusters were selected on the basis of integrated UBV colors or low metallicity. The clusters were imaged to produce c-m diagrams and the age estimated from the magnitude difference between the red clump and the main sequence turnoff. Observations of the ~9 Gyr old cluster ESO121-SC03 were included as a check. For the 23 candidates for which an age estimate could be obtained, *none* proved to be older than ~2.5 Gyr. This result rather strongly reinforces the status of the cluster Age Gap: there are now more than 40 clusters known, via main sequence turnoff photometry, to have ages between ~1 and 3 Gyr! Sarajedini (1998), however, has reported the serendipitious discovery of three LMC clusters somewhat older than 3 Gyr. The data are archival WFPC2 short exposure images and isochrone fits suggest ages of ~4 Gyr for the clusters, slightly older than any previously age-dated LMC intermediate-age cluster. The results require confirmation from better data, especially as the Sarajedini (1998) abundance estimates differ significantly from those of Olszewski et al. (1991) for the two clusters in common. Nevertheless, independent of the Sarajedini (1998) results, ESO121-SC03 remains a unique LMC cluster and we are yet to find (if they exist) the LMC equivalents of SMC clusters like Kron 3, NGC 361 and 416, clusters with ages between ~5 and 8 Gyr.

The search for clusters in the Age Gap is important because, without such objects, the LMC Age-Metallicity relation (AMR) remains virtually unconstrained in the epoch between ~3 Gyr and the formation of the oldest clusters. Despite this, Pagel & Tautvaisiene (1998) have used the existing LMC cluster AMR to constrain analytical models for the chemical evolution of the LMC. Their models can incorporate variations in the star formation rate and they find the cluster AMR can be fit with a model that has a star formation history quite similar to that determined observationally by Geha et al. (1998) from deep WFPC2 observations of field regions. The Pagel & Tautvaisiene (1998) result can be then interpreted as further evidence for significant LMC star formation in the ~3 – 10 Gyr interval that is apparently not reflected in the cluster age distribution. If these field star formation histories are assumed to apply to the cluster formation history also, then based on the current size of the known ~1 – 3 Gyr cluster population, we expect there should be perhaps a dozen LMC clusters with ages in the ~3 – 10 Gyr age range, even allowing for evolutionary fading and some cluster disruption. At the present time we know definitely of only one such cluster in this age range; perhaps the Sarajedini (1998) objects may represent one or two more. While searches for such clusters should continue, it is apparent that the deficiency of age $\geq$ ~3 Gyr clusters in the LMC is probably real. In that case we must address the fact that prior to the large increase in star formation ~3 Gyr ago, the LMC did form significant numbers of field stars but did not form any relatively luminous (i.e., massive) star clusters (other than the old clusters). Identification of the star formation conditions under which this can occur (or perhaps better, the conditions under which massive cluster formation *does* occur) is best left to theorists. But they should not forget that such clusters have been forming approximately continously in the SMC.

One final point deserves mention. In Fig. 1 there is a strong suggestion of a radial abundance gradient in the old LMC cluster system. It is of relevance to the LMC enrichment history to ask if such a gradient is also present among the intermediate-age cluster population which, like the old clusters, has disk kinematics. The extensive work of Olszewski et al. (1991) found a slight difference between "inner" ($r \leq 5°$) and "outer" clusters, in the sense that the outer clusters were marginally more metal-poor. There was, however, considerable overlap in the total abundance range. On the other hand, Bica et al. (1998, see Geisler, these proceedings) have provided (photometric) abundance estimates for the 13 most distant of the intermediate-age clusters studied by Geisler et al. (1997). They find generally lower abundances than Olszewski et al. (1991) for the clusters in common and their results hint at the existence of a radial abundance gradient. A comprehensive investigation of this question is clearly called for.

3. SMC

In the SMC star cluster AMR shown in Da Costa (1991), the cluster ages were determined from main sequence turnoff luminosities but the abundances were mostly based on giant branch colors. Recently, however, there have been both new abundance determinations and new age estimates for an increased sample of SMC clusters. Thus a reinvestigation of the SMC cluster AMR is appropriate.

As regards cluster abundances, Da Costa & Hatzidimitriou (1998) have published estimates for six old and intermediate-age clusters based on spectroscopy at the CaII triplet of typically five individual red giant members per cluster. Similarly, Mighell et al. (1998 and these proceedings) have determined "photometric" abundances for seven old and intermediate-age clusters by applying the "simultaneous reddening and metallicity" and the "red giant branch slope" methods to cluster c-m diagrams. Finally, de Freitas Pacheco et al. (1998) have

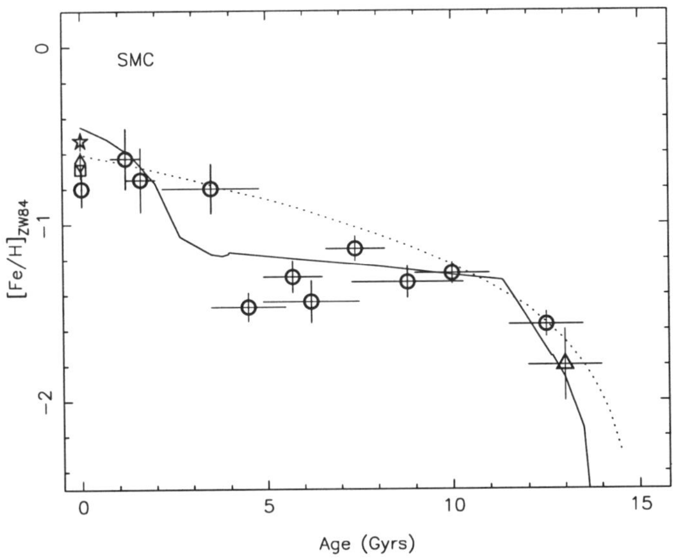

Figure 2. The SMC age-metallicity relation. Star clusters are shown as circles, the field RR Lyrae results of Butler et al. (1982) by the triangle and present-day field star results by the star, diamond and square symbols. The dotted line is the prediction of a simple (closed box) chemical evolution model while the solid line is that for the "bursting" model of Pagel & Tautvaisiene (1998). Note that this latter model has a very low star formation rate in the ~4 to 12 Gyr interval, which is in conflict with the numbers of clusters found in this age range.

determined abundances for six old and intermediate-age clusters from integrated cluster spectra. Their technique involves the measurement of Lick system indices calibrated via single stellar population models. The agreement between these independent abundance determinations is generally quite good. Thus they can be combined to produce improved abundance estimates for a total of 10 old and intermediate-age SMC star clusters.

As regards cluster ages, Mighell et al. have used archival WFPC2 short exposure images to estimate ages for several clusters using the color difference between the red clump and the red giant branch as an age indicator. Similarly, de Freitas Pacheo et al. have used their single stellar population models to derive age estimates from their integrated cluster spectra. The agreement between these data and the ages available from (ground-based) determinations of the main sequence turnoff luminosities is usually reasonable, so that the values can be combined to produce improved age estimates.

The SMC cluster AMR based on these new data is shown in Fig. 2. Somewhat surprisingly perhaps, the "morphology" of this relation isn't greatly different from that in Da Costa (1991), although the cluster data are undoubtedly more reliable. What are we to make of this relation? First, we can compare the data to the predictions of a simple (closed box) chemical evolution model; such models are often assumed to apply to dwarf galaxies like the SMC. The prediction of such a model, scaled to the present-day SMC abundance and gas fraction, and assuming an age of 15 Gyr, is shown as the dotted curve in Fig. 2. The fit is clearly not satisfactory: it might be adequate for the earliest epochs but the model generally predicts abundances that are too large for the intermediate-age clusters. An alternative approach is that adopted by Pagel & Tautvaisiene (1998). As noted above, in their models these authors allow for variations in the star formation rate (SFR). For the SMC they find that if large variations in the SFR are included, then age-metallicity relations can be generated that are more representative of the cluster data. In particular, they associate the period of little obvious enrichment with a long interval in which the SFR is very low. They then use a large increase in the SFR at $\sim$4 Gyr to "explain" the rapid rise in cluster abundance. The solid line in Fig. 2 shows the AMR for one of these "bursting" models, and it does appear to be representative of the cluster data. However, the assumed star formation history underlying this model is problematic. It assumes that the SFR is very small over the $\sim$4 – 12 Gyr interval, yet, while the current sample of SMC clusters is undoubtedly biased by various selection effects, most observed clusters in fact formed during this postulated "quiescent" period, an inconsistency with the model assumptions.

As a guide then to future modellers of these cluster data, and to SMC observers, two salient features of Fig. 2 should be noted. First, the rapid rise in cluster abundance at $\sim$3 – 4 Gyr seems inescapable. Tighter constraints on the SMC star formation history, from both field star and additional cluster studies, are required however, to gain insight into the origin of this sharp abundance rise. Second, among the SMC intermediate-age clusters there is a definite abundance range, but with little obvious correlation between age and abundance. For example, contrast the clusters NGC 339 and NGC 416 with [Fe/H] ≈ -1.45 at $\sim$5 Gyr with Kron 3 at [Fe/H] ≈ -1.15 at $\sim$7.5 Gyr. The existence of this abundance range is somewhat unexpected given that in dwarf galaxies lacking significant systematic rotation, such as the SMC, enrichment products are expected to be well-mixed over galaxy-wide scales on timescales considerably less than a Hubble time. Certainly the present-day abundances in the SMC are homogeneous: studies of HII regions and young stars suggest the present-day abundance dispersion is $\lesssim$0.1 dex. In this context it is also worth noting that there is no apparent spatial – abundance correlation in these SMC cluster data; e.g., there is no evidence for a radial abundance gradient.

These results suggest that the chemical evolution of the SMC was (and is) quite complex. In particular, it is becoming increasingly clear that we must relax the simplfying assumption that the SMC has evolved as an isolated system, and no longer ignore the effects of interactions with the LMC and the Galaxy.

4. Summary

The latest data for old and intermediate-age clusters in the Magellanic Clouds seem to have raised as many questions as they have supplied answers as regards the star and cluster formation histories and the chemical evolution of these galaxies. Clearly a variety of complex processes are involved. We can but look forward to the impact the new large telescopes in the Southern Hemisphere will have on these subjects. For example, it will be possible to study quantities such as the [α/Fe] element ratio in red giants in Magellanic Cloud clusters of different ages. The results of such studies will tell us a lot about the details of the enrichment processes in these galaxies. Indeed, at the next Magellanic Clouds IAU Symposium, we'll undoubtedly be discussing the new puzzles that the new large telescope data will reveal.

References

Bica, E., et al. 1998, AJ, 116, 723

Butler, D., Demarque, P., & Smith, H.A. 1982, ApJ, 257, 592

Da Costa, G.S. 1991, in The Magellanic Clouds, IAU Symp. 148, eds R. Haynes and D. Milne (Kluwer, Dordrecht), p. 183

Da Costa, G.S. 1993, in The Globular Cluster – Galaxy Connection, eds G.H. Smith and J.P. Brodie (ASP, San Francisco), ASP Conf. Ser. 48, p. 363

Da Costa, G.S., & Hatzidimitriou, D. 1998, AJ, 115, 1934

de Freitas Pacheco, J.A., Barbuy, B., & Idiart, T. 1998, A&A, 332, 19

Geha, M.C., et al. 1998, AJ, 115, 1045

Geisler, D., et al. 1997, AJ, 114, 1920

Mighell, K.J., Sarajedini, A., & French, R.S. 1998, AJ, 116, 2395

Olsen, K.A.G., et al. 1998, MNRAS, 300, 665

Olszewski, E.W., et al. 1991, AJ, 101, 515

Pagel, B.E.J., & Tautvaisiene, G. 1998, MNRAS, 299, 535

Sarajedini, A. 1998, AJ, 116, 738

Schommer, R.A., et al. 1992, AJ, 103, 447

Suntzeff, N.B., et al. 1992, AJ, 104, 1743

Discussion

van den Bergh: In the dwarf irregular galaxy IC 1613 we see lots of star formation, but no cluster formation. This shows that the specific cluster forming frequency is $\geq 10^2$ times higher in the LMC *now* than it is in IC 1613.

Da Costa: Yes, this further illustrates the point I was trying to make in regard to the LMC. We shouldn't assume that the formation history of relatively luminous (massive) star clusters necessarily tells us about the star formation history of the general population. The formation of such clusters may require special conditions and so the lack of such clusters doesn't necessarily mean no star formation at all.

Hans Zinnecker: The jump in metallicity in the SMC some 3 Gyr ago strikingly agrees with the onset of enhanced star formation rate in the LMC! Something must have happened then, some common trigger. Perhaps a close encounter with the Milky Way?

Rebecca Elson: Are there any constraints from field star studies on the star formation history in the 4-10 Gyr gap where the Pagel et al. model shows no star formation?

Jon Holtzman: Our deep HST data near the bar of the SMC appear to be inconsistent with the predictions of the Pagel model. Comment: Note that it may still be possible for a star formation history which matches that of LMC clusters to fit the deep luminosity function if one allows for a steeper IMF.

Norbert Langer: Pagel & Tautvaisiene's model for the chemical evolution of the SMC points out that there may be a fundamental problem with the interpretation of the age-metallicity data from clusters in the SMC: a constant metallicity for a time interval of about 5 Gyr seems to imply that there was no significant star formation during that time (at least in a closed box model), in contrast with the existence of clusters with these ages.

Da Costa: Yes, the lack of enrichment shown by the SMC clusters over the $\sim$4 to 10 Gyr interval, assuming the formation of clusters is indicative of general on-going star formation throughout this period, requires either the outflow of metal-enriched gas, which seems unlikely unless the star formation was very vigorous, or the infall of primordial, or at least low abundance, gas.

New Views of the Magellanic Clouds
IAU Symposium, Vol. 190, 1999
Y.-H. Chu, N.B. Suntzeff, J.E. Hesser, and D.A. Bohlender, eds.

Young Magellanic Cloud Clusters (< 1 Gyr): Census, Properties, Star Formation History

Eva K. Grebel, Dennis Zaritsky, Jason Harris

UCO/Lick Observatory, UCSC, Santa Cruz, CA 95064, USA

Ian Thompson

Observatories of the Carnegie Institution, 813 Santa Barbara St., Pasadena, CA 91101, USA

Abstract. We report preliminary results from an automated cluster survey of the Magellanic Clouds aimed at improving the cluster census and at deriving cluster properties from their resolved stellar content. The survey is tripling the number of known clusters. The cluster age distribution shows similar peaks at 100–200 Myr in LMC and SMC, coincident with the closest encounter of the Clouds and perigalacticon.

1. Introduction

In contrast to the Milky Way, the Magellanic Clouds contain a plethora of compact star clusters of all ages. Photographic surveys led to the detection of ~ 2400 clusters in the LMC and ~ 700 clusters in the SMC, while the true number of star clusters is estimated to be more than twice as high (Westerlund 1997).

The new large-scale CCD surveys of the Magellanic Clouds provide homogeneous multi-color coverage of the Clouds with good spatial resolution and are well-suited to improve the census of star clusters. We are using data from the Magellanic Clouds Photometric Survey (Zaritsky et al. 1997) and from the Optical Gravitational Microlensing Experiment (Udalski et al. 1998) for this purpose. These surveys also allow us to derive ages and structural parameters based on the resolved stellar content of the clusters.

2. Search Methodology

Our search for clusters uses a minimum spanning tree algorithm to identify areas of higher stellar density (Zahn 1971). The algorithm finds clusters by searching for minima in the average distance between any two stars. The user specifies the allowable standard deviation of the average distance and a minimum number of stars per "cluster". Without these constraints the algorithm will find extended strings of stars as clusters, or consider any two close stars a "cluster".

In addition to the density criterion we make use of the magnitude and color information in our dataset. The data are divided into subsets covering certain magnitude ranges and color ranges representing different evolutionary phases. We require detection in at least two of these subsets to consider a density enhancement to be a star cluster candidate. For instance, a density enhancement detected in the red clump subset and the subset of faint main-sequence stars is a good candidate for an intermediate-age cluster. Similarly, a density enhancement detected in the blue and in the bright subsamples stands a good chance of indicating a young cluster. We use five such color-magnitude selections of subregions representing certain evolutionary phases plus the full data set.

After fine-tuning the cluster algorithm criteria for each subset, 94% of the known star clusters could be retrieved. Clusters that escaped detection were either objects within OB associations with displaced or poorly defined centers, or objects that were not visible in our data even when visually inspecting our frames (possibly plate errors in the earlier photographic surveys?).

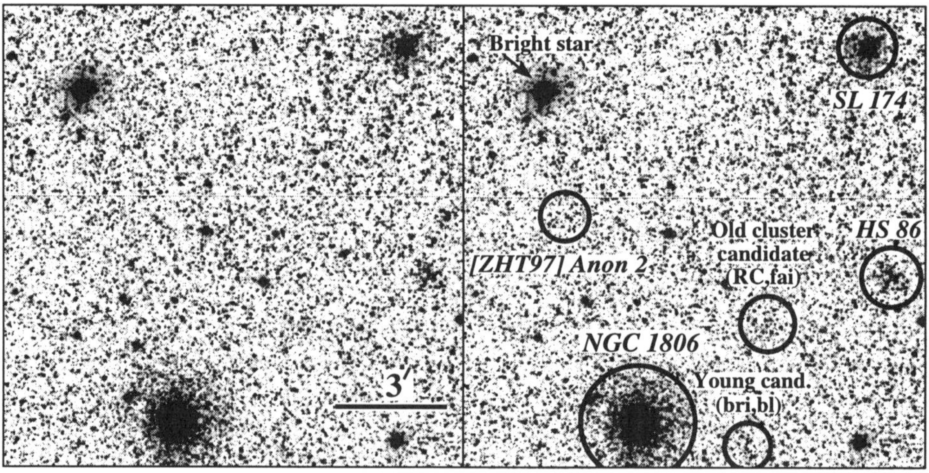

Figure 1. Cluster search results. Left: Small field from the Magellanic Clouds Photometric Survey (Zaritsky et al. 1997). Right: Same field with marked cluster detections. For new candidates the magnitude/color-selected subsets in which they were detected are specified (bri=bright, bl=blue, fai=faint, RC=red clump).

3. Monte-Carlo Simulations

Monte-Carlo simulations were run to assess the reliability of our detections on artificial fields with randomly distributed stars (10^6) and fields with artificial density gradients. The density in both Monte-Carlo sets was chosen to reproduce the observed densities in our data away from and near to the LMC bar. Furthermore, we created fields with artificial clusters (14,000). Observed magnitudes were randomly assigned to the Monte-Carlo star positions, and magnitude/color-selected subsets were extracted as before.

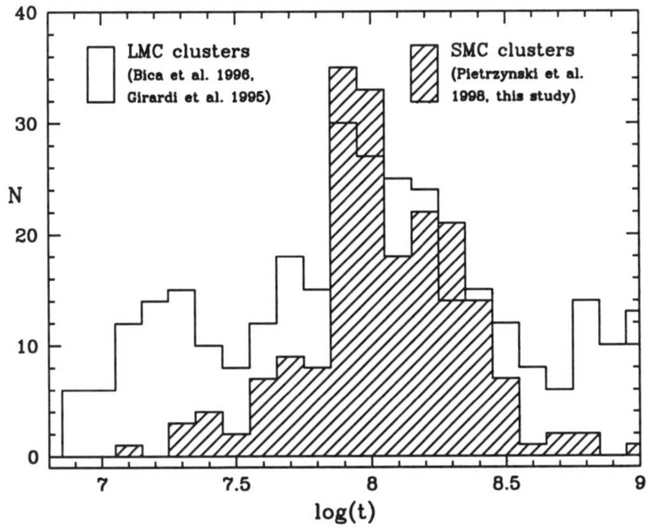

Figure 2. The age distributions of LMC and SMC clusters both peak around log(t)=8, when the Clouds had their closest encounter and last perigalacticon.

On average we found 10 – 20% spurious detections in each of the magnitude/color-selected subsets with a slight increase of 2 – 3% in fields with artificial density gradients. Spurious cluster candidate detections were reduced to 5% when we imposed the requirement of independent detection in at least two subsets. About 60% of the artificial clusters with ≥ 30 members and $\sim 100\%$ of clusters with ≥ 50 stars were detected by the minimum spanning tree algorithm, which demonstrates its high sensitivity.

4. Search Results

In the areas of the Magellanic Clouds Photometric Survey searched so far the number of cluster candidates was *tripled.* Among the new clusters are a few bright amorphous agglomerates in extended OB associations with ages of mostly < 30 Myr. Many sparse, bright concentrations of OB stars were detected as cluster candidates (ages < 40 Myr). These objects resemble the multiple, unbound compact cores of OB associations that will disperse quickly (van den Bergh, this volume). Finally, the new detections comprise many faint, sparse clusters with estimated ages > 600 Myr. *Our new detections seem to represent a continuum of open clusters in various stages of dissolution* ranging from sparse young objects that will be unobservable after > 100 Myr to initially more compact agglomerates that require $\gg 100$ Myr to disperse.

A preliminary analysis of structural parameters indicates that few loose clusters with radii $> 40''$ are detectable after 200 Myr, while slightly more compact clusters with radii $< 30''$ are easily detected until ~ 400 Myr. Older clusters tend to have higher concentration parameters.

5. Age Distribution and Interactions

Isochrone-based ages using both Geneva and Padua models were derived for a complete sample of ~ 200 populous star clusters from the catalog of Pietrzyński et al. (1998) in the central ~ 2.4 degrees^2 of the SMC. The resulting age distribution shows a pronounced peak at ~ 100 Myr and a second, smaller peak at ~ 200 Myr (Figure 2). These two peaks resemble the broad peak around 100–200 Myr in age distribution of LMC clusters found by Girardi et al. (1995) from an age calibration of integrated colors.

Geisler et al. (1997) showed that ages for intermediate-age and old clusters derived from integrated colors can be significantly (8–10 Gyr) in error due to stochastic effects by bright field stars and crowding. Comparing our isochrone ages for clusters younger than 1 Gyr to integrated-color ages for the same SMC clusters from Elson & Fall (1985) we find good overall agreement despite considerable scatter (log(Δt) $\pm$0.3), indicating that for bright young clusters the errors are less significant.

The SMC data do not show a period of enhanced cluster formation at more recent times (log(t)~ 7), nor a pronounced peak at log(t)~ 9.1, as seen in the LMC.

N-body simulations (Gardiner et al. 1994) suggest that the last close LMC/SMC encounter was ~ 200 Myr ago, and that the last perigalacticon occurred ~ 100 Myr ago. The common peaks in the cluster age distribution at log(t)~ 8 may be due to an enhanced cluster formation rate triggered by these events, while interaction between the Clouds does not seem to have been the cause of the increased cluster formation in the LMC seen at earlier and later times.

Acknowledgments. Support for this work was provided by NASA LTSA grant NAG-5-3501, NSF AST-96 19576, the Packard Foundation, and the Sloan Foundation.

References

Bica, E.L.D., Clariá, J.J., Dottori, H., Santos, J.F.C., & Piatti, A.E. 1996, ApJS, 102, 57

Elson, R.A.W., & Fall, S.M. 1985, ApJ, 299, 211

Gardiner, L.T., Sawa, T., & Fujimoto, M. 1994, MNRAS, 266, 567

Geisler. D., Bica, E., Dottori, H., Clariá, J.J., Piatti, A.E., & Santos, J.F.C. 1997, AJ, 114, 1920

Girardi, L., Chiosi, C., Bertelli, G., & Bressan, A. 1995, A&A, 298, 87

Pietrzyński, G., Udalski, A., Kubiak, M., Szymański, M., Woźniak, P., & Żebruń, K. 1998, AcA, 48, 175

Udalski, A., Szymański, M., Kubiak, M., Pietrzyński, G., Woźniak, P., & Żebruń, K. 1998, AcA, 48, 147

Westerlund, B.E. 1997, The Magellanic Clouds, Cambridge University Press

Zahn, C.T. 1971, IEEE Trans. on Computers, C-20, 68

Zaritsky, D., Harris, J., & Thompson, I. 1997, AJ, 114, 1002

Discussion

John Gallagher: With regard to the long lives of star-forming regions, it might be interesting to note that Robbie Dohm-Palmer found a similar behaviour in HST observations of the dI galaxy Sextans A (Dohm-Palmer et al. 1998, AJ). He used blue loop stars for his age tracers and showed that star-forming regions have lifetimes of > 100 Myr.

Grebel: Indeed this is very similar to what we find in the LMC from supergiants and Cepheids. We see long-lived chains of star formation that were active for ≈ 200 Myr. Interestingly these areas appear to correspond to the earlier proposed asymmetric "spiral arms" in the LMC. Gardiner and collaborators carried out N-body simulations to study the effect of an off centered bar on a rotating disk and showed that this leads to the formation of a non-density wave "spiral arm" and sites of star formation along this feature. The resulting picture looks very similar to the observed global star formation patterns across the face of the LMC disk. It would seem possible that the bar of Sextans A, which appears to be off centered, may have a similar effect in Sex A by confining gas along what develops into long-lived chains of star formation.

David Graff: What was the source for your Cepheids?

Grebel: The catalogue of Payne-Gaposhkin and the General Catalogue of Variable Stars (Kholopov 1995, 1997).

David Graff: How does the Drift-scan survey compare with the OGLE survey?

Grebel: The Magellanic Clouds Photometric Survey will cover $8^\circ \times 8^\circ$ of the LCM and the central $4^\circ \times 4^\circ$ of the SMC in $UBVI$. The goal of this drift-scan survey is to provide multicolor photometry down to $V = 21$ mag for stellar population and star formation history studies.

The OGLE survey concentrates on the bar regions of the Magellanic Clouds and covers them in BVI searching for microlensing events and other signs of variability.

New Views of the Magellanic Clouds
IAU Symposium, Vol. 190, 1999
Y.-H. Chu, N.B. Suntzeff, J.E. Hesser, and D.A. Bohlender, eds.

Stellar Associations in the LMC

E. Kontizas, M. Kontizas[1], D. Gouliermis[1], A. Dapergolas

Astronomical Institute, National Observatory of Athens, P.O. Box 20048, GR-118 10 Athens, Greece

R. Korakitis

National Technical University of Athens, GR-157 80 Zografos, Greece

D.H. Morgan

Royal Observatory of Edinburgh, Scotland EH9 3HJ, UK

Abstract. The stellar associations are defined as loose unbound concentrations of young stars with a bright OB stellar component, mainly located at the most recent star forming regions, representing the smallest units in the hierarchy of stellar systems in galaxies. The definition of the associations is discussed and the main properties of their stellar content are summarized. Using plates taken with the 1.2m UK Schmidt Telescope a method is developed to detect all stellar associations in the LMC and to find their spatial distribution relative to the LMC's recent star formation history. In 40% of the LMC region, the number of "single peak" associations has increased considerably ($\sim 3\times$). A very interesting result is that the faint limit, where the identified associations are revealed, varies by about 3 mag. This indicates not only extinction differences but also the presence of pre-main sequence stars at a wide range of masses.

1. Introduction

The identification and definition of a stellar association is a difficult task. Any loose group of stars within an area where HII regions and bright O and B spectral type stars exist, giving evidence of common origin, was accepted as a stellar association (Morgan et al. 1952). In an extended review, Blaauw (1964) described the properties of the associations, which include a variety of objects with a wide range of sizes, often comprising more than one concentration of stars and, in some cases, connected to a star cluster as a bound part of the association. Their dimensions range from those of ordinary galactic clusters to a few hundred parsecs. However "the matter is moreover complicated by the fact that, in some associations, distinct subsystems can be distinguished".

[1]Department of Physics, University of Athens, GR-157 83 Zografos, Greece

The need for a more explicit definition of a stellar association was pointed out when such objects were searched in nearby galaxies (Hodge 1986; Efremov et al. 1987; Ivanov 1987; Efremov 1995). It was then realized that different authors use different criteria, which give conflicting sizes of a wide range. It has therefore become clear that various authors do not always refer to the same kind of object and a new definition would be necessary to compare the same kind of stellar system in galaxies (Ivanov 1987; Kontizas et al. 1994; Elmegreen & Efremov 1996; Bresolin et al. 1998; Elmegreen & Efremov 1998; Brown 1998).

2. Current Definition of Stellar Associations

Stellar associations are single, unbound concentrations of early-type luminous stars, embedded in the most recent star formation regions in a galaxy. The stellar mass density in these systems is accepted to be $\rho \leq 0.1\ M_{\odot}/pc^3$ (Blaauw 1964; Lada & Lada 1991). Taking into account this definition, we consider the stellar association as a single concentration of stars that can be directly compared to star clusters, either open or globular, providing one extreme end in the various stellar systems in a galaxy. Table 1 illustrates the properties of the main structural parameters of the stellar systems. The open clusters are the intermediate objects and, quite often, they may be confused with stellar associations (Batinelli & Capuzzo-Doltcetta 1991; Brown 1998).

	Gravitational connection	density ρ	No of Stars N	Total Mass M_t	Top end of LF	half mass radius r_h	max size
STELLAR ASSOCIATIONS	unbound (positive energy)	$\rho \leq 0.1 \frac{M_{\odot}}{pc^3}$ ↓	↓	↓	↑	↑	within THE SAME limits
OPEN CLUSTERS	unbound / bound						depends on age ↓ due to disruption and / or relaxation
GLOBULAR CLUSTERS	bound (negative energy)	$\rho \geq 1.0 \frac{M_{\odot}}{pc^3}$					

Table 1. Properties of single peak stellar systems

3. Stellar Associations in the LMC

The first catalogue of OB associations in the LMC was published by Lucke & Hodge (1970, hereafter LH), who listed 122 objects with sizes from 15 to 300 pc and number of stars from a few to 225, down to a limit of V=14.7 mag. They pointed out that the most extended associations had star clusters as their members and those would better be called "star clouds". It was also found that 90% of them coincided with HII regions. Five of those "star clouds" were examined by Kontizas et al. (1994) using star counts and spectral classification of stars, in order to isolate the "single stellar groups".

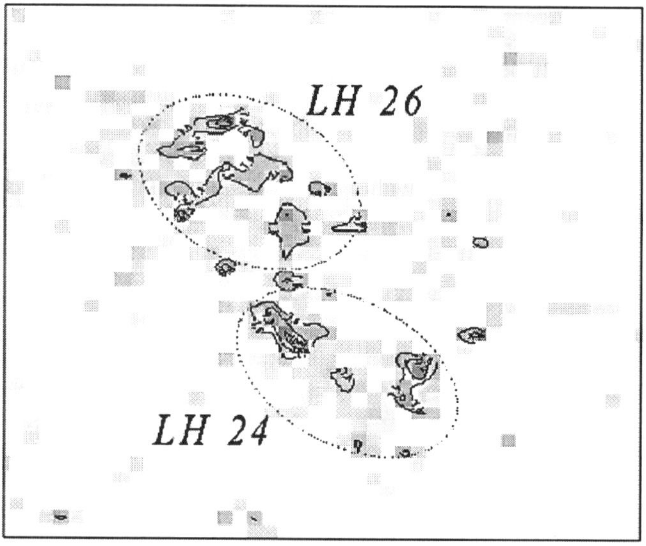

Figure 1. Example of the detection of several associations revealed in the region of the objects LH 24 & 26.

3.1. Stellar Content

Initial Mass Function. The stellar content of the LMC stellar associations has been investigated by several authors. Being young objects, a major part of their stellar content consists of massive stars, so the age of such systems and the IMF become very important sources of information on the recent star formation and the bright end of the luminosity function (LF).

Garmany (1991) gives an extended review of the previous work, whereas Westerlund (1997) summarizes all recent progress in this direction. The IMF shows a slope (γ) from -1.0 to -1.3 (Hill et al. 1994; Massey et al. 1989a,b, 1995), whereas in this meeting Massey (this volume) reports that the slope γ is -1.3 for the massive stars in associations of all three galaxies (SMC, LMC and the Milky Way), with no evidence of upper mass cut off.

Stellar Population - Pre-main Sequence Stars (PMS). The associations have been well-known as the primary hosts of very massive stars but, being young in stellar content, they are expected to contain a significant number of PMS which are not visible in the optical wavelengths. Brandl et al. (1996) have observed a region in the core of 30 Doradus (R136), where they detected 110 extremely red sources which are low- or intermediate-mass PMS candidates. The PMS stars were revealed in R136 by Siriani et al. (this volume) from observations with the HST. They found that the red population in the CMD is well tracked by PMS isochrones and it consists of low-mass stars down to 0.8 $M_{\odot}$. These stars have ages from 1 to 10 Myr, with a flattening from 3 $M_{\odot}$ to 1$M_{\odot}$. HST/NICMOS H-band images (Zinnecker, this volume) have shown that the LF rises at least until H $= 20$ (~ 2 $M_{\odot}$) and there are stars much less massive, down to 0.4 $M_{\odot}$

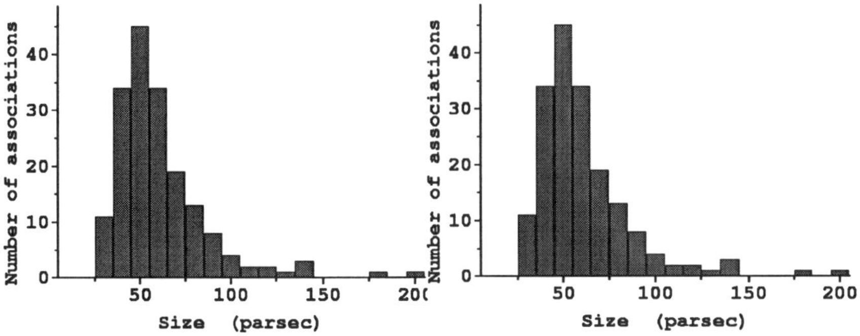

Figure 2. (left) Size distribution and (right) volume density distribution of the detected associations.

in 30 Doradus. There is evidence that PMS stars are identified in this region (Brandner et al., this volume).

Latest observations of the Galactic "associations" and their "subgroups", summarized by Brown (1998), have revealed that the classical membership down to spectral type B8 is corrected and extended to include stars beyond F type, PMS, T-Tauri and evolved F8, K3 and Wolf-Rayet stars.

Age. The age of the associations is difficult to determine because of the uncertainties in the upper end of the luminosity and its transformation to mass. Depending on the way the mass is determined (photometrically or spectroscopically), for the very massive stars it is particularly sensitive to the observations. In any case, the associations are not found older than a few times 10^7 yr. Often, there is evidence for sequential star formation within a group of nearby associations, such as those of the LH58 which were found to exhibit age differences from 4 - 9 My for the various spatial groups of stars (Hill et al. 1994). The region of LH58 is found to comprise four discrete components (associations), with sizes of about 50, 70, 100, and 40 pc respectively, showing four different episodes of star formation in space and most probably in time. 30 Doradus, as observed with HST/WFPC2, revealed that "two-stage starbursts" may be characteristic of the massive star formation on this scale (Rubio et al. 1998).

3.2. Search for "Single" Stellar Associations in the LMC

We have further developed our method (Kontizas et al. 1994) of detecting all "single peak" stellar OB associations in the LMC, from star counts in very large areas of the LMC at two wavebands: U and HeII (4686Å). Digitized photographic plates, taken with the 1.2m UK Schmidt Telescope, have been used and star counts were performed on the image of stars produced by APM and SuperCOSMOS, in order to list all LMC associations according to the current definition.

The magnitudes of the stars detected by the plate scanning procedure were calibrated using as standards the CCD measurements by Will et al. (1996), so that images at various magnitude ranges were produced. Star counts were performed at each magnitude range and statistical tests enabled us to determine the

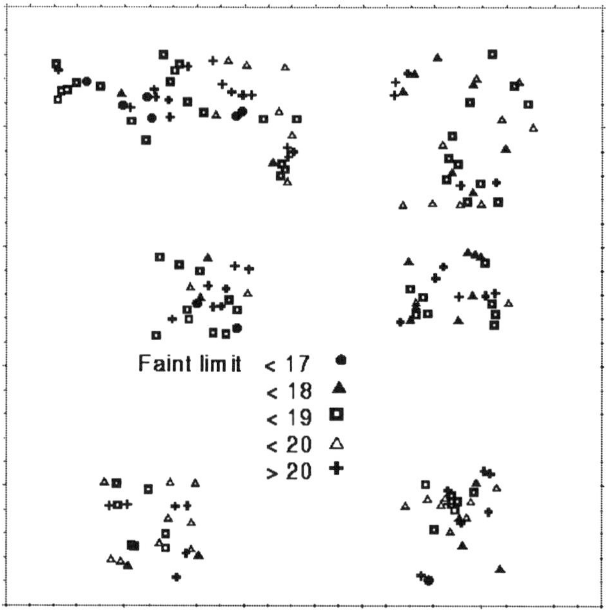

Figure 3. Spatial distribution of the lower detection limit.

background density (number of stars per unit area) and its standard deviation. Using the IRAF & MIDAS packages, regions with a high star concentration were revealed.

Isodensity contours were produced in each area and the ones corresponding to densities $\geq 3\sigma$ illustrate the regions where an enhancement of stars is observed. Such isodensity contours in the various magnitude ranges indicate the groupings of the stars at the corresponding magnitudes. This investigation shows the limits at which these stellar groupings are detected (the upper and lower magnitude limit of detection) and whether the bright stars (since the search is for OB associations) show a distinct concentration compared to the background (Fig. 1).

The size adopted for each association is the largest dimension of the outermost isodensity contour (3σ), as described above. Many associations were revealed, bringing their number much higher than in the LH catalogue. This was expected, since these objects were selected using the new definition of the stellar association. The size distribution of the detected associations is shown in Fig. 2a. This result agrees well with the distribution found by Ivanov (1996) for M31 and M33, and Bresolin et al. (1998) for the nearby galaxies.

The limit of detection of each stellar association at our wavebands showed a wide range of magnitudes, indicating that the low mass stellar component is hidden at different magnitudes, even for the same region on the selected areas (Fig. 3). This effect can be due to extinction but the reddening is never high enough to allow for such a large difference. However, considering that PMS stars are expected in these objects, the difference in the limiting magnitude may indicate the existence of PMS stars of different masses in the various associations.

Finally, the detection of the associations in the various magnitude ranges permits the estimation of a mean mass range of the stars found in each object. A total number of stars and, therefore, a total mass within the half mass radius, is calculated. The volume density of each association (Fig. 2b) estimated by this method, even allowing for the existence of low mass stars and/or PMS stars for the adopted IMF (Massey 1999; Zinnecker 1999), gives strong evidence that these objects are in principle loose and unbound.

Acknowledgments. The authors would like to express their sincere thanks to the 1.2m U.K Schmidt Telescope Unit for loan of the material and the British Council and the General Secretariat of Research and Technology of Greece for financial support of this project.

References

Batinelli, P., & Cappuzzo-Doltcetta, R. 1991, MNRAS, 249, 76.

Blaauw, A. 1964, ARA&A, 2, 213

Brandl, B., et al. 1996, ApJ, 466, 254.

Bresolin, F. , et al. 1998, AJ 116, 119.

Brown, A. 1998, NATO Advanced Study Institute, May 1998, Crete Greece

Efremov, Y. 1995, AJ, 110, 2757

Efremov, Y., Ivanov, G., & Nikolov, N. 1987, A&SS, 135, 119

Elmegreen, B., & Efremov, Y. 1996, ApJ, 466, 802

Garmany, C. 1991, PASP Conf. Ser., vol. 13

Hodge, P. 1986, IAU Symp. No 116, p. 369

Hill, R.J., Madore, B.F., & Freedman, W.L. 1994, ApJS, 91, 583

Ivanov, G. 1987, A&SS 136, 113

Ivanov, G. 1996, A&A, 305, 1990

Kontizas, M., et al. 1994, A&AS, 107,77

Lucke, P., & Hodge, P. 1970, AJ, 75, 171

Massey, P., 1999, this volume

Massey, P., Lang, C.C., Degioia-Eastwood, K., & Garmany, C.D. 1995, ApJ, 438, 188

Massey, P., Parker, J.W., & Garmany, C.D. 1989b, AJ, 98, 1305

Massey, P., Silkey, M., Garmany, C.D., & Degioia-Eastwood, K. 1989a, AJ, 97, 107

Morgan, W.W., Sharpless, S., & Osterbrock, D. 1952, AJ, 57, 3

Rubio, M., Barba, R.H., Walborn, N.R., Probst, R.G., Garcia, J., & Roth, M.R. 1998, AJ, 116, 1708

Will, J-M., Bomans, D.J., & de Boer, K.S. 1995, A&A, 295, 54

Will, J-M., Bomans, D.J., Vallenari, A., Schmidt, J.H.K., & de Boer, K.S. 1996, A&A, 315

Westerlund, B. 1997, "The Magellanic Clouds", Cambridge University Press.
Zinnecker, H. 1999, this volume

Discussion

Philip Massey: Two comments: First, as a user of OB associations I think that a critical test of whether or not an association or group is "physical" is to what extent the member stars are coeval. Although there are a few exceptions, most of the "A1" associations of Lucke and Hodge (1970, AJ, 75, 171) turn out to be highly coeval (Massey et al. 1995, ApJ, 454, 151). You cite the Hill et al. photometric survey of LH58 showing sequential star formation and subclumps of distinctly different ages, but this is simply NOT what we find if we actually take spectra of the bright blue stars and place them on the HRD that way. The cluster was analyzed by Garmany et al. (1994) and rediscussed by Massey et al. (1995, ApJ, 438, 188) along with comparison to the Hill et al. results.

My second comment concerns how reproducable these "subjective" delineations of OB associations are. I've been blessed with a number of excellent undergraduate REU students over the years, and the current one, Liz Waterhouse, is doing *UBV* photometry of some of the Lucke-Hodge associations that contain WR stars. On our $15' \times 15'$ CCD frames I always have her look at the frame and draw where she sees the OB association. So far her outlines agree very well with those in Lucke's PhD thesis (Lucke 1972, U. Washington). So these A1 associatons are easily distinguished and at some level must be real.

Kontizas: I do not disagree totally since we also found that many of LH associations are single objects, so these must be coeval. To have smaller, single "subgroups" does not necessarily mean they are not coeval too. On the other hand in some areas we expect second generation star formation from one group to the other.

Jan Palouš: If the dissolution time of OB associations derives from (the) spatial density of stars, this number should be different from a critical density for the Milky Way. You give a rather short dissolution time, which may be in contradiction (with) the conclusion of Eva Grebel on the unchanging patterns of star clusters over the past 600 Myr.

Kontizas: The clusters as bound systems are not expected to dissolve in time scales of the same order as associations do. The associations are a lot less dense (often 10^3 times less) so their disruption time is much shorter.

Frank Winkler: You mentioned the excellent correlation between Lucke-Hodge associations and HII regions. Have you looked at correlations between your own associations and $H\alpha$ emission?

Kontizas: No. We have not done this yet.

New Views of the Magellanic Clouds
IAU Symposium, Vol. 190, 1999
Y.-H. Chu, N.B. Suntzeff, J.E. Hesser, and D.A. Bohlender, eds.

Deep STIS Luminosity Functions for LMC Clusters

R. Elson, N. Tanvir, G. Gilmore, R.A. Johnson, S. Beaulieu

Institute of Astronomy, Madingley Rd., Cambridge CB3 0HA, England

Abstract. We present deep luminosity functions derived from HST STIS data for three rich LMC clusters (NGC 1805, NGC 1868, and NGC 2209), and for one Galactic globular cluster (NGC 6553). All of the LMC cluster luminosity functions are roughly consistent with a Salpeter IMF or with the solar neighbourhood IMF from Kroupa, Tout & Gilmore (1993). They continue to rise at least to $0.7M_{\odot}$. NGC 1868 shows evidence for mass segregation which may be primordial. A comparison of deep luminsoisty functions for seven Galactic globulars shows that the luminosity functions are eroded at low masses by amounts that are strongly correlated with distance from the Galactic plane.

1. Introduction and Data

The LMC is an ideal laboratory for studying the formation and evolution of rich star clusters. We describe the first results from a 95-orbit Cycle 7 HST project (No. 7307) with this as its primary aim (Elson et al. 1997). Briefly, we are using WFPC2, NICMOS2 and STIS (in imaging mode with the F28X50LP filter) to obtain deep $(V - H)$ CMDs and $\sim R$-band luminosity functions for eight clusters: NGC 1805 and NGC 1818 ($\sim 10^7$ yr), NGC 1831 and NGC 1868 ($\sim 10^8$ yr), NGC 2209 and Hodge 14 ($\sim 10^9$ yr), and NGC 2210 and Hodge 11 ($\sim 10^{10}$ yr). We also have data for the Galactic globular cluster NGC 6553, primarily for calibration purposes.

These data will allow us to determine age spreads in the young clusters and identify pre-main-sequence stars (and thus investigate the timescale and sequence of star formation), to quantify the binary population and trace its evolution and the development of mass segregation. We will also be able to investigate the universality of the IMF, which is the focus of this presentation. NGC 1868 and NGC 2209 are particularly interesting in this regard. They have similar ages but very different core radii, and one possibility is that different IMFs ($x \sim 1$ as opposed to $x \sim 2$, where $x = 1.35$ is the Salpeter value) have caused their cores to expand (through mass loss due to stellar evolution) at different rates (Elson et al. 1989).

We determined magnitudes using PSF fitting, with PSFs constructed from stars in the images. Completeness was determined using artificial star tests, and background contamination using STIS images of a field near each cluster. STIS magnitudes were transformed to absolute magnitudes in the HST system using a zero-point of $K_{STIS} = 23.4$ (H. Ferguson, 1998, private communication), a distance modulus of 18.5, and an absorption $A_{STIS} = 0.18$.

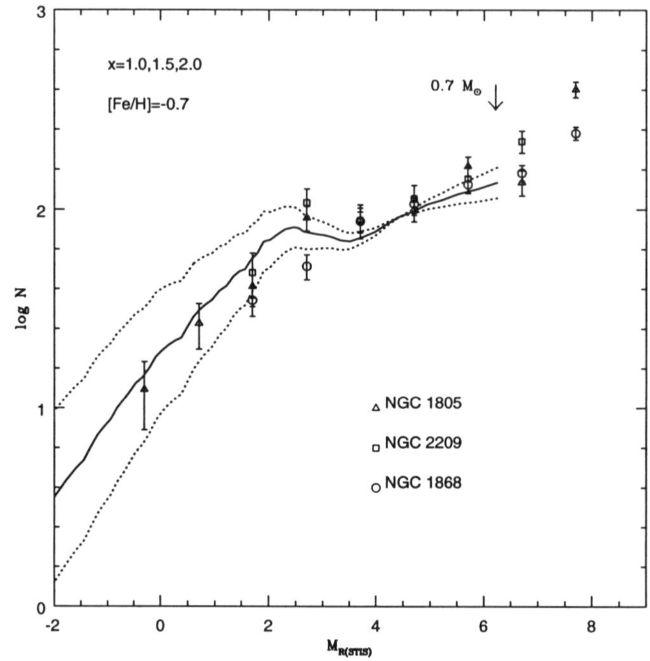

Figure 1. Luminosity functions for three LMC clusters compared to model power-law IMFs with $x = 1.0, 1.5, 2.0$. Arbitrary shifts in $\log N$ have been applied.

2. Results and Discussion

Figure 1 shows luminosity functions for NGC 1805, NGC 1868 and NGC 2209. The LFs are for stars near the half-mass radius where any dynamical evolution should not affect the mass function. All three are similar. Also shown are LFs corresponding to power-law IMFs with slopes $x = 1.0, 1.5, 2.0$. All the luminosity functions continue to rise at masses $< 0.7 M_{\odot}$. Variations in IMF slope do not appear to be responsible for the differences in core radii of NGC 1868 and NGC 2209. The LFs appear to be similar to those of Galactic globular clusters. The rising LF at low masses suggests that the clusters will survive disruption through evaporation, and evolve to objects like the old LMC clusters. They also suggest that 30 Doradus, which is probably a younger counterpart of these clusters, will reveal a similar population of low mass stars (cf. H. Zinnecker, this volume). This is important, as 30 Dor has often been cited as an archetypal starburst, assumed to have an IMF truncated at relatively high mass.

Figure 2 shows luminosity functions for NGC 1868 derived at three different radii. There is clear evidence for mass segregation. N-body modelling, which is an integral part of our project, will indicate to what extent this is primordial or due to dynamical evolution.

Figure 3 shows the LF for NGC 6553 and those of six other Galactic globulars from the literature (Elson et al. 1995; Santiago et al. 1996; Piotto et al. 1997).

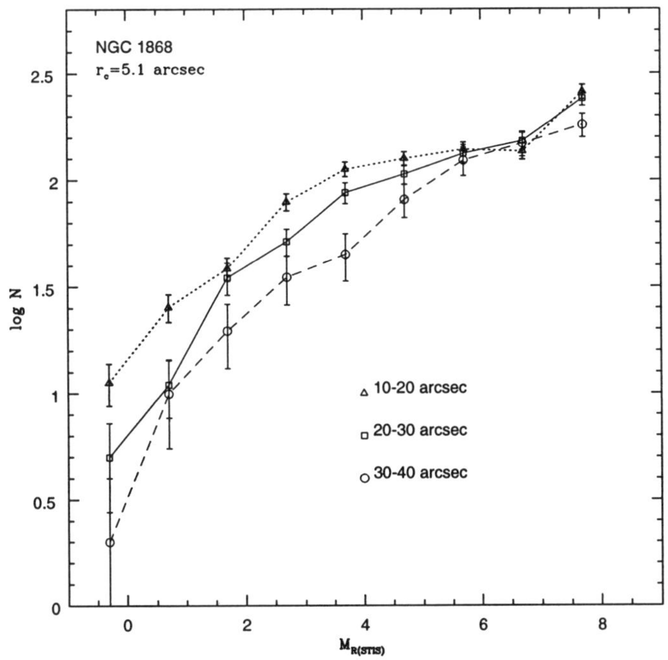

Figure 2. Luminosity functions for NGC 1868 at three different radii.

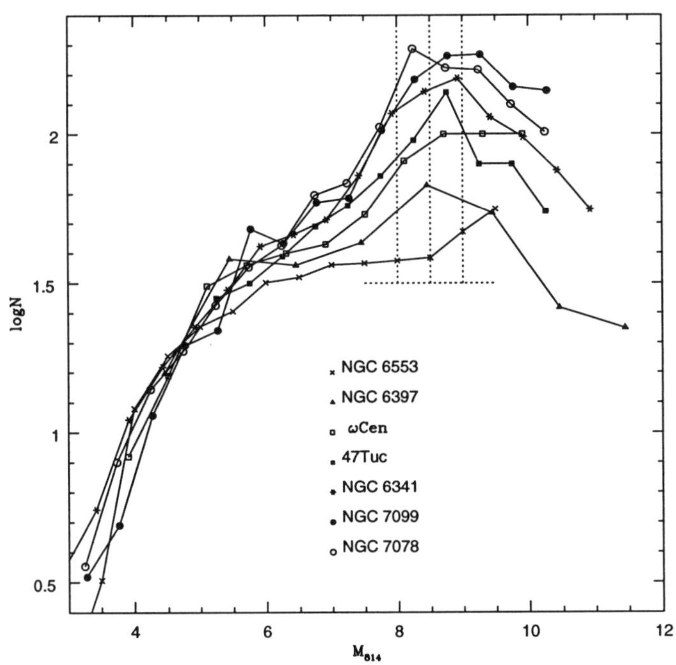

Figure 3. Luminosity functions for 7 Galactic globulars.

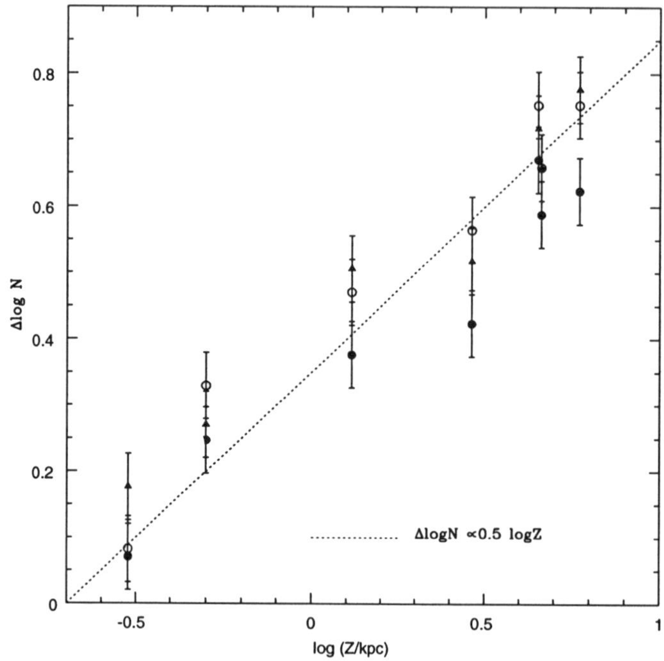

Figure 4. Relative flattening of the faint end of the luminosity functions in Fig. 3, measured at $M_{814} = 8.0, 8.5, 9.0$ (filled circles, open circles, triangles) plotted against distance above the Galactic plane.

A transformation from STIS magnitude to M_{814} was derived from the Padova isochrones (G. Worthey, 1998, private communication). While the luminosity functions agree well at brighter magnitudes, the faint ends differ markedly. Such differences have been noted before, and have been attributed to either metallicity effects, evaporation of low mass stars, or stripping of low mass stars by tidal shocking (cf. Piotto et al. 1997). We quantified the difference among the luminosity functions as the increment $\Delta \log N$ at $M_{814} = 8.0, 8.5, 9.0$. Figure 4 shows $\Delta \log N$ plotted against the distance, Z, of each cluster from the Galactic plane. There is a striking correlation, which suggests that tidal shocking is primarily responsible for the differences among the luminosity functions. A plot of $\Delta \log N$ against metallicity ($-0.3 <$ [Fe/H] < -2.5 for this sample) shows no correlation.

We are awaiting STIS images of four more clusters, NGC 1818, NGC 1831, Hodge 14, and Hodge 11. Also, background data still to be acquired, as well as improved data reduction methods, will allow us to push our luminosity functions $1-2$ mag fainter than at present (and to verify the upturn at $M_{814} > 8.5$ in the LF of NGC 6553).

References

Elson, R., Freemam, K., & Lauer, T. 1989, ApJ, 347, L69

Elson, R., Gilmore, G., Davies, M., Sigurdsson, S., Aarseth, S., & Santiago, B. 1997, STScI Newsletter, 14, 1

Elson, R., Gilmore, G., Santiago, B., & Casertano, S. 1996, AJ, 110, 682

Kroupa, P., Tout, C., & Gilmore, G. 1993, MNRAS, 262, 545

Piotto, G., Cool, A., & King, I. 1997, AJ, 113, 1345

Santiago, B., Elson, R., & Gilmore, G. 1996, MNRAS, 281, 1363

Discussion

John Pritchard: What binary star creation models are you using in your model clusters, and how did you determine the binary star fraction in NGC 1818?

Elson: Our models are N-body models, so at present do not include star formation. We input a binary frequency and distribution of mass ratios, and watch how it evolves.

The binary star frequency was determined from the CMD - there was a binary sequence that stood out clearly from the single star main sequence. We did some simple modelling to estimate the range of mass ratios of the binaries in our sample.

Hans Zinnecker: Congratulations on your very nice HST results! Just one comment: for your youngest cluster (NGC 1805) you need pre-MS tracks to deal with subsolar masses, ce n'est pas?

Elson: Merci! We will need pre-MS tracks to interpret our STIS LF's at the faint end. Also, our NICMOS and WFPC data (V-H CMD's) will allow us to identify pre-MS stars and quantify their numbers and ages.

Jan Palous: This is a comment: I assume that GCs follow radial orbits. It may imply that the individual clusters did suffer from tides to different extents since they come to the bar from various sites.

Tammy Smecker-Hanes: Since you have such striking evidence that the change in the IMF at the low mass end (loss of low mass stars) is a function of scale height, z, and hence is due to disk shocking, are there predictions for changes in cluster shape or concentration, etc., that you could look for also?

Elson: I haven't looked into other correlations, but it would certainly be an interesting thing to do.

New Views of the Magellanic Clouds
IAU Symposium, Vol. 190, 1999
Y.-H. Chu, N.B. Suntzeff, J.E. Hesser, and D.A. Bohlender, eds.

Spontaneous and Induced Star Formation in the LMC

Yu.N. Efremov

Sternberg Astronomical Institute, MSU, Moscow 199899, Russia

B.G. Elmegreen

IBM Watson Researcher Center, Yorktown Heights, NY 10598, USA

Abstract. Spontaneous and triggered star formation in the LMC is discussed with data on star clusters ages and positions. The supershell LMC4 and the stellar arcs in the same region are suggested to be triggered by GRBs, the progenitors of which might have escaped the old elliptical cluster NGC1978, close to which are a number of X-ray binaries and the SGR/SNR N49.

1. Stochastic Star Formation

The LMC is the best site to study large-scale features of star formation as displayed by mutual positions and ages of star clusters. The ages are now available for about 600 clusters.

Efremov & Elmegreen (1998a) demonstrated recently that the larger the mutual distance between these clusters, the larger their age difference. This relation may be considered part of a general correlation between the size of a star formation region and the duration of star formation there. For example, OB subgroups each form in $\sim$ 3 My, the whole OB associations that surround them form in $\sim$ 10 My, and the star complexes, such as Gould's Belt, that include the OB associations, form in $\sim$ 30 My; each scale is larger than the previous by a factor of 10 (e.g., 3 pc, 30 pc, and 300 pc). Thus the duration of star formation scales approximately with the square root of the size of the region in this example (see review in Elmegreen & Efremov 1999).

Such scaling is expected for star formation in turbulent gas, because star formation usually operates on a time scale proportional to the cloud or clump crossing time, and this crossing time scales with the square root of region size. Thus the formation of hierarchical young stellar groups results from hierarchical cloud structure. There is no preferred scale for cluster formation.

2. Triggered Stellar Arcs in the LMC4 Region

The LMC harbours what is often considered to be the classical region of triggered star formation, the LMC4 supershell identified and photographed in Hα by Meaburn (1980). This region is sometimes erroneously called Shapley's Constellation III, but that title should go to the region around the NGC 1974 cluster (see Efremov & Elmegreen 1998b). Three or four gigantic arcs of stars and clusters are known in the LMC4 region (Hodge 1967). The most obvious arc was considered by Westerlund & Mathewson (1966): it is inside the southern part of the supergiant HI hole/shell, found by McGee & Milton (1966), Domgörgen et al. (1995), and Kim et al. (1997). Westerlund & Mathewson (1966) ascribed the origin of the stellar arc and HI features to the outburst of a Super-Supernova, following Shklovsky (1960). Later on, the region was considered the best manifestation of triggered self-propagated star formation (Dopita et al. 1985). However, there are serious difficulties with this interpretation if the brightest stars in the center of the HI ring are considered to be the triggering cluster, because these stars are too young (Olsen et al. 1997; Braun et al. 1997).

Recently, Efremov & Elmegreen (1998b) suggested that two well-shaped arcs in this region formed by triggered star formation in gas that was swept-up by centralized sources of pressure. The strictly circular shapes of both arcs are the strongest evidence for this. Six coeval A I stars near the center of the larger arc (called Quadrant) were suggested to be the remnants of an association, including O-stars, which swepted up the gas in the larger region starting $\sim$ 30 My ago. A small and younger cluster near the center of the smaller arc (Sextant) was shown to be responsible for that one.

These centralized stellar sources of pressure could produce both young stellar arcs at the right time and position, as Efremov & Elmegreen (1998b) demonstrated, yet the general picture is still not satisfactory. The main questions remaining are:

1. Why are there no giant stellar arcs or rings around other, even more rich, clusters in the LMC?

2. Why are all of the stellar arcs in the LMC close to each other in only this place?

3. Why are there just arcs and not full stellar rings?

4. Why are the arcs in the region of the largest and deepest HI hole in the LMC?

The recent identification of Gamma-ray burst (GRB) afterglows in distant galaxies revived the possibility that single super-explosions can produce large shells and trigger star formation (Efremov, Elmegreen & Hodge 1998; Perna & Loeb 1998). There may not even be a central cluster or evidence of an extragalactic cloud impact in the triggered region. Indeed, the arcs in the LMC4 region, and the whole HI supershell as well, could be produced by GRB-like explosions (Efremov, Elmegreen & Hodge, 1998).

This explains question (1) above, yet not the other three. Why would all the peculiar outbursts occur in the same region of the LMC? An answer to this question might come if we consider the common assumption that explosions of some GRBs are the result of merging components of close binaries that include a neutron star, black hole or white dwarf. This hypothesis was recently advanced by Efremov (1998) to account for the stellar arcs and LMC4 supershell; the binaries could have escaped from NGC 1978, which is a massive cluster of intermediate age in the same area of the LMC, up and right of $2''$ within the circle in Figure 1 by Efremov & Elmegreen (1998b).

3. NGC1978 as the Origin of GRB Progenitors

The high rate of occurrence of X-ray binaries (with one component a neutron star) inside dense globular clusters is well known (e.g., Baylin 1996). It was explained long ago as consequence of the high probability of formation of close binaries after tidal captures in the dense cluster (e.g., Shklovsky 1982; Davies 1995). It was also shown (McMillan 1986) that a large number of tidally captured binaries may escape a dense old cluster as the result of three-body collisions. Recently Hanson & Murali (1998) suggested that stellar encounters in globular clusters were able to produce not only millisecond pulsars, but also binaries that evolve into GRBs.

The formation of a neutron star after a SN explosion in a binary system leads to a high recoil velocity, the most likely value of which is 150 - 200 km^{-1} (Lipunov et al. 1997). Such high velocities would spread out any future GRB over very large distances around the paternal cluster. Even smaller velocities would disperse the GRB progenitors significantly, because the binaries may take 100 My before they merge to give a GRB (Lipunov et al. 1997). We therefore suggest that the relics of GRB might be observed in the kiloparsec-scale regions surrounding old dense clusters.

NGC 1978 is a rich and old cluster, though much younger than classical globulars in the Milky Way and the LMC. Its age is about 2×10^9 (Bomans et al. 1995), and it is the richest such cluster in the LMC. Indeed, it has a few hundred red giants with masses of around 1 $M_\odot$!

NGC 1978 is also unusual in its extremely flattened shape (Geisler & Hodge 1980). This may indicate a formation process involving the merger of two clusters, especially because no rotation has been detected (Fisher et al. 1992). Besides the shape, there is no other evidence for merging, yet Kravtsov (1999) found some signs of abundance differences in two parts of the cluster. The process of merging surely increases the probability of stellar encounters, leading both to the formation of close binaries and to the escape of many stars from the resulting cluster (e.g., de Oliveira et al. 1999). The observation of a large number of blue stragglers in NGC 1978 (Cole et al. 1997) might also indicate a high rate of stellar encounters there.

There are also other objects near NGC 1978 that are binary stars with a compact component and are therefore related to GRB progenitors if they are not progenitors themselves. Three X-ray binaries are within 20′ of NGC 1978 and more are in a wider surrounding, as is evident from Haberl et al. (1999). These objects are ascribed mostly to high-mass binaries, yet some classified as Be/X-ray binaries might contain white dwarfs (see Neguerula 1998).

Moreover, the famous GRB of March 5, 1979 is also at about 20′ to the NW of the cluster. It is now known as the Soft Gamma Ray (SGR) repeater SGR 0526-66, and it is an X-ray binary, i.e., a stellar remnant of a SN inside the young remnant N49 (Danner et al. 1998).

This is consistent with the conclusion of Nakamura (1998) that low energy GRBs leave behind SGR repeaters. Nakamura explained in this way the properties of SN1998bw (of b/c type), which was a bright Supernova and also a GRB with low energy (Woosley et al. 1999). All three well-studied SGRs are indeed connected with young SNRs (Kouvelitou et al. 1998). The connection between SGRs and GRBs was recently suggested by Wang & Wheeler (1998).

We propose that the outburst that produced SGR 0526-66 plus SNR N49 was similar to the SN1998bw/GRB980425 event, and also that events like these produced the stellar arcs near LMC4. Indeed, the Quadrant and Sextant arcs might be produced by the stellar winds and supernovae from one to five dozen O-type stars (Efremov & Elmegreen 1998b), which corresponds approximately to 10 to 50 common SNs. The energy from this is comparable to or a bit below that of common GRBs. An energy 10-100 times higher was necessary to form the entire LMC4 HI hole; this is within the energy range observed for GRBs. We noted that a second, rich, intermediate-age cluster in the LMC, NGC1806, is inside the HII supershell LMC6, in the region of low HI density.

4. Conclusions

Stellar arcs near LMC4, and the LMC4 HI hole itself, might have been produced by GRBs whose progenitors originated in the intermediate-age, nearby, globular cluster NGC 1978. GRBs like this may occur anywhere in a galaxy near such a dense cluster, whereas sequential SNs in young clusters might only occur in the spiral arms and near the planes of galactic disks. Indeed, the whole LMC4 region and other similar patches of star formation with giant stellar arcs (as in M83 and NGC6946 – see Efremov, Elmegreen & Hodge 1998) might be produced by GRBs. The arcs might be only parts of circles (e.g., Hodge 1967; Efremov 1998) because the GRBs explode outside the galactic plane. The abundance of SNs in both M83 and NGC 6946 might be one more indication of a connection between certain SNs and GRBs.

One important issue that is still unresolved is why other dense globular clusters in the LMC and elsewhere have no similar concentrations of GRB relicts. For example, Ciardullo et al. (1990) suggested that the large number of X-ray binaries in the bulge of M31 might be the result of ejection from the globular clusters there. The unique properties of NGC 1978 that might have led to the observed stellar arcs are its elliptical shape (suggesting a merger) and its high abundance of red giants (which are more massive stars than in classical globulars). These properties are plausibly connected with the occurrence of surrounding GRB relics and X-ray binaries. Perhaps the compact binaries ejected from this cluster evolve not only to X-rays sources, but to SN/SGR and GRBs as well.

References

Baylin, C.D. 1996, ASP Conf. Ser., 90, 320

Bomans, D.J., Vallenari, A., & de Boer, K.S. 1995, A&A, 298, 427

Braun, J.M., Bomans, D.J., Will, J.-M., & de Boer, K.S. 1997, A&A, 328, 167

Ciardullo, R., Tamblyn, P., & Phillips, A.C. 1990, PASP, 102, 1113

Cole, A.A., et al. 1997, AJ, 114, 1945

Danner, R., Kulkarni, S.R., & Trumper, J. 1998, BAAS, 192, No. 43.09

de Oliveira, M., Dottori, H., & Bica, E. 1999, this volume
Domgörgen, H., Bomans, D.J., & de Boer, K.S. 1995, A&A, 296, 523
Davies, M. 1995, MNRAS, 276, 887
Dopita, M.A., Mathewson, D.S., & Ford, V.L. 1985, ApJ, 297, 599
Efremov, Yu.N. 1999, Pis'ma Astron. Zh., in press
Efremov, Yu.N., & Elmegreen, B.G. 1998a, MNRAS, 299, 588
Efremov, Yu.N., & Elmegreen, B.G. 1998b, MNRAS, 299, 643
Efremov, Yu.N., Elmegreen, B.G., & Hodge P.W. 1998, ApJ, 501, L163
Elmegreen, B.G., & Efremov, Yu.N. 1999, in The Orion Complex Revisited, ed. M. J. McCaughrean & A. Burkert, ASP Conference Series, 1999, in press.
Geisler, D., & Hodge, P. 1980, ApJ, 242, 66
Haberl, F., Pietsch, W., & Filipovic, M. 1999, this volume
Meaburn, J. 1980, MNRAS, 192, 365
Hanson, B.M.S., & Murali, C. 1998, ApJ, 505, L15
Hodge, P.W. 1967, PASP, 79, 29
Kim, S., Staveley-Smith, L., Sault, R.J., Kesteven, M.J., McConnell, D., & Freeman, K.C. 1997, PASA, 14, 119
Kravtsov, V. 1999, this volume
Kouvelitou, C. et al. 1998, Nature, 393, 235
Lipunov, V.M., Postnov, K.A., & Prokhorov, M.E. 1997, MNRAS, 288, 245
Loeb, A., & Perna, R. 1998, ApJ, 503, L35
MacMillan, S.L. 1986, ApJ, 307, 126
McGee, R.X., & Milton, J.A. 1966, Aust. J. Phys., 19, 343
Nakamura, T. 1998, submitted to Prog. Theor. Phys. (astro-ph/9807087)
Negueruela, I. 1998, A&A, 338, 505
Olsen, K.A.G., Hodge, P.W., Wilcots, E.M., & Pastwick, L. 1997, ApJ, 475, 545
Shklovsky, I.S. 1960, Soviet Astron. - AJ, 4, 355
Shklovsky, I.S. 1982, Comments on Astrophys., 9, 261
Wang, L., & Wheeler, J.C. 1998, ApJ, 504, L87
Westerlund, B.E., & Mathewson, D.S. 1966, MNRAS, 131, 371
Woosley, S.E., Eastman, R.G., & Schmidt, B.P. 1999, ApJ, 516, 788

Discussion

Lance Gardiner: There is evidence of asymmetrical spiral structure in the LMC from the analysis of the spatial distribution of young star clusters by Doffer et al. Numerical simulations have shown that an asymmetric spiral structure can be induced by an off center bar, as in the LMC. Is the hierarchical distribution of young stars you found consistent with the existence of spiral structure in the LMC?

Efremov: The star-forming spiral arms are just the highest level in the hierarchy of star-forming regions, so if they exist in the LMC, our conclusions on the hierarchical star formation are not affected. The density wave arms do surely not exist in the LMC, and even the sheared spiral fragments are seen probably only in the outer HI disk, as Kim's data revealed.

Nolan Walborn: Do you think a configuration of alignments such as you propose would be maintained for 10^8 years?

Efremov: The supergiant HI shells may live up to 100 Myrs or so if there is no shear and the HI disk thick enough. This is why these supershells are found usually in irregular galaxies and in the outer parts of spiral galaxies. As concerns stellar arcs, the dynamical evaluations of their lifetime could be of great use. Surely the arcs cannot live long, the Quadrant arc is already more dispersed than the Sextant arc (ages are ≈ 15 and ≈ 7 Myrs respectively).

During the banquet, Rebecca Elson, Jay Gallagher, and Bengt and Vivi Westerlund contemplate whether Hans Zinnecker wears his World Cup cap while sleeping, too.

Part 6. Stellar Clusters

Section B. Poster Papers

Tammy Smecker-Hane and Jay Gallagher enjoy a moment following the Symposium banquet.

New Views of the Magellanic Clouds
IAU Symposium, Vol. 190, 1999
Y.-H. Chu, N.B. Suntzeff, J.E. Hesser, and D.A. Bohlender, eds.

Ages and Metallicities of LMC Clusters and Their Surrounding Fields

Doug Geisler

Kitt Peak National Observatory, NOAO, Tucson AZ, 85719 USA

Eduardo Bica and Horacio Dottori

Instituto de Física, UFRGS, Porto Alegre, RS, BRAZIL

Juan J. Clariá and Andrés E. Piatti

Observatorio Astronómico, Córdoba, ARGENTINA

João F.C. Santos Jr.

Dpto. de Física, UFMG, Belo Horizonte, MG, BRAZIL

Abstract. We report on a survey of 25 candidate old LMC clusters. Washington photometry was obtained on the CTIO 0.9m in $< 1^h$, reaching below the turnoff. Ages based on the magnitude difference δT_1 between the giant branch clump and the turnoff revealed that no new old clusters were found. The candidates all turned out to be of intermediate age (1-3 Gyr). We confirm that there was apparently no cluster formation in the LMC from 3-9 Gyr ago, and that there was a pronounced epoch of cluster formation beginning 3 Gyrs ago that peaked at $\sim$ 1.5 Gyrs ago. We also determine ages for the surrounding fields, as well as metallicities for both the clusters and fields from the color of the giant branch. In most cases the stellar population of each cluster is quite similar to that of its field. The mean metallicity for the intermediate age outer disk clusters is -0.65. A few clusters stand out in the age–metallicity relation in the sense that they are intermediate age clusters at relatively low metallicity. In the northern part of the LMC disk 3 fields all have a secondary clump $\sim$0.45 mag fainter than the dominant clump, suggesting a component located behind the LMC at a distance comparable to that of the SMC.

1. Introduction

There are only a handful of known LMC clusters that are genuinely old, similar to the globular clusters (GCs) in the Milky Way. LMC clusters show a pronounced age gap between a large number of intermediate age clusters (age $\sim 1-3$ Gyr, hereafter IACs) and the classical GCs, with ages $>$ 12 Gyr (Olszewski et al. 1991), with the sole exception of ESO121-SC03 with an age of $\sim$ 9 Gyr. This prevents us from using clusters to tell us any details about chemical evolution in the LMC over most of its history, despite our ability to determine accurate ages and metallicities for them. Finding even a single additional cluster in this age gap, or more old GCs, would significantly enhance our knowledge of the chemical evolution in the LMC. Several studies have uncovered a number of interesting old cluster candidates. Our primary source was the catalog of integrated UBV photometry for 624 clusters by Bica et al. (1996). We obtained Washington photometry for the 25 best candidate old clusters from this study.

2. Results

Our results have been published in Geisler et al. (1997) and Bica et al. (1998). Our CMD for ESO121-SC03 reaches well below the turnoff in this ≈ 9Gyr cluster and therefore shows that our technique can easily distinguish between IACs and older clusters. The magnitude difference (δT_1) between the giant branch clump in IACs (the horizontal branch in GCs) and the turnoff is well correlated with age, and is independent of reddening and distance.

Our intriguing result is that all the old cluster candidates turned out to be IACs!. This confirms with a larger and more homogeneous dataset previous indications that the LMC apparently did not form clusters between 3–9 Gyrs ago, a period covering $\sim 1/2$ of its life, and that it only formed a single (surviving) cluster between 3 Gyrs and the epoch of formation of the classical GCs. We have also found additional evidence that the onset of IAC formation ocurred at 3 Gyr, with a peak at ~ 1.5 Gyrs.

Da Costa and Armandroff (1990) showed the utility of the $(V - I)$ color of the giant branch for measuring metallicities in old stellar populations. Geisler and Sarajedini (1999) introduced a similar technique using the $(C - T_1)$ color. This technique **has three times the metallicity sensitivity of** $(V - I)$. We have used their calibration (corrected for age effects) to derive metallicities for both clusters and their fields, as well as their age-metallcity relation.

Three of our fields present a unique feature: a populated secondary clump ≈0.45 mag fainter than the prominent main clump seen in all other fields. Intriguingly, these 3 fields are all located within about 1° of each other! This feature might be interpreted as a depth structure consisting of a layer of stars located approximately 10 Kpc behind the LMC, at a distance comparable to that of the SMC, or perhaps a dwarf companion to the LMC. More studies are needed to investigate this intriguing feature.

References

Bica, E., et al. 1996, ApJS, 102, 57

Bica, E., et al. 1998, AJ, 116, 723

Da Costa, G.S., & Armandroff, T.E. 1990 AJ, 100, 162

Geisler, D., et al. 1997, AJ, 114 1920

Geisler, D., & Sarajedini, A. 1999, AJ, 117, 308

Olszewski, E.W., et al. 1991, AJ, 101, 515

Discussion

David Graff: Do you have a program to map out the velocities of this area? What is the density of stars in your field?

Geisler: Yes, radial velocities wil be very important to understand the origin of this phenomenon. I would guess we roughly have 5,000-10,000 stars in our $13.6' \times 13.6'$ fields.

Despina Hatzidimitriou: Do you have a rough estimate of the number of old red HB stars compared to the total number of red clump stars?

Geisler: We are currently assessing such values. Although not yet available in detail, the ratio in these dual clump fields is very small (maybe 10%).

New Views of the Magellanic Clouds
IAU Symposium, Vol. 190, 1999
Y.-H. Chu, N.B. Suntzeff, J.E. Hesser, and D.A. Bohlender, eds.

Nonuniform Star Population of the LMC Cluster NGC 1978?

V. Kravtsov

Sternberg Astron. Inst., 13, University Ave., Moscow 119899, Russia

Abstract.
The LMC cluster NGC 1978 shows different properties in the two parts divided by its minor axis. These two parts have different color distributions of the core He-burning (CHB) stars and giant branch (GB) positions on the CMDs at and below the level of the CHB stars. These differences may be caused by spread in metallicity.

1. Introduction

The LMC shows the presence of pairs and even trios of clusters (Bhatia et al. 1991; Bica et al. 1992). A study of such objects is important to achieve understanding of the formation and evolution of star clusters. Binary nature of the LMC cluster NGC 1978 is one of the explanations of its high ellipticity (Fischer et al. 1992). We pay attention to a possible nonuniformity of its star population (see Alcaino et al. (1999) for details).

2. Results and Conclusions

Quadrants A and B (with $a_{int} \approx 0'.8$ (150 pixels) and $a_{ext} \approx 3'.3$; $e = (a - b)/a$ is equal to 0.3) separated by the cluster's minor axis are to the northeast of the cluster's major axis. They belong to the cluster's northwestern half (NWH) and the southeastern half (SEH), respectively. Figure 1 shows that the color distributions of CHB stars for quadrant B are more symmetrical, with a maximum near their centers, whereas the distributions for quadrant A have an excess at their red ends compared to B. These features are also traced on corresponding color distributions of the whole sample of the CHB stars. Note that the visible position of the lower GB on CMDs of the cluster's SEH seem to be systematically bluer than that of the cluster's NWH. The differences discussed *are apparent for colors formed using B magnitudes, but are practically absent for* $V - I$.

The color-position diagram (CPD) shows the above-mentioned differences on the Figure 2. In the color range $0.80 < B - V < 1.05$ most of the CHB and the lower GB stars (upper gians are excluded) with $X < 0$ (SEH) are to the left of the vertical reference line in contrast to position of such stars having $X > 0$ (NWH). The maximum difference in $B - V$ color between them may be as large as $0^m.05 - 0^m.07$.

The isochrones by Bertelli et al. (1994) with values of age and [Fe/H] which are close to those of NGC 1978, show that the permitted age spread of a few

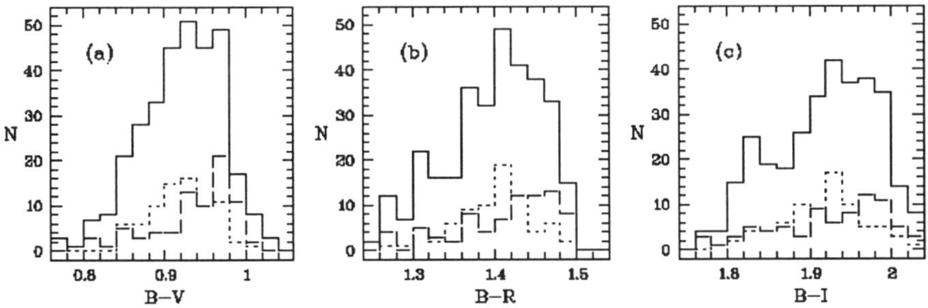

Figure 1. The color distributions (a) $B-V$, (b) $B-R$, and (c) $B-I$ of the CHB stars in different zones of NGC 1978: the dashed line for quadrant A (78 stars), the dotted line for quadrant B (77 stars), and the solid line for the whole ($5' \times 5'$) cluster field (319 stars).

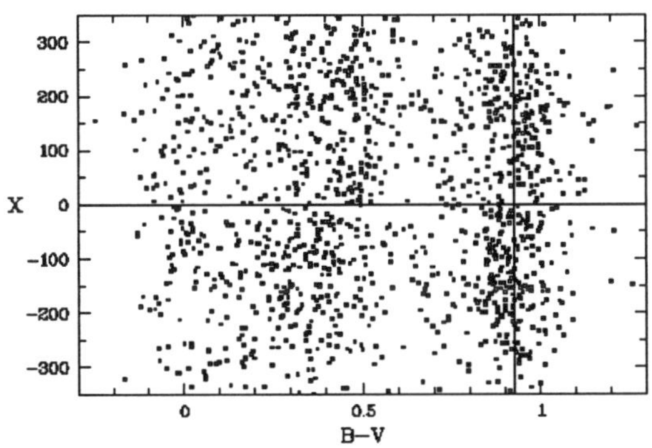

Figure 2. The CPD presents the plot of color $B-V$ versus position (X, in pixels) of stars along the major axis of NGC 1978. Positive and negative values of X correspond to NWH and SEH of the cluster. Stars near the vertical reference line are the CHB stars and the lower giants.

tenths of a Gyr (there is no obvious variation of the turnoff magnitude) cannot cause the differences discussed. A metallicity spread leads to a scatter of the *lower giants* and of the CHB stars in $B-V$ to be greater than that in $V-I$ by a factor of ~ 2.0. The difference in metallicity between stars in NWH and SEH of NGC 1978 may be as large as Δ[Fe/H] $\sim 0.1-0.2$.

References

Alcaino, G., Liller, W., Alvarado, F., et al. 1999, A&AS, 135, 103

Bertelli, G., Bressan, A., Chiosi, C., et al. 1994, A&AS, 106, 275

Bhatia, R. K., Read, M. A., Hatzidimitriou, D., et al. 1991, A&AS, 87, 335

Bica, E., Claria, J. J., & Dottori, H. 1992, AJ, 103, 1859

Fischer, P., Welch, D. L., & Mateo, M. 1992, AJ, 104, 1086

New Views of the Magellanic Clouds
IAU Symposium, Vol. 190, 1999
Y.-H. Chu, N.B. Suntzeff, J.E. Hesser, and D.A. Bohlender, eds.

Young Populous Clusters in the Magellanic Clouds - Implications for Stellar Evolutionary Models

Stefan C. Keller, Michael S. Bessell, and Gary Da Costa

Mount Stromlo and Siding Spring Observatories, Australia.

Abstract. We present new UV, visual and Hα photometry obtained with the WFPC2 of NGC 330, NGC 1818, NGC 2004 and NGC 2100, four young populous clusters in the MCs. We present observational evidence for a degree of convective core overshoot in excess of that currently applied in standard models.

1. Introduction

One of the most important and actively debated issues in stellar evolution is that of the presence and amount of extension of the convective core due to internal mixing. A sizable number of studies have revealed the need for some degree of core extension (Langer & Heger, in this volume). The present study focuses on four young populous clusters in the mass range 10–20 $M_\odot$.

2. Discussion

We have imaged the clusters in F160BW, F555W and F656N. The F160BW-F555W color provides an excellent temperature discriminator for the bright MS members. The clusters are known to be rich in Be stars (Keller et al. in preparation), given the discordant photometric characteristics of Be stars we obtained a F656N image to distinguish those stars with significant Hα emission. Further details can be found in Keller et al. (1999, A&AS, 134, 489).

It is within the mass range of 10–20 $M_\odot$ that the effects of internal mixing are most apparent as evolution progresses. The mechanism for core extension that we consider here is that brought about by convective core overshoot (CCO). CCO brings more H to the central core, the MS lifetime is increased and the star evolves further to the red and is more luminous. Subsequent evolution occurs at a more rapid pace, that is the ratio of core H to core He burning lifetimes (t_H/t_{He}) is reduced. It is only through observational studies that the degree of overshoot within theoretical models can be constrained. To ascertain the degree of CCO present we compare our sample to that produced synthetically from SEMs incorporating varying degrees of CCO. We test how well the models match the data in two ways: by examination of the normalized integrated luminosity function (nILF) and the morphology of the MS.

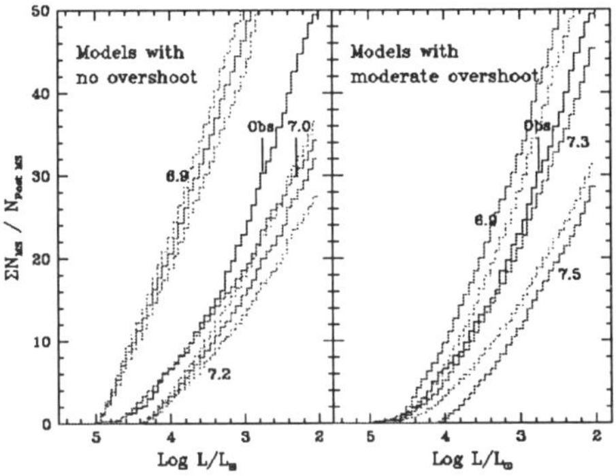

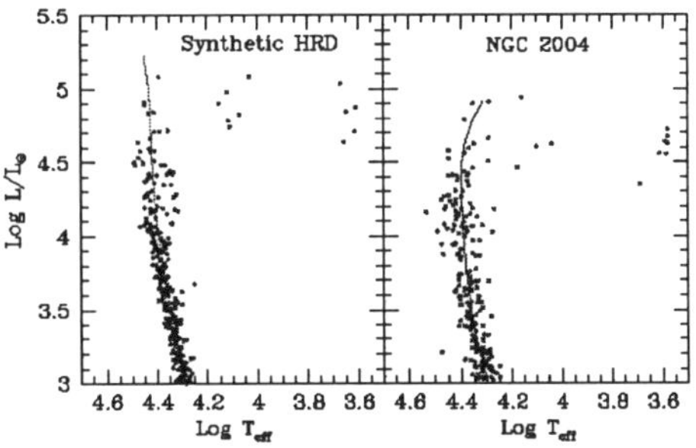

Left - Figure 1. The left panel shows the nILF for NGC 2004 and models without overshoot. The dotted bounding lines represent 1σ uncertainties estimated from repeated trials. Ditto in the right panel from Padua models.
Right - Figure 2. The left panel shows the population synthesized from the best fitting model of Fig. 1a. The curve represents cubic spline fit derived from repeated trials. The right panel shows the observed HR diagram for NGC 2004.

Figure 1 shows the ILF for MS stars normalized to the number of post-MS stars in NGC 2004. The nILF reflects the t_H/t_{He} which is a function of the amount of CCO. Overlaid in figure 1 are the nILF predicted from models of various ages containing no overshoot. All fail to match the observations. These models predict too many evolved stars from the number of MS progenitors, i.e. t_H/t_{He} of no overshoot models is too high to match the observations. nILFs derived from the "moderate" overshoot SEM of the Padova and Geneva groups present a much better fit to the observed nILF with the omission of the brightest portion. To achieve a better match with the bright tail of the nILF we include a slight age spread. This tells us that the t_H/t_{He} of these models is appropriate for the population.

As we have accurate T_{eff} for these stars we examine the morphology of the MS band. We see that the model with moderate overshoot and age spread described above fails to match the shape of the MS band. The inclusion of an age spread produces the observed number of stars at high luminosities but at higher temperatures than observed. The observations show a distinct turnover to the red consistent with extended MS evolution. The amount of extension of the MS to the red is seen to increase with increasing turnoff mass amongst the sequence of four clusters.

3. Conclusions

The present study has highlighted a number of problems for standard SEMs. We find strong evidence for a degree of CCO in excess of that incorporated in standard models. In addition, increasing amounts of CCO are required in higher mass stars.

New Views of the Magellanic Clouds
IAU Symposium, Vol. 190, 1999
Y.-H. Chu, N.B. Suntzeff, J.E. Hesser, and D.A. Bohlender, eds.

Outer Halo Star Clusters: Comparisons for Cloud Clusters

J.E. Hesser, P.B. Stetson, S. van den Bergh

National Research Council of Canada, Herzberg Inst. of Astrophysics, Dominion Astrophysical Obs., Victoria, BC V8X 4M6, Canada

M. Bolte, J.A. Johnson

U. of California, UCO/Lick Observatory, Santa Cruz, CA 95064, USA

W.E. Harris

McMaster University, Department of Physics and Astronomy, Hamilton, ON L8S 4M1, Canada

D.A. VandenBerg

University of Victoria, Department of Physics and Astronomy, Victoria, BC V8W 3P6, Canada

R.A. Bell

University of Maryland, Department of Astronomy, College Park, MD 0742-2421, USA

H.E. Bond, L.E. Fullton

STScI, 3700 San Martin Dr., Baltimore, MD 21218, USA

G.G. Fahlman, H.B. Richer

University of British Columbia, Department of Physics and Astronomy, Vancouver, BC V6T 1Z4, Canada

Abstract. Color-magnitude diagrams reaching from the giant branches to $V_{\rm lim} \sim 27.0$, or about three magnitudes fainter than the turnoff, have been obtained in V and I with WFPC2 on HST for NGC 2419, Pal 3, Pal 4 and Eridanus, whose relative ages are discussed.

1. Formation of the Milky Way Halo

The ages, chemistry and other properties of outer Galactic halo star clusters make them powerful objects for comparison with the oldest star clusters of the Magellanic Clouds (e.g., see the individual papers by Geisler, Mighell, Da Costa, Olsen, and Johnson in these proceedings) and other Local Group galaxies. They also constrain the chronology and duration of early cluster formation by contributing to understanding the relative roles of rapid collapse and multi-Gigayear accumulation of many fragments. We seek to understand if the more remote globular clusters with unusually red horizontal branches (HBs) for their [Fe/H] — the "second-parameter" clusters — are systematically younger than inner-halo objects, where the second-parameter phenomenon is weaker. Direct age measurements can be made only with color-magnitude diagram (CMD) photometry reaching well below the main sequence turnoff, thus avoiding dependence upon HB morphology, which is sensitive to many parameters besides age.

WFPC2 F555W ('V') and F814W ('I') photometry processed through the suite of DAOPHOT/ALLSTAR/ALLFRAME codes cleanly defines CMDs for the following clusters from their giant and HBs down through their upper main sequences: Pal 3: ($M_V = -5.2$, $R_\odot \sim 87$ kpc), [Fe/H] ~ -1.6; Pal 4: ($M_V = -5.8$, $R_\odot \sim 98$ kpc), [Fe/H] ~ -1.3; Eridanus: ($M_V = -4.8$, $R_\odot \sim 78$ kpc), [Fe/H] ~ -1.4; and NGC 2419: ($M_V = -9.5$, $R_\odot \sim 88$ kpc), [Fe/H] ~ -2.1. Each cluster exhibits a population of blue stragglers. Pal 3 contains several RR Lyrae variables.

2. Relative Ages

Accurate knowledge of abundances, and abundance ratios, is essential for both relative and absolute age determinations. Unfortunately, even in nearby Galactic globular clusters (whose giants are sufficiently bright for high-dispersion analyses), there is considerable dispersion among studies. For the outer-halo objects, abundance information is qualitatively different, as it is based upon low-dispersion spectra (often only of the calcium triplet in a few giants) and various indirect photometric indicators.

NGC 2419 and the nearby cluster M92 have indistinguishable ages, providing there are no unusual ratios of abundances (e.g., [α/Fe]) in either of these clusters (Harris et al. 1997). Thus, the earliest star (or globular cluster) formation began at essentially the same time everywhere in the Galactic halo throughout a region now almost 200 kpc in diameter.

Two relative age indicators, $\Delta V_{\rm TO}^{\rm HB}$ and $\Delta(V - I)$, suggest that all three second-parameter clusters are younger than their nearby counterparts by 1.5 to 2 Gyr (Stetson et al. 1999), which is in the same sense and approximately the same size predicted by HB star modeling. However, if we ask what other parameters would have to be different for the "second parameter" clusters to have the same ages as the inner halo clusters, we find that an overestimation of the former's abundances by only 0.3 dex (either in [Fe/H] or [α/Fe] would be required. If Pal 3, Pal 4 and Eri were to have primordial abundances of [α/Fe] ~ 0.0, as Brown et al. (1997) found for Rup 106 and Pal 12, that would erase half of the inferred age difference between them and the inner halo clusters.

References

Brown, J.A., Wallerstein, G., & Zucker, D. 1997, AJ, 114, 180

Harris, W.E., et al. 1997, AJ, 114, 1030

Stetson, P.B., et al. 1999, AJ, 117, 247

New Views of the Magellanic Clouds
IAU Symposium, Vol. 190, 1999
Y.-H. Chu, N.B. Suntzeff, J.E. Hesser, and D.A. Bohlender, eds.

Binary Clusters in the Magellanic Clouds

A. Dieball

Sternwarte der Universität, Auf dem Hügel 71, 53121 Bonn, Germany

E. K. Grebel

UCO/Lick Observatory, UCSC, Santa Cruz, CA 95064, USA

1. Introduction

Binary clusters account for more than 10% of the cluster population in the Magellanic Clouds. Statistically fewer than 50% of the found pairs are expected to be chance superpositions. We are studying binary cluster candidates in the Magellanic Clouds to investigate whether the cluster pairs are of common origin and if they show evidence for interaction. We determine ages for the cluster components through isochrone fitting (isochrone models based on Schaerer et al. 1993, and Bertelli et al. 1994) to the colour magnitude diagrams (CMDs). We investigate the stellar spatial density in the surroundings of the clusters to look for signs of possible interactions between the components.

2. The Large Magellanic Cloud

We present first results for three cluster pairs which are located in the bar of the LMC. The data were obtained with EFOSC 2 at the ESO/MPI 2.2-m telescope at La Silla using Washington and Hα filters.

SL 353 & SL 349 are located in the outer northwestern part of the LMC bar. Ages from isochrone fits to the core-He-burning giants of the two clusters are slightly higher than main-sequence-turnoff ages. We adopt a mean age of 520 ± 80 Myr for both clusters which is in agreement with Vallenari et al. (1998). Radial velocities are similar for both clusters and different from the surrounding field. This strongly supports that this cluster pair was formed at the same time from the same giant molecular cloud (GMC).

SL 385 & SL 387 are located in the eastern part of the LMC bar. The best fitting isochrones result in ages of 220 ± 40 Myr for SL 385 and 280 ± 60 Myr for SL 387. Our data, however, do not support an age difference as large as suggested by Vallenari et al. (1998) (150 and 500 Myr). The ages of both clusters are sufficiently similar to agree with near-simultaneous formation but are no evidence whether this cluster pair is a binary cluster or not.

NGC 1971 & NGC 1972 & NGC 1969 are located close to the geometrical LMC centre in the western part of the bar. A curved bridge of enhanced stellar density connects NGC 1971 and NGC 1972. We do not see such sign of possible interaction towards NGC 1969. Isochrone fits result in similar ages for all three clusters: 63 ± 8 Myr for both NGC 1971 and NGC 1969, and 56 ± 6

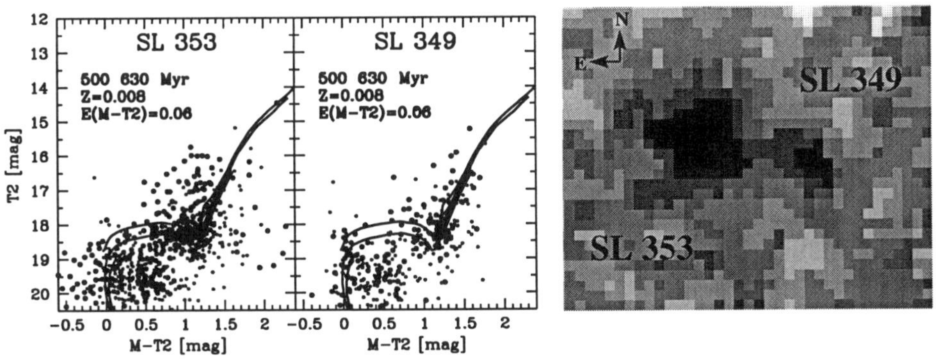

Figure 1. As an example we present the CMDs (left) and the stellar density plot (right) for SL 353 & SL 349.

Myr for NGC 1972. We suggest that NGC 1971 & NGC 1972 are a true binary cluster and all three clusters may have been formed in the same GMC.

3. The Small Magellanic Cloud

The data for six SMC cluster pairs were extracted from the SMC *BVI* database of the Optical Gravitational Lensing Experiment (OGLE; e.g., Udalski et al. 1998). Our results are presented in the following table:

Three of our SMC binary cluster candidates have sufficiently similar ages to indicate near-simultaneous formation. Chance superposition in the crowded SMC bar region (most likely) and gravitational capture (possible, but rare) may explain the age differences between the remaining three cluster pairs. We suggest that the last ones can be removed from the list of SMC binary cluster candidates. Kinematic studies are needed to validate this suggestion.

Table 1. Binary cluster data

cluster names	sizes [″]	ages [Myr]	separation [′]
H 99 & H 100	12 & 20	140 ± 20 & 140 ± 20	0.507
H 78 & H 76	18 & 7	72 ± 8 & 250 ± 40	0.812
H 179 & H 175	29 & 7	35 ± 5 & 400 ± 60	0.542
B 73 & H 143	11 & 18	72 ± 8 & 180 ± 20	1.301
IC 1612 & H 186	27 & 9	80 ± 8 & 90 ± 10	0.480
B 65 & H 134	29 & 9	72 ± 8 & 80 ± 8	0.617

References

Bertelli, G., Bressan, A., Chiosi, C., Fagotto, F., & Nasi, E. 1994, A&AS 106, 275

Schaerer, D., Meynet, G., Maeder, A., & Schalle,r G. 1993, A&AS 98, 523

Udalski, A., Szymanski, M., Kubiak, M., Pietrzynski, G., Wozniak, P., & Zebrun, K. 1998, Acta Astron. 48, 147

Vallenari, A., Bettoni, D., & Chiosi, C. 1998, A&A 331, 506

New Views of the Magellanic Clouds
IAU Symposium, Vol. 190, 1999
Y.-H. Chu, N.B. Suntzeff, J.E. Hesser, and D.A. Bohlender, eds.

Star Cluster Encounters in the Magellanic Clouds

M.R. de Oliveria, H. Dottori, and E. Bica

Instituto de Fisica - UFRGS, Caixa Postal 15051, CEP 91501-970, Porto Alegre, RS, Brasil

1. Cluster Pairs: Sample and Numerical Simulation

The study of star cluster pairs in the Magellanic Clouds (MCs) has been explored extensively in the past years (Bhatia & Hatzidimitriou 1988; Rodrigues et al. 1994; de Oliveira et al. 1998). The evolution of these pairs can provide fundamental insight into the past history of cluster formation in the MCs.

We have selected MC clusters (Bica & Schmitt 1995; Bica et al. 1998) with morphologies resembling those of the present models. The images in this selection of pairs were obtained from the Digital Sky Survey (DSS). We have also performed N-body simulations of star cluster encounters under different initial conditions, with the total number of particles per simulation ranging from 1024 to 20480.

We analyze the morphology of these selected cluster pairs in the MCs and compare them to the results of our numerical simulations of star cluster encounters. Evidence is found that during the interactions, transient morphological effects such as expanded halo, isophotal deformation and isophotal twisting can occur due to tidal effects and dynamical friction. These simulations, when compared to isodensity maps of cluster pairs in the MCs, can reproduce a wide range of morphological structures.

2. Examples

From the analysis of the time evolution of each model, it is possible to observe in the beginning of the encounter the trend to form bridges. As time goes by the smaller cluster is disrupted by the massive one, partly merging into a single cluster at the end of the simulation, partly being ejected to the field.

In Fig. 1 we compare the isodensity map of the pair HS319/BRHT52b with a suitable model. This suggests that the small cluster is enduring a disruption by dynamical friction.

In Fig. 2 we show the isolated LMC cluster NGC1783 isophotes and ellipticity compared with a final merger cluster model. The similarities between the model and the observation suggests that NGC1783 could be the result of a cluster pair encounter.

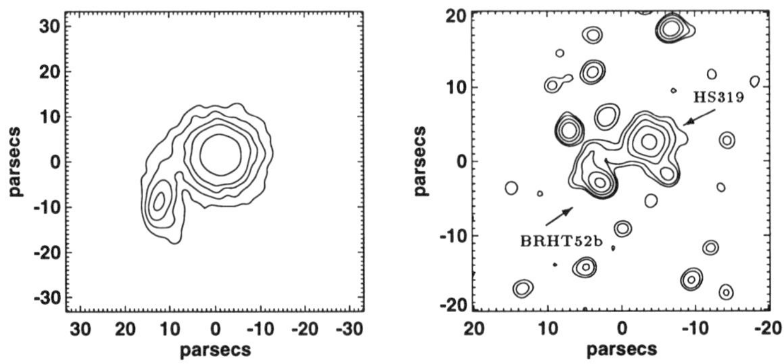

Figure 1. Comparison of a suitable model (left panel) with the LMC pair HS319/BRHT52b (right panel). In this model the projected plane is XY and the time of simulation is t = 19.5 Myr.

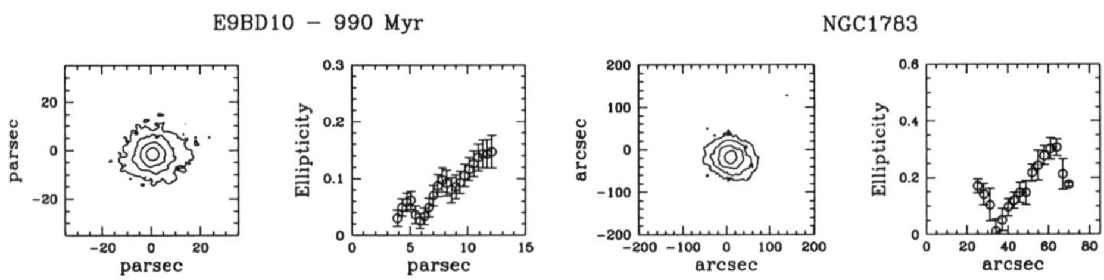

Figure 2. Comparison of a final stage model (left two panels), in a XZ projection with the LMC cluster NGC1783 (right two panels).

Acknowledgments. We thank Dr. Hernquist for allowing us to use TREE-CODE and the CESUP-UFRGS for alloted time in the CRAY YMP-2E computer. We acknowledge support from the Brazilian Institutions CNPq, CAPES and FINEP. The DSS images are copyright by the Royal Observatory Edinburgh and the Anglo-Australian Observatory.

References

Bhatia, R.K., & Hatzidimitriou, D. 1988, MNRAS, 230, 215

Bica, E., & Schmitt, H. 1995, ApJS, 54, 33

Bica, E., Schmitt, H., Dutra, C.M., & Oliveira, H.L. 1999, AJ, 117, 238

de Oliveira, M.R., Dottori, H., & Bica, E. 1998, MNRAS, 295, 921

Rodrigues, I., Rodriguez, A., Schmitt, H., Dottori, H., & Bica, E. 1994, Proceedings of the IIIrd ESO/CTIO Workshop on *The Local Group: Comparative and Global Properties*, (eds.) A. Layden, R.C. Smith & J. Storm

New Views of the Magellanic Clouds
IAU Symposium, Vol. 190, 1999
Y.-H. Chu, N.B. Suntzeff, J.E. Hesser, and D.A. Bohlender, eds.

The Oldest Star Clusters in the Small Magellanic Cloud

Kenneth J. Mighell[1]

Kitt Peak National Observatory, National Optical Astronomy Observatories, P. O. Box 26732, Tucson, AZ 85726-6732, USA

Ata Sarajedini[2]

Department of Physics and Astronomy, San Francisco State University, 1600 Holloway Avenue, San Francisco, CA 94132, USA

Rica S. French[3]

Middle Tennessee State University, Physics & Astronomy Department, WPS 219, P. O. Box 71, Murfreesboro, TN 37132, USA

Abstract. We present our analysis of archival *Hubble Space Telescope* Wide Field Planetary Camera 2 (WFPC2) observations in F450W ($\sim B$) and F555W ($\sim V$) of the intermediate-age populous star clusters NGC 121, NGC 339, NGC 361, NGC 416, and Kron 3 in the Small Magellanic Cloud. We use published photometry of two other SMC populous star clusters, Lindsay 1 and Lindsay 113, to investigate the age sequence of these seven star clusters in order to improve our understanding of the formation chronology of the SMC. We analyzed the V vs $B-V$ and M_V vs $(B-V)_o$ color-magnitude diagrams of these populous Small Magellanic Cloud star clusters using a variety of techniques and determined their ages, metallicities, and reddenings. These new data enable us to improve the age-metallicity relation of star clusters in the Small Magellanic Cloud. In particular, we find that a closed-box continuous star-formation model does not reproduce the age-metallicity relation adequately. However, a theoretical model punctuated by bursts of star formation is in better agreement with the observational data. The full details of this analysis are reported in Mighell, Sarajedini, & French (1998, AJ, 116, 2395).

[1]This research was supported by a grant from the National Aeronautics and Space Administration (NASA), Order No. S-67046-F, which was awarded by the Long-Term Space Astrophysics Program (NRA 95-OSS-16).

[2]Hubble Fellow

[3]Based on research conducted at NOAO as part of the Research Experiences for Undergraduates program.

New Views of the Magellanic Clouds
IAU Symposium, Vol. 190, 1999
Y.-H. Chu, N.B. Suntzeff, J.E. Hesser, and D.A. Bohlender, eds.

AM-3: An Intermediate-Age Star Cluster in the Extreme Outskirts of the SMC

G. S. Da Costa

Mt. Stromlo & Siding Spring Observatories, Private Bag, Weston Creek Post Office, ACT 2611, Australia

1. Introduction

The faint star cluster AM-3 is one of three such objects discovered by Madore & Arp (1979). AM-1 has since been shown to be one of the most distant of the Galaxy's halo globular clusters, while AM-2 is now recognized as a distant, reddened Galactic open cluster. AM-3, however, has been largely ignored despite the fact that it lies only ~4.5 degrees from the center of the SMC making association with the SMC a possibility. This possibility was strengthened by the results of Irwin (1990) who showed that AM-3 lies (in projection) just within the outermost density contours of the SMC field population.

AM-3 was observed with the CTIO 4m prime focus CCD camera using a TI CCD as detector. Three 300 sec V integrations in 1.″3 seeing and three 500 sec B integrations with 1.″5 images were obtained in photometric conditions. The images were analyzed using ALLSTAR and calibrated in the standard manner. The resulting color-magnitude (c-m) diagram for stars within 22″ of the cluster center is shown in Fig. 1.

2. Results

The c-m diagram of Fig. 1 shows a sparse clump of red giants at V = 19.6 ± 0.1 and the top of the cluster main sequence at V ≈ 22. The magnitude for the clump stars in AM-3 is comparable to those in old and intermediate-age SMC star clusters. For example, in the SMC clusters studied by Da Costa & Hatzidimitriou (1998), the clump magnitudes vary from 19.2 (Lindsay 1) to 19.6 (NGC 121). On this basis the association of AM-3 with the SMC appears confirmed, though a radial velocity for the cluster is required to put the question beyond doubt. The projected distance of AM-3 from the center of the SMC is then ~4.5 kpc, larger than the previous western-most cluster identified (Lindsay 1 at ~3.4 kpc in projection) and larger also than the projected distance (~4.2 kpc) of the eastern-most cluster Lindsay 113.

The age of AM-3 can be constrained through the use of isochrone fits. Fig. 1 shows isochrones from Bertelli et al. (1994) for Z = 0.001 and ages of 4, 6 and 10 Gyr fitted assuming a reddening of E(B–V) = 0.03 mag. The vertical fit of the isochrones was set by matching the V magnitude of the clump stars and the resulting modulus, $(m-M)_V$ = 19.00, again supports SMC membership for AM-3. If a higher abundance isochrone set is used (Z = 0.004), the fit is less satisfactory and the implied age is somewhat younger than the ~6 Gyr suggested

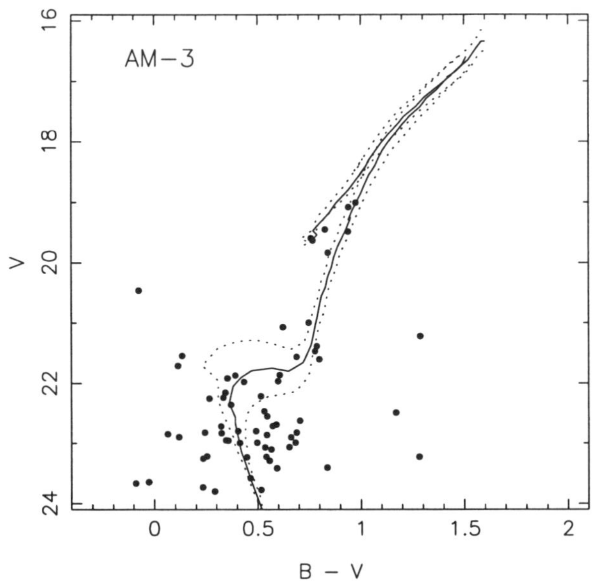

Figure 1. A color-magnitude diagram for AM-3. Superposed are isochrones from Bertelli et al. (1994) for Z = 0.001 placed assuming E(B-V) = 0.03 and $(m-M)_V$ = 19.00. The solid line is for an age of 6 Gyr, while the dotted lines are for 4 (upper) and 10 (lower) Gyr.

by Fig. 1. The best estimate for the age of AM-3 is then ~5 – 6 Gyr; a more exact determination requires deeper and more precise photometry. Thus AM-3, despite its extreme location relative to the SMC center, is an intermediate-age object. Whether it formed at this radius or is seen near apogalacticon in an elliptical orbit (or even whether it is bound to the SMC) cannot be answered at this time, though the cluster does lie at the outermost HI contour in the map of Mathewson & Ford (1984). Based on comparing the number of AM-3 clump stars to the number of red horizontal branch stars in clusters such as Eridanus or Pal 4, and on adding up the luminosities of the stars in Fig. 1 together with a correction for fainter stars, we estimate $M_V \approx -3.5 \pm 0.5$ mag for AM-3. It is quite probable that this cluster would not have been noticed if it lay closer to the main body of the SMC.

References

Bertelli, G., et al. 1994, A&AS, 106, 275

Da Costa, G.S., & Hatzidimitriou, D. 1998, AJ, 115, 1934

Irwin, M.J. 1990, in IAU Symp. 148, eds R. Haynes and D. Milne, (Kluwer: Dordrecht), p. 453

Madore, B.F., & Arp, H.C. 1979, ApJ, 227, L103

Mathewson, D.S., & Ford, V.L. 1984, in IAU Symp. 108, eds S. van den Bergh and K. de Boer, (Kluwer: Dordrecht), p. 125

New Views of the Magellanic Clouds
IAU Symposium, Vol. 190, 1999
Y.-H. Chu, N.B. Suntzeff, J.E. Hesser, and D.A. Bohlender, eds.

Results from *HST* Observations of Six LMC Globular Cluster Fields

K.A.G. Olsen

Cerro Tololo Inter-American Observatory, Casilla 603, La Serena, Chile

P.W. Hodge

Astronomy Department, Box 351580, University of Washington, Seattle, WA 98195, USA

M. Mateo

Department of Astronomy, University of Michigan, 821 Dennison Building, Ann Arbor, MI 48109, USA

E.W. Olszewski

Steward Observatory, University of Arizona, Tucson, AZ 85721, USA

R.A. Schommer, N.B. Suntzeff, and A.R. Walker

Cerro Tololo Inter-American Observatory, Casilla 603, La Serena, Chile

Abstract. We present deep *HST* color-magnitude diagrams of fields centered on the six old LMC globular clusters NGC 1754, NGC 1835, NGC 1898, NGC 1916, NGC 2005, and NGC 2019. Separate cluster and field star CMDs are shown. The time of formation of the LMC is studied from an analysis of the cluster CMDs. Based on a comparison of the CMDs with sequences of the Milky Way clusters M3, M5, and M55, we suggest that the LMC formed its first stars at the same time as the Milky Way to within 1 Gyr. We find additional evidence that these LMC globular clusters are as old as the oldest Milky Way clusters through a comparison of our data with the horizontal branch evolutionary models of Lee, Demarque, & Zinn (1994).

The evolution of the LMC following its formation is studied through an analysis of the field star CMDs. Through an automated comparison with stellar evolution models, we extract the star formation histories implied by the CMDs. Our best-fit star formation histories imply that the LMC has been actively forming stars over the last 4 Gyr, in agreement with previous field star studies. The four fields that lie in the Bar also contain significant numbers of stars with ages of 4–8 Gyr in the best-fit cases. The most notable disagreement between the best-fit models and observed CMDs is in the color of the red giant branch.

1. Observations

Observations of the clusters NGC 1754, NGC 1835, NGC 1898, NGC 1916, NGC 2005, and NGC 2019 were taken with *HST* during Cycle 5. Each cluster was observed for one orbit with WFPC2 through the F555W and F814W filters.

2. Analysis

After performing photometry with DoPHOT 2.5, calibrating, and statistically subtracting field stars from the cluster CMDs, we analyzed the ages of the clusters in two ways. First, we matched the LMC cluster CMDs with fiducial sequences of Milky Way clusters having similar metallicities. Ages of the LMC clusters were then measured relative to those of the Milky Way fiducials from the differences in the colors of their RGBs (cf. VandenBerg, Bolte, & Stetson 1990). Compared with the Milky Way clusters M3, M5, and M55, we find $\langle \mathrm{Age_{LMC}} - \mathrm{Age_{MW}} \rangle = 1.0 \pm 1.3$ Gyr. Second, we compared the HB morphologies and metallicities measured from the RGBs with the HB evolutionary tracks of Lee, Demarque, & Zinn (1994). Within the errors, five of the LMC clusters fall on the track associated with the old Milky Way halo. The sixth LMC cluster, NGC 1916, was not analyzed because of its high degree of differential reddening.

Star formation histories were extracted from the field star CMDs by fitting Bertelli et al. (1994) isochrones through a technique similar to that described by Dolphin (1997). For these fits, we adopted the age-metallicity relation suggested by LMC clusters and tested different IMFs, distances, and reddenings. For a Salpeter IMF, we find that the NGC 1754 field, which lies in the LMC disk, has a star formation rate that is low until 4 Gyr ago, after which it increases steadily. The four fields lying in the LMC Bar, however, also show high star formation activity in the age range 4–8 Gyr. As this activity appears unique to the Bar, we suggest that the Bar may be an old feature.

The detailed star formation histories may be incorrect because of some mismatch of the models and observations. We find that the old, metal-poor models are unable to fit both the lower main sequence and RGB simultaneously. This suggests that either the adopted age-metallicity relation is incorrect, in which case the LMC globular clusters studied here are ~0.5 dex more metal-poor than our measurements, or the model RGBs are too red by ~0.08 magnitudes. Despite this problem, the large differences in the star formation histories of the NGC 1754 field and the Bar fields should remain.

Acknowledgments. We thank M. Bolte, J. Johnson, and G. Mandushev for their unpublished MW fiducials, and D. VandenBerg for valuable discussion. This work was supported by STScI grant GO-05916 to NBS.

References

Bertelli, G. et al. 1994, A&AS, 106, 275

Dolphin, A. 1997, New Astronomy, 2, 397

Lee, Y.-W., Demarque, P., & Zinn, R. 1994, ApJ, 423, 248

VandenBerg, D., Bolte, M., & Stetson, P. 1990, AJ, 100, 445

New Views of the Magellanic Clouds
IAU Symposium, Vol. 190, 1999
Y.-H. Chu, N.B. Suntzeff, J.E. Hesser, and D.A. Bohlender, eds.

Three Old LMC Globular Clusters

Jennifer A. Johnson

University of California, Santa Cruz, CA 95064, USA

Michael Bolte

UCO/Lick Observatory, UCSC, Santa Cruz, CA 95064, USA

Howard E. Bond

Space Telescope Science Institute, Baltimore, MD 21218, USA

James E. Hesser

National Research Council, Victoria, B.C. V8X 4M6 Canada

Claudia M. de Oliveira

European Southern Observatory, Garching, Germany

Harvey B. Richer

University of British Columbia, Vancouver, BC V6T 1Z4, Canada

Peter B. Stetson

National Research Council, Victoria, B.C. V8X 4M6 Canada

Don A. VandenBerg

University of Victoria, Victoria, BC V8W 3P6, Canada

Abstract. We used deep HST imaging of three LMC clusters, NGC 2257, NGC 1466 and Hodge 11, to determine their ages relative to Milky Way clusters. They all have similar ages to within $\pm$ 1 Gyr. For NGC 2257 and NGC 1466, we measured the blue straggler specific frequencies and find them to be similar to those of Galactic clusters.

1. Introduction

With the advent of HST, it is possible to study the LMC globular clusters in similar detail to those in our own Galaxy. We can then probe the formation history of the LMC, placing constraints on the epoch of initial star formation and the duration and nature of globular cluster formation in the LMC halo. Recent results by Olsen et al. (1998), Mighell et al. (1996) and Brocato et al.

(1996) show that the LMC clusters are close in age to the Milky Way clusters, while Testa et al. (1995) argues that NGC 2257 is $\sim$ 2-3 Gyrs younger.

2. Observations and Results

We have V and I WFPC2 photometry extending to $\sim$ 3 magnitudes below the main-sequence turnoff for three old, metal-poor LMC clusters: NGC 1466, NGC 2257 and Hodge 11. We compared their color-magnitude diagrams to those of local Galactic globulars both qualitatively and quantitatively using the ΔV method. We find that these three LMC clusters are as old as the Milky Way clusters. Our photometry probes faint enough magnitudes that the blue straggler sequences are easily visible. We calculated the specfic frequency of the blue stragglers (F_{BSS}) (Bolte et al. 1993) for two of our clusters. We find a F_{BSS} for NGC 1466 of 0.10 and for NGC 2257 of 0.14. These numbers are similar to MW clusters such M3 (0.09) and NGC 5053 (0.14) (see Cohen et al. (1997) and references therein).

3. Discussion

We used the results of the semi-analytic models of galaxy formation of Somerville & Primack (1999) to compare our results with the predictions of hierarchial clustering in a CDM universe. The models gave us the merging and star-formation histories of galaxies inside 10 Local Group-sized halos. We compared the times of the first large-scale star formation of the parent galaxies with those of their satellite galaxies. Although the first subclumps of satellite galaxies cool and collapse later on average than their counterparts for the parent galaxies, a substantial number of LMC-like objects collapse and begin their first starburst within 1 Gyr of their parent's first burst. Thus, the similar ages to Galactic globulars and the LMC clusters are consistent with hierarchial clustering models of galaxy formation, if we assume globular clusters form in the first major starburst. Our blue straggler results show that the blue straggler populations are not affected a large amount if the clusters belong to a different galaxy and that other, internal properties must be important.

References

Bolte, M., Hesser, J.E., & Stetson, P.B. 1993, ApJ, 408, L89

Brocato, E., Castellani, Ferraro, F.R., Piersimoni, A.M., & Testa, V. 1996, MNRAS, 282, 614

Cohen, R.L., Guhathakurta, P., Yanny, B., Schneider, D.P., & Bahcall, J.N. 1997, AJ, 113, 669.

Mighell, K.J., Rich, R.M., Shara, M., & Fall, S.M. 1996, AJ, 111, 2314

Olsen, K.A.G., Hodge, P.W., Mateo, M., Olszewski, E.W., Schommer, R.A., Suntzeff, N.B., & Walker, A.R. 1998, MNRAS, 300, 665

Somerville, R.S. & Primack, J.R. 1999, MNRAS, in press (astro-ph/9802268)

New Views of the Magellanic Clouds
IAU Symposium, Vol. 190, 1999
Y.-H. Chu, N.B. Suntzeff, J.E. Hesser, and D.A. Bohlender, eds.

The Stellar Complexes in the Large Magellanic Cloud

F. Maragoudaki and M. Kontizas

Section of Astrophysics Astronomy & Mechanics, Department of Physics, University of Athens, GR-157 83 Athens, Greece

E. Kontizas and A. Dapergolas

Astronomical Institute, National Observatory of Athens, P.O. Box 20048, GR-118 10 Athens, Greece

D.H. Morgan

Schmidt Telescope Unit, Royal Observatory of Edinburgh, EH9 3HJ, UK

Abstract. A method has been developed for the detection of stellar complexes and the derivation of their fundamental properties in the LMC. We have used digitized UKST plates in order to perform star counts and produce isodensity contour maps in various colours and luminosity slices. About 50 large stellar groupings have been revealed. Their properties favour a "top-down" mechanism acting in star forming regions.

1. Observational material and detection of stellar complexes

We use direct photographic plates taken with the UK 1.2m Schmidt Telescope, in various wavebands: U, R and HeII ($\lambda = 4686$ Å) down to a magnitude limit of 19-20 mag. The plates were digitized with the fast measuring machines APM and SuperCosmos and the derived data give for the detected images positions, magnitudes etc.

The detection of the stellar complexes is based on:

1. *Star counts and isodensity contour maps*
   Star counts were performed on the U detected stellar images and the isodensity contour map of the field was produced. The regions with enhanced stellar density were considered as large stellar groupings candidates. The local backgrounds were taken into consideration and were subtracted from the studied areas.

2. *Luminosity slices*
   The instrumental magnitudes have been calibrated by us using known CCD magnitudes from the literature (Will et al. 1997). Star counts and isodensity contour maps performed for various luminosities allowed us to study the distribution of stars at the various luminosity slices.

3. *Various wavebands*
   Steps 1 and 2 were performed on U, R and HeII plates. In the U plates,

the blue stars of the upper main sequence stand out against the older, background stellar populations, while the R plates cover a wider range of stellar populations. The HeII waveband covers part of the B magnitude and the obscuration by the gas is minimal.

The structures found from the above three steps (larger than 150 pc) were accepted as stellar complexes under the following criteria:

1. The stellar density of the structure is enhanced by at least 3σ above the mean density of the surrounding field in more than two slices in the U plate. At least one of these slices must contain stars brighter than 17.5 mag.
2. The structure is also revealed in the isodensity maps of the HeII data.
3. If a structure is embedded in a larger one its density should be at least 5σ above the mean density of the surrounding field.

2. Discussion

The above criteria have been applied to a region of the LMC of about $6° \times 7°$ centered at RA $\sim 05^h26^m$ and DEC $\sim -69°$.

About 50 large stellar groupings have been revealed showing: (1) hierarchical structure, where the smallest are found within the large ones; (2) their size distribution has peaks at 250± 50pc (aggregates) and 600± 50pc (complexes), there are also a few cases with size 1200± 50pc (supercomplexes); (3) the luminosity slices show that these structures dissapear at U ~17.5 mag corresponding to spectral types of about B8 and mass around $3M_\odot$; (4) there is evidence that at the fainter magnitudes these structures are aligned to a general trend whereas at the brighter end (B1, O spectral types) they become more clumpy and symmetrical in shape.

This hierarchical picture of stellar groupings and their preferred dimensions which has been also found in other galaxies (M31, M33), favours a "top-down" scenario for the star formation mechanism in these galaxies (Efremov & Elmegreen 1998).

Acknowledgments. The authors would like to express their sincere thanks to the 1.2m U.K Schmidt Telescope Unit for loan of the material, the British Council for financial support and to the General Secretariat of Research and Technology for financial support of this project.

References

Battinelli, P., Efremov, Yu.N., & Magnier, E. 1996, A&A, 314, 51

Efremov, Yu.N. 1995, AJ, 110, 2757

Elmegreen B.G., & Efremov Yu.N. 1996, ApJ, 466, 802

Kontizas, M., Maravelias, S.E., Kontizas, E., Dapergolas, A., & Bellas-Velidis I. 1996, A&A, 308, 40

Will, J.M., Bomans, D.J., & Dieball, A. 1997, A&AS, 123, 455

New Views of the Magellanic Clouds
IAU Symposium, Vol. 190, 1999
Y.-H. Chu, N.B. Suntzeff, J.E. Hesser, and D.A. Bohlender, eds.

Relaxation of Star Clusters and Color Gradients

Christian Boily[1]

Queen Mary & Westfield College, University of London
Mile End Road E1 4NS, England.

Abstract. Large interstellar distances imply that protostars are dynamically decoupled from gas ram pressure (see, e.g., Aarseth, Lin & Papaloizou 1988) and are effectively isolated self-gravitating objects. With this in mind we apply Newtonian gravity to collapsing cold spheroids of point masses to model the Lin-Mestel-Shu (1965) solution for dust. Incomplete violent relaxation (e.g., van Albada 1982) means that traces of the initial orbit distribution in energy remain. We show here that furthermore the degree of orbit mixing achieved correlates well with initial morphology, by collapsing two-mass component uniform spheroids of different aspect ratio. We setup a test problem to seek out a relation between orbit survival and initial morphology which we then solve numerically.

1. N-Body Integration.

Our toy model consists of a uniform sphere containing two mass components, one inside, the other outside, radius $r = \pi$; the bounding radius $= 5$. Each mass bin is sampled with 5,000 bodies. With $m_1/m_2 = 1/3$ the mean density is everywhere the same, hence the free-fall time is unique. With the more massive stars confined to $\pi < r < 5$, the intention is to favour mass ejection from, or otherwise disturb, the much lighter central distribution. This is a robust test of orbit survival in energy space under violent relaxation.

The 10,000-particle simulation was performed with the code NBODY1 on the HARP computer. The length l of the smoothed potential $= 3 \times 10^{-4}$ (in system units), well below close-encounter deflection impact-parameter values (see Boily et al. 1999 [BCM]; Theis & Spurzem 1998). The simulation ran for a total of 6 initial crossing times. A small velocity was given to each star such that the virial ratio $Q \equiv -T/W = 1/200$ initially. Two more simulations were performed with the same random seed distributions, but now flattening the z-axis by factors of 2 (hamburger) and 10 (pancake): this covers an appropriate range of spheroid aspect ratios.

The short timescale of the simulations matches age estimates of the youngest LMC star clusters (Elson et al. 1987).

[1]Now at: Astronomisches Rechen-Institut, Mönchhofstrasse 12-14, Heidelberg D-69120, Germany.

We measured orbit-mixing by monitoring in time the ranking $\mathcal{R}_{t,i}$ of each (i^{th}) star relative to one another, introducing

$$\Delta \equiv \left(\frac{1}{2N} \sum_{S} \sum_{i_{\rm bin}} [\mathcal{R}_{t,i} - \mathcal{R}_{o,i}]^2 \right)^{1/2} .$$

Completely mixed orbits give $\Delta \simeq 57.7$ when the number of rank bins $S = 200$.

2. Discussion.

We found that while both spherical and 'pancake' collapses erase memory of the initial vertical (z) structure rather well, this is not so for the 'hamburger' simulation (aspect ratio = 1/2): we find post-collapse $\Delta_z \approx 42$, well below the expected value for fully-mixed systems.

The discovery of young star clusters with color gradients (see Fischer et al. 1998; Hunter et al. 1998; Kontizas et al. 1998) suggests stellar formation processes may have been at work throughout their formation history. When star formation proceeds on a timescale short compared with a cluster free-fall time, color (mass) gradients may develop early on and not be deleted by violent relaxation. Memory of the initial particle distribution is not completely obliterated by cold collapse, particularly when collapse proceeds from a distribution of initial aspect ratio $\sim 1/2$.

Combining this result with the findings of BCM, we would anticipate color gradients be present in roundish clusters, and absent for those with large (on the mean) ellipticities. These results obtained for axisymmetric distributions are to be applied cautiously, given that (young) LMC star clusters are likely triaxial (e.g., Han & Ryden 1994).

Acknowledgments. Warm thanks to C.J. Clarke and S.D. Murray for discussions, and the organisers of the conference for agreeing to this (last minute) contribution.

References

Aarseth, S.J., Lin, D.N.C., & Papaloizou, J.C.B. 1988, ApJ, 324, 288

Boily, C.M., Clarke, C.J., & Murray, S.D. 1999, MNRAS, 302, 339

Elson, R.A.W., et al. 1987, ApJ, 323, 54

Fischer, P., et al. 1998, AJ, 115, 92

Han, C.H., & Ryden, B.S. 1994, ApJ, 433, 80

Hunter, D.A., et al. 1998, ApJ, 478, 124

Kontizas., M., et al. 1998, A&A, 336, 503

Lin, C.C., Mestel, L., & Shu, F.H. 1965, ApJ, 142, 1431

Theis, C., & Spurzem, R. 1998, A&A, 341, 361

van Albada, T.S. 1982, MNRAS, 201, 939

New Views of the Magellanic Clouds
IAU Symposium, Vol. 190, 1999
Y.-H. Chu, N.B. Suntzeff, J.E. Hesser, and D.A. Bohlender, eds.

Bow Shock Induced Star Formation in the LMC

K.S. de Boer, J.M. Braun

Sternwarte, University of Bonn, D-53121 Bonn, Germany

A. Vallenari

Osservatorio Astronomico, Vic. Osserv. 5, I-35122 Padova, Italy

U. Mebold

Radioastronomisches Institut, Univ. Bonn, D-53121 Bonn, Germany

1. Facts

The supergiant shell LMC 4 is known for its ring of H II regions and the deep minimum in HI in the interior (Rohlfs et al. 1984). LMC 4 shows X-ray emission detected by ROSAT and highly ionised gas (C IV) detected with HST (Bomans et al. 1994, 1996). The velocities of its neutral gas indicate it cannot have been formed by the collision with a high velocity cloud (Domgörgen et al. 1995).

The age of the association NGC 1948 at the NE edge of LMC 4 (Vallenari et al. 1993, Will et al. 1996) is 5-10 Myr, while the stars in the interior are 10-15 Myr old (Braun et al. 1997). The similarity of all the ages derived for star groups inside and at the edge of LMC 4 (Braun et al. 1997) pointed toward the need to identify a large scale star formation triggering event.

ROSAT X-ray data of the LMC were analysed by Blondiau et al. (1997) and they found near the dark cloud south of 30 Dor an anticorrelation between X-ray emission and neutral hydrogen + IRAS dust emission. They noted also that this region is close to and aligned with the steep gradient in H I near the leading edge, a region from which the X-ray intensity is lower than the general background of the X-ray sky (see map of Snowden & Petre 1994).

2. Model

Collectively, the described facts led to the following scenario. Gas is compressed at the leading edge of the LMC because of the bow shock of the LMC formed in the interaction with the galactic halo. Such compressed gas starts to form stars, likely on a large scale. Due to the rotation of the LMC these regions move away from the leading edge, and a sequence of such events will lead to a sequence of structures along the eastern and northern edge on the LMC increasing in age in a clockwise fashion (de Boer et al. 1998).

The kinematics of the LMC are as follows. Models for the Magellanic Clouds and Stream (Heller & Rohlfs 1994) indicate that the LMC moves towards the East, confirmed by Hipparcos data of LMC stars (Kroupa & Bastian 1997). The full space velocity is $\simeq 265\ \mathrm{km\,s^{-1}}$. Combining the radial and the tangential rotation curve, the true rotation speed of the LMC at a radius of 2 kpc is $\simeq$ 150 $\mathrm{km\,s^{-1}}$. Given the above, the maximum velocity difference between LMC gas and gas of the Milky Way halo is $\simeq$ 450 $\mathrm{km\,s^{-1}}$, to be found near the SE edge.

3. Predictions

The clockwise progression of age, starting at the leading edge as given by the fields listed by de Boer et al. (1998), should continue along the edge of the LMC and populations older than 50 Myr should be found at the western and southern edges. The star formation at the leading edge of the LMC need not be a continuous process. Interruptions in the piling up of gas may occur when a region devoid of gas is moving up to the leading edge, such as a 'former LMC 4'. The NE side of the LMC shows such gaps.

The bow shock induced star formation model predicts a clock-wise increase in the age of superstructures, in step with the rotation velocity of $\simeq$ 150 $\mathrm{km\,s^{-1}}$. At a radius of about 2 kpc this means one revolution in $\simeq$ 100 Myr. One might therefore expect at each location along the outer edge of the LMC stellar populations with ages of modulo 100 Myr. Looking for the sequence of ages, only *accurate* ages are of use, with precision better than 20%. Such a precision can be reached with star groups, but hardly when using individual objects. Other mechanisms of star formation also operate in the LMC (de Boer et al. 1998), and hence the basic bow shock structure may be overlain with secondary star formation and thus with other ages.

The inclination of the LMC with respect to the line of sight implies that the vector of motion of the LMC lies in the plane of the LMC itself. Thus the LMC slices its way as a disk through the MW halo. The radial rotation curve of the LMC indicates solid body rotation out to 2 kpc. This means that the geometry of the star forming regions near the leading edge may stay the same during the further rotation. Is it significant that the *shape* of the dark cloud south of 30 Dor is the same as that of association LH 77 inside LMC 4?

References

Blondiau, M.J., Kerp, J., Mebold, U., & Klein, U. 1997, A&A, 323, 585
Bomans, D.J., Dennerl, K., & Kürster, M. 1994, A&A, 283, L21
Bomans, D.J., de Boer, K.S., Koornneef, J., et al. 1996, A&A, 313, 101
Braun, J.M., Bomans, D.J., Will, J.-M., & de Boer, K.S. 1997, A&A, 328, 167
de Boer, K.S., Braun, J.M., Vallenari, A., & Mebold, M. 1998, A&A, 329, L49
Domgörgen, H., Bomans, D.J., & de Boer, K.S. 1995, A&A, 296, 523
Heller, P., & Rohlfs, K. 1994, A&A, 291, 743
Kroupa, P., & Bastian, U. 1997, New Astronomy, 2, 77
Rohlfs, K., Kreitschmann, J., Siegman, B.C., et al. 1984, A&A, 137, 343
Snowden, S.L., & Petre, P. 1994, ApJ, 436, L123
Vallenari, A., Bomans, D.J., & de Boer, K.S. 1993, A&A, 268, 137
Will J.-M., et al. 1996, A&A, 315, 125

New Views of the Magellanic Clouds
IAU Symposium, Vol. 190, 1999
Y.-H. Chu, N.B. Suntzeff, J.E. Hesser, and D.A. Bohlender, eds.

Numerical Simulation of the Global Star Formation Pattern in the LMC

L.T. Gardiner

Dept. of International Education, Sun Moon University, Rep. of Korea

C. Turfus

Dept. of Mathematics, Sun Moon University, Rep. of Korea

Abstract. We have constructed a dynamical model of the LMC to examine the effects of an off-center bar perturbation on the global distribution of the gas and star formation activity. We find that a weak rotating bar potential, whose center is displaced by 0.6 kpc from the disk center, gives rise to an asymmetric spiral structure which mimics the observed chains of recent star formation as well as activity in the bar region.

1. Numerical Model

Recently, Dottori et al. (1996) proposed that the large-scale gravitational potential due to an off-centered bar has induced the formation of spiral features in the distribution of young star clusters in the LMC. We have conducted numerical simulations to examine this scenario in greater detail. Our numerical model of the LMC was based on the N-body/cellular automaton scheme of Gardiner et al. (1998a), which incorporates star formation based on local gravitational instability and propagation within a cloud complex. We simulated the gas disk of the LMC with 7000 particles, representing a total gas mass of $4.2 \times 10^8 M_\odot$, initially distributed uniformly on the surface of a disk of radius 3.5 kpc. The particles were subjected to their mutual gravitational forces, as well as to external forces due to the dynamical mass of the SMC associated with the stellar component. These external forces were considered to arise from the superposition of an axisymmetric potential and a weak rotating bar-like perturbation whose center was offset from the disk center by 0.6 kpc. The bar was assumed to spin about its own axis at the same rate as it rotates about the disk center with a pattern speed of $\Omega_b = 50$ km/s/kpc.

2. Results

Figure 1 shows the distribution of star particles up to 30 Myr old produced by the simulation at $T = 300$ Myr. The simulation produces a trailing spiral arm pattern which corresponds to the spiral arms A and B defined by Smith et al. (1987). The pattern also resembles the distribution of young supergiant stars found by Grebel & Brandner (1998, these proceedings). Although the

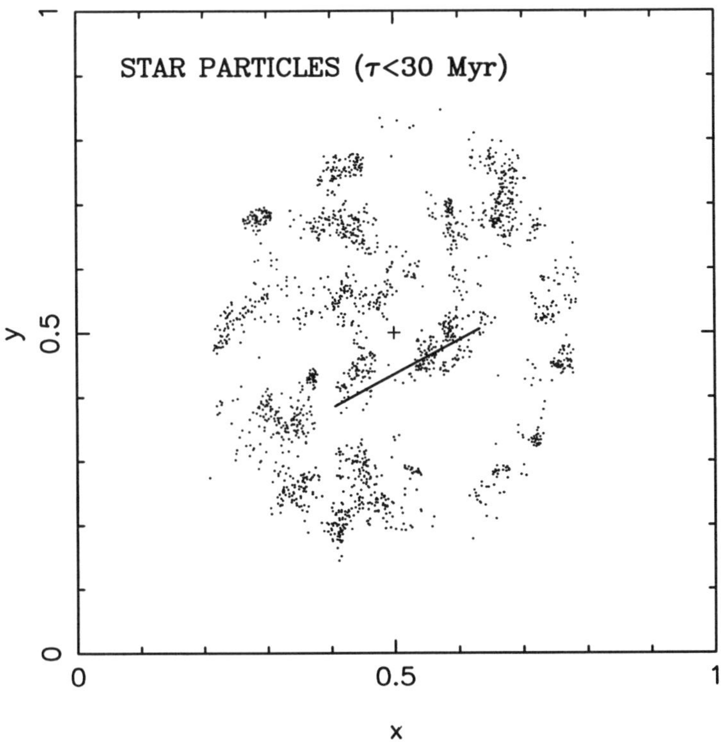

Figure 1. The star particle (age less than 30 Myr) distribution generated by the model at $T = 300$ Myr after the start of the simulation. The disk center and the major axis of the bar are shown. Units are in tens of kpc.

simulation code incorporates a mechanism for propagating star formation, the large-scale spiral features are produced by the off-center bar potential (Colin & Athanassoula 1989), not by the combination of propagating star formation plus shear. Nevertheless, the mechanism of propagating star formation is responsible for the generation of stellar complexes and gaseous structures up to 1 kpc in size which are probably associated with the Shapley Constellations and supergiant shells. Refer to Gardiner et al. (1998b) for a more detailed account of this work.

References

Colin, J., & Athanassoula, E. 1989, A&A, 214, 99

Dottori, H., Bica, E., Claria, J.J., & Puerari, I. 1996, ApJ, 461, 742

Gardiner, L.T, Turfus, C., & Wang, M. 1998a, PASJ, 50, 375

Gardiner, L.T, Turfus, C., & Putman, M.E. 1998b, ApJ, 507, L35

Smith, A.M., Cornett R.H., & Hill R.S. 1987, ApJ, 320, 609

New Views of the Magellanic Clouds
IAU Symposium, Vol. 190, 1999
Y.-H. Chu, N.B. Suntzeff, J.E. Hesser, and D.A. Bohlender, eds.

The Formation and Evolution of LMC Globular Clusters: The Database

S.F. Beaulieu, R. Elson, G. Gilmore, R.A. Johnson, N. Tanvir

Institute of Astronomy, University of Cambridge, Madingley Road, Cambridge, CB3 0HA, UK

B. Santiago

Instituto de Fisica, Universidade Federal do Rio Grande do Sul, Av. Bento Goncalves 9500, Porto Alegre RS, CEP: 91501-970, Brazil

Abstract. We present details of the database from a large Cycle 7 HST project to study the formation and evolution of rich star clusters in the LMC (see Elson et al., this volume). Our data set, which includes NICMOS, WFPC2 and STIS images of 8 clusters, will enable us to derive deep luminosity functions for the clusters and to investigate the universality of the stellar IMF. We will look for age spreads in the youngest clusters, quantify the population of binary stars in the cores of the clusters and at the half-mass radii, and follow the development of mass segregation.

1. Introduction

The Large Magellanic Cloud (LMC) is unique in containing massive star clusters at all stages of evolution. We are presently obtaining very deep HST images of 8 optimally selected clusters. This large observing program will enable us to probe the cluster's stellar content down to 0.2 $M_\odot$. In Cycle 7, a total of 95 orbits have been granted for this project. We are using WFPC2 (F555W,F814W), NICMOS (all three cameras) (F110W,F160W) and STIS (CCD/F28X50LP) instruments in primary mode for each cluster, with parallel observing of surrounding fields as well, giving a total of 285 orbits of imaging data. These observations, in conjunction with state-of-the-art N-body models being calculated by the Institute of Astronomy's N-body group, will quantify the origin and evolution of rich star clusters in the Milky Way, the LMC and beyond.

2. The LMC clusters

Our clusters are grouped in four pairs, with ages of $10^7 - 10^{10}$ yr (see Table 1) and are observed at their core and half-mass radius. They are among the richest clusters in the LMC and have masses $\approx 10^4$ $M_\odot$. In the youngest clusters, the fraction of primordial binaries, the stellar initial mass function (IMF), and the degree of primordial mass segregation will allow us to investigate the process of star formation in a protocluster. Age spreads and a search for pre-MS stars will reveal both the timescale for star formation, which has important implications

Table 1. The LMC globular cluster sample

Cluster	log *age* (yr)	[Fe/H][a]	R_g[b] (°)	R_c[c] ('')	Recent CMD
NGC 1805	7.2	...	3.7	...	...
NGC 1818	7.3	-0.8	3.4	7.9	Will et al. (1995)
NGC 1831	8.7	-0.3	4.6	14.2	Mateo (1998)
NGC 1868	8.8	-0.6	5.4	5.1	Corsi et al. (1994)
NGC 2209	9.1	-1.0	5.7	20.0	Corsi et al. (1994)
H14=SL506	9.2	...	4.4	10.3	none available
NGC 2210	10.1	-2.2	5.0	7.1	Brocato et al. (1996)
H11=SL868	10.1	-2.1	5.2	19.8	Mighell et al. (1996)

[a]The values of metallicity are quite uncertain; for many of the clusters widely differing values are quoted in the literature

[b]Distance from the LMC centre

[c]Core radius

for its trigger, and whether the low and high mass stars form sequentially or together. In the intermediate and old clusters, we will trace the development of mass segregation and the binary fraction in the core and at the half-mass radius; binaries play a crucial role in the structural evolution of a cluster, and in particular, affect the onset of core collapse. A comparison of the IMFs in our clusters will help answer the far-reaching question of whether there is such a thing as a universal IMF.

In addition to the 8 LMC clusters, we are observing two LMC background fields in order to correct for contamination from LMC field stars and unresolved background galaxies. Also from these background fields and the parallel data, we will study the LMC field star population and the star formation history. We are also observing one galactic globular cluster of similar metallicity to the LMC clusters, NGC 6553, in order to provide an empirical calibration of the NICMOS and STIS passbands relative to absolute V magnitude and mass. We have similar data for 47 Tuc from the HST project 7419 (PI: R.F.G. Wyse).

The data reduction and analysis is performed using STSDAS and photometry is performed using the DAOPHOT package. Some preliminary results can be found in Elson et al. (this volume), Johnson et al. (1998), and at our webpage: `http://www.ast.cam.ac.uk/LMC`.

References

Brocato, E., et al. 1996, MNRAS, 282, 614

Corsi, C.E., et al. 1994, MNRAS, 271, 385

Mighell, K.J., et al. 1996, AJ, 111, 2314

Will, J.-M, Bomans, D.J., De Boer, K.S. 1995, A&A, 295, 54

Johnson, R.A., et al. 1998, in *NICMOS and the VLT*, edited by Wolfram Freudling & Richard Hook, ESO Conference & Workshop Proceedings No 55, 140

New Views of the Magellanic Clouds
IAU Symposium, Vol. 190, 1999
Y.-H. Chu, N.B. Suntzeff, J.E. Hesser, and D.A. Bohlender, eds.

Recent Migratory Star Formation in Nearby Magellanic Irregulars

Robert Ruotsalainen

Physics Dept., Eastern Washington University, Cheney, WA 99004

Abstract. UBV measurements of resolved stars in the associations of two Magellanic irregulars, Leo A and Sextans B, are compared with synthetic color-magnitude diagrams based upon stellar evolutionary models.

Image-tube exposures of the galaxies Leo A and Sextans B were obtained at the 2.2 m telescope of Mauna Kea Observatory. Color-magnitude diagrams were produced from the resolved images of some 590 stars in Leo A and 556 stars in Sextans B. These observations were compared with synthetic color-magnitude diagrams based upon stellar evolutionary models of Schaller et al. (1992), supplemented by increased mass loss models of Meynet et al. (1994), as well as by models of de Loore & Vanbeveren (1994) for massive close binaries with Magellanic Cloud metallicities, and of De Greve (1993) for close binary systems of intermediate mass.

Point-spread functions were fit to the digitized photographic data for the measurement of apparent blue magnitudes and $U-B$ colors. Figure 2 illustrates a model adjusted to match the observations, seen in Figure 1, for 306 stars resolved in the central region of Leo A. Modeled distributions depend on values assigned for the intervals of star formation over ages t, and also on values assigned for the initial mass function index α (of the form: $N \sim M^{-\alpha}$ for N number of stars of mass M), the metallicity Z, and the distance modulus μ. The Salpeter (1955) value of $\alpha = 2.35$ was used. Reduced metallicities ($Z = 0.001$ or $Z = 0.002$) were chosen for the modeling of single stars and massive binaries; only solar composition ($Z = 0.02$) models were available for the binary stars of intermediate mass ($3 < M/M_{\odot} < 8$).

For the mixed binary/single model seen in Figure 2, every fourth tested case included a close binary system, of randomly selected primary mass ($3 < M/M_{\odot} < 40$) and age, alternately chosen to have a mass ratio of either 0.9 or 0.6. A distance modulus of $\mu = 26.0$ is reflected in the plate limit assigned to the modeled distribution. Magnitude-dependent smoothing was applied to the model for the simulation of random photometric errors.

Matches, of the computed distributions of points to the observations, are consistent with a migration in the recent formation of massive and intermediate mass stars, from west to east, with stars having formed most actively in the central region of each galaxy. Such migratory activity is similar to modeled results obtained earlier for the Magellanic irregulars, DDO 53 and GR 8.

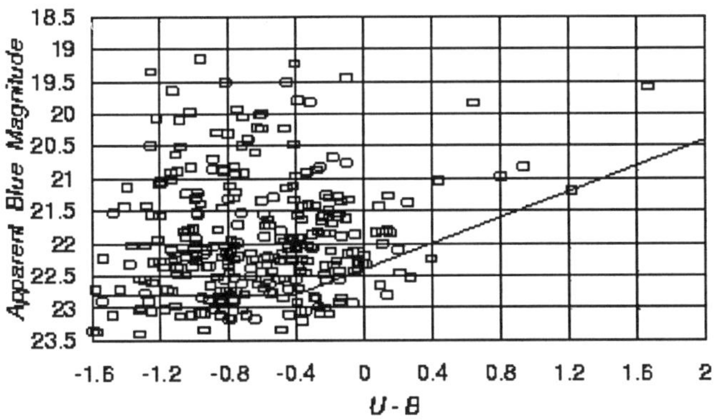

Figure 1. Observations of 306 Stars in Central Region of Leo A.

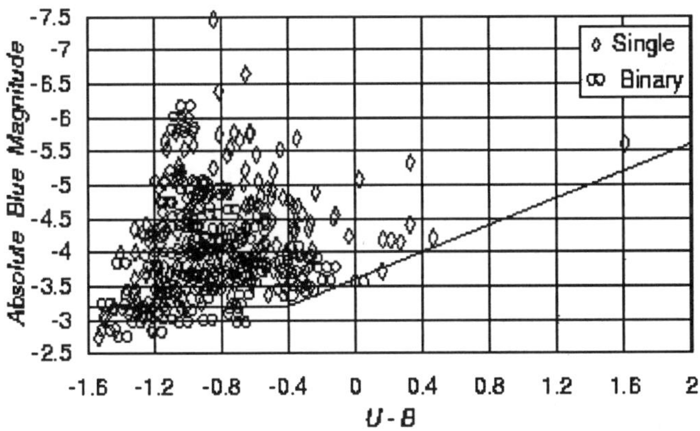

Figure 2. Mixed binary/single model of 306 points: 5<t (Myr)<70.

References

De Greve, J. P. 1993, A&AS, 97, 527

de Loore, C., & Vanbeveren, D. 1994, A&AS, 103, 67

Meynet, G., Maeder, A., Schaller, G., Schaerer, D., & Charbonnel, C. 1994, A&AS, 103, 97

Salpeter, E. E. 1955, ApJ, 121, 161

Schaller, G., Schaerer, D., Meynet, G., & Maeder, A. 1992, A&AS, 96, 269

New Views of the Magellanic Clouds
IAU Symposium, Vol. 190, 1999
Y.-H. Chu, N.B. Suntzeff, J.E. Hesser, and D.A. Bohlender, eds.

The Recent Star Formation History of the LMC Southwest Quadrant

Paolo Battinelli

Osservatorio Astronomico di Roma, Italia

Serge Demers

Université de Montréal, Canada

1. Introduction

Recently, Demers & Battinelli (1998) have shown that the young intercloud stellar aggregates are 10 to 25 Myr old, thus younger than the estimated age (200 Myr) of the last close LMC–SMC encounter (see, e.g., Irwin et al. 1996). These stellar aggregates show a distance gradient that confirms the existence of a "link" between the two Magellanic Clouds. Numerous blue stars have been detected by Demers & Irwin (1991) east of the aggregate ICA76 (the closest to the LMC). These blue stars, that extend to the southwestern periphery of the LMC, may be an aftermath of the close LMC–SMC encounter. Our present study is aimed to survey this SW part of the LMC and to determine the origin of these blue stars.

2. Observations

Twenty one 15′×15′ fields were observed, in B and V, with the CTIO 1.5m telescope. The detector was a Tek 2K CCD. The data were obtained in December 1997, during three consecutive photometric nights. For each field two 450 s B and V exposures were taken. Data were analyzed with DAOPHOT II and ALLFRAME. This combination of detector (large field) and exposure times allows us to produce CMDs down to V=22 ±0.05, i.e., well below the populous turnoff of the bulk of the old population of the LMC.

We selected our target fields mainly looking at the distribution of the blue stars on F033 and F032 ESO/SERC survey fields. The plate pairs have been scanned by the APM in order to produce a CMD of the area. A certain number of targets were also selected in order to survey the LMC radially (these fields are mostly along the 5^h hour circle).

3. Results and Discussion

Our survey indicates that there are essentially no young stars (age <25 Myr) in the western periphery of the LMC. The stellar aggregate ICA 76 – at 7° from the center and with more than 80 blue stars – is to be considered unique. We have, however, found a region, about one degree in size, where numerous stars younger than ~800 Myr are seen. Some of them may be as young as 100 Myr. This region is not as close to the central part of the LMC as expected, but some 6 degrees from its center. These young stars are found in three of our CCD fields located near each other and close to ICA 76. The three fields with a substantial number of blue stars are just east of ICA 76 and between −74° and −75° (see Fig. 1). A detailed study of the stellar populations of this unusual area will be presented elsewhere.

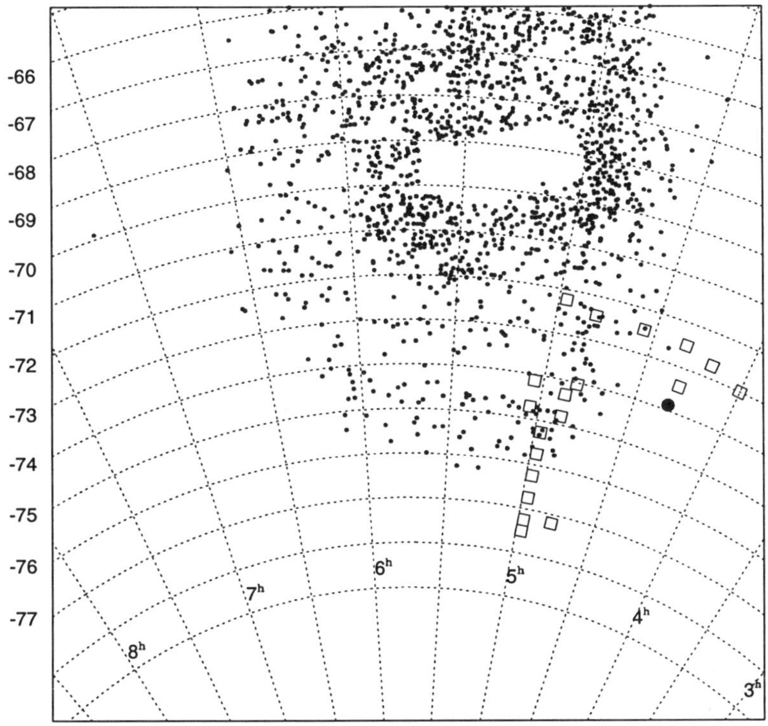

Figure 1. Dots: 1627 clusters from Kontizas et al. 1990; squares: CCD fields of the present survey; big black dot: ICA 76. The three *unusual* fields are just East of ICA 76.

References

Demers, S., & Battinelli, P. 1998, AJ, 115, 154

Demers, S., & Irwin, M.J. 1991, A&AS, 91, 171

Irwin, M.J., Demers, S., & Kunkel, W.E. 1996, BAAS, 28, 932

Kontizas. M., Morgan, D.H., Hatzidimitriou, D., & Kontizas, E. 1990, A&AS, 84, 527

New Views of the Magellanic Clouds
IAU Symposium, Vol. 190, 1999
Y.-H. Chu, N.B. Suntzeff, J.E. Hesser, and D.A. Bohlender, eds.

HST Study of the Stellar Populations Within 30 pc of SN1987A

Martino Romaniello[1] and Nino Panagia[2]

STScI, 3700 San Martin Drive, Baltimore, MD 21218, USA

Salvatore Scuderi

Osservatorio di Catania, Viale A. Doria 6, Catania, Italy

Abstract. We present the results of a study of the stellar population around SN1987A in the Large Magellanic Cloud based on an analysis of multi-band HST-WFPC2 images.

We recover the intrinsic luminosity and temperature, as well as the reddening, of individual stars by fitting the observed fluxes in six bands, from the UV to the I band, to the ones computed from the model atmospheres by Bessel et al (1998). T Tauri stars, i.e., low-mass Pre-Main Sequence stars, are identified through their Hα (488 stars with $EW(\mathrm{H}\alpha) > 8$ Å) and/or U-band (850 stars with $(U-B)_0 < -0.3$) excesses. The resulting HR diagram is shown in the left panel of Figure 1, together with the theoretical Zero Age Main Sequence for $Z = 6 \times 10^{-3}$. A number of different generations of stars are required to explain the observations. In particular, the most luminous stars and the bulk of T Tauri stars indicate that a burst of star formation took place roughly 12 Myr ago, i.e., when the progenitor of SN1987A was born.

The *spatial distribution* of the stars belonging to the same young generation is shown in the right panel of Figure 1. The stars more massive than 6 $M_\odot$ are mainly concentrated in a cluster, whereas the low mass ones ($M < 2\ M_\odot$) are more evenly distributed. In fact, *the surface density of T Tauri stars is highest where that of the massive stars is lowest.* For comparison, the spatial density of Red Giant clump stars is uniform, as expected for the LMC field.

The Star Formation Rate (SFR) as a function of time is shown in the left panel of Figure 2. The right panel shows the IMF derived if T Tauri stars are identified through their Hα ($\Gamma = -1.55$) or their U-band ($\Gamma = -1.87$) excesses.

Full account of this work can be found in Panagia et al. (1999) and Romaniello (1998).

[1] Also Scuola Normale Superiore di Pisa.

[2] Affiliated to the Astrophysics Division, Space Science Department of ESA.

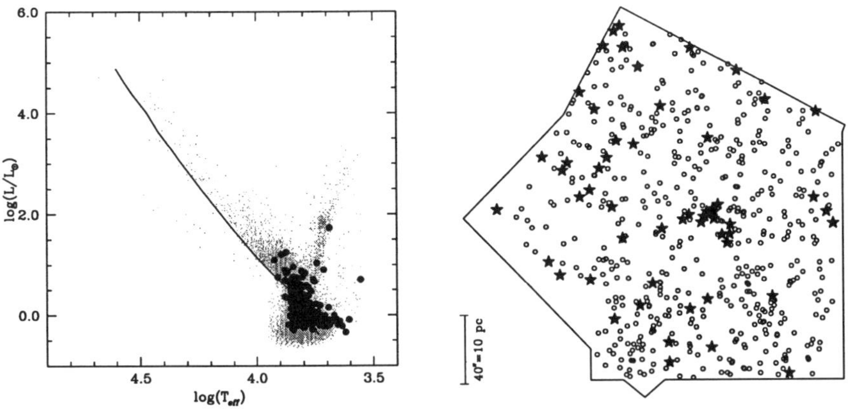

Figure 1. *Left panel:* HR diagram; T Tauri stars identified through their Hα excess are shown as dots. *Right panel:* Spatial distribution of massive (star symbol) and Pre-Main Sequence (circles) stars.

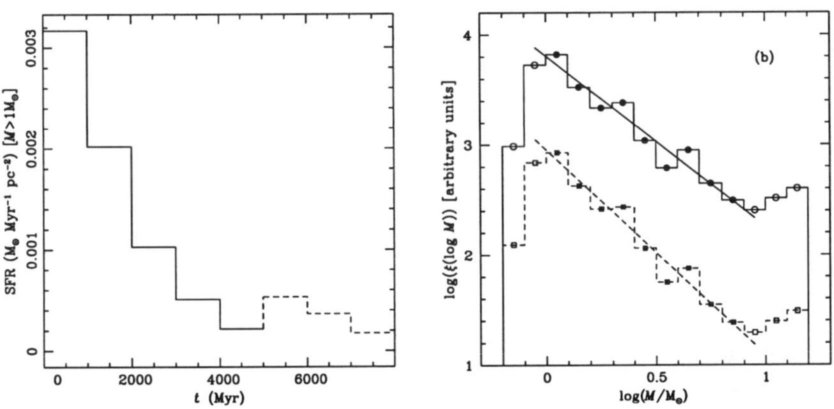

Figure 2. *Left panel:* SFR for the field around SN1987A. Look back times greater than 5 Gyr are affected by incompleteness (dashed histogram). *Right panel:* IMF including only stars with Hα excess (full line) or also those with U-band excess (dashed line).

References

Bessell, M.S., Castelli, F., & Plez, B. 1998, A&A, 333, 231

Panagia, N., Romaniello, M., Scuderi, S., & Kirshner, R.P. 1999, in preparation

Romaniello, M. 1998, PhD Thesis, Scuola Normale Superiore, Pisa

New Views of the Magellanic Clouds
IAU Symposium, Vol. 190, 1999
Y.-H. Chu, N.B. Suntzeff, J.E. Hesser, and D.A. Bohlender, eds.

Star Clusters in the Magellanic Type Irr Galaxy NGC 4449

Patrick Seitzer

University of Michigan, Dept. of Astronomy, 818 Dennison Bldg, Ann Arbor, MI 48109, USA

Eva Grebel

Lick Observatory, Univ. of California, Santa Cruz, CA 95064, USA

Abstract. We compare the luminosity function of star clusters in the LMC and NGC 4449: a IBm/Sm type galaxy very similar in appearance and luminosity ($M_B = -18.5$) to the LMC ($M_B = -18.4$). The brightest of several hundred clusters in NGC 4449 have $M_V = -10$ (assuming a distance of 3.9 Mpc), comparable to the brightest clusters in the LMC.

How unique is the population of star clusters in the LMC? Do all Magellanic type galaxies have a similar system of star clusters? Or does it depend on environment, and any possible interactions the parent galaxy has endured? Although there have been many studies of clusters in members of the Local Group (NGC 6822, WLM, and others), most of these systems are considerably less luminous than the LMC. Indeed, in Harris' (1991) review of globular cluster systems in galaxies, there is no mention made of any clusters in galaxies similar to the LMC. Thus it is impossible to determine if the LMC cluster system is unique.

For comparions, we studied the clusters in NGC 4449, which is classified as IBm/Sm in the Second Reference Catalog of Galaxies system, and is at a distance of 3.9 Mpc (Melisse & Israel 1994) with $M_B = -18.5$. Visually, it appears very similar to the LMC. This galaxy has been studied frequently because of a burst of recent star formation (Hunter & Gallagher 1997; Hill et al. 1994; Bothun 1986). It is isolated on the sky: the nearest large galaxy is NGC 4258 which is over 3° projected separation. But Hunter et al. (1998) show that gas in the system is highly disturbed in the outer parts, evidence perhaps for a recent interaction.

Analysis of WFPC2 archival images showed several hundred extended objects centered on the galaxy, which we interpret to be the star cluster population of NGC 4449. A total of 97 such objects brighter than $V = 21$ were classified as star clusters based on their radial profiles. The sample will be incomplete at fainter magnitudes, in the center of the galaxy, and for those clusters with half light radii smaller than several WFC pixels (2 pc at 3.9 Mpc).

It is evident that NGC 4449 has a substantial number of rich star clusters, the brightest of which are comparable to the brightest in the LMC. The figure shows the observed luminosity function of the NGC 4449 clusters along with two estimates of the LF for the LMC clusters. Because the NGC 4449 sample is much more incomplete than the LMC samples, there will be a scaling factor in log(N) which still needs to be determined.

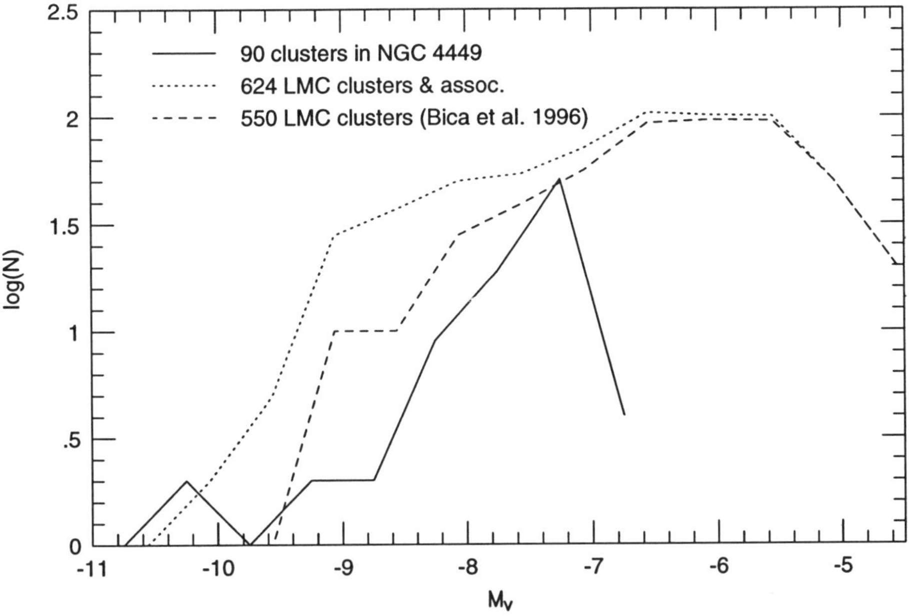

Figure 1. V band luminosity functions for clusters in NGC 4449 and the LMC.

The luminosity function is a reflection of the underlying mass function of the clusters, their age distribution, and metallicities. For the moment (lacking colors) we can say nothing about the ages and metallicities. But it is safe to conclude that NGC 4449 has a substantial population of bright clusters like the LMC. To within a scaling factor the cluster luminosity function is much like the LMC. On the basis of the recent starburst activity, we predict that the majority of these clusters will be found to have ages less than 1 Gyr.

References

Bothun, G.D. 1986, AJ, 91, 507

Harris, W.E. 1991, ARA&A, 29, 543

Hill, R.S., Home, A.T., Smith, A.M., Bruhweiler, F.C., Cheng, K.P., Hintzen, P.M.N., & Oliversen, R.J. 1994, ApJ, 430, 568

Hunter, D.A., & Gallagher, J.S. 1997, ApJ, 475, 65

Hunter, D.A., Wilcots, E.M., van Woerden, H., Gallagher, J.S., & Kohle, S. 1998, ApJ, 495, 47

Melisse, J.P., & Israel, F.P. 1994, A&AS, 103, 391

New Views of the Magellanic Clouds
IAU Symposium, Vol. 190, 1999
Y.-H. Chu, N.B. Suntzeff, J.E. Hesser, and D.A. Bohlender, eds.

Recent Star Formation History of the Magellanic Clouds

Eva K. Grebel

UCO/Lick Observatory, UCSC, Santa Cruz, CA 95064, USA

Wolfgang Brandner

JPL/IPAC, Mail Code 100-22, Pasadena, CA 91125, USA

Abstract.
A new age calibration of Cepheids and supergiants is used to study the large-scale recent star formation history of the LMC and the SMC. We find evidence for migration of star formation along the LMC bar as well as for the existence of long-lived (≈ 200 Myr) extended star-forming features.

1. Introduction

Due to their proximity the Magellanic Clouds are well suited to study large-scale star formation processes based on resolved stellar populations. The ongoing CCD surveys such as the Magellanic Clouds Photometric Survey and the EROS, MACHO, and OGLE microlensing projects are only partially available, yet the databases maintained at the Centre des Données astronomiques de Strasbourg (CDS) and at the Astrophysical Data Center (ADC, NASA) contain already a wealth of information useful for this kind of study.

We are using data on Cepheids (Artiukhina et al. 1997) and non-variable supergiants in both Clouds to trace their recent large-scale star formation history. We carried out a new period-age calibration based on Galactic and Magellanic Cepheids in clusters by fitting Geneva isochrones to the cluster main sequences. The *UBV* photometry of the non-variable supergiants (Rousseau et al. 1978; Isserstedt 1979, 1982a; Feitzinger & Isserstedt 1983; Schmidt-Kaler et al. 1998) was dereddened using Isserstedt's (1982b) intrinsic colors for Magellanic supergiants. The supergiants were age-calibrated through Geneva evolutionary tracks.

The Cepheid age calibration covers an age range of roughly 230 to 30 Myr, and the supergiants trace the past 30 to 2 Myr. 230 Myr correspond to one rotation period of the LMC. We visualized the resulting star formation history through animations to study its spatial variations with time.

2. Results for the LMC

Our data show that star formation was active in the bar region at distinct regions at certain times, while other areas were inactive. Star formation along the bar progressed from the southeast to the northwest.

Only two large activity regions appear to have formed within the past 30 Myr: 30 Dor and the surrounding supershells LMC 3 (simultaneously) and LMC 2 (a little later), and LMC 4 & 5. LMC 4 is the only supershell with a radial stellar age gradient as expected from self-propagating star formation in a spherically expanding shell. Our data contradict the suggested formation of 30 Dor through shock-shock collisions of LMC 2 and LMC 3 (Chernin et al. 1995).

Many of the areas coincident with today's supergiant shells have been active throughout the past 200 Myr forming long-lived chains of star formation. There is no clear evidence of progression of star formation along these chains, which are vaguely reminiscent of spiral arms. The large-scale age structure does not support the clock-wise age increase suggested by the bow-shock scenario of de Boer et al. (1998). The long-lived chains are best reproduced by models that include the effect of an off-centered bar on a rotating disk (Gardiner et al. 1998).

3. Results for the SMC

The SMC shows a much more complex picture since it is not seen nearly face-on like the LMC, is strongly disrupted, and has a large depth extent along the line of sight. Our data show that the most recent star formation occurred mostly along the bar of the SMC, while slightly older populations show a more widespread distribution. Recent star formation is also visible along the beginning of the bridge between the Magellanic Clouds.

For both Clouds ages and positions of young star clusters agree very well with the star formation history as outlined by Cepheids and other supergiants.

Acknowledgments. We wish to thank Jörg Isserstedt whose impressive movie of the star formation history of the LMC inspired the current study. Terry Figel's (UCSC Vizlab) help in transferring our movie to video tape is gratefully acknowledged. EKG thanks Dennis Zaritsky for support through NASA LTSA grant NAG-5-3501.

References

Artiukhina, N.M., Durlevich, O.V., Frolov, M.S., et al. 1995, General Catalogue of Variable Stars, 4th ed., vol.V, Kosmosinform, Moscow

Chernin, A.D., Efremov, Y.N., & Voinovich, P.A. 1995, MNRAS, 275, 313

de Boer, K.S., Braun, J.M., Vallenari, A., & Mebold, U. 1998, A&A, 329, L49

Gardiner, L.T., Turfus, C., & Putman, M.E. 1998, ApJ, 507, L35

Feitzinger, J.V., & Isserstedt, J. 1983, A&AS, 51, 505

Isserstedt, J. 1979, A&AS, 38, 239

Isserstedt, J. 1982a, A&AS, 50, 7

Isserstedt, J. 1982b, A&A, 115, 97

Rousseau, J., Martin, N., Prévot, L., Rebeirot, E., Robin, A., & Brunet, J.P. 1978, A&AS, 31, 243

Schmidt-Kaler, Th., Gochermann, J., Östreicher, M.O., et al. 1999, MNRAS, 306, 279

Eva Grebel would likely have won any award for most colourful poster!

Part 7. Tidal Interactions

Section A. Invited Reviews

(Top) Allen Sweigart and Peter Bergbusch ponder the modelling of those pesky observations that (bottom) Bill Harris and Jon Holtzman make.

New Views of the Magellanic Clouds
IAU Symposium, Vol. 190, 1999
Y.-H. Chu, N.B. Suntzeff, J.E. Hesser, and D.A. Bohlender, eds.

Proper Motions of the Clouds

Claudio Anguita

Dept. de Astronomía, Univ. de Chile, Casilla 36-D, Santiago, Chile

Abstract. The proper motion of the LMC relative to three background QSOs has been determined. 125 CCD frames taken from 1989.0 to 1997.2 at the Cassegrain focus of the CTIO 1.5 m telescope were used. The method of observation and reduction is breafly presented. The results are compared with those obtained by other authors who use different methods and proper motion reference systems. We find a good agreement in $\mu_\alpha \cos\delta$, but a rather large discrepancy in μ_δ. Our LMC proper motion seems to indicate that the LMC is not leading the Magellanic Stream.

1. Introduction

Since Mathewson et al.'s (1974) discovery of the Magellanic Stream (MS), all models of the Magellanic System assume that the Large Magellanic Cloud (LMC) and the Small Magellanic Cloud (SMC) lead the MS in their orbital motions. Given that the MS forms a near great circle perpendicular to the Galactic plane, the Clouds have orbits which run over the Galactic poles. For the LMC, the models of Murai & Fujimoto (1980), Lin & Lynden Bell (1982), Shuter (1992), and Gardiner et al. (1994) predict a proper motion in the range of 1.5 to 2.0 mas yr^{-1} with a position angle $\theta \approx 90°$, measured in the celestial coordinate system in the sense NESW. In the case of the SMC, only Gardiner et al. (1994) predict its proper motion. In their numerical simulations, they compute the LMC and SMC current space velocities and proper motions that give the best reproduction of observational features of the Magellanic System. One of the results they obtain is $\mu = 1.6$ mas yr^{-1} with $\theta = 150°$ for the SMC proper motion.

Three quasars behind the LMC have been reported: Q0557−6713 (Blanco & Heathcote 1986), Q0558−6707, and Q0615−6615 (Maza 1989), at 3°.9, 4°.1, and 5°.9 from the center of the LMC. These quasars provide an excellent reference system for us to measure the proper motion of the LMC. From our experience with the parallax programs at CTIO, we have been able to measure the centroid of a point-like object with a precision of 3 mas (s.d.) and the mean position with a standard error of 1–2 mas. We have therefore decided to measure the proper motion of the LMC, in order to compare with the aforementioned predictions.

2. Observations and Results

The QSOs used as inertial fiducial points to measure the absolute proper motion of the LMC were observed with the CTIO 1.5 m telescope, using CCDs attached to the f/13.5 cassegrain focus. Observations were carried out in the course of the Universidad de Chile parallax program at CTIO, over an eight year period, using a Johnson R bandpass to reduce the effect of refraction.

During the first part of the program (1989 and 1990), we used an RCA (312 × 508) chip that gives a scale of $0\farcs3$ pixel^{-1} and a field of view of $1\farcm6 \times 2\farcm5$; for the remainder of the program, two TEK (1024 × 1024) chips, giving a scale of $0\farcs24$ pixel^{-1} and a $4\farcm1 \times 4\farcm1$ field of view, were used. Nevertheless, for the last epoch of Q0557−6713 and Q0558−6707 observations, we had to used a TEK (2048 × 2048) that gives the same scale as the smaller TEK chips. All the chips we used are thinned and back-side illuminated. Ideally this kind of program should use only one chip, but this is not possible because chips wear off and are replaced by improved chips on a rapid timescale.

The LMC stars defining the stellar reference frame of each QSO field were chosen according to the following criteria: they had to be on all frames taken with a seeing $< 1\farcs7$ and to have a magnitude near R = 2.5 mag, with the QSOs at the bright part of it. Only frames with a seeing less than $1\farcs7$ were used. In order to use the frames of Q0557−6713 and Q0615−6615 taken at the beginning of the program, we limited the stellar reference frames of these QSO fields to LMC stars within the smaller field of view of the RCA CCD chip.

In order to select only LMC stars as members of the reference frame, B,V,R photometry was done for the three QSOs fields. Their color-magnitude diagrams show that the typical stellar population is that of the LMC with little contamination. These fields are almost completely dominated by LMC stars.

Coordinates (x, y) of the images centroids were calculated using DAOPHOT (Stetson 1987). On each frame, barycenter coordinates $(x_{(b)} = x - \overline{x}, y_{(b)} = y - \overline{y})$ of the reference stars and of the QSO were computed. For each QSO, the average of three frames taken in succession on a night of good seeing and with good orientation of the CCD relative to the cardinals points was chosen as the standard frame; the coordinates of the reference stars were then used to make linear transformations such that all frames have the same orientation and scale as the standard frame.

All the observations of a QSO were combined in the standard frame by

$$x_{(b,t_i)} = x_{(b,t_o)} + \mu\ x_{(b)} \times (t_i - t_o),$$

where $x_{(b,t_i)}$ is the barycenter coordinate of an object at epoch t_i, $\mu\ x_{(b)}$ is the proper motion in the coordinate x of the object relative to the barycenter of the stellar reference system, and t_o is the beginning of the year when the observations started. A similar equation is used to combine $y_{(b,t_i)}$. Before solving this set of equations, we applied corrections for color refraction which were determined empirically by taking frames at different hour angles, from three hours east to three hours west. Nevertheless, only frames taken with hour angle less than one hour were used to determine the proper motions of the objects.

To compute the proper motions of the QSOs and of LMC stars belonging to the stellar reference frame, we use the method called ordinary least-squares (OLS) regression by Isobe et al. (1990). Table 1 is a summary of the project; it gives (1) the name of the quasar; (2) and (3) the right ascension and declination components of the LMC proper motion, which are the "reflex" motion of the QSOs relative the LMC stars of the area, and their standard deviations; (4) the number of frames; (5) the number of epochs and (6) the observation period.

Table 1. Proper Motion of the LMC

relative to	LMC $\mu_\alpha\cos(\delta)$ (mas yr^{-1})	μ_δ	Number of: frames	epochs	observation period
Q0557−6713	+1.4 ± 0.3	+3.1 ± 0.2	61	12	1989.02−1996.86
Q0558−6707	+1.3 ± 0.4	+3.4 ± 0.6	32	7	1992.81−1996.86
Q0615−6615	+1.7 ± 0.2	+2.8 ± 0.3	32	8	1989.90−1997.17
The three QSOs (weighted mean)	+1.6 ± 0.2	+3.0 ± 0.2			

3. Comparison with Previous Measurements

In the upper part of Table 2, we give the results of all measurements known of the LMC proper motion with precisions less than 1 mas yr^{-1} in both μ components, and the reference system used in the project. In the case of Jones et al. (1994) and Anguita et al. (1999), the proper motion components are given for the field where the observations were done, and for the center of rotation of the Cloud. The last values include corrections for rotation of the LMC plane and for solar motion. Similar information is given for the SMC in the lower part of Table 2.

Kroupa & Bastian (1997) calculate the weighted mean of the three independent LMC proper motion measurements known at the moment, and obtained $\mu_\alpha\cos(\delta)$ = 1.61 ± 0.19 mas yr^{-1} and μ_δ = -0.06 ± 0.21 mas yr^{-1}, which correspond to μ = 1.61 ± 0.28 mas yr^{-1} with $\theta = 92°$, in agreement with model predictions. Comparing those values with our measurements we find that the R.A. components are quite similar, but that there is a large discrepancy between the two values of the declination components. The LMC proper motion we calculate is μ = 3.3 ± 0.2 mas yr^{-1} with $\theta = 31°$. The direction of our proper motion vector seems to indicate that the LMC is not leading the Magellanic Stream. In relation to this point, it is interesting to note that the LMC proper motion (μ = 1.7 mas yr^{-1}, $\theta = 50°$) measured by Kroupa et al. (1994), using stars of the PPM Catalogue, is another vector that does not have the direction predicted by theoreticians.

What makes these results indeed interesting is that our value of the LMC proper motion is larger by a considerable amount than any prediction from models considering the Clouds gravitationally bound or unbound to the Galaxy, yet the fact that the same telescope was used during a relative short span between initial and final epochs (eight years) and that a QSO is used as reference should make the outcome highly reliable.

CCD relative astrometry of ten QSOs behind the SMC given by Tinney et al. (1997) is needed to address the discrepancy found in this project and to increase our knowledge of the kinematic properties of the Clouds. Unfortunately, the SMC proper motion, $\mu = 1.73 \pm 0.80$ mas yr^{-1} (Kroupa & Bastian 1997) calculated with 11 stars from data obtained with the Hipparcos satellite, has an uncertainty too large to give valuable information that could be used to analyze

Table 2. Proper Motion of the Magellanic Clouds

Source		LMC $\mu_\alpha \cos(\delta)$ (mas yr^{-1})	μ_δ	Reference System
Kroupa, et al. (1994)		+1.3 ± 0.6	+1.1 ± 0.7	PPM
Jones, et al. (1994)	field	+1.20 ± 0.28	+0.26 ± 0.27	Galaxies
	(LMC center)	+1.37 ± 0.28	−0.18 ± 0.27	
Kroupa & Bastian (1997)		+1.94 ± 0.29	−0.14± 0.36	Hipparcos
Anguita, et al. (1998)	field	+1.6 ± 0.2	+3.0 ± 0.2	Quasars
	(LMC center)	+1.7 ± 0.2	+2.8 ± 0.2	
Source		**SMC $\mu_\alpha \cos(\delta)$ (mas yr^{-1})**	**μ_δ**	**Reference System**
Kroupa, et al. (1994)		+0.5 ± 1.0	−2.0 ± 1.4	PPM
Kroupa & Bastian (1997)		+1.23 ± 0.84	−1.21± 0.75	Hipparcos

the results obtained with Hipparcos and future ground-based CCD observations of SMC regions.

References

Anguita, C., Loyola, P., & Pedreros, M. 1999, in preparation
Blanco, V.M., & Heathcote, S. 1986, PASP, 98, 635
Gardiner, L.T., Sawa, T., & Fujimoto, M. 1994, MNRAS, 206, 565
Isobe, T., Feigelson, E.D., Akritas, M.G., & Babu, G.J. 1990, ApJ, 364, 104
Jones, B.F., Klemola, A.R., & Lin, D.N.G. 1994, AJ, 107, 1333
Kroupa, P., & Bastian, U. 1997, New Astron., 2, 77
Kroupa, P., Röser, S., & Bastian, U. 1994, MNRAS, 266, 412
Lin, D.N.C., & Lynden-Bell, D. 1982, MNRAS, 198, 707
Mathewson, D.S., Cleary, M.N., & Murray, J.D. 1974, ApJ, 190, 291
Maza, J. 1989, private communication
Murai, T., & Fujimoto, M. 1980, PASJ, 32, 581
Shuter, W.L. 1992, ApJ, 386, 101
Stetson, P.B. 1987, PASP, 99, 191
Tinney, C.G., Da Costa G.S., & Zinnecker, H. 1997, MNRAS, 285, 111

Discussion

Steven Majewski: How do you account for color terms in your astrometric solution? Because the mean color and shape of the spectral energy distribution is very different between your QSOs and the Magellanic stars at the same magnitude (i.e. red giants) this is a critical correction. Have you checked your solution against the reference frame provided by redder, compact galaxies that you should have in your frames (which go to $V = 22$)?

Anguita: For several years I have been measuring CCD trigonometric parallaxes with the same telescope, so I know quite well how to determine and apply a correction we know as Differential Colour Refraction.

No, I have not checked our solution against the reference frame you mentioned. It could be an interesting test to do it.

William Kunkel: What assumptions did you employ about the rotation of the LMC?

Anguita: I measured the proper motions of three different LMC fields. To obtain the corresponding values of the LMC center of rotaton, I used the assumptions employed by Lin in 1994. In any case, the corrections were, at most, 0.0002 mas yr^{-1}.

Lance Gardiner: Are you aware that Mike Irwin has recently measured the proper motion of the SMC using CTIO 4m plates and found that the SMC is leading the Magellanic Stream?

Anguita: I am sorry, but I have no information about the SMC proper motion which, as you say, has been recently measured.

New Views of the Magellanic Clouds
IAU Symposium, Vol. 190, 1999
Y.-H. Chu, N.B. Suntzeff, J.E. Hesser, and D.A. Bohlender, eds.

Review of *N*-Body Models of Tidal Interactions

L.T. Gardiner

Dept. of International Education, Sun Moon University, Rep. of Korea

Abstract. Considerable progress has been made in the current decade with the help of *N*-body simulations towards a deeper understanding of the nature of the dynamical relationship between the Large and Small Magellanic Clouds and the Galaxy. The origin of such features as the Magellanic Stream, the inter-Cloud Bridge, the Wing and large extension in depth of the SMC has come to be interpreted in the context of the tidal interactions among the members of the LMC-SMC-Galaxy triple system. The inclusion of gas-dynamical effects and star formation in the latest models has added further refinements to this picture and confirmed the enhancement of the star formation rate in the SMC as a result of the recent LMC-SMC encounter.

1. Introduction and Numerical Models

Test particle models of the Magellanic Clouds system (e.g., Murai & Fujimoto 1980; Gardiner et al. 1994) have been successful in searching a large parameter space for orbits leading to simulation of the gross features related to the tidal disruption of the Clouds. However, in order to obtain a more realistic picture of the effects of tidal interactions on the Magellanic Clouds, it has been necessary to construct *N*-body models in which the global galactic potential responds dynamically to the evolution of the matter distribution.

There have been only a limited number of studies specifically of the Magellanic Clouds which have employed *N*-body techniques. Moore & Davis (1994) carried out a high-resolution simulation of tidal stripping of the LMC, and found that the tidal debris defined a thick plane centered on the Magellanic orbital plane. On this basis, they argued against the validity of a tidal origin for the Magellanic Stream (MS). Gardiner & Noguchi (1996) produced an *N*-body model of the SMC, Magellanic Stream and inter-Cloud region, which has been extended by Yoshizawa (1998) to include gas-dynamical effects and star formation. Kunkel et al. (1997) carried out smaller scale *N*-body simulations of the LMC-SMC system. Their work focussed mainly on the interpretation of the kinematics of carbon stars observed at the periphery of the LMC. Li & Thronson (1999, these proceedings) have recently developed a self-consistent smoothed-particle hydrodynamics (SPH) model of the Galaxy-LMC-SMC system. Their work emphasises the importance of accurate treatment of the gaseous components responsible for the production of the MS and the bridge between the Clouds.

In the remainder of this presentation, I review the major results produced by the Magellanic system models of Gardiner & Noguchi (1996, he reafter GN96) and Yoshizawa (1998, hereafter Y98), which represent the most extensive completed studies of the tidal interactions of the Magellanic system to date. In these models, which are based on the same assumptions (see GN96 for details), the Galaxy and the LMC were represented by fixed potentials, whereas the SMC was constituted as a self-gravitating system of 15000 collisionless (GN96) or 25000 gas/star particles (Y98) with a disk/halo mass ratio of 1:1 (GN96) or 3:7 (Y98). The Magellanic orbits used for the models were derived from test particle modeling of the Magellanic system by Gardiner et al. (1994). The latter models predicted polar orbits with the Clouds leading the MS and gave current space velocities consistent with subsequent proper motion measurements (Kroupa & Bastian 1997). The inclusion of gas-dynamics and star formation in the Y98 model was based on the use of the 'sticky-particle' method for gas cloud collisions together with the application of a Schmidt Law for star formation above some specified volume density threshold. The models were computed from $T = -2$ Gyr to the present, $T = 0$.

2. The Magellanic Stream and Leading Arm Feature

In the above-mentioned Magellanic system models, the MS is a tidal tail torn from the SMC as a result of the interaction between the Magellanic Clouds and the Galaxy. More specifically, a bridge-tail structure emerges from the SMC as a result of a relatively close encounter with the LMC 1.5 Gyr ago, and these features become extended by the tidal force of the Galaxy into a long trailing tail and a weaker leading counterpart (see Fig. 6 of GN96). Some major characteristics of the MS were reproduced by the simulations, including its emergence from a region between the Clouds to form a long ($> 100°$) arc in the sky, its mass of a few $10^8 M_\odot$, and its velocity profile.

The models address two major criticisms of the tidal theory for the origin of the MS: (1) the lack of observational evidence for an equally prominent leading bridge counterpart to the striking trailing tail, and (2) the absence of stars in the gaseous MS. The GN96 model predicted the existence of a relatively weak leading arm feature (LAF) on the opposite side of the Clouds to the MS whose overall geometry resembled the distribution of scattered high-velocity clouds in the region $260° < l < 310°$, $-30° < b < 30°$. The HI survey results of Putman et al. (1998) have confirmed that the section of the proposed LAF with $b < -8°$ comprises a continuous feature. These observations, combined with a metallicity determination for HVC 287.5+22.5+240 by Lu et al. (1998), strongly suggest that the material in the proposed LAF originated in the Clouds. The weakness of the LAF is explained by the models as resulting from the asymmetry introduced by the LMC-SMC binary interaction. The lack of stars in the MS, considered enigmatic since tidal forces should affect stars and gas equally, is likely to be explained by the fact that the material in the simulated MS was torn from the outer part of the SMC disk beyond 3 kpc from the SMC center. The Y98 model, whose initial configuration comprised a compact stellar disk (radius ~ 3 kpc) embedded in a more extended gaseous component (radius ~ 5 kpc), introduced very few stars into the MS. Furthermore, the low gas density of material in

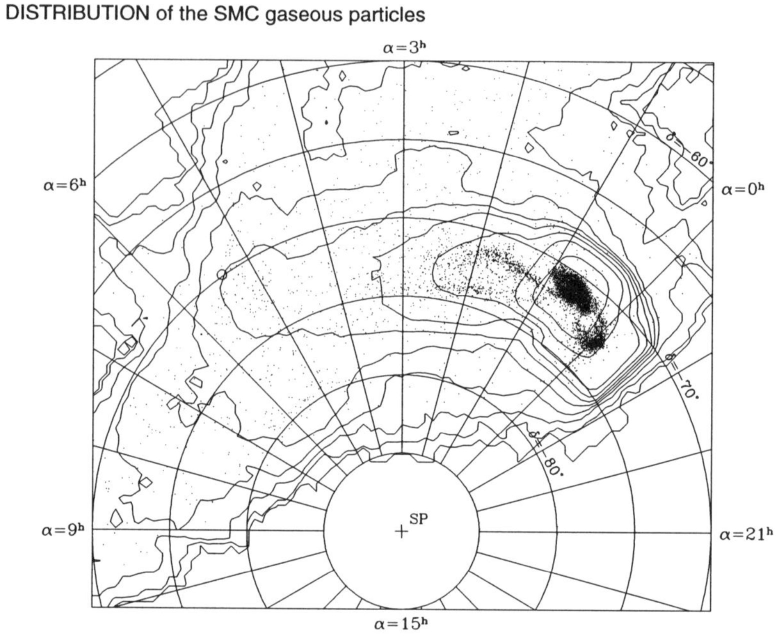

Figure 1. Distribution of SMC gas particles projected on the sky in the Y98 simulation. Contours denote the surface number density of particles weighted by the inverse square of distance.

the simulated MS accounts for the absence of star formation activity since its creation.

3. Structure of the SMC and the Inter-Cloud Region

The orbits employed in the models predict a close encounter between the LMC and the SMC occurring 0.2 Gyr ago at a separation of 7 kpc. This close passage drew out a prominent bridge-tail feature which can account for many characteristics of the structure and kinematics of the SMC and the inter-Cloud region (ICR). Fig. 1 shows the gas particle distribution produced by the Y98 model projected on the sky in the region of the Clouds. This distribution shows the beginnings of the MS in the north and an extension of material from the SMC towards the LMC forming the inter-Cloud bridge. The gas forms a narrower distribution in the ICR than in the GN96 model owing to the inclusion of gas dissipation, which keeps the SMC gas disk dynamically cold.

The tidal bridge and tail structures produced by the simulation are seen roughly superposed along the line-of-sight, leading to a large proposed depth for the ICR. In Fig. 2, the distances of particles from the sun given by the GN96 model are shown plotted against right ascension. The figure clearly indicates the separation between the bridge (closer) and tail (farther) structures along the line-of-sight, and suggests that stellar associations in the ICR may be related to one or other of these structures. Future accurate distance and velocity measurements

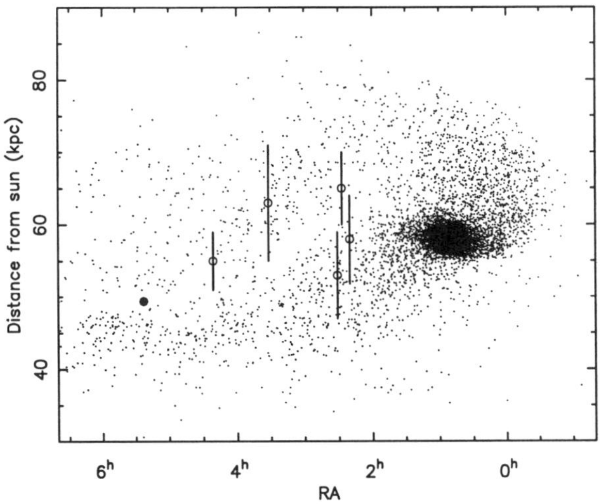

Figure 2. Line-of-sight structure in the GN96 simulation. The distances of the particles have been plotted against right ascension. The LMC is represented by a large dot. Also shown (large open circles) are the distances of five stellar associations in the inter-Cloud region with measurement errors (Grondin et al. 1992).

for the ICR associations can provide important data for evaluation of current models.

The existence of the bridge-tail structure originating in the SMC also gives rise to a large SMC depth, as indicated by Cepheid (Caldwell & Coulson 1986) and red clump star observations (e.g., Hatzidimitriou & Hawkins 1989). The simulations suggest that the SMC is elongated along the line-of-sight with material spanning a distance range of 50 kpc to nearly 80 kpc. The largest depths are predicted in the eastern regions, in agreement with red clump star studies.

The structure and kinematics of the central regions of the SMC appear to result directly from the existence of a strong bar ~ 5 kpc in length seen almost end-on (see GN96; note, however, that the Y98 model did not develop a bar due to the stabilising influence of the heavier halo). In particular, the highly non-circular motions of particles in the SMC bar result in a steep gradient in the radial velocity–right ascension plane (refer to Fig. 10 of GN96).

4. Star Formation in the SMC

The inclusion of a star-formation algorithm in the numerical code employed by Y98 allows useful insight to be gained into the effects of tidal interactions on star formation activity in the SMC. Fig. 3 shows the star formation rate (SFR) as a function of time. The plot indicates that the SFR in the SMC is strongly enhanced by the tidal encounters with the LMC occurring 1.5 Gyr (14 kpc separation) and 0.2 Gyr ago (7 kpc separation) by up to a factor of $\sim 3-4$. Interestingly, the peak of the SFR following the first interaction occurs 200-300

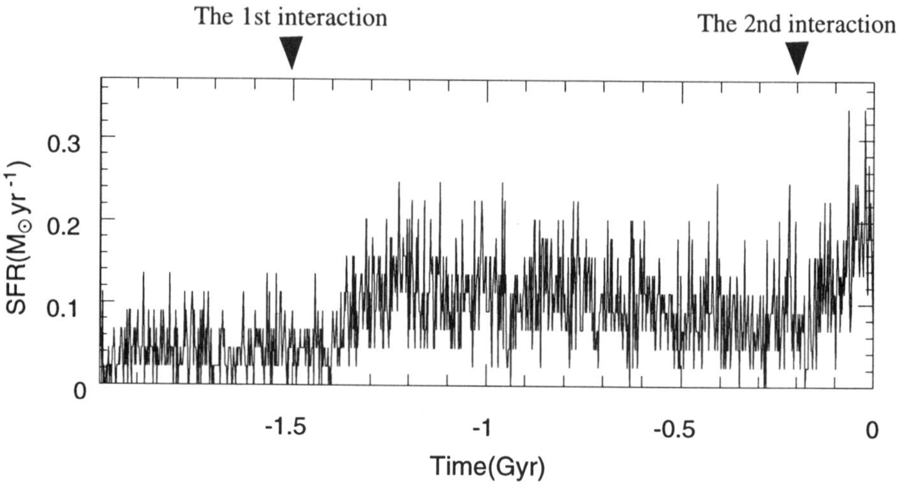

Figure 3. Star formation history (star-formation rate versus time) from the Y98 model.

Myr after the encounter event, while the SFR may still be rising at the present epoch 200 Myr after the second interaction. These enhancements and delays are consistent with the results of interacting galaxy models by Noguchi (1986), which suggest that star formation is triggered in spiral features produced by the tidal interaction.

The SMC Wing, containing numerous young stellar associations, is associated with star formation activity occurring less than 100 Myr ago. There is a correspondingly strong feature in the gas distribution in the model between $1^h < \alpha < 2^h$ in Fig. 1. In the Y98 model, a significant enhancement of star formation activity within the past 300 Myr is found in this region, confirming that the SMC Wing is a direct product of tidally induced star formation. The model did not give rise to star formation further east of $\alpha = 2^h$ because of the low gas density in the region between the Clouds. Nevertheless, a small number of faint stellar aggregates have been detected between the Clouds (Irwin et al. 1990).

5. Future Work

The continued accumulation of observational data related to the Magellanic Clouds system offers new challenges for future modelling of tidal interactions. The spectacular observational confirmation by Putman et al. (1998) of the existence of the leading counterpart to the Magellanic Stream which was predicted by the GN96 model testifies to the importance of numerical simulations in understanding the dynamical interactions of the Magellanic Clouds–Galaxy system. Recent work (see Gardiner 1999) indicates that the inclusion of a weak drag force due to ram pressure exerted by diffuse gas in the Galactic halo may lead to improved modelling of the leading arm feature. The effects of tidal interactions on the LMC have so far been little explored. It is now clear from HI observations (Putman et al. 1998) that the LMC also makes a significant contribution to the gas in the ICR. In addition, it would be interesting to discern whether the LMC's off-center bar is a result of tidal interactions or due to a spontaneous internal instability. As for numerical techniques, models are becoming increasingly detailed ($> 10^5$ particles) and sophisticated, involving accurate treatment of the gaseous component and/or star formation. However, there is still an important role for relatively straightforward (and less CPU-intensive) collisionless N-body models and test particle simulations in surveying the parameter space of possible orbits and model configurations that are hitherto only loosely constrained by observations.

Acknowledgments. I would like to thank A.Yoshizawa for his communication of research results and permission to reproduce Figs. 1 and 3 from his PhD thesis.

References

Caldwell, J.A.R., & Coulson, I.M. 1986, MNRAS, 218, 223

Gardiner, L.T. 1999, in "Proceedings from the Stromlo Workshop on High-Velocity Clouds", eds Gibson, B.K., & Putman, M.E., p.292

Gardiner, L.T., & Noguchi, M. 1996, MNRAS, 278, 191 (GN96)

Gardiner, L.T., Sawa, T., & Fujimoto, M. 1994, MNRAS, 266, 567

Grondin, L., Demers, S., & Kunkel, W.E. 1992, AJ, 103,1234

Hatzidimitriou, D., & Hawkins, M.R.S. 1989, MNRAS, 241, 667

Irwin, M.J., Demers, S., & Kunkel, W.E. 1990, AJ, 99, 191

Kroupa, P., & Bastian, U. 1997, New Astron., 2, 77

Kunkel, W.E., Demers, S., Irwin, M.J., & Albert, L. 1997, ApJ, 488, L129

Lu, L., Savage, B.D., Sembach, K.R., Wakker, B.P., Sargent, W.L.W., & Oosterloo, T.A. 1998, AJ, 115, 162

Moore, B., & Davis, M. 1994, MNRAS, 270, 209

Murai, T., & Fujimoto, M. 1980, PASJ, 32, 581

Noguchi, M. 1986, MNRAS, 219, 305

Putman, M.E., Gibson, B.K., Staveley-Smith, L. et al., 1998, Nature, 394, 752

Yoshizawa, A. 1998, PhD thesis, Tohoku University, Sendai, Japan (Y98)

Discussion

Despina Hatzidimitriou: We do not yet know how big a kinematic halo (if any) the SMC has. So, it may be a good idea to explore the possibility that the SMC is a pure disk, as a limiting case. It may be worth investigating what happens to the rotation curve, if you assume a rotation curve before the last two close encounters.

Gardiner: In future models it will be good to explore the effect of changing the disk-to-halo mass ratio. However, I believe the observational evidence does support the existence of a kinematic halo in the SMC.

K. Innanen: If the Magellanic Clouds are natural, primodial satellites of the Galaxy, one asks the question: how did they acquire the angular momentum of their orbit(s) around the Galaxy? One possibility is that they may not be primodial satellites, but have arrived by some other dynamical event in the Local Group many Myr ago.

Gardiner: Yes, I agree. The Magellanic Clouds could have been subjected to torques exerted by other Local Group galaxies.

Hans Zinnecker: What was the critical threshold gas density or surface gas density that was used to describe the star formation rate in the Yoshizawa model? Secondly, what is the reason for the (model) 200 Myr delay for the onset of star formation after a close LMC-SMC encounter?

Gardiner: The critical surface density threshold was 25×10^{20} H cm^{-2} (Kennicutt et al. 1995). A scale height of $\approx$ 450 pc was assumed for the HI layer in the SMC, giving a critical volume density of 0.04 $M_{\odot}pc^{-3}$.

The delay appears to be related to the time taken for the tidal bridge and tail to form and produce density enhancements in the gas.

New Views of the Magellanic Clouds
IAU Symposium, Vol. 190, 1999
Y.-H. Chu, N.B. Suntzeff, J.E. Hesser, and D.A. Bohlender, eds.

The Aftermaths of the SMC/LMC Encounter

Serge Demers

Département de Physique, Université de Montréal, Montreal Qc, Canada H3C 3J7

W. E. Kunkel

Las Campanas Observatory, Carnegie Institution of Washington, La Serera, Chile

Abstract. We present structural and kinematic evidence pointing to a gravitational interaction between the two Magellanic Clouds. We discuss, in particular, the results of our kinematic survey of carbon stars in the periphery of the LMC. Stars with peculiar velocities, in the LMC south-west, are identified as part of debris left on the SMC orbit around the LMC.

1. Introduction

The determination of the proper motion of the SMC by Irwin et al. (1996) implies that it is most certainly bound to the LMC and that it is currently moving away from it. The close encounter presumed to have taken place some 200 Myr ago is once more supported by observations. The analysis of the radial velocities of nearly 1000 carbon stars (Kunkel et al. 1997a) in the periphery of both Clouds brings new facts, not only about the mass of the LMC (Kunkel et al. 1997b) but also about the geometry of the interaction. Some topics of this analysis are discussed here.

2. The Inter-cloud Young Population

Young star aggregates, extending east of the SMC wing, were discovered by Irwin et al. (1990). These stellar groups were first investigated by Grondin et al. (1992) and more recently by Demers & Battinelli (1998) who showed that the distance to the five major aggregates decreases from the SMC side to the LMC side. They are indeed part of a "bridge" linking the two Clouds. Furthermore, the stellar population making up these aggregates was found to be very young: 10 to 25 Myr. This implies that some delayed star formation mechanisms must be at work in this region, if we believe that the tidal impact took place $\sim$ 200 Myr ago.

The kinematics of the young inter-Cloud population is still barely known. It is, for example, not known if these stars are all part of a tidal bridge or if some of them are in the tidal tail which, according to interaction models, would be, seen from earth, behind the bridge. Clues concerning this come from the radial velocity of the two brightest stars in aggregate ICA 76 - the eastern most one, 8° from the LMC center, which differs by 80 km s^{-1} (Demers et al. 1991). We are dealing here with main sequence stars of V > 17 thus requiring a large telescope to obtain spectra.

The further extension of the young inter-Cloud population toward the LMC has recently been investigated by Battinelli & Demers (1999) who showed that blue stars are present in the western periphery of the LMC, 6° from its center, but no stars as young as the ones of the inter-Cloud aggregates are seen. The origin of this patch of blue stars is not currently known.

3. The Carbon Stars

In order to increase substantially the number of kinematic probes we directed our interest toward carbon stars. Such red stars are easy to identify from color magnitude diagrams but they require spectra to confirm their nature. Carbon stars are more numerous and brighter than inter-Cloud upper main sequence stars. Furthermore, they are much more numerous than clusters to define the kinematics at large distance from the center of the LMC or the SMC. They are known to be of intermediate age, which means that they did not witness the first stages of the LMC or SMC formation. The few hundred carbon stars observed by us in the LMC have a velocity dispersion, (taking into account the translation and the rotation of the LMC), of 16 km s^{-1} which compared to the velocity dispersion of LPV (Hughes et al. 1991) must be 1 to 2 Gyr old, old enough to have experienced the last encounter.

As explained by Kunkel et al. (1997a), the red candidates for spectroscipic follow-ups are selected from UKST plate pairs scanned by the APM. As we reach smaller radial distances from the center of the LMC, the number of candidates becomes quite high. For this reason, in the inner zone, only very red stars were spectroscopically observed. Our survey of carbon stars is complete in the LMC periphery outside a deprojected radius of 6.2°. For the deprojection, we adopt the HI rotation center for the LMC center, an inclination of 33° and a line of nodes at 168°.

As we can see from Figure 1, the distributions of the LMC carbon stars and of clusters are quite different in the periphery. The carbon stars extend to larger distances, one in the north east is as far as 20° from its center. There are few clusters outside of the 6.2° circle while there are many carbon stars. The reason for this different distribution is not known. Tidal interaction may have more facility to redistribute individual stars than clusters. It is also obvious that the cluster and carbon star distributions are asymmetric relative to the HI rotation center. The distributions are displaced toward the east.

4. Radial Velocities

Carbon stars as well as clusters, LPVs and HI can be used to investigate the orbital velocity of LMC members as a function of their radial distance. This has been described by Kunkel et al. (1997b) who used the data to estimate the LMC mass. Figure 2 presents the orbital velocities of the above mentioned probes for two annulii centered on the LMC. These velocities are corrected for the solar motions and for the translation motion of the LMC. Panel a) corresponds to the 4 to 20 degree zone and b) represents the 3 to 5 degree zone. One can easily see, in the upper panel, the points near PA = 220 which do not follow the sinusoidal curve. These are, mostly hydrogen, in the inter-Cloud area. In the inner zone, displayed on panel b), one can identify points which do not follow the rotation. Carbon stars are among the points which "do not belong".

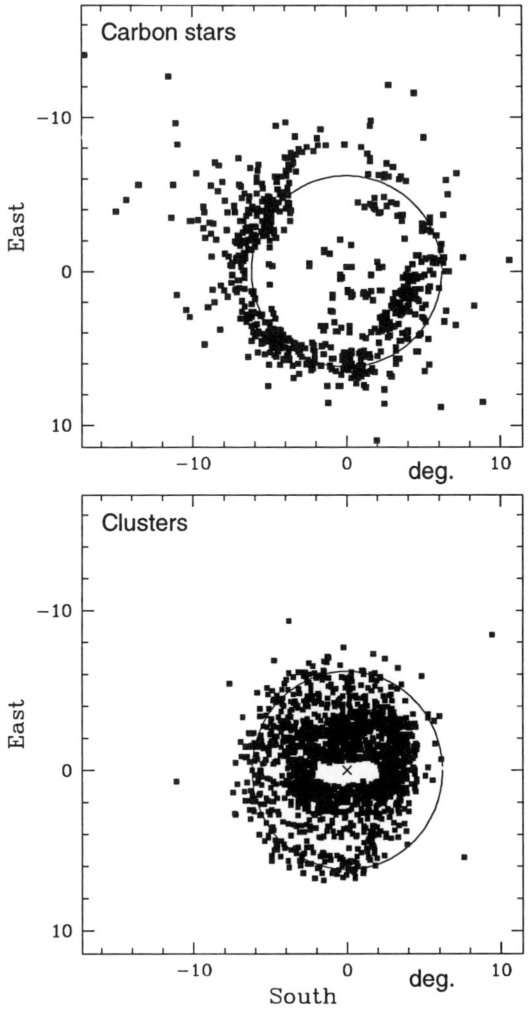

Figure 1. Distribution of carbon stars and clusters on the deprojected plane of the LMC. The circle has a radius of 6.2°.

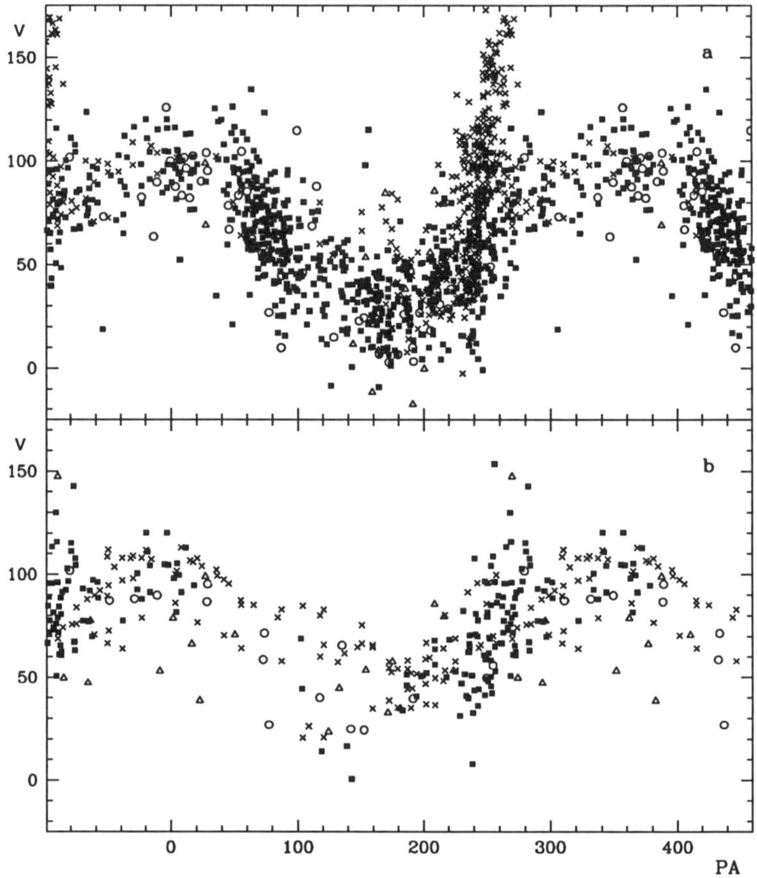

Figure 2. Radial velocities as a function of the deprojected position angles. Squares: carbon stars; circles: clusters; triangles: LPVs; crosses: HI. a) is the 4° to 20° zone. b) is the 3° to 5° zone

5. Conclusion

We believe that certain carbon stars, in the southwest of the LMC, do not belong to the LMC but are debris left along the SMC orbit. The peculiar radial velocity of these stars made them debris candidates. The ill-defined absolute magnitude of carbon stars prevents us from establishing if these stars lie in front or behind the LMC. One should search for RR Lyrae, red clump stars or globular clusters associated with this kinematic peculiarity to map in 3D the suspected orbit of the SMC.

Acknowledgments. This project was financially supported, in parts (S. D.) by the Natural Sciences and Engineering Research Council of Canada.

References

Battinelli, P., & Demers, S. 1999, this volume

Demers, S., Grondin, L., Irwin, M.J., & Kunkel, W.E. 1991, in The Magellanic Clouds, R. Haynes & D. Milne, Dordrecht: Kluwer, 473

Demers, S., & Battinelli, P. 1998, AJ, 115, 154

Grondin, L., Demers, S., & Kunkel, W.E. 1992, AJ, 103, 1234

Hughes, S.M.G., Wood, P.R., & Reid, N. 1991 in The Magellanic Clouds, R. Haynes & D. Milne, Dordrecht: Kluwer, 81

Irwin, M.J., Demers, S., & Kunkel, W.E. 1990, AJ, 99, 191

Irwin, M.J., Demers, S., & Kunkel, W.E. 1996, BAAS, 28, 932

Kunkel, W.E., Irwin, M.J., & Demers, S. 1997a, A&AS, 122, 463

Kunkel, W.E., Demers, S., Irwin, M.J., & Albert, L. 1997b, ApJ, 488, L133

New Views of the Magellanic Clouds
IAU Symposium, Vol. 190, 1999
Y.-H. Chu, N.B. Suntzeff, J.E. Hesser, and D.A. Bohlender, eds.

The Magellanic Cloud Interaction: Interpretation and Constraints from Observations of Carbon Stars as "Test Particles"

William Kunkel

Las Campanas Observatory, Carnegie Institution of Washington, La Serena, Chile

Serge Demers

Université de Montréal, Montréal, P.Q., Canada

Abstract.
Since gravity alone acts on them, and their evolutionary age restricts sampling to "mid-life", carbon stars are examined as tracers of recent dynamical evolution of the Magellanic Clouds. Ablation processes modify the distribution of HI, and velocity outliers reflect recent tidal interactios.

1. Introduction

Relying on simulations to understand the dynamical evolution of the Magellanic Clouds depends on our understanding the behavior of test particles, and when our expectations are not met, errors of interpretation may ensue. The present investigation centers on carbon stars as test particles, given their favorable properties, including the assurance that, unlike neutral hydrogen streams surrounding the Clouds, gravity alone acts on stars.

Carbon stars are luminous, and can be observed spectrographically to $(V - M_v) \sim 22$ with a 2.5-m aperture. Their evolutionary phase sets in after at least half a crossing time has elpased, and does not persist beyond six Gyr (Frogel, Mould, & Blanco 1990). Their kinematic information is decoupled from recent star forming environments, nor is it masked by the oldest collapse and accretion processes that gave birth to galaxies. The equivalent width of the CN absorption features in raw spectra is high, good correlation peaks can be measured for signal-to-noise ratios in raw spectra as low as five. Carbon stars are abundant, readily observed in large numbers, even on the outermost periphery of these systems. And gravity alone determines their motion. Hence they serve as a benchmark against which the reliability of other diagnostics can be evaluated.

2. The SMC

Figure 1 shows a sky projection of the SMC combining 150 carbon stars from Kunkel et al. (1997a, 1999), and an additional 48 stars from Hatzidimitriou et al. (1997). An annulus of 2.2° width and 7° radius about the photocenter ($l = 302.8°, b = -44.3°$) samples a zone of Galactocentric radial velocities, corrected

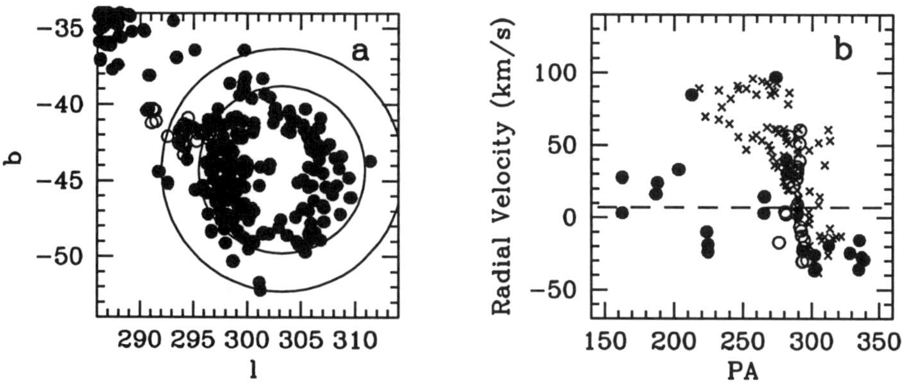

Figure 1. SMC periphery objects in Galactic coordinates, left panel. Galactocentric radial velocities vs. *PA* within the annulus, right panel. Carbon stars are dots, early type stars are open symbols, and HI are crosses. The dashed line represents the mean SMC velocity

for motion in the plane of the sky toward $PA\sim150°$ of 180 km s^{-1} (Irwin 1996), shown in Figure 1b. If the outer portions of the SMC are currently breaking up, with an expanding velocity gradient of 8 km $s^{-1}kpc^{-1}$ (Hatzidimitriou et al. 1993); further, if these detached outermost components conserve their orbital angular momentum about the LMC, then a velocity difference in the line of sight of 120 km s^{-1} corresponds to 15 kpc where Hatzidimitriou et al. measured, and somewhat more, closer to the LMC. In Figure 1b the slope of the upper envelope of the HI velocity with *PA* then implies that we are seeing points separated by 14 kpc in the line of sight, projected on but 4.4 kpc of the sky ($272° < PA < 316°$): a sheet inclined at an angle to the line of sight of $18\pm3°$, with the more remote edge toward the North. Carbon stars lie predominantly at the lower velocities, approaching the observer. The HI dominates the more positive velocities and lower *PA*'s than either the stars of early type or the majority of carbon stars. The latter lie at *PA*'s furthest South in a continuation of the same sheet. Since the ratio of carbon stars to HI varies with *PA*, and orbital motion is to the right, or increasing *PA*'s, the simplest hypothesis explaining the dearth of gas at larger *PA*'s is that neutral gas on the near side has disappeared, most probably by collisional processes involving the hot Galactic corona, analogous to those observed via Fabry-Perot spectrophotometry of Hα emission on the leading edges of Magellanic Stream fragments observed by Weiner & Williams (1996).

3. The LMC

Carbon stars on the LMC periphery are seen to be asymmetrically distributed with respect to the rotation center (at $l = 279.9°, b = -33.2°$), (Figure 2). In the plane of the LMC disk, inclined at 33° with a line of nodes at 168° (Feitzinger et al. 1977), the furthest extent of this periphery lies to the NE, near $PA \sim 55°$, opposite the SMC, while the shortest radius lies toward the SMC. The effect was first shown in the isopleths of the LMC envelope in Figure 3 of Irwin (1991). The HI shows quite the opposite orientation (McGee & Milton 1966). The shortest radius to the HI periphery is toward the NE in all reported surveys, including that of Putman, Gibson, and Stavely-Smith (1999).

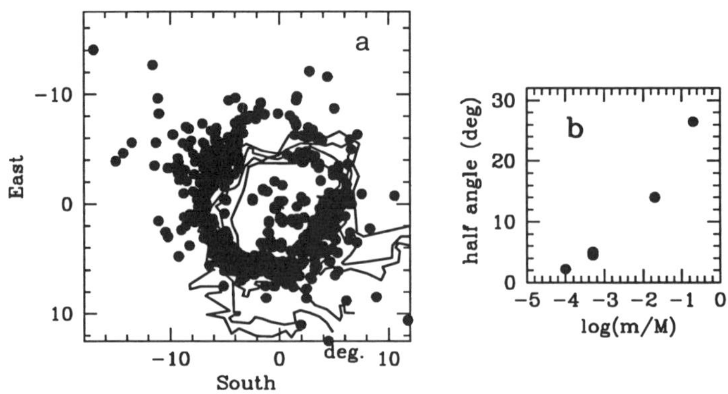

Figure 2. Map of LMC periphery carbon stars (RA, dec), with HI contours from McGee and Milton (1966), left panel. Half angle of debris stream vs. $log(m/M)$, right panel.

No single mechanism can generate the dissimilarity in these asymmetries. On stars gravity alone acts, so an additional, second mechanism must act on the gas, in addition to gravity. The clue common to both the LMC and SMC is the depletion of the gas which is in the same sense in both Clouds: in the direction of orbital motion. These observations lead to the conclusion that by itself the dynamical behavior of the gas is an unreliable guide to the orbital motion of the Magellanic Clouds, obscuring the details of their origin.

4. Closing remarks

In earlier work Kunkel et al. (1997b) have shown the usefulness of PP $N-$body codes for interpreting the data, relying on the *minimization* of the parameter count as the dominant constraint. Tidal phenomena seen in those simulations were found sufficient to account for the major perturbations observed. Toomre & Toomre (1972), Moore & Davis (1994) and Piatek & Prior (1995) have shown that a strong tidal impulse from a remote massive perturber forms a bridge and tail almost symmetrically on opposite sides of a victim. A small perturber, on the other hand, penetrating the LMC disk with pericenter toward the NE at 5 to 7 kpc leading the LMC in orbit almost always produces an asymmetric periphery toward the NE, often leaving an off-center bar, in the manner observed (Kunkel et al. 1999).

These simulations also demonstrate a "slingshot" acceleration of a tiny fraction of LMC stars, under one percent, flung forward 15 kpc and more, ahead in orbit, by the SMC penetration, with Galactocentric velocities between +20 and +100 km s^{-1}. To two carbon stars in this class from the original survey new data from Majewski et al. (1999) report six K giants in this category.

Besides the six "slingshots" in the six fields to the NE, the survey of Majewski et al. (1999) found some 20 K giants in the field NE5, all with negative velocities between –230 and –50 km s^{-1}, peaking near –170 km s^{-1}. These return to the Galaxy, appearing to belong to a population of stars in the Magellanic Plane (Kunkel and Demers 1976), whose origin in a prior encounter of the Magellanic Clouds with the Galaxy was suggested by Kunkel (1979). Their narrow distribution on the sky points to an early separation from the Magellanic progenitor. The infalling K giants belonged predominantly to NE5 and were not strongly evident in the adjacent fields NE4 and NE6, each some 3° away. By applying equation (5) of Kunkel (1979), calibrated with a series of simulations shown in Figure 2, their distribution suggests that the mass of the tidal fragment to which they belong was less than 0.05% the mass of the Galaxy. The positive velocity "slingshot" stars, on the other hand, seen with a broader distribution covering all six NE fields, track the more energetic disruptions to which the current survivors (i.e., the MC's) of a Searle & Zinn (1978) scenario are subjected today.

References

Feitzinger, J.V., Isserstedt, J., & Schmidt-Kaler Th., 1997, A&A, 57, 265

Frogel, J.A., Mould, J., & Blanco, V.M. 1990, ApJ, 352, 96

Hatzidimitriou, D., Cannon, R.D., & Hawkins, M.R.S. 1993, MNRAS, 261, 873.

Hatzidimitriou, D., Croke, B.F., Morgan, D.H., et al. 1997, A&AS, 122, 507

Irwin, M.J. 1991, IAU Symp. 148, The Magellanic Clouds, eds. R. Haynes & D. Milne (Dordrecht: Kluwer), p.453

Irwin, M.J. 1996, personal communication

Kunkel, W.E. 1979, ApJ, 228, 718

Kunkel, W.E., & Demers, S. 1976, ROB No. 182, p.241

Kunkel, W.E., Demers, S., & Irwin, M.J. 1997a, A&AS, 122, 463

Kunkel, W.E., Demers, S., & Irwin, M.J. 1997b, ApJ, 488, L129

Kunkel, W.E., Demers, S., & Irwin, M.J. 1999, AJ, submitted

Majewski, S.R., Ostheimer, J.C., Kunkel, W.E., et al. 1999, this volume

McGee, R.X., & Milton, J.A. 1966 Austr. J. Phys. 19, 515

Moore, B., & Davies, M. 1994, MNRAS, 270, 209

Piatek S., & Prior, C. 1995, AJ, 109, 1071
Putman, M.E., Gibson, B.K., & Staveley-Smith, L. 1999, this volume
Searle, L., & Zinn, R. 1978, ApJ, 225, 357
Toomre, A., & Toomre, J. 1972, ApJ, 178, 623
Weiner, B.J., & Williams, T.B. 1996, AJ, 111, 1156

Discussion

Nicholas Suntzeff: Does your model explain the existence of the Cowley & Harwick carbon stars in the LMC which are apparently young, blue and have negative velocities with respect to the LMC center?

Kunkel: The young carbon CH stars show strong negative velocity excesses to -40 kms^{-1} below the HI velocities, and this effect is shared by the carbon stars we observed and the LPV's. The fact that the 5 globular clusters also show this feature suggests that either the entire LMC disk responded to the SMC impact 0.2 Gyr ago, or the HI did so. No SMC debris appears to be involved in this feature. But 0.2 Gyr ago, the SW piece of the LMC lay where the SMC penetrated.

Mary Putman: Can you show me the regions which were observed on your map of the carbon star distribution and the HI contours? Do you have an idea of what mechanism is leaving the carbon stars behind (or in front of) the HI?

Kunkel: The limits for the LMC lie at -76° to the South, -58° to the North (except for SRC fields 161 and 162), and fields 55 and 58 to the West and East, respectively. For the SMC the radius is approximately 10°.

The mechanism leaving carbon stars to the East of the LMC would appear to be one of two: (1) either ram pressure is active, or (2) collisional processes ionize the hydrogen "out of sight." The presence of carbon stars on the single velocity envelope of the inter-cloud hydrogen suggests that the latter dominates. Moore & Davis (MNRAS 1994) seem to offer the best treatment of such physics.

Mary Kontizas: Do you think that the SMC when its orbit passes by the LMC could produce an infall of the outer region's gas to the inner parts of the LMC? I have in mind the young inner LMC clusters and what effect has been responsible for their formation.

Kunkel: Toomre and Toomre (1972 ApJ) teaches us such debris should "get trapped" roughly in an inclined plane through the LMC disk. But I cannot suggest where to look for these. If we could find reliable spectral signatures identifying the origin of a cluster (r-to-s process ratio?) I should not be surprised if clusters such as NGC 1818 were "acquired" by the LMC that way.

Part 7. Tidal Interactions

Section B. Poster Papers

Animated conversations at the opening reception.

New Views of the Magellanic Clouds
IAU Symposium, Vol. 190, 1999
Y.-H. Chu, N.B. Suntzeff, J.E. Hesser, and D.A. Bohlender, eds.

The Magellanic Stream and the History of the Tidal Interaction between the LMC and SMC

T. Sawa

Department of Physics and Astronomy, Aichi University of Education, Kariya 448–8542, Japan

M. Fujimoto

Uedayama 4-104, Tenpaku, Nagoya 468–0001, Japan

Y. Kumai

Faculty of Commerce, Kumamoto Gakuen University, Kumamoto 862–8680, Japan

Abstract. Adopting observed proper motions of the MC's and cluster formation histories of the LMC as observational constraints, we found a new binary orbit for the LMC and SMC in which the geometry and kinematics of the Magellanic Stream are well reproduced. In this orbit, the LMC and SMC experienced very close encouters 0.2 and 3 Gyr ago.

1. Introduction

It is widely accepted that the Magellanic Stream (MS) was produced by tidal interaction between the LMC, the SMC and the Galaxy (Murai & Fujimoto 1980; Gardiner et al. 1994; Gardiner & Noguchi 1995). However, the orbit in which the MS was formed has not been uniquely determined because of the lack of observational constraints on theoretical models. Here, we introduce two new observational constraints: the proper motion of the LMC and SMC observed by Hipparcos satellite (Kroupa & Bastian 1997) and the age distribution of star-clusters in the LMC (Girardi et al. 1995) which indicates two enhanced star-cluster formation epoch in the LMC at 0.1 and 3 Gyr ago.

2. Binary Orbits of the LMC and SMC

We searched for binary orbits of the LMC and SMC which endure around the galaxy for more than 10 Gyr, and then investigated the geometry and kinematics of the MS formed in each orbit by using test-particle simulations. We assumed a flat rotation curve for the Galaxy and used the present transverse velocities of the Clouds and the direction of the spin axis of the SMC as free parameters.

We first selected binary orbits of the LMC and SMC in which the geometry and kinematics of MS are well reproduced. Next, we checked whether or not the adopted model values of the present transverse velocities of the Clouds coincide

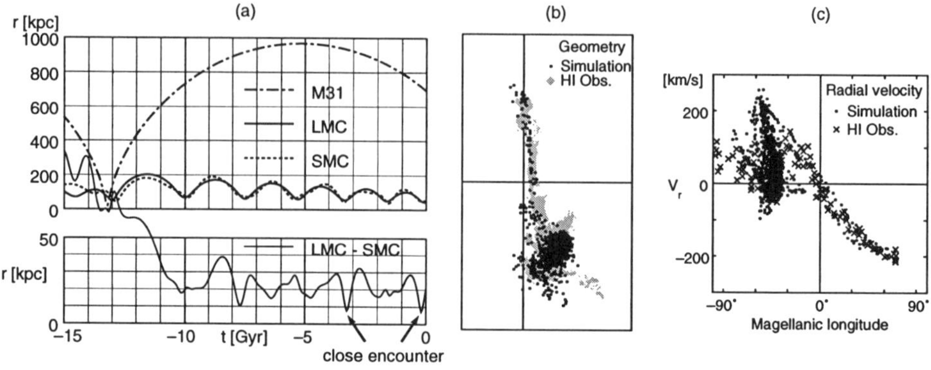

Figure 1. Our best model. (a) The time variation of the Galaxy-M31 and Galaxy-Magellanic Clouds distances. The LMC-SMC distance is given in the lower panel. (b) The geometry of the MS. (c) The radial velocity distribution of the MS along the Magellanic longitude.

with proper motions measured by Hipparcos, and accepted only orbits which satisfy this criterion. Finally, we examined time-variations in LMC-SMC separation for the past 15 Gyr, and chose the best model based on the coincidence between epochs of very close approaches in the model and those of enhanced cluster formation in the LMC. Figure 1 shows our best model: the LMC and the SMC experienced two very close encounters 0.2 Gyr and 3 Gyr ago, which gives satisfactory agreement with the cluster age distribution in the LMC. At these epochs, the considerable disturbance experienced by the LMC could have led to the formation of a huge number of clusters.

3. Effect of the Expanding Universe

Finally, we have traced back in time our best orbit until the early phase of the expanding universe when the Galaxy makes contact with M31 (Fujimoto et al. 1997). We find that the LMC and the SMC were already in a binary state in the Galaxy potential of the early universe, suggesting that they formed together and were then bound to the Galaxy.

References

Fujimoto, M., Sawa, T., & Kumai, Y. 1997, IAU Symp. 186, ed. D. Sanders, 6

Gardiner, L. T., & Noguchi, F. 1995, MNRAS, 278, 191

Gardiner, L. T., Sawa, T., & Fujimoto, M. 1994, MNRAS, 266, 567

Girardi, L., Chiosi, C., Bertelli, G., & Bressan, A. 1995, A&A, 298, 87

Kroupa, P., & Bastian, U. 1997, New Astronomy, 2, 77

Murai, T., & Fujimoto, M. 1980, PASJ, 32, 581

New Views of the Magellanic Clouds
IAU Symposium, Vol. 190, 1999
Y.-H. Chu, N.B. Suntzeff, J.E. Hesser, and D.A. Bohlender, eds.

The Interstellar Medium in the Magellanic Bridge

N. Lehner, F.P. Keenan, J.V. Smoker, P.L. Dufton, W.R.J. Rolleston, F.C. McKenna

Department of Pure and Applied Physics, The Queen's University of Belfast, Belfast, BT7 1NN, Northern Ireland

K.R. Sembach

The John Hopkins University, Department of Physics and Astronomy, Bloomberg Center, 3400 N. Charles Street, Baltimore, MD 21218, USA

J.E. Tohline

Department of Physics and Astronomy, Louisiana State University, Baton Rouge, LA 70803-4001, USA

G. White, S. Stanimirović

Department of Physics, University of Western Sydney Nepean, Kingswood, NSW, Australia

Abstract. HST ultra-violet and ground-based optical spectroscopy are presented for DI 1388, an early-type star in the Magellanic Bridge. These data have been supplemented with H I radio mapping of its environment. The spectroscopy shows interstellar components at LSR velocities of 0, 90 and 200 km s^{-1} – corresponding to the local ISM, a high velocity cloud and the Magellanic Bridge ISM, respectively. The last is the first absorption line detection of this material and shows the presence of both low (e.g., N I) and high (Si IV) ionization material. Further optical and HST UV observations of DI 1338 and another Bridge star, DGIK 975, are scheduled to investigate the chemical homogeneity and star formation history of the Bridge.

1. Introduction

The Magellanic Bridge probably originated some 200 Myr ago as a result of a tidal interaction during an LMC/SMC close approach, which resulted in material being pulled from the wing of the SMC. Detailed studies of the interstellar lines towards stars in the Bridge should give valuable information on its composition and origin.

Previously, we have obtained AAT spectra for five B-type stars in the Bridge (Hambly et al. 1994; Rolleston et al. 1998). Although these were principally used to measure stellar features, they also provided the first detection of interstellar absorption lines from the Bridge gas. Here we present HST UV spectroscopy and H I radio mapping of DI 1388 (RA: 02 57 11.94, Dec: 02 57 11.94 (J2000), $V = 14.39$, $B - V = -0.26$, MK Type: B0V, $v \sin i = 180 \pm 30$ km s^{-1}).

2. Observations and Results

HST Observations: STIS with the E140 grating provided wavelength coverage from 1150 to 1700 Å at a resolution of $\sim$ 3 km s^{-1}pixel^{-1} and S/N of $\sim$ 10. Absorption components with $v_{\rm lsr} \sim$ 0, 90, 200 km s^{-1} are present. The disk absorption lines are normally saturated, while the HVC absorption is weak and only a few chemical elements are identified. The Bridge absorption is detected unambiguously for 16 lines of Si II, Si III, Si IV, O I, S II, N I, C II, Fe II, Al II, Ni II.

ATCA Observations: Seven sightlines were observed, including one towards DI 1388. The peak brightness temperature of 11.2 K compares with the nearest single-dish Parkes values of 4 K (McGee & Newton 1986). The H I column density map from the ATCA data has a maximum value of n_H=1$\times$10^{20} cm^{-2} compared with single-dish values of $\sim$5$\times$10^{20} cm^{-2}. The ATCA maps show no correlation between peak H I temperatures and stellar positions.

3. Discussion

HST observations of a second Bridge star, DGIK 975, will be obtained next year, while AAT time has been allocated to re-observe both stars at high resolution and S/N. Our optical, UV, and new single-dish radio data will be combined to determine accurate physical parameters for the Bridge interstellar gas, including:

- Bridge origin from chemical composition: The Bridge composition should reflect that of the SMC. We will compare our Bridge abundances with those found in both the LMC and SMC and in the Magellanic Stream to investigate the proposition that both the Stream and Bridge originated from the same material.

- Star formation in the Bridge: Although the Bridge is believed to be approximately 200 Myr old, our B-type targets have ages of $\leq$ 20 Myr. Hence star formation is still occurring in the Bridge. Given its very low H I column density, the mechanism is probably different from that operating in galactic disks. It has been suggested that collisions between cloudlets within HVCs in the halo of our Galaxy leads to star formation. Given that the H I densities inferred for the Bridge and the HVCs are similar, such a formation mechanism may be operating in the Bridge. We therefore intend to compare our Bridge results with those found for halo sightlines.

References

Hambly N.C., Dufton P.L., Rolleston W.R.J., et al. 1994, A&A, 285, 716

McGee R.X., & Newton L.M. 1986, PASA, 6, 471

Rolleston W.R.J., Dufton P.L., Mc Erlean N.D., et al. 1999, A&A, submitted

New Views of the Magellanic Clouds
IAU Symposium, Vol. 190, 1999
Y.-H. Chu, N.B. Suntzeff, J.E. Hesser, and D.A. Bohlender, eds.

New Particle Simulation of the Magellanic Clouds

P.S. Li

University of Wyoming, P.O. Box 3905, University Station, Laramie, WY 82 071, USA

H.A. Thronson

Office of Space Science, NASA, Washington D.C. 20546, USA

Abstract.
We use a modified TreeSPH code to simulate the MilkyWay-LMC-SMC system using the latest observational results from Hipparcos on the proper motion of the Clouds. More than 120,000 particles are used for the simulation. Gravitational and hydrodynamic interactions among gaseous and stellar components in the system have been considered. The Magellanic Stream, some high velocity clouds, and the barred structure of LMC are reproduced. A leading gas arm from the SMC is predicted. From the simulation, the long term binary status of LMC-SMC is questioned as frequent encounters between the LMC and the SMC will have difficulty in explaining the survival of the SMC.

1. Introduction

The Magellanic Clouds are the two largest satellite galaxies of the Milky Way. The discovery of the Magellanic Stream provides direct evidence of interaction between the Clouds and the Milky Way. The previous simulations (e.g., Murai & Fujimoto 1980; Gardiner & Noguchi 1996) on this system did not include hydrodynamics, so cannot answer all the questions about the actual mechanism of forming the Magellanic Stream, the gas bridge between the LMC and the SMC, and the origin of high velocity clouds. Also, gaseous exchange between the two Clouds is important to their structure and evolution.

We developed a parallelized TreeSPH code and used it to simulate the Milky Way-LMC-SMC system. Initial conditions of the 3 galaxies are obtained by backward integration using the latest observational results from Kroupa & Bastian (1997) for the proper motion of the Clouds. A total of 116,736 particles (1/3 are gas particles) have been used. All three galaxies in the system are assumed to be disk galaxies. The mass ratio, including dark matter, of Milky Way:LMC:SMC = 1:0.02:0.003. In the model, the Milky Way has a very massive dark halo based on the latest estimation by Kulessa & Lynden-Bell (1992). The simulation lasts an equivalent of 1.35 billion years.

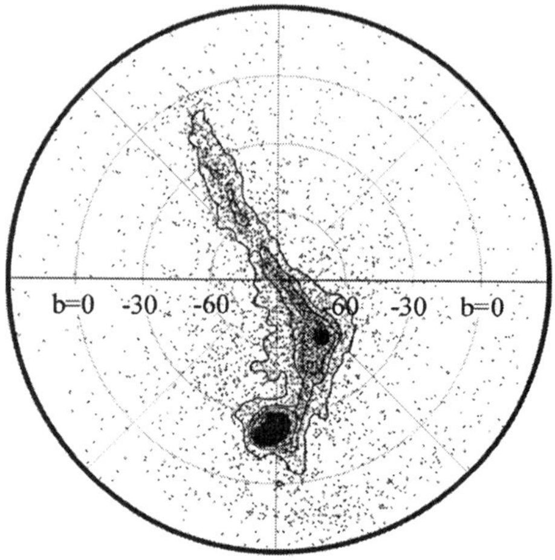

Figure 1. Column density of gas around the Magellanic Clouds and the Stream.

2. Simulation Results and Conclusion

We clearly rule out the Stream formation mechanism due to ram pressure from a halo of gas (Heller & Rohlfs 1994) and reproduce most of the observed features, including the Magellanic Stream (see Fig. 1), high velocity clouds, barred structure of the LMC, and the leading gas arm from the SMC, which appears as the material bridge between the LMC and the SMC.

Some of the high velocity clouds are formed from the gas material of the lead gas arm and the SMC. These gas clouds are scattered and accelerated by the LMC during their encounter.

From the simulation, we find that the LMC and the SMC are not in a binary orbit at the beginning of the simulation. They came within 20 kpc of each other about 550 million years ago and the encounter lasted for about 400 million years. During this encounter, more than half of the gas from the SMC is lost and the SMC is almost destroyed. The continued interaction between the LMC and the SMC will have difficulty in explaining the survival of the SMC.

Acknowledgments. This work is supported by the ISO program and the NCSA grant #MCA94P001N, and is a part of the first author's Ph.D. Thesis.

References

Murai, T., & Fujimoto, M. 1980, PASP, 32, 581

Gardiner, L.T., & Noguchi, M. 1996, MNRAS, 278, 191

Heller, P., & Rohlfs, K. 1994, A&A, 291, 743

Kulessa, A.S., & Lynden-Bell, D. 1992, MNRAS, 255, 105

Discussion

Eric Maurice: What can you say about the thickness of the LMC disk after the encounter of the two Magellanic Clouds and of the Galaxy?

Li: The thickness of the LMC disk is increased after the encounter with the SMC and the Milky Way. However, I haven't measured the actual value from the data yet.

Nolan Walborn: The LMC bar looks centered in the simulations. Can you comment on why the actual bar is eccentric?

Li: Some of the participants have asked about this question in the last few days. Unfortunately, we haven't investigated this question in detail yet. It is hard to tell if the bar is centered with the stellar disk from the animation. We will look at this problem to find out if the Milky Way is the cause of the offset of the LMC bar as soon as possible.

William Kunkel: What is the bright blob flying off the gas disks?

Li: Some of the gas from the SMC and the leading arm are scattered and accelerated by the LMC and fall back in the direction of the Milky Way. The masses of these gas blobs are in the range of 10^6 to 10^7 $M_\odot$. They could explain the origin of some of the high velocity clouds observed.

New Views of the Magellanic Clouds
IAU Symposium, Vol. 190, 1999
Y.-H. Chu, N.B. Suntzeff, J.E. Hesser, and D.A. Bohlender, eds.

The Metallicity and Dust Content of HVC287+22+240: Evidence for a Magellanic Clouds Origin

Bart Wakker, Blair D. Savage

Dept. of Astronomy, Univ. of Wisconsin, Madison, WI 53706, USA

Limin Lu, Wallace L.W. Sargent

California Institute of Technology, 105-24, Pasadena, CA 91125, USA

Kenneth R. Sembach

Department of Physics and Astronomy, Johns Hopkins University, Baltimore, MD 21218, USA

Tom A. Oosterloo

CNR, Istituto di Fisica Cosmica, Milan, Italy

We present evidence that the properties of the high-velocity cloud catalogued as #187 by Wakker & van Woerden (1991) (HVC287+22+240) can be explained if it is part of a leading arm counterpart of the Magellanic Stream, and we use this to argue that the HVC originated in the Magellanic Clouds after a tidal interaction.

The sulphur and iron abundance were measured using GHRS spectra of NGC3783, a background Seyfert galaxy. These data were fully described by Lu et al (1998). An accurate reference HI column density was found by combining data taken with the Australia Telescope Compact Array with data from the Green Bank 140 ft telescope. The particle distributions predicted by the tidal model of Gardiner & Noguchi (1996) are compared with those of a subsample of the high-velocity clouds.

We find that (S/H)=0.25$\pm$0.07 solar in the HVC, providing the first accurate measurement of the chemical enrichment level in an HVC. Note that this is similar to the interstellar abundance of S in the Magellanic Clouds: (S/H)=0.27 solar in the LMC, and 0.21 solar in the SMC (Russel & Dopita 1992).

We find that (Fe/H)=0.13$\pm$0.04 solar in the HVC. In diffuse disk gas subsolar ratios indicate that most of the iron is locked into dust ((Fe/S)$\sim$0.005 solar in cool gas, (Fe/S)$\sim$0.25 solar in warm gas; Savage & Sembach 1996). The ratio in the HVC implies that 7 out of 8 Fe atoms are in dust (if Fe/S is intrinsically solar, as is the case in the SMC). If the intrinsic ratio were that in metal-poor halo stars ($\sim$0.4, Wheeler, Sneden & Truran 1989), 2 out of 3 Fe atoms are in dust. The Fe/S ratio is similar to the interstellar ratio in the SMC, which is found by combining (Fe/Zn)$\sim$0.25 (Welty et al. 1997; Roth & Blades 1997), with (Zn/S)$\sim$1 (Russel & Dopita 1992).

HVC287+22+240 was interpreted as an extra-galactic cloud by West et al. (1985) on the basis of its low value of (Ca/H) (0.014$\pm$0.006 solar); however, this ratio suffers from an unknown ionization and depletion correction. Blitz et al. (1996) argued that many HVCs, including HVC287+22+240, are best explained as "remnants of Local Group formation", and they used the subsolar S abundance as supporting evidence, although our observed S abundance would be rather high for relatively unprocessed gas.

Gardiner & Noguchi (1996) presented an N-body simulation of the Magellanic system, modeling the Milky Way and the LMC as fixed potentials, but the SMC as a self-gravitating collection of particles. The model produces the Magellanic Stream, Magellanic Bridge and other features; especially noteworthy for our purposes is the prediction for the properties of the leading tidal arm. When we compare these predictions with the sky and velocity distribution of some HVCs, we find that they have similar global properties, although there are differences in the details.

We argue that the metallicity and depletion information for the HVC, coupled with its velocity and position on the sky, strongly suggest that the HVC (and by implication also the Magellanic Stream proper), originated from the Magellanic Clouds as the result of a tidal interaction between the LMC and the SMC. The alternative model for the origin of the Magellanic Stream is that it is the result of ram-pressure stripping by a tenuous galactic corona (Moore & Davis 1994), but in this model the presence of gas with Magellanic abundances ahead of the Stream cannot be explained.

References

Blitz, L., Spergel, D., Teuben, P., Hartmann, D., & Burton, W.B. 1996, BAAS, 189, 61.01

Gardiner, L.T., & Noguchi, M. 1996, MNRAS, 278, 191

Lu, L., Savage, B.D., Sembach, K.R., Wakker, B.P., Sargent, W.W.L., & Oosterloo, T.A. 1997, AJ, 115, 162

Moore, B., & Davis, M. 1994, MNRAS, 270, 209

Roth, K.C, & Blades, J.C. 1996, ApJ, 474, 95

Russel, S.C., & Dopita, M.A. 1992, ApJ, 384, 508

Savage, B.D., & Sembach, K.R. 1996, ARA&A, 34, 279

Wakker, B.P., & van Woerden, H. 1991, A&A, 250, 509

Welty, D.E., Lauroesch, J.T., Blades, J.C., Hobbs, L.M., & York, D.G. 1998, ApJ, 429, 627

West, K.A., Pettini ,M., Penston, M.V., Blades, J.C., & Morton, D.C. 1985, MNRAS, 215, 481

Wheeler, J.C., Sneden, C., & Truran, J.W. 1989, ARA&A, 27, 279

New Views of the Magellanic Clouds
IAU Symposium, Vol. 190, 1999
Y.-H. Chu, N.B. Suntzeff, J.E. Hesser, and D.A. Bohlender, eds.

A Search for Tidal Stellar Debris from the Magellanic Clouds: Survey Results from the First Two Years

S. R. Majewski[1], J. C. Ostheimer

University of Virginia, Department of Astronomy, Charlottesville, VA 22903-0818, USA

W. E. Kunkel

Las Campanas Observatory, Carnegie Observatories, La Serena, Chile

K. V. Johnston

Institute for Advanced Study, Princeton, NJ 08540, USA

R. J. Patterson and C. Palma

University of Virginia, Department of Astronomy, Charlottesville, VA 22903-0818, USA

1. Introduction

An important discriminant between leading models for the origin of the Magellanic Stream is the presence of a stellar counterpart to the HI gas stream: ram pressure stripping of gas by a putative hot Galactic halo would act only on Magellanic gas while gravitational tidal stripping would act on both gas and stars. Several previous attempts to find tidal stellar debris have failed to find carbon stars, A stars, or other main sequence stars in the Magellanic Stream (Mathewson et al. 1979; Recillas-Cruz 1982; Brück & Hawkins 1983; Guhathakurta & Lin 1999). However, there has long been a suggestion (Kunkel 1979; Lynden-Bell 1982) of a possible Magellanic association of satellite galaxies and globular clusters that have similar orbits and may derive from the break up of a greater Magellanic galaxy (Lynden-Bell & Lynden-Bell 1995; Majewski et al. 1997). Recent models (Moore & Davis 1994; Johnston 1998) of the tidal disruption of Large Magellanic Cloud (LMC)-like systems indicate a wide dispersal of debris, much wider than the rather confined HI stream, so that the contrast of tidal debris against the Galactic fore/background would be low. If true, this could explain some of the previous negative results for tidal debris searches.

2. Survey and Results to Date

Results of these recent dynamical models drive the strategy of our new attempt to find stellar tidal debris from the Magellanic Clouds. To detect widely dis-

[1]Visiting Associate, Carnegie Observatories; Packard Foundation Fellow; Cottrell Scholar

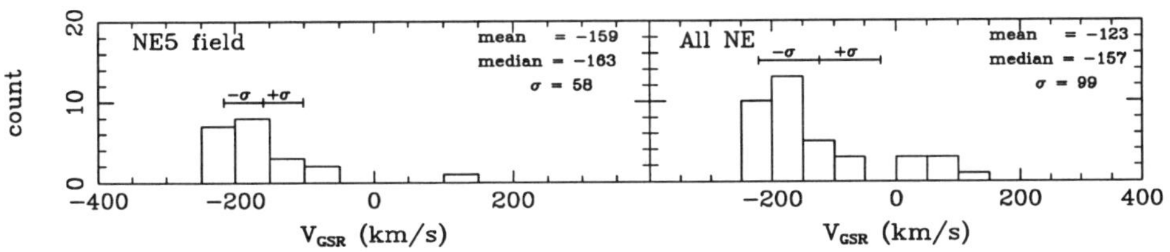

Figure 1. The radial velocity distribution of giants in a single field (left panel) northeast of the LMC ($l = 271°, b = -19°$), and for a total of five fields in the same region of the sky (right panel)

persed, low contrast debris requires that large areas be surveyed. Presently, one may cover large areas practically only with small telescopes, but to search for debris at the distance of the Magellanic Clouds then requires bright stars as targets. We have therefore begun a photometric search for giant stars in a $\sim 25°$ radius ring of fields around the Clouds. We have used the Las Campanas 1-m Swope telescope to image in the DDO51 and Washington broadband filters, a combination that effectively separates K and M giants from dwarfs, since the DDO51 filter is centered on the surface gravity sensitive Mg b triplet and MgH stellar features around 5100 Å . To date we have imaged ~ 25 deg^2 in 29 fields to Washington $M \sim 19.5$. Giant star candidates are observed with the Modular Spectrograph on the DuPont 2.5-m telescope, from which we obtain 10-15 km s^{-1} radial velocities and checks on the surface gravity.

Figure 1 shows the radial velocity distribution (corrected for Local Standard of Rest) of giants in a single field northeast of the LMC ($l = 271°, b = -19°$), and for a total of five fields in the same region of the sky. The distributions clearly do not follow expectations for a random halo field population – i.e., a Gaussian with ~ 110 km s^{-1} dispersion centered on 0; rather, they seem to show dominance by a negative velocity "moving group". Moreover, the distance-velocity distributions we obtain for these and our other fields roughly match the expected distributions for multiply-wrapped, tidal arms as predicted by our semi-analytical LMC debris model (see Johnston 1998). However, while we conclude from these first survey results that we are finding correlated streams of halo stellar debris around the Magellanic Clouds, much more work is needed before we may associate these streams definitively with the Clouds. But the apparent lack of an appreciable *random* field star population in our survey data at least suggests that tidal debris makes a predominant contribution to the outer parts of the Milky Way.

References

Brück, M. T., & Hawkins, M.R.S. 1983, A&A, 124, 216

Guhathakurta, P., & Lin, D. N. C. 1999, in preparation

Johnston, K. V. 1998 ApJ, 495, 297

Kunkel, W. E. 1979, ApJ, 228, 718

Lynden-Bell, D. 1982, Observatory, 102, 202
Lynden-Bell, D., & Lynden-Bell, R.M. 1995, MNRAS, 275, 429
Majewski, S. R., Phelps, R.L., & Rich, R.M. 1997, ASP Conf. Ser., 112, 1
Mathewson, D.S., et al. 1979, in IAU Symp. 84, 547 (ed. W. B. Burton)
Moore, B., & Davis, M. 1994, MNRAS, 270, 209
Recillas-Cruz, E. 1982, A&A, 124, 216

Part 8. Cepheids and Distances

Section A. Invited Reviews

Sidney van den Bergh shares a joke with (left to right) Johannes Andersen, Doug Welch, and Kem Cook.

New Views of the Magellanic Clouds
IAU Symposium, Vol. 190, 1999
Y.-H. Chu, N.B. Suntzeff, J.E. Hesser, and D.A. Bohlender, eds.

Cepheid Variables in the LMC and SMC

D.L. Welch, and the MACHO Collaboration[1]

Dept. of Physics & Astronomy, McMaster University, Hamilton, Ontario, L8S 4M1 Canada

Abstract. In this paper, we will review major new results regarding classical Cepheids, in the Large Magellanic Cloud (LMC) and Small Magellanic Cloud (SMC). Specifically, we discuss recent work regarding multi-mode Cepheids and describe new observations of a W Vir star (HV 5756) and a Cepheid which are each in eclipsing binary systems. An additional interesting pulsating supergiant in an eclipsing system is also identified. Ephemerides for eclipses for the three systems are provided.

1. MACHO Project Status

The MACHO Project is a wide-field photometric survey whose primary goal is to characterize the mass of non-luminous or under-luminous matter along different lines-of-sight in the Milky Way galaxy. As of July 1998, observations span 2100 nights with over 75,000 images archived. Data collection has taken place essentially continuously since 1992, making this one of the most extensive data sets for variable stars obtained in the history of variable star astronomy. It is expected that similar observations will be obtained up until at least 1999 December 31. A relatively recent review of the results of the MACHO Project can be found in Cook *et al.* (1997).

2. Classical Cepheids

At present, the MACHO Project has identified 1767 classical and first-overtone Cepheids in the LMC. EROS has reported a total of 290 Cepheids in the LMC. Both groups agree that the fraction of first-overtone Cepheids is 30%. At shorter periods, the fundamental and first-overtone mode sequences are less distinct, suggesting the possibility of a range in masses and/or ages for these stars. Such a conclusion has been suggested by Bersier *et al.* (1998) and Alcock *et al.* (1999a).

---

[1]C. Alcock, R.A. Allsman, D. Alves, T.S. Axelrod, A.C. Becker, D.P. Bennett, K.H. Cook, K.C. Freeman, K. Griest, M.J. Lehner, D.W. Kurtz, S.L. Marshall, D. Minniti, B.A. Peterson, M.R. Pratt, P.J. Quinn, A.W. Rodgers (Deceased), A. Rorabeck, W. Sutherland, A. Tomaney, and T. Vandehei

3. Beat Cepheids

Alcock *et al.* (1999b) report some of the results found by Rorabeck (1997). The second-overtone lightcurve derived from 1st/2nd-overtone beat Cepheid lightcurves is found to be very sinusoidal. Beaulieu *et al.* (1997), Welch (1998), and Rorabeck (1997) have reported beat Cepheids in the SMC.

4. Eclipsing Cepheid Variables

We have discovered three Cepheid variables in eclipsing binary systems to date. Times of primary and secondary minima for these systems are given in Tables 1, 2, and 3, respectively.

1. *HV 5756* Our first reported discovery, HV 5756 (MACHO ID 78.6338.24) was reported in Welch *et al.* (1996). A total of six primary minima have been observed at this writing (Dec 1998). Furthermore, images from the Harvard College Observatory plate archives have now been obtained to search for historical minima.

2. *6.6454.5* A finder chart with suitable comparison stars is available at the URL: `http://wwwmacho.mcmaster.ca/EclCep/`. Diluted lightcurve subtracted photometry is shown in Figure 1.

3. *5.4763.71* This system appears to contain a 4.69-day Cepheid with a slowly-varying amplitude and has an orbital period of 123.9 days. It is possibly a low-mass Cepheid (in between the tradition classifications of BL Her and W Vir variables). It is located at equinox J2000.0 coordinates 05:09:59.2 -69:58:28. Both primary and secondary minima are visible in the lightcurve phased with the orbital period. The system is also remarkable in that secondary minimum appears to take place at an orbital phase of 0.55, indicating an eccentric orbit.

These systems are likely just the tip of the iceberg, since the hotter companions are usually more luminous than the pulsator and hence are easily selected.

Table 1. Eclipse Predictions for HV 5756

Primary		**Secondary**	
JD	UT	JD	UT
2451122.6	1998 Nov 5.1	2451332.6	1999 Jun 3.1
2451542.6	1999 Dec 30.1	2451752.6	2000 Jul 27.1
2451962.6	2001 Feb 22.1	2452172.6	2001 Sep 20.1
2452382.6	2002 Apr 18.1	2452592.6	2002 Nov 14.1
2452802.6	2003 Jun 12.1	2453012.6	2004 Jan 8.1

Table 2. Eclipse Predictions for 6.6454.5

Primary		Secondary	
JD	UT	JD	UT
2450867.06	1998 Feb 22.56	2451065.81	1998 Sep 9.31
2451264.56	1999 Mar 27.06	2451463.31	1999 Oct 11.81
2451662.06	2000 Apr 27.56	2451860.81	2000 Nov 12.31
2452059.56	2001 May 30.06	2452258.31	2001 Dec 14.81
2452457.06	2002 Jul 1.56	2452655.81	2003 Jan 16.31
2452854.56	2003 Aug 3.06	2453053.31	2004 Feb 17.81

Table 3. Eclipse Predictions for 5.4763.71

Primary		Secondary	
JD	UT	JD	UT
2451270.40	1999 Apr 1.90	2451338.54	1999 Jun 9.04
2451394.30	1999 Aug 3.80	2451462.44	1999 Oct 10.94
2451518.20	1999 Dec 5.70	2451586.34	2000 Feb 11.85
2451642.10	2000 Apr 7.60	2451710.24	2000 Jun 14.75
2451766.00	2000 Aug 9.50	2451834.14	2000 Oct 16.65
2451889.90	2000 Dec 11.40	2451958.04	2001 Feb 17.54
2452013.80	2001 Apr 14.30	2452081.94	2001 Jun 21.44
2452137.70	2001 Aug 16.20	2452205.84	2001 Oct 23.35
2452261.60	2001 Dec 18.10	2452329.74	2002 Feb 24.25
2452385.50	2002 Apr 21.00	2452453.64	2002 Jun 28.15

Acknowledgments. Work performed at Lawrence Livermore National Laboratory (LLNL) is supported by the Department of Energy (DOE) under contract W7405-ENG-48. Work performed by the Center for Particle Astrophysics (CfPA) on the University of California campuses is supported in part by the Office of Science and Technology Centers of the National Science Foundation (NSF) under cooperative agreement AST-8809616. Work performed at MSO is supported by the Bilateral Science and Technology Program of the Australian Department of Industry, Technology and Regional Development. DLW and AJR were supported, in part, by a Research Grant from the Natural Sciences and Engineering Research Council of Canada (NSERC) during this work.

References

Alcock, C. *et al.* 1999a, AJ, in press

Alcock, C. *et al.* 1999b, ApJ, in press

Beaulieu, J.-P. *et al.* 1997, A&A, 321, L5

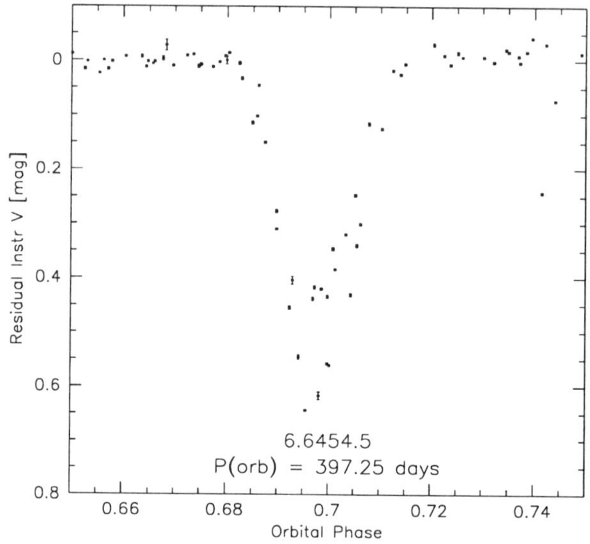

Figure 1. Five consecutive primary minima of the system containing the 4.97-day classical Cepheid 6.6454.5 phased with an orbital period of 397.25 days. The scatter in the minima is not observational in origin but is due to the removal of the contaminated Cepheid lightcurve from the data. A full solution of the system is required to remove the variation during eclipse properly. An epoch of mid-eclipse is HJD 2449277.1. Primary eclipses typically last about 12 days.

Bersier, D. 1998, A.S.P. Conf. Series, 135, 24

Cook, K.H. *et al.* 1997, Proc. 12th I.A.P., 17

Rorabeck, A. 1997, M.Sc. thesis, McMaster University

Welch D.L., 1998, A.S.P. Conf. Series, 135, 355

Welch D.L., *et al.* 1998, I.A.U. Circular 6802

Welch D.L., *et al.* 1996, I.A.U. Circular 6434

Discussion

Sidney van den Bergh: Do amplitude-modulated Cepheids fall on the normal period-luminosity relation?

Welch: The amplitude-changing Cepheids are all around a period of $1\overset{d}{.}25$. As I recall, their colours are not distinct from the population of "normal" Cepheids.

Carme Gallart: Any news about anomalous Cepheids?

Welch: An excellent question. It is likely that the shortest-periods objects in our selections are "anomalous" Cepheids, since, at some point, intermediate-mass Cepheid progenitors no longer can enter the Cepheid instability strip. However, the stars, at that point, can be fed from older and lower mass stars.

New Views of the Magellanic Clouds
IAU Symposium, Vol. 190, 1999
Y.-H. Chu, N.B. Suntzeff, J.E. Hesser, and D.A. Bohlender, eds.

The MACHO Project LMC Variable Star Inventory: Classical Cepheids, AGB Variables, and the Nine Million Star Color-Magnitude Diagram

D.R. Alves

University of California, Davis, CA 95616, USA

A. Basu, K.H. Cook

Lawrence Livermore National Lab., Livermore, CA 94550, USA

D.L. Welch

McMaster University, Hamilton, ON L8S 4M1, Canada

and the MACHO Collaboration

Abstract. We measure the ratio of $\sim 5M_\odot$ blue and red LMC supergiants (representative of classical Cepheids) using the MACHO Project's 9 million star color-magnitude diagram (9M CMD) of the LMC bar. We find $b/r = 0.39$, which favors the Z=0.008, $5M_\odot$ theoretical model of Schaerer et al. (1993) over that of Fagatto et al. (1994). Next, we examine the low mass (old) and low metallicity LMC field population (Pop. II). Features in the 9M CMD and properties of LMC field RRab variables are consistent with a mean iron abundance of [Fe/H] ≈ -1.5 dex for this population. Newly discovered post-HB/early-AGB Pop. II variables are identified in order to delineate the instability strip (IS). Good agreement with the theoretical IS of Bono et al. (1997) is found. We then compare the field RRab with newly identified RRab variables in the LMC clusters NGC 1898 and NGC 1835. We find the mean colors of these cluster RRab lie near the red and blue edges of the IS, respectively, which is similar to their respective (overall) red and blue HB morphologies. Since the field RRab lie on the red side of the IS, we infer the LMC field Pop. II is likely to have a red HB morphology.

1. Introduction

MACHO Project two-color photometric data have been calibrated to Kron-Cousins V and R with a precision $\sigma_V = \sigma_R = 0.020$ mag for ~9 million stars in and nearby to the LMC bar (Alves 1998; Alcock et al. 1999a). We use the composite "9 million star color-magnitude diagram" (9M CMD) along with our

[1] C. Alcock, S. Marshall, T. Axelrod, A. Drake, K. Freeman, B. Peterson, R. Allsman, A. Becker, A. Tomaney, K. Griest, T. Vandehei, M. Lehner, D. Bennett, D. Minniti, M. Pratt, W. Sutherland, & P. Quinn

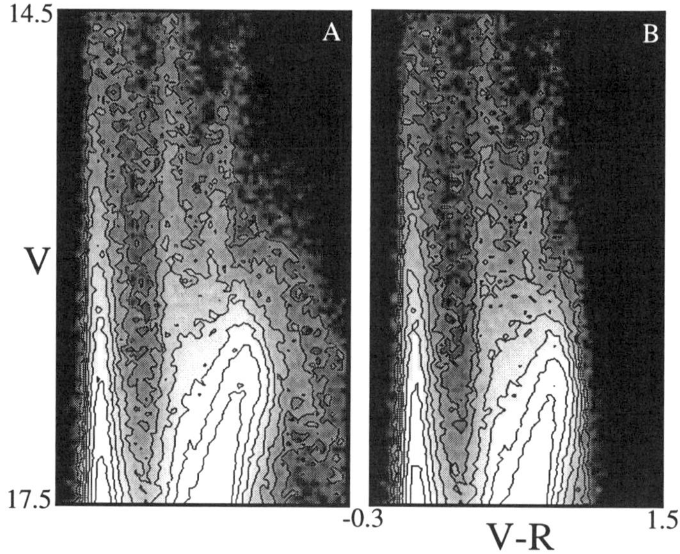

Figure 1. Log-scaled Hess diagrams representing **(A)** 261805 LMC stars with $(V - R) = -0.3$ to 1.5 mag, $V = 17.5$ to 14.5 mag, and **(B)** the same diagram but with 35169 candidate variable stars removed. The variable stars are defined by poor fits to constant brightness lightcurves. Intensity and contours (1.0 to 2.5 dex in 0.25 steps) indicate the logarithmic surface density of stars (no. per ~10 sq. deg. and per 0.03 mag square color-mag bin).

photometric lightcurve data to make new tests of stellar evolution theory and to study LMC stellar populations.

Figure 1 shows the bright end of the 9M CMD. The prominent red giant branch (RGB) and asymptotic giant branch (AGB; see also Alves et al. 1998) extend up to $V \sim 16.5$ mag. The striking difference between panels is on the AGB where virtually all stars appear to be variables. The bright and nearly vertical plumes are (from blue to red): the upper main sequence, Galactic foreground disk stars, and red supergiants (SGs). Careful examination reveals that many stars lying on the red edge of the upper-main sequence are also variables. The blue SGs lie between the upper main sequence and foreground disk star plumes.

2. Classical Cepheids: the LMC Blue/Red Supergiant Ratio

In this section, we summarize our measurement of the blue to red supergiant ratio in the 9M CMD and compare with theory. We report the following ridge line sequences for blue (b) SGs: $(V - R) = 0.09$, 0.18, 0.30, and 0.39, and for red (r) SGs: 0.76, 0.71, 0.66, and 0.64 at $V = 14.75$, 15.25, 15.75, and 16.25 mag, respectively. After consideration of the foreground stars, and shifting $z = 0.008, 5M_{\odot}$ model sequences to match these SG ridge lines, we count b/r SGs in $\Delta(V - R, V) = 0.2, 0.5$ mag rectangular "apertures" centered at $(V - R, V)$

= 0.18, 15.25 and 0.66, 15.75. The average ratio of non-variable $\sim 5M_{\odot}$ LMC SGs is $b/r = 0.39$, $(b+r) = 2869$. Adopting the Cepheid instability strip at $logT = 3.76$ to 3.70 (cf. Alcock et al. 1999b) to demark the non-variable b and r SG phases, the $z = 0.008$, $5M_{\odot}$ model of Fagatto et al. (1994) yields $b/r = 0.92$ while the model of Schaerer et al. (1993) yields $b/r = 0.34$. The predicted ratios of Cepheids to non-variable SGs are $c/(b+r) = 0.08$ and 0.06, respectively. A count of MACHO-discovered LMC Cepheids in a parallelogram in the CMD connecting the b & r apertures yields $c = 300$, $c/(b+r) \approx 0.10$. We note that these MACHO-discovered LMC Cepheids show an excess at $\sim 5M_{\odot}$ (pulsation masses) over model predictions for a constant star formation rate and Salpeter IMF (Alves et al. 1997, 1998; Alcock et al. 1999b).

3. Population II and AGB Variables in the LMC Bar

Next we examine the low mass (old) and low metallicity LMC field population, which we call "Population II." Well known Pop. II variable stars are the RR Lyrae (RRab are the fundamental mode pulsators). Pop. II horizontal branch (HB) stars (like RRab) evolve from the zero-age HB toward the asymptotic giant branch-bump (AGB-bump), i.e., the transition from core helium burning to shell burning. Some of these post-HB/early-AGB stars are also variables.

We find the 9M CMD giant branch is consistent with two superposed populations. Cluster CMDs representative of these dual giant branches are the $\sim$1.5 Gyr-old, [Fe/H] = -0.7 dex SMC cluster NGC 411 (Alves & Sarajedini 1999) and the Galactic globular cluster M3 with [Fe/H] ≈ -1.5 (Ferraro et al. 1997). If the LMC Pop. II is more metal rich than this, it is likely to be younger than M3. We report the color and magnitude of the AGB-bump (Gallart 1998) in the 9M CMD (not shown) at $(V-R, V) = 0.57$, 18.38 mag. We report the median color and magnitude of the LMC field RRab as $(V-R, V) = 0.31$, 19.45 mag (note the considerable scatter due to differential reddening, blends, and HB evolution in Fig. 2). ΔV(AGB-bump − RRab) depends on metallicity. We find the Z=0.0004, scaled-solar, HB models of Castellani, Chieffi, & Pulone (1991) provide a good match, although perhaps a slightly higher metallicity would be preferred. We then shift these model HB tracks in the CMD to match the location of the AGB-bump. This accounts for the uncertain LMC distance and reddening, and also recalibrates the color-temperature relation to the 9M CMD. In the right panel of Figure 2, LMC field RRab ($0.46 < P < 0.71$ day) are shown in the Bailey diagram along with the ridge line for M3 (see inset). This diagram indicates a mean metallicity for these RRab that is similar to M3. The "cloud" of RRab to the right of the prominent sequence in this diagram may indicate some LMC field RRab have lower metallicities. We also refer to the spectroscopic metallicities for a sample of these field RRab in Alcock et al. (1996) which indicate [Fe/H] ~ -1.6, albeit with considerable scatter. Thus, by four methods we arrive at [Fe/H] ≈ -1.5 dex for the LMC field Pop. II.

In the CMD panel of Figure 2, we also plot twenty-four newly identified post-HB variables. These were found in a search of between $17.5 < V < 18.8$, $0.15 < (V-R) < 0.75$, and restricted to variables with periods in the range $0.73 < P < 5$ day. These variables help us define the Pop. II instability strip (IS). The prediction of Bono et al. (1997) for the fundamental blue and red edge shows good agreement. Moreover, it lends support to our color-temperature calibration

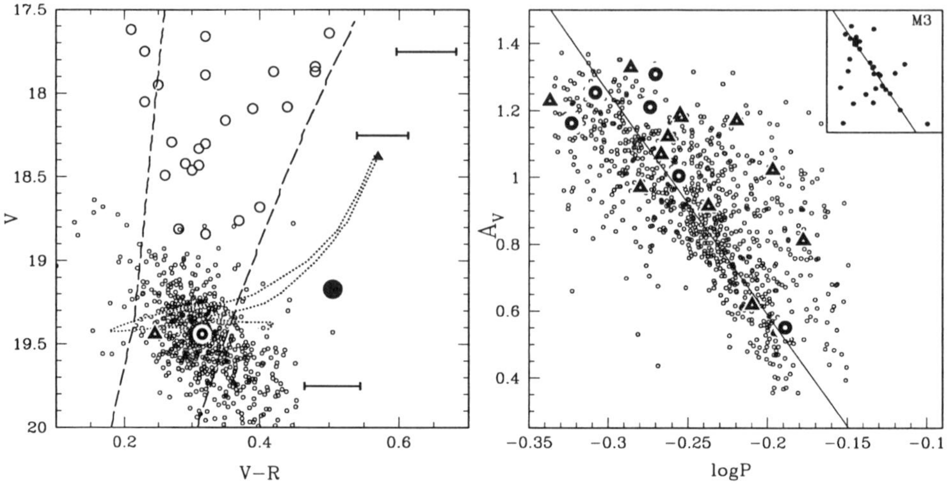

Figure 2. **(Left)** CMD showing LMC field RRab (small open circles) and Pop. II post-HB variables (larger open circles). 9M CMD fiducial sequences are indicated as follows: RGB (errorbar marks), the AGB-bump (solid triangle), and the red HB clump (solid circle). Theoretical data shown are $M = 0.70$ and $0.75\ M_{\odot}$, Z=0.0004 HB evolution tracks from Castellani et al. (1991; dotted lines), and Pop. II variable IS (F mode) from Bono et al. (1996; dashed marks) for $M = 0.60$ $M_{\odot}$. The mean location of NGC 1835 and NGC 1898 RRab are indicated with the bold-faced triangle and circle respectively. **(Right)** Bailey (period-V amplitude) diagram showing LMC field RRab (small circles), RRab from the LMC clusters NGC 1835 (bold triangles) and NGC 1898 (bold circles). The M3 ridge line is also indicated. In the inset (same axes) we plot RRab from Kaluzny et al. (1998) for M3 with ridge line.

derived from the AGB-bump and HB models. We note the discrepancy at the blue edge and high luminosities may imply some of these variables have higher masses.

In the Bailey diagram, we show newly identified RRab found near the LMC bar clusters NGC 1898 and NGC 1835. Spectroscopic metallicities for these clusters are [Fe/H] ~ -1.4 and -1.8 dex, respectively (Olszewski et al. 1991). Recent *HST* CMDs (Olsen et al. 1998) show these clusters have red and blue HB morphologies respectively and also confirm their “old” age. Our selected samples of these cluster RRab are certainly not complete, but we will assume the mean colors and brightnesses are representative. We find mean colors $(V - R) = 0.31$ and 0.25 mag respectively and mean brightnesses of $V \sim 19.45$ mag for both clusters. The RRab mean color for these two clusters distribute in the IS in a manner similar to their overall HB morphologies. Since the LMC field RRab lie in the red portion of the IS, we infer the LMC field Pop. II has a red HB morphology.

Acknowledgments. Work at LLNL is supported by DOE contract W7405-ENG-48. Work at CfPA is supported by NSF AST-8809616 and AST-9120005. Work at MSSSO is supported by the Australian Dept. of ITRD. WJS thanks PPARC-AF, KG thanks DOE-OJI, Sloan, and Cottrell awards.

References

Alcock, C., et al. 1996, AJ, 111, 1146

Alcock, C., et al. 1999a, in preparation

Alcock, C., et al. 1999b, ApJ, 117, 920

Alves, D. 1998 Ph.D. dissertation University of California, Davis

Alves, D., et al. 1997, in *Proc. of IAU Symposium 180 Planetary Nebulae*, eds. Habing & Lamers, p.468

Alves, D., et al. 1998, in Pulsating Stars - Recent Developments in Theory and Observation, eds. Takeuti & Sasselov, p.17

Alves, D., & Sarajedini, A. 1999, ApJ, 511, 225

Bono, G., Caputo, F., & Santolamazza, P. 1997, A&A, 317, 171

Castellani, V., Chieffi, A., & Pulone, L. 1991, ApJS, 76, 911

Fagatto, F., Bressan, A., Bertelli, G., & Chiosi, C. 1994, A&AS, 105, 29

Ferraro, F., et al. 1997, A&A, 320, 757

Gallart, C. 1998, ApJ, 495, 43

Kaluzny, J., Hildich, R. Clement, C., & Rucinski, S. 1998, MNRAS, 296, 347

Olsen, K.A., et al. 1998, MNRAS, 300, 665

Olszewski, E., Schommer, R., Suntzeff, N., & Harris, H. 1991, AJ, 101, 515

Schaerer, D., Meynet, G. & Maeder, A., & Schaller, G. 1993, A&AS, 98, 523

Discussion

Lance Gardiner: I'm interested in a couple of structural paprameters for the LMC. Is it possible to estimate (1) the total mass of the LMC bar and (2) the inclination of the LMC disc, from the data in the 9 million star CMD?

Alves: I have not estimated the structural parameters from all of the stars in the 9M-CMD. However, using the type ab RR Lyrae and the surface density profiles that Kem Cook showed during his talk, I find an inclination of 35°, and a new center very near the optical center.

Assuming the RRab stars trace a luminous halo population in the LCM, I estimate the mass contribution from this halo is a few percent of the total mass of the LMC.

John Dickel: Do you have a handle on how such a small fraction of highly reddened upper main sequence stars might affect your blue supergiant counts?

Alves: Keep in mind that the 9M CMD I showed you is log-scaled in the density. Therefore, the number of upper main sequence stars which might suffer heavy reddening and contaminate my counts of the blue supergiants is really quite small. I would think the effect is negligible.

Jim Hesser: In your CMD for NGC 2065 you said you were able to do a good statistical field star correction because of your excellent field star samples. However, there seemed to be many small black dots to the red of the main sequence.

Alves: That's correct. My field star subtraction was not perfect and I made some cuts in color and magnitude to futher remove field stars. I should be able to revisit the field star subtraction procedure and improve on this preliminary analysis.

New Views of the Magellanic Clouds
IAU Symposium, Vol. 190, 1999
Y.-H. Chu, N.B. Suntzeff, J.E. Hesser, and D.A. Bohlender, eds.

Recent Results on Cepheids in the Magellanic Clouds from the EROS Survey

J.B. Marquette (The EROS Collaboration)

Institut d'Astrophysique de Paris, 98bis Bd Arago, 75014 Paris, France

Abstract.

Between October 1996 and February 1997 a specific campaign of the EROS experiment was devoted to Cepheids in central regions of the LMC and the SMC. The Period–Luminosity (PL) relations were determined in the EROS photometric system, and a significant slope change of the period–luminosity relation has been observed for the SMC fundamental mode Cepheids with periods shorter than 2 days.

1. Introduction

The different microlensing surveys have the ability to monitor millions of stars, therefore they are suitable to study Cepheid variables. In that way the EROS 1 collaboration has already reported the discovery of 97 Cepheids towards the bar of the LMC (Beaulieu et al. 1995), and 450 Cepheids towards the SMC (Beaulieu & Sasselov 1996). The present paper reports the results of a dedicated Cepheid campaign performed between October 1996 and February 1997. Other topics concerning microlensing survey as well as astrophysical returns of the EROS experiment can be found in this volume (see Beaulieu, Graff et al., Milsztajn, Ripepi et al.).

2. Observations and Data Reduction

The observations were done using the new EROS2 experimental setup which consists of a 1m F/5 Ritchey–Chretien telescope, with two 2 × 4 CCD mosaic cameras mounted in different focal planes. The simultaneous imaging is done in a two-color passband, 420–650 nm ("V_{EROS}" band) and 620–920 nm ("R_{EROS}" band). It has to be stressed that these bands are *not* standard. The available field of view covers a surface of 0.7 deg. (right ascension) × 1.4 deg. (declination) on the sky. A detailed description of the experiment is available elsewhere (Bauer 1997; Bauer et al. 1997).

Two fields per Magellanic Cloud (see Table 1) were monitored about once per night with an exposure time of 20 seconds. When possible, the four fields were imaged almost simultaneously at the same airmass. A total of ~110-160 images was obtained for each field.

Table 1. Coordinates (J2000) of the field centres.

field	LMC1	LMC2	SMC1	SMC2
α	5 23 34	5 15 36	0 51 54	0 41 54
δ	-69 44 22	-69 44 22	-73 36 32	-73 42 32

3. Data Analysis and Discussion

After photometric reduction using the standard EROS 2 photometry package (Peida++, Ansari 1996), the systematic search for periodic variable stars was performed (for $P > 0.5$ day), using an algorithm proposed by Scargle (1982), among 1,134,000 light curves in LMC and 504,000 in SMC. A 5th-order Fourier series was used to fit the light curves. This allow to distinguish classical (F) Cepheids from first overtone (1-OT) Cepheids as done in previous study (Beaulieu et al. 1995) : 1-OT Cepheids have a lower content of second and third harmonics than F Cepheids. A second visual inspection of Cepheid light curves was then performed for the few stars that lie in-between the F or 1-OT Cepheid samples. As a result we obtained a new EROS 2 Cepheid catalog, still under construction (Bauer et al. 1999) which consists of 590 Cepheids towards the SMC (351 F and 239 1-OT Cepheids) and 290 Cepheids towards the LMC (177 F and 113 1-OT Cepheids).

The PL relations for the F and 1-OT Cepheids are shown in Fig. 1. For LMC Cepheids, and SMC 1-OT Cepheids as well, the relations are compatible with a linear behaviour. In contrast, that of the SMC F Cepheids displays a change in the slope for periods smaller than 2 days, visible in both colours. The magnitude deviation for these short–period Cepheids, with respect to an extrapolation of the PL relation for longer period Cepheids, reaches 0.2–0.3 mag at $P = 1$ day. Hence, a simple linear fit to the full SMC F Cepheid sample would result in a biased slope and thus to an incorrect estimate of the distance to the SMC.

Note that the shorter period Cepheids are 1-OT objects, while one finds only F Cepheids at the longest periods. This fact is related to the physics of the pulsation itself and can be explained theoretically (for recent advances on Cepheid theory, see Bono, this volume). The comparison of the LMC and SMC F Cepheids population shows that there are no such stars in LMC for the shorter periods. This reflects the metallicity effect, e.g., the blue loops in LMC are not allowed to cross the instability strip at these shorter periods. In order to quantify the significance of the non–linearity in Fig. 1, we fit the PL relations (see Table 2) in two different ways: (1) using a standard linear regression for the full data; (2) using two straight lines that cross at a "break–period" $P_{\rm break}$,

$$f(x) = \alpha + \beta_{\rm i} \log(P/P_{\rm break}) \tag{1}$$

where i is an index for $P < P_{\rm break}$ (resp. $P > P_{\rm break}$). We find the break–period to be 2.0 days for F Cepheids and use 1.4 day for 1-OT Cepheids from the known ratio of F and 1-OT Cepheid periods. It is beyond the scope of the present paper to give a firm explanation of the reported effect. Three hypotheses based

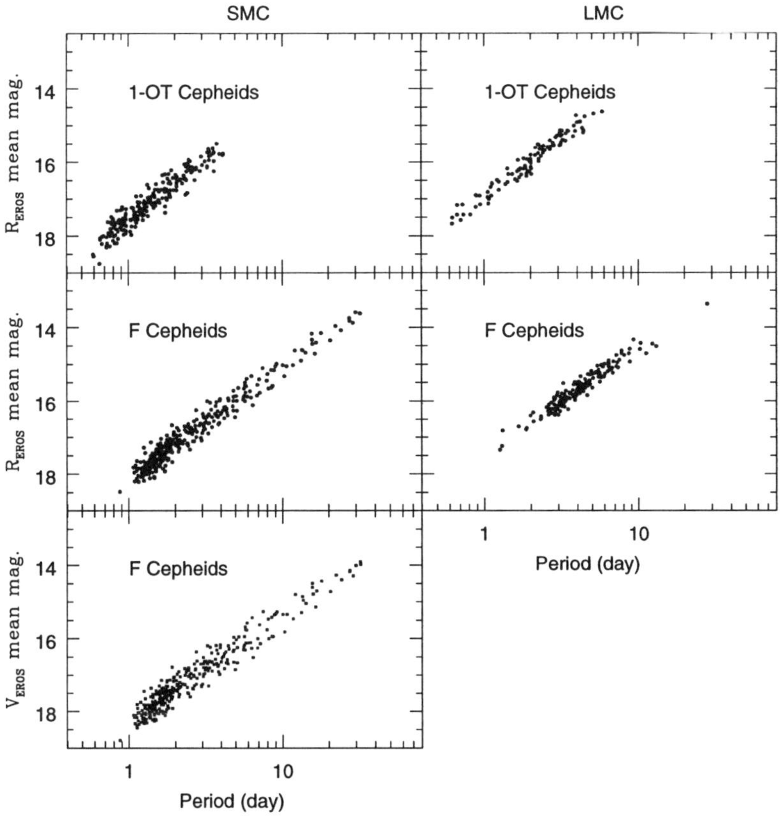

Figure 1. PL relations for first overtone Cepheids (top) and fundamental mode Cepheids (center) towards the SMC (left) and the LMC (right) seen in R_{EROS}. The figure at bottom left shows the PL relation for SMC F Cepheids seen in V_{EROS}. The five panels are drawn to the same scale in order to facilitate comparison.

on qualitative considerations could be proposed : (i) the observed non-linearity is due to the fact that for a given mass and metallicity the evolution tracks do not cross the entire instability strip, which would be non-uniformly populated. This is unlikely as we observe 1-OT Cepheids which would be affected in a similar way; (ii) the mixing of two stellar populations, for example, metal-poor anomalous Cepheids. This very tentative hypothesis would need further observations to be confirmed; (iii) the intrinsic shape of the instability strip becomes non-linear for small masses. The last explanation seems to be the most likely.

Acknowledgments. This work is based on observations made at ESO, La Silla, Chile. I thank the members of the EROS group for great fun, as I am back in astrophysics after six years spent in software industry. I am grateful to the IAU 190 organization committee for both invitation and travel grant.

Table 2. PL relation fits described in Eq.(1), given in the non-standard EROS system. N is the number of stars in the sample, and σ_{res} the dispersion of the fit residuals.

Population	Cloud	Method	N	α	β	σ_{res}
F Cepheids	LMC	all	177	V_{EROS} : 17.63 ± 0.04	2.77 ± 0.07	0.18
				R_{EROS} : 17.44 ± 0.04	2.89 ± 0.06	0.15
		P > 2 d	170	V_{EROS} : 17.60 ± 0.03	2.72 ± 0.07	0.17
				R_{EROS} : 17.43 ± 0.02	2.89 ± 0.06	0.14
		P < 2 d	7	V_{EROS} : 17.86 ± 0.03	3.59 ± 0.57	0.18
				R_{EROS} : 17.46 ± 0.02	2.99 ± 0.51	0.17
	SMC	all	351	V_{EROS} : 18.29 ± 0.02	2.91 ± 0.04	0.24
				R_{EROS} : 18.12 ± 0.02	3.04 ± 0.03	0.20
		P > 2 d	164	V_{EROS} : 18.20 ± 0.02	2.80 ± 0.06	0.25
				R_{EROS} : 18.05 ± 0.02	2.95 ± 0.05	0.20
		P < 2 d	187	V_{EROS} : 18.40 ± 0.02	3.47 ± 0.18	0.22
				R_{EROS} : 18.20 ± 0.02	3.49 ± 0.16	0.20
1-OT Cepheids	LMC	all	113	V_{EROS} : 17.11 ± 0.03	3.12 ± 0.07	0.18
				R_{EROS} : 16.93 ± 0.03	3.18 ± 0.07	0.16
		P > 1.4 d	87	V_{EROS} : 17.13 ± 0.03	3.15 ± 0.12	0.19
				R_{EROS} : 16.96 ± 0.03	3.23 ± 0.11	0.16
		P < 1.4 d	26	V_{EROS} : 17.11 ± 0.03	3.03 ± 0.21	0.16
				R_{EROS} : 16.93 ± 0.03	3.07 ± 0.20	0.16
	SMC	all	239	V_{EROS} : 17.70 ± 0.02	3.06 ± 0.08	0.25
				R_{EROS} : 17.55 ± 0.02	3.21 ± 0.07	0.21
		P > 1.4 d	103	V_{EROS} : 17.55 ± 0.03	2.71 ± 0.26	0.25
				R_{EROS} : 17.43 ± 0.03	2.91 ± 0.23	0.21
		P < 1.4 d	136	V_{EROS} : 17.70 ± 0.03	3.20 ± 0.12	0.25
				R_{EROS} : 17.55 ± 0.03	3.32 ±, 0.1	0.21

References

Ansari, R. 1996, Vistas in Astronomy, 40, 519

Bauer, F. 1997, PhD thesis, Université Paris VII, or Saclay Report Dapnia–SPP 97/10

Bauer, F., et al. (the EROS coll.) 1997, in Optical Detectors for Astronomy, ESO: Garching

Bauer, F., et al. (the EROS coll.) 1999, in preparation

Beaulieu, J.P., et al. (the EROS coll.) 1995, ApJ, 303, 137

Beaulieu, J.P., & Sasselov D. (the EROS coll.) 1996, in Variable Stars and the Astrophysical Returns of Microlensing Surveys, R. Ferlet, J.P. Maillard, & B. Raban, Frontières: Paris, 1996, 193

Scargle, J.D. 1982, ApJ, 263, 835

New Views of the Magellanic Clouds
IAU Symposium, Vol. 190, 1999
Y.-H. Chu, N.B. Suntzeff, J.E. Hesser, and D.A. Bohlender, eds.

On the Period-Luminosity-Color Relation of Classical Cepheids

G. Bono[1]

Osservatorio Astronomico di Roma, Via Frascati 33, 00040 Monteporzio, Italy

M. Marconi[2]

Osservatorio Astronomico di Napoli, Via Moiariello 16, 80100 Napoli, Italy

Abstract.
We compare the theoretical and empirical PL relations in V and I bands for Cepheids in the LMC. We found that, within the current intrinsic dispersions, the theoretical predictions are in remarkable agreement with observational data. We also discuss the PLC relations in (V,B-V) and (V,V-I) as well as the dependence of distance determinations on uncertainties in colors, in reddening corrections and in metal content.

1. Introduction

Classical Cepheids are the most popular primary distance indicators. The key role played by this group of radial variables in estimating the cosmic distance scale is soundly confirmed by the large number of both observational and theoretical investigations aimed at improving both accuracy and reliability of distance measurements. At least three are the main reasons for the widespread use of the Cepheid distance scale:

- Cepheids are bright intermediate-mass objects with visual magnitude range from M_V=-3 at short-periods ($logP \approx 0.6$) to M_V=-6 at long-periods ($logP \approx 2.0$). This feature and the luminosity variations over the pulsation cycle make Cepheids an appealing observational target, since they can be detected and measured in a large number of Local Group galaxies. With the HST, the distance scale key project has identified a good number of Cepheirds in two dozens galaxies out to the Virgo cluster.
- The physical mechanisms which govern the pulsational instability of these objects have been firmly established. This notwithstanding, the pulsation behavior and the modal stability of Cepheids are often outlined by adopting näive arguments. Moreover, up to now no general theoretical consensus on the dependence of the Cepheid luminosity on the metal abundance has been reached.

---

[1] On leave from Osservatorio Astronomico di Trieste, Via G.B. Tiepolo 11, 34131 Trieste, Italy

[2] Dipartimento di Fisica Università di Pisa, Piazza Torricelli 2, 56126 Pisa, Italy

In fact, theoretical predictions based on linear, nonadiabatic, radiative models suggest that the blue edge of the instability strip presents a negligible dependence on metal content (Saio & Gautschy 1998). On the other hand, recent theoretical predictions which account for both blue and red edges of the instability strip support the evidence that Cepheid luminosity depends on chemical composition. In particular, it turns out that metal-poor Cepheids are, at fixed period, brighter than metal-rich ones. This prediction is at odds with their empirical behavior, and indeed a trend opposite to this has recently appeared in the literature (Sasselov et al. 1997; Kennicutt et al. 1998 and references therein).

• Some of the long-standing questions on distance determinations can be properly addressed within the Cepheid scenario. The intrinsic width of the instability strip is substantially narrower in near infrared (NIR) bands than in optical. Metallicity dependence and reddening corrections give the same outcome, and indeed NIR PL relations are less affected by these uncertainties than optical PL relations. The calibration of the PL zero-points can be accomplished by adopting several independent methods such as the Baade-Wesselink method and its progeny (Krockenberger et al. 1997; Di Benedetto 1997), the trigonometric parallaxes (Feast & Catchpole 1997), and the main sequence fitting (Gieren & Fouquè 1993). Therefore a straightforward analysis of both systematic and intrinsic errors affecting distance determinations may be undertaken.

Cepheids are the primary extragalactic distance indicators, and affect directly the evaluation of the Hubble constant -H_0. As noted by Trimble (1997), over the years H_0 estimates decreased by one order of magnitude, but the error bars remained almost constant! Clearly, we need a comprehensive theoretical investigation of the deceptive errors which could affect distance measurements.

In §2 we discuss the comparison between theoretical and empirical PL relations in V and I bands for Cepheids in the Large Magellanic Cloud (LMC), as well as the so called Wesenheit function. Theoretical prediction concerning the Period-Luminosity-Color (PLC) relations are presented in §3 together with a brief analysis of the uncertainties due to both metallicity and reddening.

2. PL Relations for LMC Cepheids

LMC Cepheids are the backbone of cosmic distance derivation since the PL relations of the target galaxy are compared with the LMC PL relations when placing extragalactic distances on an absolute scale. Even though the Cepheid metallicity is not firmly constrained (see Luck et al. 1998) and the reddening estimates still present some uncertainties, LMC Cepheids cover a wide period range and are relatively close objects, thus enabling a proper sampling of the instability strip.

In a recent paper Tanvir (1997) performed a thorough analysis of the intrinsic and systematic uncertainties which affect Cepheid distance scale. In particular, the author derived new LMC PL relations based on current available data in V and I bands, as well as a PL relation for the reddening free Wesenheit function, i.e. $W_{VI} = V - R[V - I]$ where $R \approx 2.45$ is the adopted extinction parameter (Cardelli et al. 1989). These empirical relations were calibrated by adopting an LMC distance modulus of 18.5 and a reddening of $E_{B-V} = 0.1$ mag, respectively. In order to supply a theoretical framework for a proper comparison with observational data we derived analytical PL relations by adopting several sequences of Cepheid models constructed with fixed chemical composition (Y=0.25 Z=0.008) and a wide range of stellar masses and effective temperatures (see Bono et al. 1999a,b for further details). The mass-luminosity relation adopted for fixing the luminosity of these models is based on evolutionary tracks which neglect the convective core overshooting during the hydrogen burning phase.

In deriving these relations we anchored the period at $logP = 1.4$. We did not restrict the period range to $logP < 1.8$ (Tanvir 1997), since we are interested in testing theoretical predictions for long-period Cepheids. Due to the well known bending of the PL relation in the long-period range we performed a quadratic fit and the results we obtained are the following:

$$
\begin{array}{lllll}
< M_V >= & -5.18 & -2.01\,[logP - 1.4] & +0.88\,[logP - 1.4]^2 & (\sigma_{rms} = 0.25) \\
< M_I >= & -6.09 & -2.43[logP - 1.4] & +0.67[logP - 1.4]^2 & (\sigma_{rms} = 0.18) \\
< W_{VI} >= & -7.41 & -3.02\,[logP - 1.4] & +0.38\,[logP - 1.4]^2 & (\sigma_{rms} = 0.08)
\end{array}
$$

Figure 1 shows the comparison between theoretical and empirical PL relations. From this comparison three interesting results emerge: 1) In the period range covered by empirical relations ($0.4 < logP < 1.7$) the mean PL_V and PL_I relations are, within the intrinsic dispersions, in remarkable agreement with theoretical predictions. 2) The theoretical PL relations show a linear behavior up to $logP \approx 1.6$ but toward longer periods they start to bend due to the shift of the instability strip toward redder colors. 3) Theoretical and empirical dispersions in the I band are, as expected, smaller than those of the V band, but they are still too large for constraining the distance modulus and the reddening values adopted for calibrating this sample. This notwithstanding, there is no plausible reason for assuming that the pulsation characteristics of LMC Cepheids are peculiar (Simon & Young 1997), thus supporting the use of these templates for estimating the absolute distance of target galaxies with similar metal contents.

The bottom panel of Figure 1 shows the comparison between theoretical and empirical PL_W relations based on Wesenheit magnitudes. The two relations are in agreement only marginally, and when moving from long to short-periods the discrepancy increases. Due to the agreement between theory and observations in the $logP - M_V$ and in the $logP - M_I$ plane, it is not clear whether this drift is caused by a systematic shift in the color-temperature relations we adopted or more likely is caused by a poor accuracy of (V-I) mean colors. In fact, in order to reduce the sampling of I light curves the V band light curves are transformed into I band light curves by adopting an empirical method (see Appendix A in Tanvir 1997). We note that the light curves are not characterized by a constant shape when moving from short to long-period Cepheids. In the short-period range they present a sawtoothed shape, then around $logP \approx 1$ they show a bump along either the rising or the decreasing branch while in the long-period range the shape becomes more sinusoidal.

It has been often pointed out that the use of these magnitudes causes a substantial decrease in the dispersion of the PL relation. This is mainly due to the fact that the Wesenheit function in two arbitrary bands (β, ξ) is the projection of the PLC (β, $\beta - \xi$) relation onto the $logP - M_\beta$ plane. As a consequence, in this plane the dispersion is much smaller since the location of

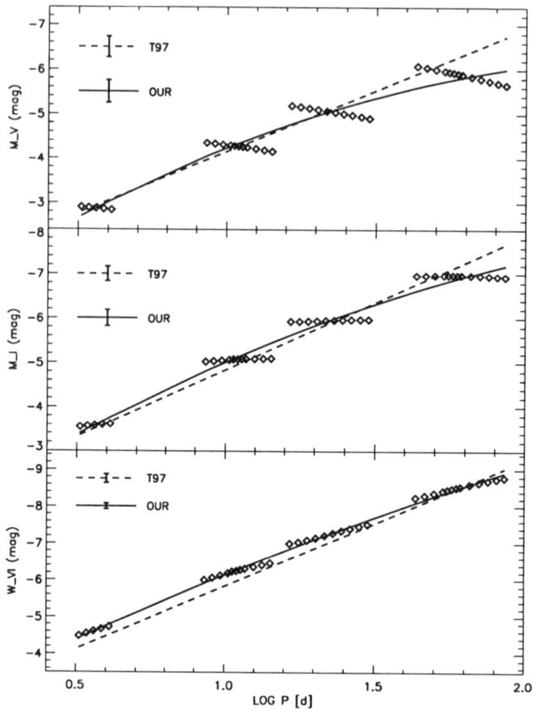

Figure 1. Comparison between our theoretical PL relations (solid lines) and the empirical relations (dashed lines) provided by Tanvir (1997). The error bars are referred to the intrinsic dispersions of analytical relations. From top to bottom the relation plotted in this figure are the PL_V, the PL_I and the Wesenheit function.

each individual Cepheid inside the instability strip is being fixed according to both period and color.

Is the Wesenheit Function Equivalent to Using a PLC Relation for Estimating Distances?

On general grounds the answer is No, because the Wesenheit function is only mimicking the behavior of the equivalent PLC relation. Due to coincidence, the value of the color coefficient in the optical PLC (V, B-V) relation is quite similar (see Bono et al. 1999a) to the extinction parameter, and thus in these bands the Wesenheit function and the PLC relation present the same behavior. However, the Wesenheit function (V,I) is derived by adopting $R \approx 2.45$ (Cardelli et al. 1989) and therefore it is quite different from the color coefficient of the PLC (V, V-I) relation (see below).

On the basis of the above results three main conclusions concerning the use of the $logP - W_{VI}$ relation can be drawn: 1) in these bands the Wesenheit function does not mimic the behavior of the PLC (V, V-I) relation and therefore it does not properly account for the intrinsic width of the instability strip. 2) This relation is sensitive to the photometric accuracy of mean colors and therefore for a sound empirical evaluation a good sampling of both V and I band light curves is needed. This means that if accurate mean colors are available the use of the PLC (V, V-I) relation for estimating distances is more physically plausible than the Wesenheit function. The interesting feature that the Wesenheit function supplies reddening free magnitudes will be discussed in a forthcoming paper (Caputo et al. 1999).

3. Discussion and Conclusions

The above discussion has been focused on the PL relations at fixed chemical composition. In order to account for the dependence of distance determinations on metallicity we adopt the PLC relations, since in this parameter space the Cepheid location is not ambiguous. We derived two analytical relations by taking into account sequences of models constructed by adopting three different chemical compositions, namely Y=0.25, Z=0.004/[Fe/H]=-0.7; Y=0.25, Z=0.008/[Fe/H]=-0.4 Y=0.25, Z=0.02/[Fe/H]=0.0. The results of this fit for (V, B-V) and (V, V-I) PLC relations are the following:

$$\begin{array}{lllll} < M_V >= & -2.83 & -3.57 logP & +2.92[< B > - < V >] & -0.35[Fe/H] \\ & \pm 0.05 & \pm 0.03 & \pm 0.05 & \pm 0.03 \end{array}$$

$$\begin{array}{lllll} < M_V >= & -3.57 & -3.59 logP & +3.80[< V > - < I >] & +0.03[Fe/H] \\ & \pm 0.03 & \pm 0.02 & \pm 0.04 & \pm 0.02 \end{array}$$

where the symbols have their usual meaning. The standard deviations of these two relations are $\sigma_{rms} = 0.05$ and 0.03 respectively. As already pointed out by Bono et al. (1999a) the PLC (V, B-V) relation suggests that metal-rich Cepheids are, at fixed period and color, brighter than metal-poor ones. On the other hand, the PLC (V, V-I) relation discloses a marginal dependence on metallicity, and thus leads strong support to the use of this relation for estimating distances of target galaxies for which the metal content is poorly known.

Since these relations are not affected by systematic errors such as reddening corrections, photometric accuracy of both magnitude and colors, and the metal content, we can estimate how plausible errors within these parameters may affect distance determinations. At first, to account for photometric and zero-point calibration uncertainties we assumed, by following Tanvir (1997), that $\sigma_B \approx \sigma_V \approx \sigma_I \approx 0.04$ mag. These uncertainties imply errors on (B-V) and (V-I) colors of the order of 0.06 mag, and in turn an uncertainty in distance of 8% and 10%. At the same time, a systematic error of the order of 0.03 mag in the reddening correction -E_{B-V}- implies errors on (B-V) and (V-I) colors of 0.03 and 0.04 mag respectively. These uncertainties imply errors in distance equal to 4% and 7%. However, if we simultaneously account for the two quoted uncertainties, the errors on the colors are equal to 0.19 and 0.26 mag, while in distances the errors are 9% and 12% respectively.

In order to account for the dependence of distance determinations on chemical composition we assumed an uncertainty on metal abundance of the order of 0.4 dex. This error takes into account not only the intrinsic errors in metallicity measurements but also the spread in metallicity of Cepheids in external galaxies. In fact, Monteverde et al. (1997) by adopting spectroscopic data of B-type supergiants in M33 estimated that the iron abundance gradient in this spiral galaxy is of the order $-0.20(\pm 0.05)$ dex kpc^{-1} The quoted uncertainty implies errors in distance modulus of 0.14 (V,B-V) and of 0.012 (V,V-I) mag and thus errors of 7% and 1% respectively.

The main outcome of this leading term error analysis are the following: 1) PLC relations based on different photometric bands present pros and cons, and indeed the PLC (V,B-V) relation, in comparison with the PLC (V,V-I) relation, is less affected by errors on both reddening corrections and colors but is more affected by a spread in metallicity. 2) The PLC (V,V-I) relation seems a promising distance indicator for target galaxies with accurate reddening and color measurements since it is marginally affected by metallicity.

Obviously a comprehensive analysis of the error budget of Cepheid distance scale based on PLC relations is not a trivial effort, since in reality it is a mixture of all previous uncertainties. A thorough analysis can be undertaken only by means of Monte Carlo simulations of Cepheids inside the instability strip, which can simultaneously account for all plausible errors on observables.

Acknowledgments. This work was partially supported by CNA through a postdoc research grant to M. Marconi, by ASI and CRA.

References

Bono, G., Caputo, F., Castellani, V., & Marconi, M. 1999a, ApJ, 512, 711

Bono, G., Marconi, M., & Stellingwerf, R. F. 1999b, ApJS, 122, 167

Caputo, F., et al. 1999, in preparation

Cardelli, J. A., Clayton, G. C., & Mathis, J. S. 1989, ApJ, 345, 245

Di Benedetto, G. P. 1997, ApJ, 486, 60

Feast, M. W. 1984, in IAU Symp. 108, Structure and Evolution of the Magellanic Clouds, eds. S. van der Bergh & K.S. de Boer, Dordrecht: Kluwer, 157

Feast, M. W. 1991, in Observational Tests of Cosmological Inflation, eds. T. Shanks et al., Dordrecht: Kluwer, C348, 147

Feast, M. W. 1995, in IAU Colloq. 155, Astrophysical Applications of Stellar Pulsation, eds. R.S. Stobie & P.A. Whitelock, San Francisco: ASP, 209

Feast, M. W., & Catchpole, R. M. 1997, MNRAS, 286, L1

Gieren, W. P., & Fouquè, P. 1993, AJ, 106, 734

Kennicutt, R. C., et al. 1998, ApJ, 498, 181

Krockenberger, M., Sasselov, D. D., & Noyes, R. W. 1997, ApJ, 479, 875

Luck, R. E., Moffett, T. J., Barnes, T. G., & Gieren, W. P. 1998, AJ, 115, 605

Monteverde, M. I., et al. 1997, ApJ, 474, L107

Saio, H., & Gautschy, A. 1998, ApJ, 1998, 498, 360

Sasselov, et al. 1997, A&A, 324, 471

Simon, N. R., & Young, T.S. 1997, MNRAS, 288, 267

Tanvir, N. R. 1997, in The Extragalactic Distance Scale, eds. M. Livio, M. Donahue & N. Panagia, Cambridge: Cambridge Univ. Press, 91

Trimble, V. 1997, in The Extragalactic Distance Scale, eds. M. Livio, M. Donahue & N. Panagia, Cambridge: Cambridge Univ. Press, 313

Discussion

David Graff: The EROS diagram showing how many stars we find as a function of period in both the LMC and the SMC does not show the standard metallicity effect. That is, it does not show that Cepheids of a particular period have a different luminosity from Cepheids with the same period, but different metallicity. It only shows that the instability strip extends to lower magnitudes for lower metallicities.

Bono: The EROS diagram shows that LMC Cepheids are systematically redder than SMC Cepheids. This effect implies that, in the $\log P - M_v$ plane, the LMC Cepheids are fainter than SMC Cepheids.

Norbert Langer: What is the improvement in the physics of your stellar models which allows you to obtain such an excellent agreement with observations, as compared to previous models.

Bono: The evolutionary tracks we adopted are canonical (i.e., no rotation, no mass-loss, no overshooting). For more details see Castellani, Chieffi, & Straniero 1992). However, to account for the difference between canonical models and models constructed by adopting a mild convective core overshooting we computed a different set of nonlinear models by increasing the canonical luminosity by about 0.25. With the exception of long period Cepheids the two sets of models present quite similar results concerning both the PL and PLC relations. Obviously, the zero-points are different. Concerning the pulsation models, the main improvement in the input physics is the inclusion of a nonlocal and time-dependant treatment of convective transport (see Stellingwerf 1982; Bono & Stellingwerf 1994).

Wolfgang Gieren: Just a comment: The best available data, both on LMC and on Galactive Cepheids, do not show any curvature of the Cepheid PL relation at the longest pulsation periods (see for example the Tauvir 1997 review paper at the Baltimore conference on the Extragalactic Distance Scale).

Bono: The EROS project in the paper by Sasselov et al. (1997) suggests that the slope of the PL relations changes toward long-period Cepheids ($\log P > 1.8$). Maybe these observations are in contradiction because the time spent inside the instability strip by high-mass stars is shorter than for lower masses. In our theoretical predictions we assume that the instability strip is uniformly filled.

Jean-Philip Beaulieu: I agree with you that one must include nonlinear, nonlocal time dependant turbulent convection in Cepheid envelopes. However, when you adopt one Mass-Luminosity relation for the metallicity of the Galaxy, LMC and SMC, you will not reproduce the position of the resonance centers (like the 10 day 2:1 resonance between the fundamental mode and second overtone) in the galaxies. You can use this constraint to have an idea about the difference in M, L between LMC, SMC, and the Galaxy.

Bono: This is a really interesting problem. We already obtained the Hertzsprung progression close to 10 days for LMC Cepheids. Owing to the small number of stellar masses we adopted we cannot say anything about the Hertzsprung progression in SMC and Galactic Cepheids.

New Views of the Magellanic Clouds
IAU Symposium, Vol. 190, 1999
Y.-H. Chu, N.B. Suntzeff, J.E. Hesser, and D.A. Bohlender, eds.

Calibration of the Cepheid Distance Scale with the Infrared Surface Brightness Technique

Wolfgang Gieren

Universidad de Concepción, Departamento de Física, Astronomy Group, Casilla 4009, Concepción, Chile

Pascal Fouqué

European Southern Observatory, Casilla 19001, Santiago 19, Chile; and Observatoire de Paris-Meudon, DESPA, F-92195 Meudon CEDEX, France

Abstract.
We review the process of determining the distances to Cepheid variables with the infrared surface brightness technique. We show that both versions of this technique (working with the V, $V - K$ and the pure infrared K, $J-K$ magnitude/color combinations) yield consistent distances to Cepheid variables. For a Cepheid with average-quality photometric and radial velocity data, the typical total uncertainty in the distance is $\pm 5\%$. We use the distances of 34 Galactic variables to calibrate period-luminosity relations in the V, I, J, H, K bands and demonstrate that these relations lead to distance moduli of the LMC which agree to ± 0.02 mag. The infrared surface brightness technique yields a best LMC distance modulus of 18.46 ± 0.02. We discuss possible metallicity corrections, such as the ones determined by Sasselov et al. from EROS Cepheids comparison in the LMC and SMC, and conclude that our distance modulus is negligibly affected by these corrections. Finally, we present a program to obtain reddening- and metallicity-independent infrared surface brightness distances of Cepheid variables in LMC clusters which will yield an even more accurate mean LMC distance independent of any metallicity correction to Cepheid absolute magnitudes.

1. Introduction

Cepheid variables continue to be the key primary objects to calibrate the extragalactic distance scale. In view of the results of the HST Key Project on the determination of H_0 from Cepheid variables in some 20 nearby galaxies out to the Virgo and Fornax clusters, it is more important than ever to provide a local and accurate calibration of the PL relation (in different bands, particularly in the V and I bands used by HST), and to establish the true distance to the LMC, whose value as derived from a variety of independent methods is still scattering between extremes of 18.1 to 18.7 (see the review of M. Feast in this volume). Also, the size of metallicity corrections to Cepheid absolute magnitudes is still under dispute (see Kennicutt et al. 1998, and references therein), and it is therefore desirable to have a method which is able to yield direct distances to Cepheid variables, in at least some of the nearest galaxies, which are independent, at least to first order, of the metallicity of the parent galaxy. Recently, such a technique has become available with the introduction of the infrared surface brightness method by Fouqué & Gieren (1997) which is capable of yielding distances to Cepheid variables nearly independent of adopted absorption corrections and the stellar metallicity, and is therefore an almost ideal tool to determine distances with very low systematic uncertainties to a set of nearby galaxies, most importantly to the LMC. The infrared surface brightness technique is of the Baade-Wesselink type and has smaller systematic uncertainties than its visual counterpart which has been extensively used by Gieren, Barnes & Moffett (1993) to study Cepheids in the Galaxy.

The use of Baade-Wesselink type methods, with their specific advantages when calibrated and applied in the near-infrared, has gained increased importance since it has been recognized from the HIPPARCOS data that the locations of the main sequences of open clusters can differ by as much as 0.6 mag (Mermilliod et al. 1997; Robichon et al. 1997; van Leeuwen & Hansen Ruiz 1997), affecting directly the classical approach to calibrate the Cepheid PL relation from the ZAMS-fitting method using Cepheid variables found in open clusters. Also, there is still considerable systematic uncertainty in the LMC distance modulus derived by Feast & Catchpole (1997) from HIPPARCOS parallaxes of nearby Cepheids, mainly due to the very small signal-to-noise ratios and the subsequent problem of correcting statistical biases, such as the Lutz-Kelker bias (Oudmaijer et al. 1998), and to the fact that a large fraction of Cepheids in the Feast & Catchpole sample are spectroscopic binaries with orbital periods in the order of the HIPPARCOS observing time interval which may systematically affect the parallaxes of these stars (Szabados 1997).

In this review we concentrate on a description of the infrared surface brightness technique and present its main features and an evaluation of its uncertainties. We then present the results of our calibration of the Galactic Cepheid PL relations from the infrared distances and apply these relations to find the distance to the LMC. Finally, we briefly discuss ongoing programs of applying the technique directly to Cepheids in the Magellanic Clouds. Among recent excellent reviews which examine other issues related to the use of Cepheids as distance indicators, we mention Feast & Walker (1987), Madore & Freedman (1991), Welch (in Jacoby et al. 1992), and Tanvir (1997).

2. The Infrared Surface Brightness (ISB) Technique

2.1. Calibration of the Technique

A detailed description of the technique is given in the paper of Fouqué & Gieren (1997). A visual surface brightness parameter F_V is defined as

$$F_V = a - 0.1\, S_V \tag{1}$$

where S_V is the visual surface brightness defined in terms of the absorption-corrected apparent magnitude V_o and the stellar angular diameter Φ as

$$S_V = V_o + 5\log\Phi \tag{2}$$

and the constant a is given by solar quantities as

$$a = 0.1\ M_{\odot}^{bol} + 1 + 0.25\log(4f_{\odot}/\sigma) = 4.2207 \tag{3}$$

The heart of the method is to calibrate F_V, or a corresponding parameter F_K defined in the K photometric band, in terms of a suitable color index, using interferometrically determined angular diameters of a set of stable giants and supergiants in the color range of Cepheid variables. Once such a calibration is available, Cepheid angular diameters, at any phase of the pulsation cycle, can be found from an absorption-corrected magnitude and color index at this phase. This angular diameter can be combined with the value of the linear displacement from the stellar mean diameter at this same phase which is found from an integration of the observed radial velocity curve of the variable which first has to be transformed to the corresponding pulsational velocity curve by multiplying with a factor (the p factor) which corrects for the effects of averaging the radial velocity over the stellar disk, and limb-darkening. The equation

$$2\ (R_o + \Delta R) = d\ \Phi \tag{4}$$

(units: Φ in arcsec, radius R in AU, distance d in pc) can then be solved for the distance and mean radius of a Cepheid variable using the observed pairs of linear displacements and angular diameters covering the pulsation cycle.

Laney & Stobie (1995) have demonstrated that the $V-K$ and $J-K$ colors are almost unaffected by the gravity and microturbulence variations of Cepheids during their pulsations and are near-ideal predictors of the surface brightnesses of these variables, as opposed to optical colors like $B-V$ or $V-R$. We therefore decided to calibrate two versions of the ISB technique using the V, $V-K$ and K, $J-K$ magnitude-color combinations. Surface brightness parameters in the V and K bands were calculated for 28 cool giant and supergiant stars from their absorption-corrected magnitudes and measured angular diameters (accurate to better than 5 percent in all cases) which were then correlated with their intrinsic $V-K$ and $J-K$ colors. Excellent linear relations were obtained for both giants and supergiants over a large color range including the one covered by Cepheid variables. There is no systematic difference in these relations between giants and supergiants. The slope of the surface brightness-color relations obtained from the giants/supergiants sample was found to be consistent with the slope obtained from the Cepheids themselves using Thompson's method (Thompson 1975). For our final surface brightness-color relations, we adopted the slopes from the Cepheids and the zero point as coming from the angular diameters of the cool giants and supergiants. This procedure leads to the following relations for the angular diameters (in mas) of a Cepheid in terms of the unreddened magnitudes and colors:

$$\Phi = \text{dex}\ (0.5474 - 0.2\ V_o + 0.262\ (V-K)_o) \tag{5}$$

$$\Phi = \text{dex}\ (0.5474 - 0.2\ K_o + 0.220\ (J-K)_o) \tag{6}$$

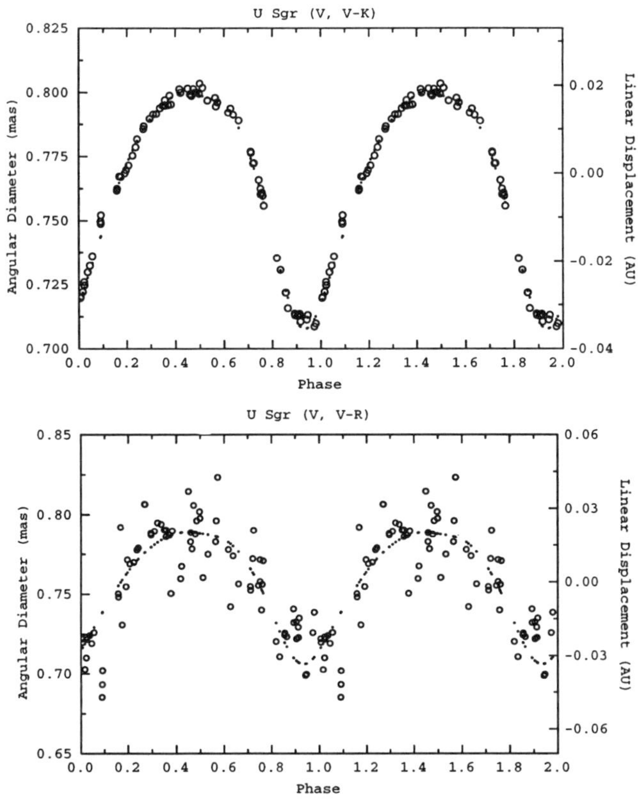

Figure 1. (top) The angular diameters (open circles) of the Cepheid U Sgr calculated from V, $V-K$ photometry according to equation (5) (left scale), plotted against phase. Overplotted (dots) is the linear displacement variation of the Cepheid (right scale). (bottom) The same adjustment from the V, $V-R_J$ photometry.

The angular diameters calculated from these equations are then combined with the linear radius displacements to solve equation (4) for the distance and mean radius of a Cepheid variable. A typical example is shown in Figure 1 where the technique is applied to the Cepheid U Sgr. For comparison, we also plot the angular diameters as obtained from a visual V, $V-R$ surface brightness technique: the dramatic increase in the accuracy of the infrared-based angular diameters over optical ones is evident.

2.2. Results and Uncertainties

The main results from an application of both versions of the ISB technique to a sample of 34 Galactic Cepheids with high-quality infrared photometry and radial velocity curves are the following:

- Both techniques produce consistent distances. The mean distance ratio is 0.983 ± 0.013.

- The most important observational constraint to the achievable distance precision is the amplitude of the color curve used in the solution.
- For Cepheids with average-quality data and color curve amplitudes larger than ~ 0.2 mag, the total uncertainty of an infrared distance is $\sim \pm 5$ percent. This uncertainty includes possible systematic errors.
- For Cepheids with very high-quality data, a distance accurate to ± 3 percent is achievable.
- The ISB distances are practically unaffected by adopted absorption corrections and the metallicity of the Cepheid, a property which makes the technique ideally suited to a direct application to extragalactic Cepheids.

A more detailed account of the effect of a variety of sources of error on the distances can be found in the papers of Gieren, Fouqué & Gómez (1997, 1998).

3. Period-luminosity Relations in the V, I, J, H and K Bands

Our approach to calibrate the PL relations is to determine the slopes of the relations, in each band, from samples of LMC Cepheids and to set the zero points from the infrared distances to the Galactic Cepheids. Due to the fact that LMC Cepheids are all at nearly the same distance, it is obvious that the slopes of the PL relations can be obtained with higher accuracy from LMC Cepheids than from Galactic ones. As the most suited LMC Cepheid data sets, we used a sample of 53 Cepheids from Tanvir (1997) for the V, I calibration, supplemented with high-quality data of Moffett et al. (1998), and a sample of 59 Cepheids for the J, H, K calibrations with mean intensity magnitudes and individual reddenings from the SAAO database (Caldwell 1997). We forced the slopes of the LMC Cepheid PL relations to the Galactic Cepheid PL relations and obtained the following absolute calibrations:

$$M_{\rm V} = -2.769\ (\pm 0.073) \times (\log P - 1.0) - 4.063\ (\pm 0.034), \quad (7)$$
$$M_{\rm I} = -3.041\ (\pm 0.054) \times (\log P - 1.0) - 4.767\ (\pm 0.029), \quad (8)$$
$$M_{\rm J} = -3.129\ (\pm 0.052) \times (\log P - 1.0) - 5.240\ (\pm 0.028), \quad (9)$$
$$M_{\rm H} = -3.249\ (\pm 0.044) \times (\log P - 1.0) - 5.628\ (\pm 0.026), \quad (10)$$
$$M_{\rm K} = -3.267\ (\pm 0.042) \times (\log P - 1.0) - 5.701\ (\pm 0.025). \quad (11)$$

The rms dispersions of these relations range from 0.209 in V to 0.173 in K. These are the most accurate PL relations as yet derived from Galactic Cepheids and can be used to derive distances to galaxies of near-solar metallicity. We note that our calibrations yield Cepheid absolute magnitudes which are on average about 0.1 mag *fainter* than the ones derived by Madore & Freedman (1991) but are consistent with these within the uncertainties. We also note that the rms dispersion of the K band PL relation (almost unaffected by reddening errors), as compared to the dispersion of the LMC Cepheid K band PL relation (0.11 mag), indicates a maximum error of ± 6 percent in the ISB distances to the Galactic Cepheids, in agreement with our error budget.

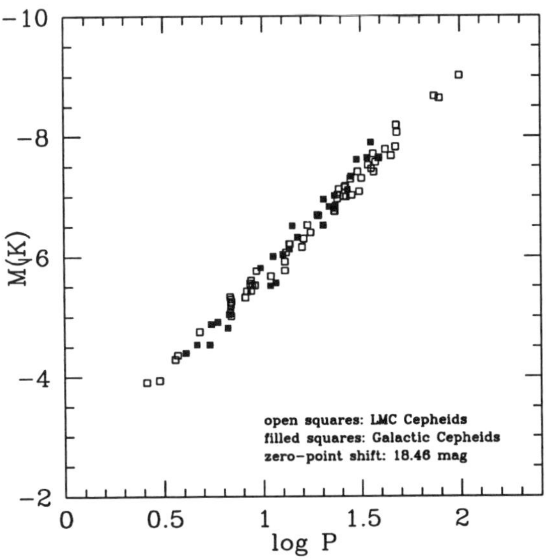

Figure 2. K band PL relations as defined by 59 LMC Cepheids, and by infrared distances of 28 Galactic Cepheids. Best agreement between the data sets occurs for a zero point shift of 18.46.

4. The Distance of the LMC

A comparison of our absolute calibrations of the PL relations to the corresponding relations as defined by the LMC Cepheids yields the distance to the LMC, as the difference of the zero points. An example is shown in Figure 2 for the K band relations. In this way we obtain the LMC distance moduli given in Table 1 from the data in the different bands. It can be appreciated that the agreement of the μ_o values is excellent, and we obtain a mean LMC true distance modulus of 18.46 with an internal uncertainty as low as ± 0.02 mag.

Recent work of the HST Key Project team on Cepheids in M 101 yields a metallicity dependence of $\delta(m-M)_o/\delta[O/H] = -0.24 \pm 0.16$ mag (Kennicutt et al. 1998) in the V, I spectral range, in the sense that metal-poor Cepheids appear fainter and more distant than more metal-rich stars. However, this correction cannot be directly applied to our V and I distance moduli, because it combines the metallicity effects on magnitude and on color, due to the fact that the HST Key Project calculates Cepheid fields reddenings from the V and I measures themselves. In our case, mean reddenings are adopted from literature, and we are only concerned by the metallicity effect on magnitude. This has been derived by Sasselov et al. (1997) from a comparison of Cepheids in the LMC and the SMC, detected in the EROS survey. They found a magnitude effect of 0.06 ± 0.01 in V and -0.01 ± 0.01 in I for the metallicity difference of 0.35 between the LMC and the SMC. As this is approximately the metallicity difference between the LMC and our Galactic Cepheids (Kennicutt et al. 1998 give [O/H] (LMC) = −0.40 and Luck et al. 1998 find [Fe/H] (LMC) = −0.3), we adopt the Sasselov et al. corrections and derive corrected V and I moduli of 18.42 ± 0.05 and

18.49 ± 0.05, respectively, giving an average VI metallicity-corrected distance modulus of 18.46 ± 0.04, in excellent agreement with our adopted value.

An obvious way to circumvent the problem of a metallicity correction is to apply the metallicity-independent ISB method directly to LMC Cepheids. Among the Cepheid variables in the LMC, those in Cepheid-rich LMC clusters as found in the surveys of Welch, Mateo & Olszewski (1993) and more recently by the MACHO collaboration are obviously the ideal targets. Studies of this kind are underway for the clusters NGC 1866 and 2031 (Gieren, Fouqué & Storm, 1999) and should yield a mean LMC distance accurate to ±2 percent which is not affected by the metallicity difference between the LMC and the Galaxy.

Table 1. LMC Cepheid Period-Luminosity Relations and LMC Distance Moduli.

Band	Slope	ZP $\log P = 1$	rms	correlation coefficient	N	μ_o (LMC)
V	-2.769	14.397	0.204	0.983	53	18.478 ± 0.049
I_C	-3.041	13.693	0.150	0.992	53	18.482 ± 0.044
J	-3.129	13.220	0.146	0.992	59	18.433 ± 0.039
H	-3.249	12.832	0.124	0.995	59	18.443 ± 0.039
K	-3.267	12.759	0.118	0.995	59	18.457 ± 0.039

Acknowledgments. This research was supported by Fondecyt grant No. 1971076 to WG which is gratefully acknowledged.

References

Caldwell, J.A.R. 1997, private communication

Feast, M.W., & Walker, A.R. 1987, ARA&A, 25, 345

Feast, M.W., & Catchpole, R.M. 1997, MNRAS, 286, L1

Fouqué, P., & Gieren, W.P. 1997, A&A, 320, 799

Gieren, W.P., Barnes, T.G., & Moffett, T.J. 1993, ApJ, 418, 135

Gieren, W.P., Fouqué, P., & Gómez, M. 1997, ApJ, 488, 74

Gieren, W.P., Fouqué, P., & Gómez, M. 1998, ApJ, 496, 17

Gieren, W.P., Fouqué, P., & Storm, J. 1999 (in preparation)

Jacoby, G.H., et al. 1992, PASP, 104, 599

Kennicutt, R.C., et al. 1998, ApJ, 498, 181

Laney, C.D., & Stobie, R.S. 1995, MNRAS, 274, 337

Luck, R.E., Moffett, T.J., Barnes, T.G., & Gieren, W.P. 1998, AJ, 115, 605

Madore, B.F., & Freedman, W.L. 1991, PASP, 103, 933

Mermilliod, J.C., Turon, C., Robichon, N., Arenou, F., & Lebreton, Y. 1997, in HIPPARCOS Venice '97, ESA SP-402, 643

Moffett, T.J., Gieren, W.P., Barnes, T.G., & Gómez, M. 1998, ApJS, 117, 135

Oudmaijer, R.D., Groenewegen, A.T., & Schrijver, H. 1998, MNRAS, 294, L41

Robichon, N., Arenou, F., Turon, C., Mermilliod, J.C., & Lebreton, Y. 1997, in HIPPARCOS Venice '97, ESA SP-402, 567

Sasselov, D., et al. 1997, A&A, 324, 471

Szabados, L. 1997, in HIPPARCOS Venice '97, ESA SP-402, 657

Tanvir, N.R. 1997, in STScI Symp. Ser. 10, The Extragalactic Distance Scale, M. Livio, M. Donahue & N. Panagia, Cambridge: Cambridge Univ. Press, 91

Thompson, R.J. 1975, MNRAS, 172, 455

van Leeuwen, F., & Hansen Ruiz, C.S. 1997, in HIPPARCOS Venice '97, ESA SP-402, 689

Welch, D.L., Mateo, M., & Olszewski, E. 1993, in IAU Coll. 139, New Perspectives on Stellar Pulsation and Pulsating Variable Stars, Cambridge: Cambridge Univ. Press, 359

Discussion

David Graff: How soon will we see results on the distances to Cepheids in LMC clusters and depth and tilt of the LMC Bar?

Gieren: Distance results for Cepheids in NGC 1866 will be available in a few months (end of 1998); results for Cepheids in other LMC clusters should emerge over the next two years, with corresponding conculsions about the depth of the LMC and tilt of the LMC bar.

John Gallagher: Have you been able to take advantage of eclipsing binaries in the MC's to provide an independant check on the angular size scales and surface brightnesses? Does this appear to be a feasible approach in the future?

Gieren: This will be an important check on the angular diameters obtained from the infrared surface brightness technique once appropriate data are available. The MACHO collaboration has turned up a Cepheid in an eclipsing binary system which will yield such a check.

Douglas Welch: One hidden assumption is that the surface brightness calibration for stable stars is the same as that for Cepheids. There is some indirect evidence that the atmospheric structure is different for Cepheids than for stable stars.

Gieren: We see in the data that the slope of the infrared surface brightness - colour relation for Cepheids and stable giants and supergiants of similar colours are the same (within very small errors). This gives us much confidence that Cepheids, in spite of their pulsating atmospheres, do indeed follow the same surface brightness-colour relations as stable yellow giants and supergiants.

Nino Panagia: I guess that some of your individual errors are larger than the total error simply because you quote first the errors on individual parameters and then you give the total error on the distance determination.

New Views of the Magellanic Clouds
IAU Symposium, Vol. 190, 1999
Y.-H. Chu, N.B. Suntzeff, J.E. Hesser, and D.A. Bohlender, eds.

The Distance to the Large Magellanic Cloud; A Critical Review

Michael Feast

Department of Astronomy, University of Cape Town, Rondebosch, 7700, South Africa

Abstract. Five independent methods of deriving the distance modulus of the Large Magellanic Cloud are reviewed and compared. A value in the range 18.60 to 18.80 is presently preferred.

1. Introduction

A remarkable range of distance moduli has been proposed for the LMC in very recent times. These moduli range from 18.06, from red clump stars, to 18.70 and greater from some discussions of Cepheids, Miras and RR Lyrae stars. This paper will briefly review five methods of determining the distance to the LMC and examine their strengths and weaknesses. These five methods are all independent of one another at least so far as their basic calibrations are concerned. Some methods have been omitted because they are not independent of the ones discussed. For instance the use of the luminosity of the red giant tip involves a calibration using horizontal branch or RR Lyrae luminosities.

2. Cepheids

The Hipparcos parallaxes of Cepheids can be used to determine the zero-points of the period - luminosity (PL) and period - luminosity - colour (PLC) relations (Feast & Catchpole 1997; Feast & Whitelock 1997). These parallaxes are mostly quite small and it is essential to analyse them in a way that avoids statistical bias. Such a method was used in the above two references (see also Feast, Pont, & Whitelock 1998, Feast 1998, and Koen & Laney 1998). Other authors have used methods which introduce bias and either ignore it (e.g., Madore & Freedman 1998) or attempt to correct for it in an approximate way (e.g., Oudmaijer, Groenewegen, & Schrijver 1998). In using a PL relation it is important to realize that even the relatively nearby (Hipparcos) stars on which the zero-point is based are generally significantly reddened. This will not cause problems so long as the systems for dereddening the programme and calibrating stars are equivalent.

In applying a PL relation to the LMC one has to take into account that the metallicity of the Cepheids there is lower than in the solar neighbourhood. This has some effect on the PL zero-point (depending on the wavelength at which one is working) and also on most methods of determining the interstellar absorption. Feast & Catchpole (1997) adopt a metallicity depletion of a factor

1.4 for the LMC, LMC reddenings from *BVI* photometry (Caldwell & Coulson 1985; Caldwell & Laney 1991) and PL metallicity corrections from Laney and Stobie (1994). They then derive an LMC modulus of 18.70±0.10. If they had adopted a metal deficiency of a factor of two as suggested by some work (see V. Smith this volume) they would have obtained a distance modulus of 18.76.

The Hipparcos proper motions of Cepheids can be used to estimate PL and PLC zero-points. The most satisfactory way to do this is to compare the value of the Oort constant of differential galactic rotation (A) derived from the proper motions (this is almost independent of distance scale) with that derived from radial velocities (this varies inversely with the distance scale) (Feast & Whitelock 1997; Feast, Pont, & Whitelock 1998). In this way one derives a zero-point which is 0.04 brighter than that derived above and an LMC modulus of 18.74±0.13 for a metal deficiency of a factor 1.4.

Until the availability of the Hipparcos results the best PL zero-point came from Cepheids in galactic open clusters. The distance scale for the clusters most commonly used was based on the parallaxes of nearby main-sequence stars using the shape of the Pleiades main sequence as a template. Using this scale Laney & Stobie (1994) obtained an LMC modulus of 18.53. A later discussion using more parallax data (Feast 1995) increased this to 18.64. Obviously there is now a great deal of data from Hipparcos which could be used to update this result. The steepness of the main sequence makes the cluster method rather sensitive to systematic errors in the photometry and the reddening correction.

One can also estimate the PL zero-point from Baade-Wesselink type analyses of Cepheids. One of the most recent such estimates has been made by Gieren, Fouqué, & Gómez(1998). They derive a steeper slope for the PL relation from their galactic stars than is observed in the LMC. If this were significant it would raise questions regarding the use of Cepheids as distance indicators. Putting this possibility on one side (i.e., attributing the difference to the selection of stars and the method of analysis), the best proceedure is probably to calibrate the LMC PL relation at the mean log P of the galactic calibrating stars. This leads to an LMC modulus of 18.55. Gieren et al. quote an uncertainty of $\sim 0\overset{m}{.}02$ for their zero-point. However this appears to be essentially an internal error. In addition there may be systematic uncertainties in the radii determinations and in the colour - surface brightness calibration (there are few suitable calibrating stars in the colour and luminosity range of the Cepheids).

3. Miras

Miras in the LMC show good PL relations in the infrared (at K) or in M_{bol} (e.g. Feast et al. 1989). van Leeuwen et al. (1997) calibrated these relations using some initial results on the Hipparcos parallaxes of Miras.This has been extended by Whitelock et al. (1997) and further work is in progress. As with the Cepheids care was taken to avoid bias in the analysis (see also Feast & Whitelock 1998). van Leeuwen et al. obtained in this way LMC moduli of 18.60 ± 0.18 (PL at

K) [1] and 18.47 ± 0.17 (PL at M_{bol}). The difference between these two estimates (which is of much higher accuracy than that of the individual values) is real and expected. This is because it is known that the period - infrared colour relations in the LMC and the Galaxy differ significantly (Whitelock et al. 1994; Glass et al. 1995). This is plausibly a metallicity effect. In that case Wood (1990) predicts that the PL(K) relation should be the least affected. Taking these results and Wood's theory at their face value, one deduces that at a given period the LMC Miras are metal deficient by Δ[Fe/H] ~ -0.4 compared with those in the Galaxy and that the LMC modulus is 18.77. Obviously the metallicity correction is quite uncertain but it does suggest that the Mira modulus of the LMC is more likely to be greater than 18.6 than less.

It is also possible to calibrate the Mira PL relations using Miras in globular clusters containing Miras, provided the cluster distances are known. Reid (1998) has recently derived a distance modulus for the globular cluster 47 Tuc using Hipparcos parallaxes of subdwarfs. Using this modulus with infrared photometry of the three Miras in the cluster, Feast, & Whitelock (1998) obtain a PL(K) zero-point which leads to an LMC modulus of 18.79 ± 0.17.

4. RR Lyraes

Reid (1997,1998) obtained distances to galactic globular clusters of various metallicities using Hipparcos parallaxes of subdwarfs. He thus derived the absolute magnitudes of RR Lyraes (or the level of the horizontal branch (HB) at the position of the RR Lyraes) in these clusters and applied these results to RR Lyraes in LMC clusters. The strength of this method is that it is based on trigonometrical parallaxes, deals exclusively with cluster RR Lyraes (or HBs) and compares cluster RR Lyraes (or HBs) in the Galaxy with those of the same metallicity in the LMC. Reid (1998) found an LMC modulus of 18.71 ± 0.13 in this way. McNamara (1998) has used Hipparcos parallaxes of SX Phoenicis stars to determine distances of some clusters and galaxies and has inferred RR Lyrae luminosities in this way. His results lead him to an LMC modulus of 18.56 ± 0.10. This is an interesting approach but depends on the parallaxes of two stars and a number of intervening steps and should therefore probably be given a lower weight than Reid's determination.

One can estimate the luminosities of field RR Lyrae variables as a function of metallicity using their Hipparcos parallaxes. The results obtained by Koen & Laney (1998) in this way lead to an LMC modulus of 18.68 ± 0.22 when applied to RR Lyraes in LMC clusters. Gratton (1998) suggested that the Hipparcos parallaxes of field HB stars could be used to infer RR Lyrae luminosities. His results have been slightly revised by Koen & Laney (1998) and lead to an LMC modulus of 18.45 ± 0.11, again using the RR Lyraes in LMC clusters. The true uncertainty of this result may well be greater than that shown since large corrections are necessary to some of the HB star results to take into account the fact that the HB is far from horizontal at the wavelength of the V band.

---

[1] Further work by Whitelock (private communication) on more extensive data give an LMC modulus of 18.64 ± 0.13 (PL(K))

Considerable interest has been aroused by the results of statistical parallax analyses of field RR Lyraes based on radial velocities and Hipparcos proper motions. The statistical parallax analysis of Tsujimoto et al. (1998) leads to an LMC modulus of 18.37 ± 0.10 when their results are applied to RR Lyraes in LMC clusters. An analysis by Fernley et al. (1998) of basically the same data leads to an LMC modulus of 18.25 ± 0.17.

The difference in the RR Lyrae moduli of the LMC obtained via the galactic globular clusters (18.71 ± 0.13) and via the statistical parallaxes (18.25 or 18.37) is rather striking. One possibility is that field RR Lyraes differ from those in clusters in some way (see Reid 1998, Sweigart 1997) in which case one cannot apply the statistical parallax results to the RR Lyraes in LMC globular clusters. The result from the trigonometrical parallaxes of field RR Lyraes (Koen & Laney 1998, see also above) is unfortunately not of quite high enough accuracy to decide clearly on this point, although this result too, favours a high value for the LMC modulus and suggests at least tentatively that there is little difference between the luminosities of the field and cluster RR Lyraes of the same metallicity.. Another possibility is that the model of the galactic halo adopted in the statistical parallax work is too simple. It is now widely believed that the halo contains a significant number of infalling streams. If the RR Lyrae populations observed in different directions contain differing proportions of stars from the various streams then the statistical parallax result might well be significantly affected. The recent work of Gould & Popowski (1998) on the statistical parallax method does not deal with this problem. It would appear best therefore to regard the statistical parallax result with reserve at the present time.

5. SN1987A

A discussion of the LMC modulus to be derived from the ring around SN1987A in the LMC is given by Panagia in the present volume. He gives reasons for adopting 18.58 ± 0.05 rather than the lower value which has sometimes been suggested.

6. Red Clump Stars

The large data bases from the microlensing surveys in the direction of the Clouds open up the possibility of using the red giant clump in the c-m diagram as a distance indicator. Because of the large number of stars involved the internal precision of the method is potentially high, though systematic errors may be introduced by uncertainties in the adopted interstellar reddenings. The LMC distance modulus is obtained by comparing the apparent magnitude of the LMC clump with an absolute magnitude derived from Hipparcos parallaxes of galactic clump stars. In this way Stanik et al. (1998) obtained $18.065 \pm 0.031 \pm 0.09$ and Udalski et al. (1998) found $18.08 \pm 0.03 \pm 0.12$, where the first uncertainty in each case is statistical and the second systematic.

However evolutionary models indicate that the absolute magnitude of the clump depends on age and metallicity. Also the colour distribution of the LMC clump stars differs from that of the Hipparcos sample. Some of these problems have been discussed by Cole (1998) and Beaulieu and Sackett (1998). Cole obtained an LMC modulus of 18.36 ± 0.17 on a model of the age and metallicity distribution of the LMC stars and using evolutionary models. According to the models the luminosity of clump stars of a given metallicity changes only modestly with age for ages greater than 10^9 years, but is a strong function of age for younger stars (changing by 0.4 mag between $10^{8.7}$ and $10^{8.6}$ years (Bertelli et al. 1994)). In view of the fact that the relative ages and metallicities of LMC and Hipparcos clump stars are not known and that the interpretation of the data depends on models which among other things have to adopt some parameterization of the mass loss on the giant branch as a function of mass and metallicity, it would appear unsafe to place too much confidence in the present results of this method. The red clump may in fact be more useful as an indicator of stellar population than as a distance indicator.

7. Conclusions

The Cepheids, Miras and cluster RR Lyraes all indicate an LMC modulus near 18.70 or slightly greater. SN1987A gives a value near 18.6. Some other methods (e.g., red clump stars, statistical parallaxes of halo RR Lyraes) give lower values. Reasons can be advanced for regarding these low values with reserve. Thus at the present time a high value is to be preferred. The most likely value being in the range 18.60 t0 18.80.

8. Acknowledgements

I would like to thank Dr. Patricia Whitelock for many helpful discussions and for unpublished information. I am also much indebted to Drs. Chris Koen and Dave Laney for discussions and preprints.

References

Beaulieu, J.P., & Sackett, P.D. 1998, AJ, 116, 209

Bertelli, G., Bressan, A., Chiosi, C., Fagotto, F., & Nasi, E. 1994, A&AS, 106, 275

Caldwell, J.A.R., & Coulson, I.M. 1985, MNRAS, 212, 879

Caldwell, J.A.R., & Laney, C.D. 1991, in Haynes, R. & Milne, D. eds., IAU Symp. 148, The Magellanic Clouds, Kluwer, Dordrecht, p.249

Cole, A.A. 1998, ApJ, 500, L137

Feast, M.W. 1995, in Stobie, R.S. & Whitelock, P.A. eds., ASP Conf. Ser. 83, The Astrophysical Applications of Stellar Pulsation. Ast. Soc Pac., San Francisco, p.209

Feast, M.W. 1998, MmSAI. 69, 31

Feast, M.W., Glass, I.S., Whitelock, P.A., & Catchpole, R.M. 1989, MNRAS, 241, 375

Feast, M.W., & Catchpole, R.M. 1997, MNRAS, 286, L1

Feast, M.W., & Whitelock, P.A. 1997, MNRAS, 291, 683

Feast, M.W., Pont, F., & Whitelock, P.A. 1998, MNRAS, 298, L43

Feast, M.W., & Whitelock, P.A. 1999, in Caputo, F. & Heck, A., eds, Post-Hipparcos Cosmic Candles, Kluwer, Dordrecht, p.75

Fernley, J., Barnes, T.G., Skillen, I., Hawley, S.L., Hanley, C.J., Evans, D.W., Solano, E., & Garrido, R. 1998, A&A, 330, 515

Gieren, W.P, Fouqué, P., & Gómez, M. 1998, ApJ, 496, 17

Glass, I.S., Whitelock, P.A., Catchpole, R.M., & Feast, M.W. 1995, MNRAS, 273, 383

Gould, A., & Polowski, P. 1998 ApJ, 508, 844

Gratton, R.G., 1998, MNRAS, 296, 739

Koen, C., & Laney, C.D. 1998, MNRAS, 301, 582

Laney, C.D., & Stobie, R.S. 1994, MNRAS, 266, 441

Madore, B.F., & Freedman, W.L., 1998, ApJ, 492, 110

McNamara, D.H. 1997, PASP, 109, 1221

Oudmaijer, R.D., Groenewegen, M.A.T., & Schrijver, H., 1998, MNRAS, 294, L41

Reid, I.N. 1997,AJ, 114, 161

Reid, I.N. 1998,AJ, 115, 204

Stanek, K.Z., Zaritsky, D., & Harris, J. 1998, ApJ, 500, L141

Sweigart, A.V. 1997, in Davis Philip, A.G. ed., Proc. Third Conf. on Faint Blue Stars, L. Davis, Schenectady, NY

Tsujimoto, T., Miyamoto, M., & Yoshii, Y. 1998, ApJ, 492, L79

Udalski, A., Szymański, M., Kubiak, M., Pietrzyński, G., Woźniak, P., & Zebruń, K. 1998, AcA, 48, 1

van Leeuwen, F., Feast, M.W., Whitelock, P.A., & Yudin, B.1997, MNRAS, 287, 955

Whitelock, P.A., Menzies, J.W., Feast, M.W., Marang, F., Carter, B.S., Robert, G., Catchpole, R.M., & Chapman, J. 1994, MNRAS, 267, 711

Whitelock, P.A., van Leeuwen, F., & Feast, M.W. 1997, in Battrick, B. et al., eds., Hipparcos, Venice '97, ESA SP-40

Wood, P.R. 1990, in Mennessier, M.O. & Omont, A. eds., From Miras to Planetary Nebulae:Which Path for Stellar Evolution?, Editions Frontières, Gif-sur-Yvette, p.67

Discussion

Nolan Walborn: Sometimes a long journey brings us back to where we started. When I started working on the Magellanic Clouds in 1973, I looked up the moduli and found 18.6 for the LMC in a paper by Sandage & Gascoigne, the value I have used consistently every since. Recently I studied 100 OB stars in 30 Doradus and found an average difference between derived and (galactic) calibration M_v's of 0.05 mag. In the meantime, there was a "revision" to 18.1 - 18.3 based on cluster main sequences, which has evidently gone away due to ambiguities among reddening, metallicity, and age, and was not even mentioned today. I've been annoyed by some of the writing about the current red clump "revision", claiming a real result or at least a long-short distance controversy. Those who obtain such results should be more critical about their uncertainties and be required to address the host of other problems they would create if real. It seems clear the LMC modulus is 18.6 ± 0.1, or at most ± 0.2, on a firmer basis than 25 years ago due to the extensive intervening work.

Serge Demers: Why did you not mention main sequence fitting as a distance indicator for the MCs? A number of clusters and associations in the LMC have already been observed.

Feast: As I said at the start of my talk I could not cover all methods. As regards main sequence fitting there are three points I can make. 1. No one has yet produced a fiducial main sequence band on Hipparcos parallaxes of field stars. 2. The main sequence fitting method is sensitive to abundance which must be known quite accurately. 3. The main sequence fitting method is sensitive to the adopted reddening (because of the steep slope of the main sequence). Nevertheless, I would expect this valuable method to be used increasingly in the future.

Nino Panagia: The study by Romaniello et al. (these proceedings) of the stellar populations around SN 1987A shows that once one can correct for reddening individually for each star, a fit of the observations with a theoretical MS for $Z = 0.3\ Z_{\odot}$ is perfectly consistent with a distance modules of 18.55 as determined for SN 1987A itself. This suggests that MS fitting may be able to provide meaningful distance determinations but that a large sample of stars, with high quality photometry and individual reddening corrections, is required.

New Views of the Magellanic Clouds
IAU Symposium, Vol. 190, 1999
Y.-H. Chu, N.B. Suntzeff, J.E. Hesser, and D.A. Bohlender, eds.

Distance to SN 1987A and the LMC

Nino Panagia[1]

STScI, 3700 San Martin Drive, Baltimore, MD 21218, USA

Abstract. Using the new reductions of the IUE light curves by Sonneborn et al. (1997) and an extensive set of HST images of SN 1987A we have repeated and improved Panagia et al. (1991) analysis to obtain a better determination of the distance to the supernova. In this way we have derived an absolute size of the ring $R_{abs} = (6.23 \pm 0.08) \times 10^{17}$ cm and an angular size $R'' = 808 \pm 17$ mas, which give a distance to the supernova $d(SN1987A) = 51.4 \pm 1.2$ kpc and a distance modulus $m - M(SN1987A) = 18.55 \pm 0.05$. Allowing for a displacement of SN 1987A position relative to the LMC center, the distance to the barycenter of the Large Magellanic Cloud is also estimated to be $d(LMC) = 52.0 \pm 1.3$ kpc, which corresponds to a distance modulus of $m - M(LMC) = 18.58 \pm 0.05$.

1. Introduction

In 1991 Panagia et al. estimated the distance to SN 1987A (51.2 ± 3.1 kpc) from a comparison of the angular size of the inner circumstellar ring as measured with the *HST-FOC* in August 1990 (Jakobsen et al. 1991), with the ring absolute size as determined from the peaks of its UV emission line light curves.

Recently, Gould re-examined the problem adopting an infinitely narrow ring geometry and retaining the Panagia et al. (1991) assumption of an exponential law for the line emissivity (Gould 1995; Gould & Uza 1998). Thus, using the same data as in Panagia et al. (1991) and Sonneborn et al. (1997), respectively, but including only the NIV] and NIII] light curves, and adopting the *average* [OIII] ring size as measured by Plait et al. (1995) over the period August 1990 - May 1993, Gould concluded that the distance to SN 1987A is less than 47 kpc.

The new reductions of the IUE spectra, done by Sonneborn et al. (1997), have produced more accurate and reliable light curves. Therefore, we have decided to repeat Panagia et al. (1991) analysis using the new data set and including a more accurate and realistic estimate of the ring angular size obtained from the study of an extensive set of HST images of SN 1987A. Here we present a brief outline of our analysis and the main results of our study. A complete account of this work will be presented in a forthcoming paper (Panagia et al. 1999).

[1] On assignment from the Astrophysics Division, Space Science Department of ESA.

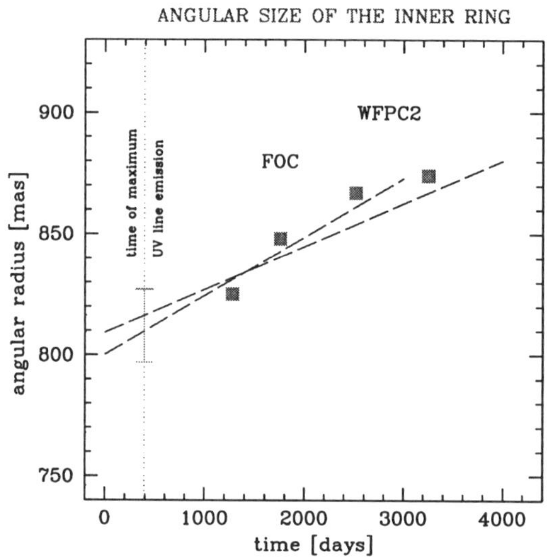

Figure 1. The average angular radius of the ring as measured in *FOC* and *WFPC*2 images taken with an [OIII] filter. Two fits to the observed points and the range of possible values of the radius at the time of the UV maximum are shown.

2. The Angular Size of the Ring

The inner circumstellar ring is clearly extended with a HPW of about 1/7 its radius (e.g., Jakobsen et al. 1991; Panagia et al. 1999). The finite width of the ring makes the definition of an average size a very delicate one, which, if done improperly, may introduce errors as large as, say, 7% or higher.

The analysis of both *FOC* and *WFPC2* images have revealed that the ring angular size, as estimated from [OIII] images, appears to increase with time (Panagia et al. 1999; see Fig. 1) while the one derived from Hβ and [NII] images remains constant in time and virtually coincides with the size measured in the earliest [OIII] image. This is the effect of both cooling and recombination of the O^{++} ion, that cause the [OIII] 5007Å line intensity to decline more quickly at the inner edge of the ring where the density is higher. Therefore, the *correct value of the ring angular size* to compare with the absolute size determined from the UV lines is an extrapolation of the observed sizes, as measured in the [OIII] band, back to the epoch of maximum UV line emission (approximately early April 1988; *cf.* Fig. 1). In this way we obtain:

$$R'' = 808 \pm 17 \text{ mas}$$

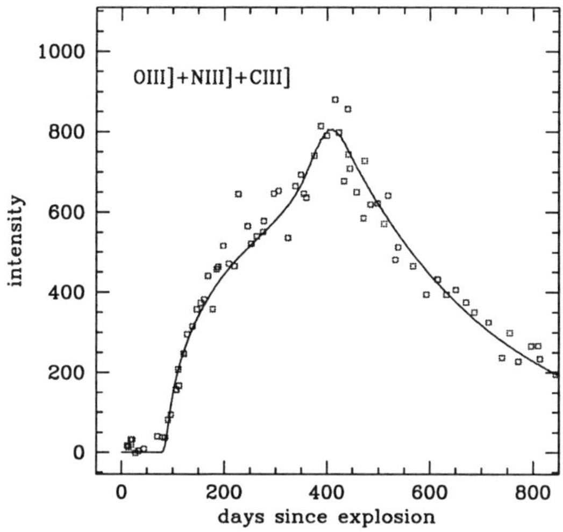

Figure 2. The sum of the observed intensities (in units of 10^{-15} erg cm^{-2} s^{-1}) of the OIII] 1666Å, NIII] 1750Å, and CIII] 1909Å lines compared to the best-fit model light curve.

3. The Absolute Size of the Ring

It has been shown (Panagia et al. 1991; Dwek & Felten 1992; Gould 1995) that under the assumption of an infinitely narrow width the *absolute radius of the ring* is simply related to the onset time of the UV line emission $t_\circ$, and the time of maximum UV line emission t_{max} as

$$R = c(t_\circ + t_{max})/2$$

Since the angular size is well measured only for O^{++}, we have limited our analysis to the UV light curves of twice ionized ions, namely OIII, NIII and CIII, and we have compared them to theoretical light curves computed under the following assumptions: (1) The ring is circular with a width of 14% the ring radius; (2) The intrinsic emission of each ion decays exponentially with time; (3) The model free parameters are: radius and inclination angle of the ring, and specific emissivity at time $t = 0$ and decay time of each line.

We computed best fits to both the individual light curves of the OIII], NIII], and CIII] lines, and the composite light curve, sum of the OIII], NIII] and CIII] line intensities (Fig. 2). The individual radius determinations fall in the range 6.0 to 6.4 $\times$ 10^{17}cm, with uncertainties of about 4% for OIII] and CIII], and slightly above 1% for NIII], resulting in an average value:

$$R = (6.23 \pm 0.08) \times 10^{17}\text{cm}$$

4. Distance Determinations

SN 1987A - Comparing the absolute and the angular sizes of the circumstellar ring, we derive a distance to SN 1987A

$$d(SN1987A) = 51.5 \pm 1.2 \text{ kpc} \qquad (m - M)_{SN\ 1987A} = 18.55 \pm 0.03$$

This value is very close to the original determination by Panagia et al. (1991) but is considerably more accurate. In the literature one finds determinations of SN 1987A distance which are all based on the analysis of UV line light curves and *HST* imaging but provide quite discrepant values, for example distance moduli ranging from 18.37 to 18.67 (Gould & Uza 1998; Sonneborn et al. 1997; Lundqvist & Sonneborn 1997; Panagia et al. 1999), all based on the same set of data. A close scrutiny of these papers reveals that all discrepancies are due to somewhat arbitrary differences in the angular size adopted and/or in the selection of UV emission lines light curves to determine the absolute size.

LMC - Allowing for the difference of the position of SN 1987A relative to the LMC barycenter (McCall 1993, 1994; also Crotts, Kunkel & Heathcote, 1995; Crotts 1998, this Conference; Panagia et al 1999) the best estimate of the distance to the center of mass of the LMC is found to be

$$d(LMC) = 52.1 \pm 1.3 \text{ kpc} \qquad (m - M)_{LMC} = 18.58 \pm 0.03$$

The error includes the uncertainties on the SN 1987A distance (± 1.2 kpc) as well as those on the depth toward SN 1987A (± 0.2 kpc) and the relative position of the LMC barycenter (± 0.3 kpc).

Our determination of the LMC distance modulus (18.58 ± 0.05) is in excellent agreement with recent determinations obtained via Hipparcos parallax calibration of both galactic Cepheids (Feast & Catchpole 1997: $(m - M) = 18.70 \pm 0.10$) and local subdwarfs (Gratton et al. 1997: $(m - M) = 18.60 \pm 0.07$). A weighted average of the three determinations gives

$$< (m - M)_{LMC} > \;=\; 18.60 \pm 0.04$$

It is apparent that the LMC distance modulus is around 18.60 and that values < 18.48 and > 18.72 are to be excluded with high confidence.

References

Crotts, A.P.S., Kunkel, W.E., & Heathcote, S.R. 1995, ApJ, 438, 724.

Dwek, E., & Felten, J.E. 1992, ApJ, 387, 551.

Feast, M.W., & Catchpole, R.M. 1997, MNRAS, 286, L1.

Gould, A. 1995, ApJ, 452, 189.

Gould, A., & Uza, O. 1998, ApJ, 494, 118.

Grattom, R., et al. 1997, ApJ, 491, 749.

Jakobsen, P., et al. 1991, ApJ, 369, L63.

Lundqvist, P., & Sonneborn, G. 1998, preprint.

McCall, M.L. 1993, ApJ, 417, L75; and *erratum* 1994, ApJ, 434, L37.

Panagia, N., Gilmozzi, R., Macchetto, F., Adorf, H.-M., & Kirshner, R.P. 1991, ApJ, 380, L23.; and *erratum* 1992, ApJ, 386, L31.

Panagia, N., Gilmozzi, Kirshner, R.P., Pun, C.S.J., & Sonneborn, G. 1999, ApJ, to be submitted.

Plait, P.C., Lundqvist, P., Chevalier, R.A., & Kirshner, R.P. 1995, ApJ, 439, 730.

Sonneborn, G., et al. 1997, ApJ, 477, 848.

Discussion

Arlin Crotts: You should subtract at least 400 pc, and better 500 pc, rather than 100 pc from the distance to LMC as determined by echo distance of SN 1987A behind the plane of LMC.

Doug Welch: 1. Can you comment on the degree to which ring clumpiness affects the distance derived?

2. Why do you favor the particular extrapolation to 0 days which you chose for the angular diameter?

Panagia: 1. Clumpiness does not affect the shape of the UV line light curves with the exception of two mysterious spikes at $\sim 200^d$ and 400^d, which indicate the existence of two very small ($\leq$ 20 light-days) and very dense ($n_e > 10^5$ cm^{-3}) clumps. All other knots are much less extreme and their effects average out completely.

2. My extrapolation of the [OIII] ring size to early times is bracketed between the smallest size measured (Aug 1990, 825 mas) in [OIII] images, and the smallest size measured in any radiation (i.e., [OI] in April 1997, $\sim$ 790 mas).

George Sonneborn: What distance to SN 1987A do you find if you assume that the [OI] diameter (Sonneborn et al. 1998, ApJ 492, L139) corresponds to the angular size of the ring at UV line maximum? This is plausible because the brightest UV emission originated from the densest gas, which recombined faster, and now would be neutral.

Panagia: The distance corresponding to the [OI] diameter you measure is 52.6 kpc or, similarly, a distance modulus of 18.60. These values are essentially my quoted 1σ upper limit to the distance to SN 1987A.

(Top) Miroslav Filipović enjoying a good story at the reception, as are (bottom) Frank Winkler, Robert Petre and Ted Stecher.

Part 8. Cepheids and Distances

Section B. Poster Papers

Ed Olszewski reviewing properties of older LMC stellar populations.

New Views of the Magellanic Clouds
IAU Symposium, Vol. 190, 1999
Y.-H. Chu, N.B. Suntzeff, J.E. Hesser, and D.A. Bohlender, eds.

The Theoretical PLC Relations of Classical Cepheids

Marcella Marconi

Osservatorio Astronomico di Capodimonte, 80131 Napoli, Italy

Abstract.
On the basis of an extensive grid of Cepheid, nonlinear, nonlocal, time-dependent convective models by Bono et al., analytical Period-Color-Luminosity (PLC) relations as a function of chemical composition were evaluated and compared with observations in LMC and SMC. I provide an analysis of PL and PLC residuals and suggestions for disentangling the effects of reddening and metallicity on the Cepheid distance scale.

1. Comparison between the Theoretical PLC Relations and the Observations in the Magellanic Clouds

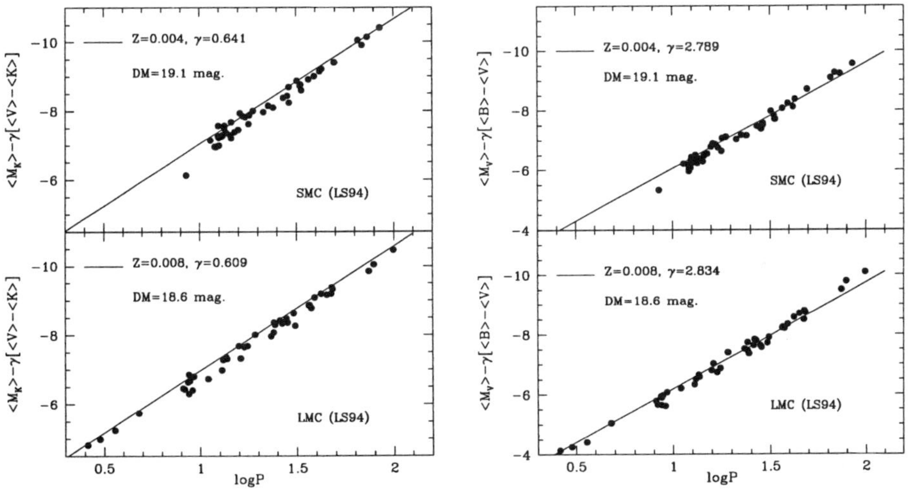

Figure 1. The comparison between the LMC (bottom) and the SMC (top) data by LS94 and our theoretical PLC relations.

The left panel of Figure 1 shows the comparison between the LMC (bottom) and the SMC (top) data by LS94 and the planar projections of the theoretical near-infrared PLC relations (Bono et al. 1998, ApJS, 122, 167, and reference therein) for Z=0.008 and Z=0.004, respectively. The right panel of the same figure shows the comparison in the optical bands. In both cases a quite good agreement is obtained adopting the labeled values of the distance moduli (18.6 mag for the LMC and 19.1 mag for the SMC).

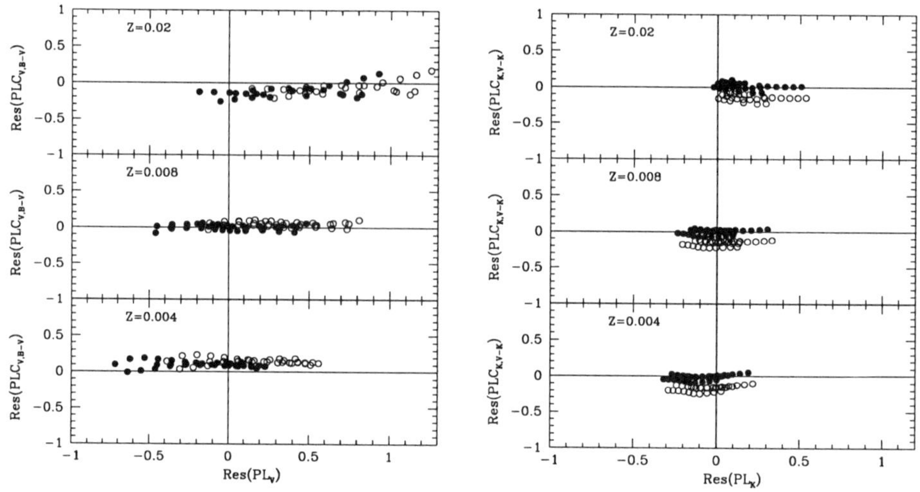

Figure 2. The optical (left) and near-infrared (right) PLC residuals versus the corresponding PL relation residuals.

2. Analysis of Residuals: A Tool to Evaluate Reddening and Metallicity

Assuming the Z=0.008 model set as the reference sample, one may compute the Z=0.008 PL and PLC residuals for model sets of different chemical compositions. In Figure 2 filled circles show the PLC residuals versus the PL relation residuals for models with Z=0.004 (bottom), Z=0.008 (middle) and Z=0.02 (top). As expected, for models with Z=0.008 the PL and PLC residuals are perfectly symmetric around the zero. For a metallicity larger (lower) than the value corresponding to the adopted PL and PLC relations, the optical PLC residuals tend to become negative (positive) and the optical PL residuals become larger (lower) than zero. Adding an artificial reddening $E(B-V) = 0.10$ mag. and a corresponding optical extinction $A_V = 0.33$ mag. (see open circles in Figure 2), the optical PL residuals increase but the optical PLC residuals are almost unchanged. In the case of infrared relations (see right panel), only the PL residuals are significantly affected by metallicity differences, while the PLC residuals are remarkably invariant. On the other hand, an artificial reddening ($E(V-K) = 0.30$ mag. and $A_K \simeq 0.03$ mag.; see open circles) now produces a vertical shift toward negative PLC residuals but leaves the PL residuals practically identical. Therefore we may conclude that, once the metal abundance is estimated through examination of the optical PLC residuals and near-infrared PL residuals, the information provided by the near-infrared PLC residuals allows us to evaluate the sample reddening.

Acknowledgments. It is a pleasure to thank Nino Panagia, who suggested a critical investigation of theoretical PL and PLC residuals.

New Views of the Magellanic Clouds
IAU Symposium, Vol. 190, 1999
Y.-H. Chu, N.B. Suntzeff, J.E. Hesser, and D.A. Bohlender, eds.

Observations of Cepheid Variables in Two LMC Clusters Using EROS 2 Data

V. Ripepi

Osservatorio Astronomico di Capodimonte (Napoli), Italy

F. Bauer

College de France, Paris, France

J.P. Beaulieu

Institut d'Astrophysique de Paris, France

1. Introduction

It is well known that Cepheid variables are very important distance indicators because of their PL and PLC relations. The distances derived through the quoted relations are affected by well known systematic effects, since, in general, we have a poor knowledge of both reddening and metallicity of the Cepheid sample. Good estimates of the quoted parameters can be achieved if the observed Cepheids are cluster members since, by comparison between cluster CMD and theoretical isochrones we can estimate the cluster reddening, metallicity and age, i.e., the evolving masses for each evolutionary phase. The ideal objects to be searched for Cepheid membership are Magellanic Cloud clusters since they are well populated in the relevant evolutionary phases.

The EROS 2 project (see Bauer et al. 1999, and other contributions in this book), among the other important results, provided excellent light curves and relevant parameters for Cepheid variables in the field of the two LMC cluster NGC 1943 and NGC 1958. Since the EROS 2 photometric system is difficult to calibrate, we coupled EROS 2 observations for Cepheids with new observations carried out in the standard BVR_JI_J filter system, in order to derive accurate CMDs for comparison with theoretical models. In this contribution we present some preliminary results of this work.

2. Observations and Preliminary Results

The observations were carried out partly in November 1996 with the 0.90 m. "Dutch" Telescope (for NGC 1958) and partly with the 1.5 m "Danish" during 1998, January, both at ESO (La Silla). Observations with the "Dutch" telescope consist of 11 observing nights spaced along 21 days, so they allow a search for stellar variability. A detailed report about reduction and calibration procedure will be done in a forthcoming paper.

Concerning some preliminary results, there are ten variable stars both in NGC 1943 and in NGC 1958. Most of these stars are Cepheid already present

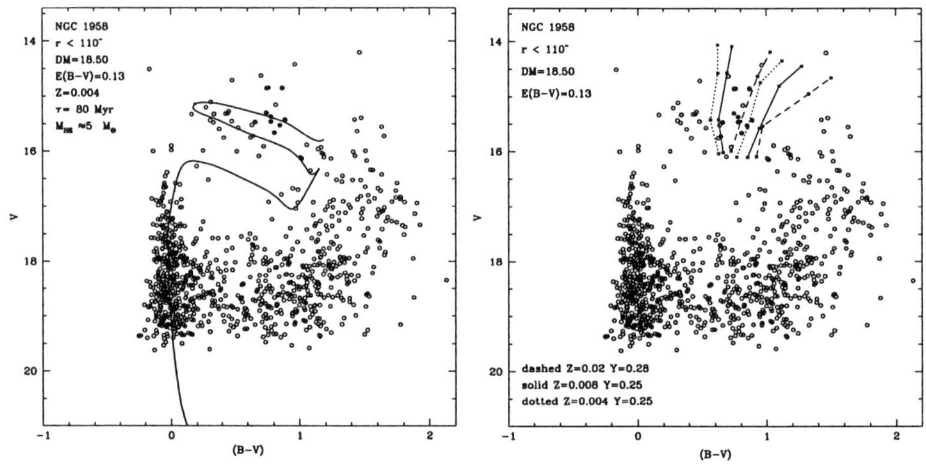

Figure 1. **Left panel:** C-M diagram for NGC 1958; filled circles represent variable stars, whereas the solid line shows the theoretical isochrone for the labelled values of age and metallicity; **Right panel:** as before, but the various lines represents the instability strips for various metallicities by Bono, Marconi & Stellingwerf (1999).

in the EROS 2 database. From the comparison between observed CMDs and theoretical isochrones (with the new input physics by Degl'Innocenti & Marconi 1998) we estimated the cluster metallicity, reddening and age as well as the Cepheid evolving masses. In Fig. 1 (left) we report the result of such a procedure for NGC 1958, but the same analysis was applied to NGC 1943. We note that adopting a set of isochrones with different input physics (e.g. Bertelli et al. 1994), the data are fitted equally well with the following values for the relevant parameters: Z=0.008, DM=18.30, $E(B-V)=0.13$, Age=100-125 Myrs. This occurrence, which would deserve a more detailed discussion, can be interpreted as a measure of the uncertainties still affecting the intermediate mass star evolutionary models. Finally we compare in Fig. 1 (right) the observed instability strip for NGC 1958 with the theoretical one (Bono, Marconi & Stellingwerf, 1999) for various assumptions about chemical composition. From the figure we notice that, at least qualitatively, the agreement between the theory of pulsations and the observations seems satisfactory.

Acknowledgments. We wish to thank Dr. Degl'Innocenti and Dr. Marconi for providing their unpublished isochrones. We thank also Dr. Bono and Dr. Marconi for providing theoretical results in advance of publication.

References

Bauer, et al. (the EROS collaboration) 1999, A&A, 348, 175

Bertelli, G., et al. 1994, A&AS, 106, 275

Bono, G., Marconi, M., & Stellingwerf, R. 1999, ApJS, 122, 167

Degl'Innocenti S, Marconi, M. 1998 (private communication)

New Views of the Magellanic Clouds
IAU Symposium, Vol. 190, 1999
Y.-H. Chu, N.B. Suntzeff, J.E. Hesser, and D.A. Bohlender, eds.

Surface Brightness Scale of Galactic Cepheids

G. P. Di Benedetto

Istituto di Fisica Cosmica, Via Bassini 15, 20133 Milano, Italy

Abstract. An accurate calibration of the surface brightness scale S_V as a function of the near-IR color $(V - K)$ has been recently measured for non-variable Galactic dwarf and giant stars. It can be shown that this correlation can be applied to the S_V scale of Galactic Cepheid variable stars, which are of major cosmological interest.

1. Introduction

Stellar surface brightness methods (Wesselink 1969; Barnes et al. 1978) can provide accurate distances for Galactic and extragalactic investigations, since the apparent brightness of a star may be measured directly by current observational techniques. I have shown the potential of the method of using accurate angular diameters available by Michelson interferometry and the near-IR colors (Di Benedetto (1993). More recently, I have used the same approach using high-precision K magnitudes and Hipparcos parallaxes for dereddening the photometric data to determine effective temperatures as accurate as $\pm$ 1% for a large sample of non-variable galactic A-F-G-K dwarfs and giants selected as standards for the Infrared Space Observatory (Di Benedetto 1998).

However, for the important class of Cepheid variable stars, the empirical S_V scale still remains poorly known due to the lack of accurate diameter measurements available for supergiants and to the uncertainties in the reddening. For the first time, Di Benedetto (1994) applied the near-IR surface brightness approach to yield Cepheid distances (Di Benedetto 1997). Steps for improving the calibration of the S_V scale have been also attempted (Di Benedetto 1995). Using the same precepts and taking advantage of the new accurate observational relationships available for the Galactic field stars, it can be shown that the S_V scale of Galactic Cepheids closely follows that of giants, enabling us to significantly improve the calibration of cosmological Cepheid-based distance indicators.

2. The Method

The surface brightness of a Cepheid variable star can be defined according to Cepheid parameters measured along the pulsation cycle of the light curve as:

$$S_V = < V_0 > + 5 \, log < R > - 0.159 - (m - M)_0 \qquad (1)$$

where $< V_0 >$ is the dereddened visual mean magnitude; $< R >$ is the mean linear radius (solar units); $(m - M)_0$ is the true distance modulus. Individual values of $< R >$ and $(m - M)_0$ for Galactic Cepheids can be derived either from direct observational data or by relationships using the pulsational period P as the sampling observable. This surface brightness is then compared to a reference sequence of non-variable giants calibrated using high-precision observational data (Di Benedetto 1998). The comparison based on the direct measurements available for Cepheids shows close agreement with the calibration of low-gravity giants. In fact, this calibration is also expected to represent the photosphere of supergiants over the color range of Cepheids according to updated model-atmosphere results (Castelli 1998). Hence, the same comparison is carried out by applying the observational relationships currently adopted as Cepheid-based distance indicators. The systematic deviations of the residuals can be observed as either slope variations or average shifts, or both. These residuals are the most sensitive signature of the type of errors affecting the calibration. In addition, they can be removed in order to improve the accuracy and reliability of the relationship itself as distance indicator.

One of the advantages in plotting the S_V data and assessing the residuals as a function of the near-IR color is that the angular/linear diameters from equation (1) in the $(V - K)$ diagram are largely insensitive to many systematics affecting the magnitude V. In fact, a change of $0\overset{m}{.}1$ in V would lead to only $0\overset{m}{.}03$ or about 1.4% of shift in the corresponding photometric stellar size. This allows us to estimate accurate Cepheid distances according to the modern Baade-Wesselink method using high-precision K magnitudes.

3. Conclusions

We are still learning how best to exploit the observed surface brightness data for extragalactic Cepheid-based distances, especially in how the results can be affected by metallicity.

References

Barnes, T.G., Evans, D.S., & Moffett, T.J. 1978, MNRAS, 183, 285

Di Benedetto, G.P. 1993, A&A, 270, 315

Di Benedetto, G.P. 1994, A&A, 285, 819

Di Benedetto, G.P. 1995, ApJ, 452, 195

Di Benedetto, G.P. 1997, ApJ, 486, 60

Di Benedetto, G.P. 1998, A&A, 339, 858

Castelli, F. 1998 private communications

Wesselink, A.J. 1969, MNRAS, 144, 297

New Views of the Magellanic Clouds
IAU Symposium, Vol. 190, 1999
Y.-H. Chu, N.B. Suntzeff, J.E. Hesser, and D.A. Bohlender, eds.

Magellanic Cloud Eclipsing Binaries: Primary Distance Indicators

J.D. Pritchard, W. Tobin

Mount John University Observatory and Department of Physics & Astronomy, University of Canterbury, Private Bag 4800, Christchurch, New Zealand

J.V. Clausen

Copenhagen University Astronomical Observatory, Julianne Maries Vej 30, DK-2100 Copenhagen Ø, Denmark

E.F. Guinan, E.L. Fitzpatrick

Department of Astronomy, Villanova University, Villanova, Pennsylvania 19085, USA

A. Giménez

LAEFF (INTA-CSIC), Apartado 50727, E-28080 Madrid, Spain

I. Ribas

Departament d'Astronomia i Meteorologia, Universitat de Barcelona, Avda. Diagonal 647, E-08028, Barcelona, Spain

Abstract. Our collaboration involves groups in Denmark, the U.S.A. Spain and of course New Zealand. Combining ground-based and satellite (*IUE* and *HST*) observations we aim to determine accurate and precise stellar fundamental parameters for the components of Magellanic Cloud Eclipsing Binaries as well as the distances to these systems and hence the parent galaxies themselves. This poster presents our latest progress.

The utility of double-lined, eclipsing binaries for the measurement of fundamental stellar parameters as well as distances, including systems in the Magellanic Clouds, has recently been reviewed by a number of authors (e.g., Guinan, Bradstreet & DeWarf 1996; Hilditch 1996; Paczyński 1997) so we shall not repeat that here except to make the following important points. The distance to an eclipsing binary system can, in principle, be determined directly, without relying on intermediary calibrating objects. The method is essentially geometric and relies on the determination of the physical stellar sizes and the intrinsic radiated flux.

The physical dimensions of the two components of an eclipsing binary are determined from light curves and the two radial velocity curves for the two components. Modern analysis tools (see Wilson 1994 for an excellent review) allow for accurate and precise modeling of well-detached, semi-detached and even complicated contact systems.

The distance is then determined simply by comparison of an apparent flux with the corresponding absolute flux. This generally relies on the estimation of the effective temperatures of the two components either via spectral classifications, multi-colour photometry or spectrophotometric flux fitting.

An ongoing collaboration between groups in New Zealand, Denmark, the U.S.A., Spain, Britain and Canada has, since 1984, been using telescopes in Chile, Australia and New Zealand as well as both the *IUE* and *HST* satellites to secure optical photometry, spectroscopic radial velocities and uv-spectrophotometry of Magellanic Cloud Eclipsing Binaries. With modern CCDs even 1m telescopes are adequate to obtain light curves with 1% or better photometric accuracy. For the radial velocities, 4m-class telescopes are the minimum requirement for groundbased observations of sufficient quality while one system (HV 2274) has been observed with the (now-defunct) GHRS on the *HST*. Flux fitting to uv-spectrophotometry obtained by satellites promises to be a useful technique for temperature determinations. Ten systems – six in the LMC and four in the SMC – have been observed with the *HST*'s (also now-defunct) FOS. All ten were also observed with the *IUE* and have modern CCD light curves, although some would certainly benefit from further observations. However only a couple have high precision radial velocity curves and HV 2274 is the only system to date for which all the requisite observations have been analysed to yield a distance determination of the highest quality – we determine a distance modulus of 18.49 ± 0.07 to HV 2274 and 18.44 ± 0.07 to the center of the LMC (Guinan et al. 1998). Our collaboration's efforts are therefore now focused on securing the photometry and spectroscopy still required for these systems plus other especially appropriate systems which have come to light as a result of recent microlensing projects such as EROS and MACHO.

The poster presented in Victoria can be downloaded from:
`http://www.phys.canterbury.ac.nz/~phys249/publications.html`

Acknowledgments. This research is supported by the Marsden Fund of New Zealand, the Danish Natural Science Research Council through its Centre for Ground-Based Observational Astronomy, NASA and NSF.

References

Guinan, E. F., Bradstreet, D. H., & DeWarf, L. E. 1996, in The Origins, Evolutions, and Destinies of Binary Stars in Clusters, E.F. Milone & J.-C. Mermilliod, Astronomical Society of the Pacific: San Francisco, 196

Guinan, E. F., Ribas, I., Fitzpatrick, E. L., & Pritchard, J. D. 1998, in Ultraviolet astrophysics: beyond the *IUE* final archive, W. Wamsteker & R. Gonzalez-Riestra, ESA Publishing division: Noordwijk, 315

Hilditch, R. W. 1996, in The Origins, Evolutions, and Destinies of Binary Stars in Clusters, E.F. Milone & J.-C. Mermilliod, Astronomical Society of the Pacific: San Francisco, 207

Panzyński, B. 1997, in Variable Stars and the Astrophysical Returns of Microlensing Surveys, R. Ferlet, J.-P. Maillard, & B. Rabam, Editions Frontières, 309

Wilson, R. E. 1994, PASP, 106, 921

Discussion

Doug Welch: What is the dominant source of uncertainty in your opinion?

Pritchard: In my opinion, the dominant source of uncertaintly arises from interstellar extinction, although the flux fitting method of Fitzpatrick & Massa is producing great results.

Russ Robb: Why not use infrared measurements to avoid extinction problems?

Pritchard: These systems are composed of hot OB-type stars. Although they will have plenty of flux in the IR, it's so far into the Rayleigh-Jeans tail of the flux distributions there is no temperature discrimination.

Smiling faces attest to an enjoyable banquet.

Part 9. Summary

Sidney van den Bergh presents the Symposium summary. That's Gretchen Harris in the "front row, right-hand side, on the aisle" seat that Sidney habitually occupied during colloquia while Gretchen was a University of Toronto graduate student (see page 2).

New Views of the Magellanic Clouds
IAU Symposium, Vol. 190, 1999
Y.-H. Chu, N.B. Suntzeff, J.E. Hesser, and D.A. Bohlender, eds.

The Magellanic Clouds, Past, Present and Future - A Summary of IAU Symposium No. 190

Sidney van den Bergh

Dominion Astrophysical Observatory
Herzberg Institute of Astrophysics
National Research Council of Canada
5071 West Saanich Road
Victoria, British Columbia
Canada V8X 4M6

1. Introduction

Important problems to which we would like to find answers are:

- What are the distances to the Large Magellanic Cloud (LMC) and the Small Magellanic Cloud (SMC)?
- What is the present distribution of stars, gas and dark matter in the Clouds, and how did it evolve?
- How, and where, did the Magellanic Clouds form, and how have their orbits evolved?
- Finally the recent discovery of numerous microlensing events in the Clouds provides answers to questions that we have only recently started to ask.

For background material on the Clouds of Magellan the reader is referred to IAU Symposium No. 108 (van den Bergh & de Boer 1984), IAU Symposium No. 148 (Haynes & Milne 1991), and to the recent monograph The Magellanic Clouds (Westerlund 1997). The sections of this summary of the conference proceedings are given approximately in the order in which they were presented at the symposium.

2. Interstellar Matter

2.1. Optical Imaging

A number of authors presented beautiful narrow-band images of both the LMC and the SMC, which showed an intricate network of hot bubbles produced by the stellar winds of giant associations, and shells formed by exploding supernovae. From the close association between supernova remnants (SNRs) and emission nebulosity Petre (NASA/GSFC) concluded that the majority (35 out of 48) of the SNRs observed in the Large Cloud had been produced by supernovae of Type II (SNe II). However, the fact that the SNR N 103B [which was produced by a SN Ia (Hughes et al. 1995)] is associated with an H II region shows that individual type assignments based on environment may be incorrect. In the SMC the presently available data are insufficient to derive the relative numbers of SNe Ia and SNe II. It would be of interest to compare recent narrow-band images of the Magellanic Clouds with digitized versions of similar images obtained a few decades ago, to search for the light echoes of prehistoric supernovae in the Clouds. Sonneborn (NASA/GFSFC) et al. reported on the recent development of a hot spot associated with the forward shock produced by SN 1987A, that is now beginning to enter the dense gas on the inner-most edge of the ring surrounding this supernova. This may be the beginning of the "second coming" of SN 1987A!

2.2. Millimeter Radio Observations

Beautiful new ^{12}CO ($J = 1 - 0$) observations obtained with the NANTEN telescope on Las Campanas were shown by various Japanese groups. Due to low metallicity molecular clouds are found to be rarer in the LMC than they are in the Galaxy. Not unexpectedly such clouds are seen to be even scarcer in the SMC. Over the mass range 10^5 $M_\odot$ to 10^6 $M_\odot$ Mizuno (Nagoya) et al. find that these molecular clouds have a power law mass spectrum of the form $dN/dM \propto M^{-1.5}$. Saito (Nagoya) et al. note that the positions of H II regions are strongly correlated with those of giant molecular clouds (GMCs). The positions of young star clusters are slightly less correlated with those of GMCs. Finally there is little evidence for a correlation between the positions of SNRs and GMCs. This suggests that most GMCs have lifetimes that are short compared to the $\geq 1 \times 10^7$ yr time-scale for the evolution of SNe II.

2.3. 21-cm Observations

A number of interesting new results, based on observations obtained with the Australia Telescope Compact Array (ATCA), were presented at the meeting. The neutral hydrogen observations by Dickey (U. Minnesota) et al. show that the cool atomic phase is quite abundant in the Clouds. Possibly such cool gas would have been transformed into molecular clouds in the dust-rich Galactic environment. Alternatively the Galactic gas may have been warmed by photoelectric heating of small grains. The hydrogen gas in the LMC forms a well-defined rotating disk, whereas that in the SMC appears chaotic. The LMC rotation curve implies a total disk mass of 2.5×10^9 $M_\odot$ within 4 kpc of the kinematic center of the Large Cloud. The total amount of neutral gas (including He) in this disk is 5×10^8 $M_\odot$. Observations with the ATCA show a number of features in the outer regions of the LMC that have been stretched into transient spiral arm-like structures by differential rotation. From their $\sim 20°$ pitch angle Staveley-Smith (ATNF) estimates an age of $\sim$ 90 Myr for these features. The ATCA survey shows an intricate system of bubble-like structures in both the LMC and SMC. A few of these bubbles appear to have no central cluster/association or supernova remnant. The origin of such bubbles is a mystery.

2.4. Absorption-line Spectroscopy

Hot ($T > 10^5$ K) gas is best observed in the ultraviolet absorption lines of highly-ionized atoms, such as O^{+5}, N^{+4}, C^{+3} and Si^{+3}. Such lines may form near O3-O7 stars in the LMC disk, or in a hot gaseous corona around the Large Cloud. Wakker (U. Wisconsin) has been able to use the *Hubble Space Telescope* (HST) to show that some lines of sight, which do not pass close to any early O-type stars, show strong C IV absorption. A comparison of the velocities of such hot gas with that of Hα emission shows that it is not co-spatial with disk material. This suggests that this hot gas is located in a corona around the LMC. The LMC is found to contain an extensive distribution of hot plasma which radiates $\sim 10^{38}$ ergs s^{-1} in the 0.5-2.0 keV band. The SMC emits only $\sim 10^{37}$ ergs s^{-1} in this energy band. De Boer et al. (1998) have detected H_2 in absorption along various lines of sight towards the LMC. They derive excitation temperatures $\leq$ 50 K for levels J $\leq$ 1 and $\sim$ 470 K for levels 2 $\leq$ J $\leq$ 4. From these observations they conclude that moderate UV pumping affects even the lowest levels of excitation towards the LMC. Richter et al. (1998) have detected an H_2 cloud with an excitation temperature $\simeq$ 70 K at +120 km s^{-1} towards the SMC, while another feature at +160 km s^{-1} has an excitation temperature > 2300 K, which must be due to strong UV radiation from its energetic environment.

3. The History of Star Formation in the Clouds

3.1. Massive Stars

The determination of the slope of the mass spectrum of star formation in the Clouds of Magellan has a long and sordid history. Massey (KPNO) pointed out that many of the problems encountered by previous investigators had been due to the fact that they only used photometric data to determine the slope of the mass function Γ. Massey finds that much more consistent results can be obtained by combining spectroscopic and photometric data on massive stars. He finds that $\Gamma = -1.4 \pm 0.2$ in the LMC, $\Gamma = -1.3 \pm 0.1$ in the SMC, and $\Gamma = -1.3 \pm 0.4$ in the Milky Way. These results suggest that the mass spectrum of star formation may be universal. Massey called attention to the fact that $\Gamma = -1.4 \pm 0.1$ in the super-compact cluster R 136, which is situated at the center of 30 Doradus complex. This value is identical to that for other regions of star formation in which the stellar density is two orders of magnitude lower. Walborn (STScI) pointed out that populations of differing age can be recognized in the 30 Dor region: (1) The central ionizing cluster R 136, which has an age of 2-3 Myr, (2) a younger generation of early-O stars and IR sources that are embedded in bright nebular filaments to the west and northeast of R 136, (3) an older population of late-O and early-B supergiants, having an age of 4-6 Myr that is scattered throughout the 30 Dor region, (4) a still older compact cluster 3' northwest of R 136 that contains A and M supergiants having ages of $\sim$ 10 Myr, and (5) a 4-6 Myr old association, that includes the luminous blue variable R 143 in the southern part of the nebula. In other words it appears that star formation in the 30 Dor region is an ongoing process, rather than a single event.

Langer & Heger (U. Potsdam) have compared theoretical predictions of the frequency distribution of massive evolved stars (Hess diagram) with observations and conclude that (1) theory yields a gap to the right of the main sequence that is not observed, and (2) theory predicts a dependence of [Fe/H] on the ratio of luminous blue stars to M-type supergiants that is in the opposite sense to that which is actually observed. They suggested that these problems might be resolved by using both [Fe/H] *and* rotation in stellar models. Such rotational mixing will destroy the onion-like internal structure of non-rotating stellar models. In the discussion following their presentation Massey suggested that both (1) and (2) might be partly accounted for by observational selection effects.

Element ratios are potentially important sources of information on evolutionary history. Smith (U. Texas) showed that the abundance ratio of the r-process element europium to that of the s-process element barium is higher in the Magellanic Clouds, than it is for stars of similar metallicities in the Galaxy. Since the mass spectrum of star formation appears to be universal this effect cannot be due to differences in the frequency with which SNe II having different progenitor masses occur.

A problem of long standing (Richtler, Spite & Spite 1991, and references therein) is that stars in the young SMC cluster NGC 330 appeared to be more metal deficient than similar stars in the SMC field. Such a difference would be difficult to explain. However, new work by Hill (Paris Obs.) appears to show that the abundance differences between NGC 330 and the SMC field, and between NGC 1818 and NGC 2100, and the LMC field are not significant.

De Boer et al. (1997) have recently proposed that star formation may be triggered as gas is compressed in a bow-shock formed at the leading edge of the LMC. Such a scenario would favor star formation at the SE edge of the Large Cloud, where the cumulative effects of the space motion and rotation of the Large Cloud are greatest. However, a recent study of the distribution of 2138 supergiants and 1170 Cepheids by Grebel (Lick Obs.) & Brandner (JPL) does not appear to support the bow-shock scenario. This suggests that the Galactic halo gas density is negligible at $R_{Gc} \sim 50$ kpc.

3.2. History of Star Formation in the Large Cloud

Observations of field stars in the LMC by Butcher (1977) showed that a burst of star formation, that has continued to the present day, started in the Large Cloud 3-5 Gyr ago. This conclusion has been supported by all subsequent studies. However, the magnitude of this increase in the rate of star formation, that occurred $\sim$ 4 Gyr ago, remains controversial. Using HST observations in three outer fields Geha et al. (1998) find that the rate of star formation increased by only about a factor of three. Gallagher (U. Wisconsin) detected no evidence for significant differences between the evolutionary histories in different LMC fields. However, Romaniello (STScI/Pisa) et al. find that the rate of star formation in the field surrounding SN 1987A is presently an order of magnitude higher than it was $\sim$ 5 Gyr ago. On balance, it appears that available data from Large Cloud field stars (Geha et al. 1998) indicate that half of all LMC stars formed during the last 4 Gyr, with the other half having formed during the preceding $\sim$ 10 Gyr. Da Costa (MSSSO) showed that this contrasts dramatically with the situation for star clusters in the Large Cloud. Elaborating on earlier work (Da Costa 1991) showed that the rate of cluster formation in the LMC was close to zero between $\sim$ 4 Gyr ago, and the era of globular cluster formation $\sim$ 14 Gyr ago. Taken at face value his data suggest that the rate of cluster formation had increased by almost two orders of magnitude $\sim$ 4 Gyr ago. This conclusion was strengthened by Sarajedini (1998) who found only 3 new clusters with ages $\sim$ 5 Gyr in the $\sim$ 10 Gyr age gap between the era of globular cluster formation and the burst that started $\sim$ 4 Gyr ago. The existence of these clusters might indicate that the burst "ramped-up" for $\sim$ 1 Gyr before reaching maximum intensity $\sim$ 4 Gyr ago. *The dramatic contrast between the history of cluster formation and that of field stars suggests that star clusters cannot be used as proxies for star formation.* In fact, it would have been very difficult to understand how stars with elevated [Fe/H] values could have formed ~ 4 Gyr ago if star (and supernova!) formation rates had remained depressed during the preceding "dark ages". The observation that the Local Group galaxy IC 1613 is forming young stars but few (if any) star clusters (Baade 1963; van den Bergh 1979), provides a clear demonstration of the fact that the rate of star formation in galaxies does not need to be closely correlated with the rate of cluster formation. In fact, Hodge (1998) finds that the rate of cluster formation, normalized to similar rates of star formation, is presently $\geq$ 600 times greater in the LMC than it is in IC 1613. The observation (Whitmore & Schweizer 1995) that star cluster are being formed very actively in the violently interacting galaxies NGC 4038/39 ("the antennae") suggests that strong shocks might favor the formation of clusters. Finally, it is of interest to note that the rate of star and cluster formation in the SMC (Mighell, Sarajedini & French 1998) appears to have proceeded at a more-or-less constant rate over the last ~ 10 Gyr. This shows that tidal interactions between the Clouds cannot be invoked to account for the LMC starbursts that started ~ 4 Gyr ago.

3.3. Early Star Formation

Olsen (U. Washington) et al. used deep HST observations to show that the LMC globular clusters NGC 1835, NGC1898, NGC 1916, NGC 2005, and NGC 2019 have ages that differ by $\leq$ 1 Gyr. Similar results were obtained for NGC 1466, NGC 2259 and Hodge 11 by Johnson (UCSC) et al. Furthermore the absolute ages of these clusters are found to be similar to those of the Galactic globulars M 3, M 5 and M 55. These results suggest that the Population II component of the Large Cloud formed during a short, but intense, burst of star and cluster formation. On the other hand deep HST observations of clusters in the SMC by Mighell (NOAO) et al. show a rather more gradual onset of cluster formation, with the oldest SMC cluster NGC 121 having an age of only 10.6 ± 0.7 Gyr. Hesser (DAO) et al. also used HST observations to show that the luminous outer halo Galactic globular NGC 2419, which is located at $R_{Gc} \sim 100$ kpc, has an age similar to that of the globular clusters in the LMC and in the main body of the Galactic halo. On the other hand they find that the fainter outer Galactic halo clusters Palomar 3, Palomar 4 and Eridanus have lower ages that are similar to that of NGC 121 in the SMC. The picture that emerges from these results is one in which massive globular clusters formed almost simultaneously in the LMC, and the inner and outer halo of the Galaxy, i.e. throughout a region with a radius of ~ 100 kpc. Less luminous massive clusters formed in the outer Galactic halo and in the SMC during the next few Gyr.

Grebel (Lick Obs) et al. strengthened observational data which show that both Magellanic Clouds shrank as they evolved, with the oldest Population II component occupying a larger volume than younger stars belonging to Population I. Van den Bergh (2000) has noted a similar effect in all other Local Group dwarf irregulars in which detailed population studies have so far been made.

4. The Distances to the Magellanic Clouds

4.1. Distance to the Large Cloud

Distance determinations to the LMC prior to 1996 have been reviewed by Westerlund (1997). An unweighted mean of these data yields a mean distance modulus $< (m-M)_o >= 18.48 \pm 0.04$. A tabulation of 16 more recent distance determinations, which was handed out at the Symposium by van den Bergh, yields an unweighted mean value $< (m-M)_o >= 18.50 \pm 0.04$. Three of the determinations listed in this table give results that are inconsistent with this mean: Using *Hipparcos* parallaxes Feast & Catchpole (1997) find $(m-M)_o = 18.70 \pm 0.10$. However, after applying Lutz-Kelker corrections Oudmaijer et al. (1998) obtain a smaller value $(m-M)_o = 18.56 \pm 0.08$. The appropriateness of these corrections was, however, disputed by Feast. A second discordant distance to the LMC was found by Udalski et al. (1998), who obtained $(m-M)_o = 18.08 \pm 0.03 \pm 0.12$ (systematic) from the mean magnitude of red clump giants in the LMC. A rediscussion of these results (Cole 1998) yields a larger modulus $(m-M)_o = 18.36 \pm 0.17$. Even more recently Paczyński (1998) has questioned the validity of using the magnitudes of clump stars as distance indicators. Finally Reid (1997) has obtained a discordant modulus of $(m-M)_o = 18.71 \pm 0.06$ by fitting the RR Lyrae variables in the LMC globulars NGC 1466 and NGC 2257 to those in the Galactic globular cluster NGC 6397. From this fit Reid obtains $(m-M)_o = 18.71 \pm 0.06$. The weak link in this chain might be the fit of the main sequence of NGC 6397 to the *Hipparcos* parallaxes of nearby subdwarfs.

Perhaps the most direct determinations of the LMC distance are those which are purely geometrical in nature. At the meeting Pritchard (Mt. John Obs.) et al. presented observations of the detached early-type eclipsing binary HV 2274 from which they obtained $(m-M)_o = 18.44 \pm 0.07$. Five additional detached early-type binaries in the LMC have yet to be observed and might strengthen this distance determination. A second purely geometrical method of distance determination is based on ultraviolet observations of the SN 1987A ring. Panagea (STScI/ESA) showed that these data yield $(m-M)_o = 18.55 \pm 0.05$ for the supernova, and $(m-M)_o = 18.58 \pm 0.05$ for the centroid of the LMC. In view of these results it seems safest to continue to use the canonical value $(m-M)_o = 18.5 \pm 0.1$, corresponding to D(LMC) = 50 kpc.

4.2. Distance to the Small Cloud

The SMC distance determinations prior to 1996 have been reviewed by Westerlund (1997). An unweighted mean of these data yields $< (m-M)_o >= 19.94 \pm 0.05$. After excluding the red clump modulus of Udalski et al. (1998), for the reasons discussed above, one finds a mean unweighted value of $< (m-M)_o >= 18.85 \pm 0.04$ for the SMC from distance determinations that have been published recently. It is, however, a source of grave concern that the four Cepheid-based determinations yield $(m-M)_o = 18.93 \pm 0.03$, which is inconsistent with $(m-M)_o = 19.73 \pm 0.02$ that is derived from five distance determinations based on RR Lyrae variables. Kunkel (private communication) has suggested that this difference may be due to the great depth along the line of sight of the SMC. Perhaps many of the young Cepheids in the Small Cloud are located in the tidal tail behind main body of the SMC. In summary it appears that the distance modulus of the Small Cloud is probably $(m-M)_o = 18.85 \pm 0.1$ (corresponding to D = 59 kpc), with the caveat that young stars might, on average, be more distant than the main body of this galaxy. Caldwell & Coulson have pointed out that the distance modulus of the Wing of the SMC is probably ~ 0.3 mag smaller than that of the main body of the Small Cloud.

5. The Orbit of the Magellanic Clouds

5.1. Evidence for Recent Interactions

The discovery of the Magellanic Stream by Mathewson, Cleary & Murray (1974) provided the first dramatic evidence for a strong tidal interaction between the Magellanic Clouds $\sim$ 1.5 Gyr ago (Gardiner & Noguchi 1996). A second encounter 0.2-0.3 Gyr ago is believed by Demers (U. Montreal) & Kunkel (OCIW) to have resulted in the formation of the Bridge between the LMC and SMC, and of the tidal tail behind the Small Cloud. Venn (Macalester College) found [Fe/H] ~ -1.0 for supergiants in the Bridge. This low metallicity suggests that gas in the Bridge was mainly drawn from the SMC. At the Symposium Putman (MSSSO) et al. reported the discovery of a narrow continuous gaseous tail, which leads the direction of motion of the Clouds, i.e., in the direction opposite to that of the Stream. Majewski (U. Virginia) et al. have searched for stars associated with the Magellanic Stream and reported that they have found a number of giant stars concentrated at distances expected for tidal debris from the Magellanic Clouds. Wakker (U. Wisconsin) et al. suspect that the high velocity cloud 287+22+240 may represent metal-poor material that originated in the Magellanic Stream.

5.2. The Orbit(s) of the Magellanic Clouds

Byrd et al. (1994) have modelled the interactions between the members of the Local Group from which they concluded that the Magellanic Clouds may have left the neighborhood of the Andromeda galaxy $\sim$ 10 Gyr ago, and were subsequently captured by the Galaxy $\sim$ 6 Gyr ago. Simulations by Sawa (Aichi U.) et al. also suggest that the LMC and SMC have formed a bound pair for $\sim$ 15 Gyr. On the other hand detailed simulations by Li (U. Wyoming) & Thronson (NASA) show that the Small Cloud lost so much matter during its two most recent interactions with the Large Cloud that it cannot have survived many such interactions. They therefore concluded that the LMC and SMC must have captured each other fairly recently. Only greatly improved proper motions for the Clouds will place significant constraints on the evolution of their orbital history. Unfortunately the presently available data on their proper motions, which are listed in Table 1, show inconsistent motions.

6. Gravitational Lensing

Paczyński (1986) wrote that "Monitoring the brightness of a few million stars in the Magellanic Clouds over a time scale between 2 hr and 2 yr may lead to the discovery of "dark halo" objects in the mass range $10^{-6} - 10^{+2}$ $M_\odot$ or it may put strong upper limits on the number of such objects." This prediction has

Table 1. Proper Motion of LMC in Milliarcseconds

$\mu_\alpha \cos\delta$	μ_δ	Method	Reference
$+1.20 \pm 0.28$	$+0.26 \pm 0.27$	Relative to galaxies	Jones et al. (1994)
$+1.94 \pm 0.29$	-0.14 ± 0.36	Hipparcos	Kroupa & Bastian (1997)
$+1.6 \pm 0.2$	$+3.0 \pm 0.2$	Relative to quasars	Anguita[1](1998)

[1] These proceedings

been brilliantly confirmed by observations of the EROS, MACHO and OGLE consortia. Particularly exciting results, on the lensing event that took place in the SMC less than a month before the symposium, were reported by Alves (LLNL) et al. The lightcurve of this event showed that it was produced by a binary. Such data on microlensing of binaries can break the degeneracy between mass, location, and transverse velocity, that occurs in the standard gravitational microlensing model for single stars. In the case of the June 1998 event seen in the direction of the SMC, it was found that the observed proper motion of the lensing object is so small that there is only a 0.15% probability that it was produced by a foreground object in the Galactic halo. Previous observations of lensing events in the direction of the LMC had already excluded objects in the mass range 10^{-7} $M_\odot$ to 1 $M_\odot$ as significant contributors to the mass of the Galactic dark halo.

The EROS, MACHO and OGLE surveys have also produced a massive and homogeneous database on variable stars in the Clouds of Magellan. Investigations based on these new data on variable stars were reported by Welch (McMaster), Alves (LLNL) et al., Marquette (Inst. d'Ap.) and Bono (Trieste) et al.

7. Conclusions

- Geometrical distance determinations, based on observations of SN 1987A and of the detached eclipsing binary HV 2274, yield distance moduli of $(m - M)_o = 18.58 \pm 0.05$ and $(m - M)_o = 18.44 \pm 0.07$, respectively for the Large Magellanic Cloud. These values are both compatible with the canonical value $(m - M)_o = 18.5 \pm 0.1$, which corresponds to a distance of 50 kpc.

- The great burst of cluster formation that started in the LMC 3-5 Gyr ago is only weakly reflected in the rate at which field stars were formed. This strongly suggests that the rate of cluster formation is not a good diagnostic for the overall rate of star formation. The observation that the present rate of cluster formation, normalized to the rate of star formation in the LMC, is more than two orders of magnitude greater than it is in the Local Group dwarf irregular IC 1613 supports this conclusion.

- Tidal interactions between the LMC and SMC that occurred $\sim$ 0.2 Gyr and $\sim$ 1.5 Gyr ago produced the Bridge and the Magellanic Stream, respectively. It is presently not clear if the LMC and SMC were closely bound between 3 Gyr and 13 Gyr ago. Improved proper motions are urgently required to constrain their orbital history.

- Observations of microlensing events strongly suggest that they are not produced by objects located in the Galactic halo. The enormous data base provided by the EROS, MACHO and OGLE consortia is proving to be a gold mine for the study of variable stars in the Clouds of Magellan.

References

Baade, W. 1963, Evolution of Stars and Galaxies (Cambridge: Harvard Univ. Press), p. 231

Butcher, H. 1977, ApJ, 216, 372

Byrd, G., Valtonen, M., McCall, M., & Innanen, K. 1994, AJ, 107, 2055

Caldwell, J.A.R., & Coulson, I.M. 1986, MNRAS, 218, 223

Caldwell, J.A.R., & Maeder, D.L. 1992, in Variable Stars and Galaxies = ASP Conference Series No. 30, Ed. B. Warner (San Francisco: ASP), p. 173

Cole, A.A. 1998, ApJ, 500, L137

Da Costa, G.S. 1991, in The Magellanic Clouds = IAU Symposium No. 148, Eds. R. Haynes & D. Milne (Dordrecht: Kluwer), p. 183

de Boer, K.S., Braun, J.M., Vallenari, A., & Mebold, U. 1997, A&A, 329, L49

de Boer, K.S., Richter, P., Bomans, D.J., Heithausen, A., & Kuornneef, J. 1998, A&A, 338, L5

Feast, M.W., & Catchpole, R.M. 1997, MNRAS, 286, L1

Gardiner, L.T., & Noguchi, M. 1996, MNRAS, 278, 191

Geha, M.C., et al. 1998, AJ, 115, 1045

Haynes, R., & Milne, D. 1991, The Magellanic Clouds = IAU Symposium No. 148 (Dordrecht: Kluwer)

Hodge, P.W. 1998, Colloquium given at Victoria, B.C., April 21

Hughes, J.P., et al. 1995, ApJ, 444, L81

Jones, B.F., Klemola, A.R., & Lin, D.C.N. 1994, AJ, 107, 1333

Kroupa, P., & Bastian, U. 1997, New Astr., 2, 77

Mathewson, D.S., Cleary, M.N., & Murray, J.D. 1974, in The Formation and Dynamics of Galaxies = IAU Symposium No. 58, Ed. J.R. Shakeshaft (Dordrecht: Reidel), p. 367

Mighell, K.J., Sarajedini, A., & French, R.S. 1998, AJ, 116, 2395

Oudmaijer, R.D., Groenwegen, M.A.T., & Schrijver, H. 1998, MNRAS, 294, L41

Paczyński, B. 1986, ApJ, 304, 1

Paczyński, B. 1998, AcA, 48, 405

Reid, I.N. 1997, AJ, 114, 161

Richter, P., et al. 1998, A&A, 338, L9

Richtler, T., Spite, M., & Spite, F. 1991, in The Magellanic Clouds = IAU Symposium No. 148, Eds. R. Haynes & D. Milne (Dordrecht: Kluwer), p. 374

Sarajedini, A. 1998, AJ, 116, 738

Udalski, A., Szymański, M., Kubiak, M., Pietrzyński, G., Woźniak, P., & Żebruń, K. 1998, AcA, 48, 147

van den Bergh, S. 1979, ApJ, 230, 95

van den Bergh, S., & de Boer, K.S. 1984, Structure and Evolution of the Magellanic Clouds = IAU Symposium No. 108, (Dordrecht: Reidel)

van den Bergh, S. 2000, The Galaxies of the Local Group (Cambridge: Cambridge Univ. Press), in preparation

Westerlund, B.E. 1997, The Magellanic Clouds (Cambridge: Cambridge Univ. Press)

Whitmore, B.C., & Schweizer, F. 1995, AJ, 109, 960

New Views of the Magellanic Clouds
IAU Symposium, Vol. 190, 1999
Y.-H. Chu, N.B. Suntzeff, J.E. Hesser, and D.A. Bohlender, eds.

End Quotes

I have a feeling that one can tweak the model to make it fit. D. Hartwick

This is certainly not an exhausting list of references (of carbon star surveys). C. Loup

If you put both diagrams on top of each other, I think that you will agree that it is a scientific, as well as an artistic, mess. M. Dopita

If someone asks me nicely, I will run some models. M. Dopita

He has kind of given me permission to report this work. L. Gardiner

A question from the audience. *May I say something?* Reply by D. Welch. *I'm sure you will say something anyway.*

When I say they are not there, I really mean they are not there. G. Da Costa

These numbers include errors and not opinions. G. Bono

No one will be satisfied with plausible reasons: what we want are real reasons. M. Feast

Sorry Nino, no one is interested in the ring distances anymore. D. Welch

All this rubbish ... is just rubbish. N. Panagia

Afterwards, anyone can pitch in, ... provided they are polite. M. Feast.

We thank our speaker from Eastern France. J. Gallagher to H. Zinnecker (Prompted by the rather unusual World Cup outfit of the speaker in honour of the win by France in the World Cup).

I asked the question in their exam and they couldn't answer it. But I couldn't answer it either, so I passed them. J. Dickel

Anyone who has been in astronomy long enough knows that differencing large numbers with systematic errors yields differences with systematic errors. J. Gallagher

I'm under attack here. J. Gallagher

The star formation rate in the LMC is consistent with the HST data. J. Gallagher

Microlensing is just an excuse to do great stellar work. S. Beaulieu

When Cassinelli proposed that there were 2000$M_\odot$ stars in R136, theoreticians could suddenly produce stable models. P. Massey

Abundance analyses are where stars earn their keep. K. Venn

The core kinda goes "poof". V. Smith describing an aspect of AGB evolution.

This isn't the real world and I'm an astronomer. V. Smith

Unfortunately, the thesis was written in German, but at least I can read the tables. D. Garnett

These are the conclusions which you will have to read because the session chair is approaching rapidly ... N. Walborn

At first Phil thought he kept on getting the same spectrum from the HST archives. D. Hunter

Most of my science fiction is read in the Astrophysical Journal. P. Hodge (editor of the Astronomical Journal)

But it is certain that it is not known to any one man to know all things... (a quote on a viewgraph). *Certainly he did not know Jay Gallagher.* P. Hodge

I have no prejudices - only inclinations. P. Hodge

... and last, and also least, planets. P. Hodge

First you eliminate all the uninteresting objects in the LMC - the stars. C. Smith

The ISM in the LMC is a mess. K. de Boer

Introductory remarks, like skirts, are better shorter. Y.-H. Chu

If I am rushed, it is only because my thesis advisor will get mad at me if I overrun the five minutes. R. Williams

It is easier to talk about what you do than what others do, because you understand what you do. J. Bregman

One little clump of stars does not an OB association make. P. Massey

Starburst is a phrase which, as far as I can tell, is used to describe any star formation more vigorous than Orion. P. Massey

New Views of the Magellanic Clouds
IAU Symposium, Vol. 190, 1999
Y.-H. Chu, N.B. Suntzeff, J.E. Hesser, and D.A. Bohlender, eds.

29 Doradus

John Dickel

Astronomy Department, University of Illinois, Urbana, IL 61801, USA

At IAU Symposium 190 on the Magellanic Clouds held in Victoria BC, 12-17 July 1998, I challenged the assembled multitude with the question:

"What is 29 Doradus?"

You give me:
1) kind of object it is
2) coordinates (preferably J2000)
3) a reference so I can check it.

I give you:
A bottle of Dickel Tennessee Sipping Whiskey. [editors note: due to the stress the preparation of these proceedings induced on several of the editors, the actual prize runs the risk of being reduced to a fraction of a bottle of whiskey. The remains may be claimed on your next visit to Victoria.]

After several false leads, I am pleased to report on 29 July 1998 –

Congratulations to Jim Kaler assisted by Al Calder who took me over to the Rare Book Room of the University of Illinois and showed me page 90 of the publication "General Description and Information on the Stars" by Johann Bode 1801.

29 Doradus, RA=$84°15'33''$, Dec=$-66°40'04''$ (1801), and m=6

Precession to J2000 gives RA=$05^{h}37^{m}01.7^{s}$, Dec=$-66°33'24.7''$ = HR1960 which is a B9.5Ve star with a magnitude of 6.3.

Second prize goes to Ed Olszewski and Ron Webbink who steered us to the correct publication.

Third prize goes to Doug Welch who steared Ed in the right direction.

Many thanks to all and I hope others had as much fun in the search as I did.

Author Index

Subject Index

INTERNATIONAL ASTRONOMICAL UNION SYMPOSIA

Published by the Astronomical Society of the Pacific

Vol. No. 190 NEW VIEWS OF THE MAGELLANIC CLOUDS
eds. You Hua Chu, Nicholas B. Suntzeff, James E. Hesser, and David A. Bohlender
ISBN: 1-58381-021-8

Vol. No. 191 ASYMPTOTIC GIANT BRANCH STARS
eds. T. Le Bertre, A. Lebre, and C. Waelkens
ISBN: 1-886733-90-2

Vol. No. 192 THE STELLAR CONTENT OF LOCAL GROUP GALAXIES
eds. P. Whitelock and R. Cannon
ISBN: 1-886733-82-1

Vol. No. 193 WOLF-RAYET PHENOMENA IN MASSIVE STARS AND STARBURST GALAXIES
eds. K. A. van der Hucht, G. Koenigsberger, and P. R. J. Eenens
ISBN: 1-58381-004-8

Vol. No. 194 ACTIVITY IN GALAXIES AND RELATED PHENOMENA
eds. Y. Terzian, E. Khachikian, and D. Weedman
ISBN: 1-58381-008-0

Please note volumes above have been printed

Complete lists of proceedings of past IAU Meetings are maintained at the IAU Web site at the URL: http://www.iau.org/publicat.html

Volumes 32 - 189 in the IAU Symposia Series may be ordered from
Kluwer Academic Publishers
P. O. Box 117
NL 3300 AA Dordrecht
The Netherlands

EDITORIAL/PUBLISHING OFFICE:
Managing Editor
PO Box 24463
211 - KMB
Brigham Young University
Provo UT 84602-4463
USA

(801) 378-2298 Phone
(801) 378-2265 Fax
pasp@astro.byu.edu E-mail

CATALOG/BOOK ORDERS:
IAU Publications
390 Ashton Avenue
San Francisco CA 94112-1722
USA

(415) 337-1100 Phone
(414) 337-5205 Fax
catalog@aspsky.org E-mail